SOLAR POLARIZATION

SOLAR POLARIZATION

*Proceedings of an International Workshop
held in St. Petersburg, Russia, 8–12 May, 1995*

Edited by

J. O. STENFLO

*Institute of Astronomy,
ETH Zentrum, CH-8092 Zurich,
Switzerland*

and

K. N. NAGENDRA

*Indian Institute of Astrophysics,
Bangalore 560 034, India*

Reprinted from Solar Physics, Volume 164, Nos. 1–2, 1996

SPRINGER-SCIENCE+BUSINESS MEDIA, B.V.

Library of Congress Cataloging-in-Publication Data

A C.I.P. Catalogue record for this book is available from the Library of Congress.

ISBN 978-94-010-6586-3   ISBN 978-94-009-0231-2 (eBook)
DOI 10.1007/978-94-009-0231-2
Printed on acid-free paper

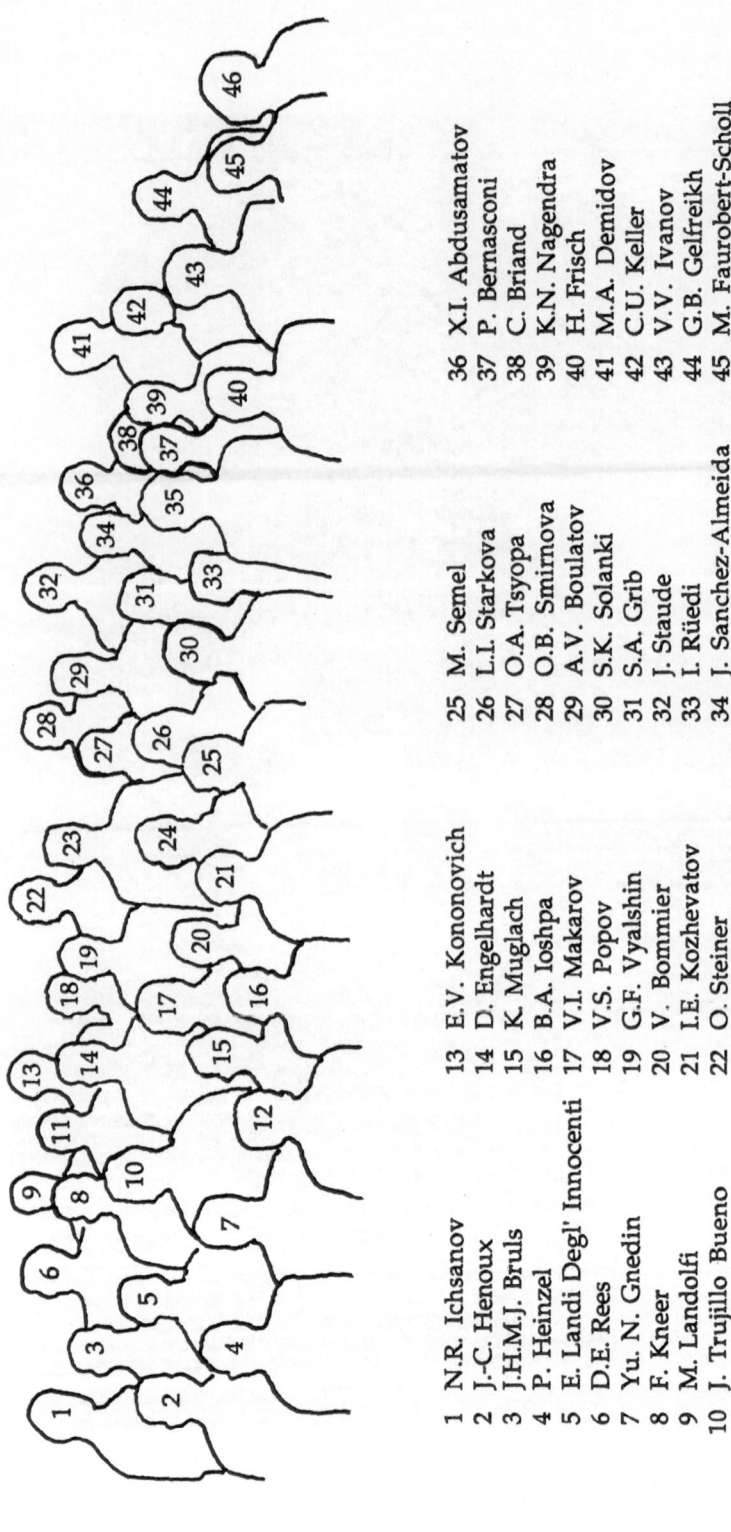

1 N.R. Ichsanov
2 J.-C. Henoux
3 J.H.M.J. Bruls
4 P. Heinzel
5 E. Landi Degl' Innocenti
6 D.E. Rees
7 Yu. N. Gnedin
8 F. Kneer
9 M. Landolfi
10 J. Trujillo Bueno
11 Yu. A. Nagovitsyn
12 V.N. Karpinsky

13 E.V. Kononovich
14 D. Engelhardt
15 K. Muglach
16 B.A. Ioshpa
17 V.I. Makarov
18 V.S. Popov
19 G.F. Vyalshin
20 V. Bommier
21 I.E. Kozhevatov
22 O. Steiner
23 J. Carlos del Toro Iniesta
24 A.P. Skumanich

25 M. Semel
26 L.I. Starkova
27 O.A. Tsyopa
28 O.B. Smirnova
29 A.V. Boulatov
30 S.K. Solanki
31 S.A. Grib
32 J. Staude
33 I. Rüedi
34 J. Sanchez-Almeida
35 J.O. Stenflo

36 X.I. Abdusamatov
37 P. Bernasconi
38 C. Briand
39 K.N. Nagendra
40 H. Frisch
41 M.A. Demidov
42 C.U. Keller
43 V.V. Ivanov
44 G.B. Gelfreikh
45 M. Faurobert-Scholl
46 T.M. Natsvlishvili

TABLE OF CONTENTS

TABLE OF CONTENTS

(*Solar Polarization*)

PREFACE

Much progress has been made in recent years in understanding the complex physics of polarized radiation in the sun and stars. This physics includes vector radiative transfer and spectral line formation in the presence of magnetic fields, scattering theory and coherence effects, partial redistribution and turbulent magnetic fields, numerical techniques and Stokes inversion, as well as concepts for polarimetric imaging with a precision only limited by photon statistics. Since these various aspects have in the past often been dealt with in a fragmented way, there was a great need to organize an international conference that could address the whole problem area in a comprehensive way. With this aim an international Workshop on *Solar Polarization* took place in St. Petersburg 8-12 May 1995. The workshop format had the advantage of allowing the participants to penetrate the various topics in greater depth, and the venue in St. Petersburg was chosen to stimulate closer collaboration between Russia and western countries.

There is a long and outstanding tradition in Russia in the theory of radiative transfer. The next logical step for this theory is to move from scalar problems to "vector transfer", i.e., to the transfer of polarized radiation. During a visit to Nice in 1993 our host there, Helène Frisch, and the visitors Seva Ivanov (St. Petersburg) and Jan Stenflo (Zurich), decided to form a "troika" who would organize an international Workshop in St. Petersburg on the topic "Solar Polarization". With this topic the area of vector radiative transfer gets embedded in a broader context that includes all the various magnetic field effects and diagnostic problems that are so central to contemporary solar physics. The various theoretical and experimental results for the "solar laboratory" also serve as benchmarks for the rest of astrophysics.

48 participants from 12 countries attended the Workshop. The Central Astronomical Observatory at Pulkovo kindly hosted the conference. We are grateful to the Observatory Director, Dr. V.K. Abalakin, and the Vice Director, Dr. Yu.N. Gnedin (who both served as Co-Chairmen of the Local Organizing Committee) for the marvellous hospitality shown to us. However, the key role among our Russian

hosts who made the conference such a success was played by Prof.
V.V. Ivanov, who from the very start was our main contact person
for all matters concerning the Workshop. The other members of the
Local Organizing Committee were S.F. Elesin, Yu.A. Nagovitzin,
V.S. Popov, and O.A. Tsiopa. To all of them we wish to express our
sincere thanks.

All the foreign participants were accomodated in the Guest
House of the Pulkovo Observatory, where a specially brought in
kitchen staff was serving us meals of outstanding quality and cater-
ing to all our wishes. The very extensive social program for the
participants and for the accompanying persons made the week in
St. Petersburg particularly memorable. The high point was our par-
ticipation in the celebration of the end of World War II on May 9.
A number of small boats had been rented, which took the Workshop participants along the canals of this beautiful city, such that
we emerged at 9 pm on the river Neva right in front of the Win-
ter Palace among the battle ships and submarines when the canon
salutes and fireworks started.

According to the new rules of SOLAR PHYSICS for the publication of proceedings from a scientific conference, all papers to be
published have to go through a regular refereeing process, and pure
reviews should be avoided, since all contributions have to deal with
new results. We are grateful for the hard and thorough work done by
all our referees, who made it possible for us to produce a high-quality
volume on "Solar Polarization" in a short time. The new rules of SO-
LAR PHYSICS also stipulate that a member of the Editorial Board
of the journal may serve as a Guest Editor for the proceedings, and
that a Proceedings Organizer has to be nominated. One of the chief
editors of SOLAR PHYSICS is designated as a Supervisor for the
proceedings. In our case the Supervisor is C. de Jager, the Guest
Editor J.O. Stenflo, and the Organizer K.N. Nagendra.

Much of Nagendra's work on the proceedings was carried out
while he spent an extended visiting period at Observatoire de la
Côte d'Azur in Nice, following the Workshop. We are grateful to
the Observatory in Nice for supporting this extensive work with its
infrastructure. Our special thanks go to Hélène Frisch for her crucial
and dedicated support as a member of the "troika" for the Workshop,
and for organizing the support from Nice for work on the proceedings.

The following served as Chairpersons for the various scientific
sessions: M. Faurobert-Scholl, H. Frisch, V.V. Ivanov, J. Sánchez-
Almeida, E. Landi Degl'Innocenti, D.E. Rees, S.K. Solanki, and J.O.
Stenflo. During much of the last day the Workshop split up in

two separate groups for in-depth discussions on two main topics of the meeting, namely "Scattering Physics" (group leader: J.O. Stenflo) and "Magnetic Field Diagnostics" (group leader: S.K. Solanki), which was followed by a concluding, summarizing discussion with both groups together.

Finally we wish to thank all the participants of the conference. Thanks to their important oral and written scientific contributions and the lively discussions they generated we could achieve the objectives set for this successful Workshop.

Zurich and Bangalore
October 1995

J.O. Stenflo
K.N. Nagendra

SCATTERING PHYSICS

J.O. STENFLO

Institute of Astronomy, ETH Zentrum, CH-8092 Zurich, Switzerland

Abstract. The theory of polarized scattering in a stellar atmosphere is formulated, first within the framework of classical physics, then in terms of quantum mechanics. The expression for the redistribution matrix that describes partial redistribution in polarization and frequency is derived for the general case when the magnetic field is of arbitrary strength. The special cases of weak fields (the "Hanle limit") and zero fields (non-magnetic scattering) are discussed. Observational examples of spectral signatures in linear polarization are presented, which show effects of hyperfine structure, interference between fine structure components, and molecular scattering.

Key words: Coherent scattering – Polarization – Solar spectrum

1. Observational Background

Radiation becomes polarized if it is scattered. Thus the polarization of the blue sky is due to molecular Rayleigh scattering. On the sun scattering processes contribute to the formation of both the continuous and line spectrum. The solar spectrum is therefore polarized, even in the absence of magnetic fields.

The degree of scattering polarization depends on the anisotropy of the radiation field that is incident on the scattering particle. Maximum polarization is obtained for 90° scattering, while no polarization results if the illumination is isotropic. If we ignore the local inhomogeneities in the solar atmosphere and consider an ideal, spherically symmetric sun, then the anisotropy of the radiation field expresses itself in the form of limb darkening (or limb brightening in the case of the EUV). This anisotropic illumination makes the scattering polarization increase from zero at disk center to a maximum at the solar limb.

Early attempts to map the linear polarization close to the solar limb throughout the whole visible solar spectrum, from 3200 Å to the infrared (Stenflo et al., 1980, 1983a, b), have revealed a Stokes Q/I spectrum with a wealth of spectral structures. V.V. Ivanov (private communication) has aptly refered to it as "the second solar spectrum", since it is so different in appearance and information content from the well-known unpolarized spectrum. We have to start over again to identify and classify what we are seeing, and to work out the basic physics that may explain how the various spectral structures are formed.

Due to the decreasing limb darkening the magnitude of the polarization generally decreases as we go upwards in the spectrum. The earlier survey

Solar Physics **164**: 1–20, 1996.

could identify polarized features that stuck up above the noise level, which was typically 0.1 % . Such features seemed to be fairly rare above 4300 Å, so it could not be properly judged how rich or interesting the spectrum at these wavelengths might really be. With the advent of 2-D imaging polarimetry that achieves demodulation by synchronous charge-shifting to build up polarized image planes within a single CCD chip and thereby eliminates spurious seeing and flat-field effects (Povel et al., 1991; Keller et al., 1992; Stenflo et al., 1992), a polarimetric accuracy of 10^{-5} in Stokes Q/I is now routinely reached, only limited by photon statistics. Such observations with the polarimetric system ZIMPOL (**Z**urich **Im**aging Stokes **Pol**arimeter) have lifted the veil over the hidden treasures of the "second solar spectrum" and found them to be richer than ever anticipated (Stenflo and Keller, 1996).

A now "classic" line in the analysis of scattering polarization is the Sr I 4607 Å line. Figure 1 shows a 2-D Stokes I and Q/I spectrum recorded with ZIMPOL I at a disk position of $\mu = 0.1$. This line has an unusually strong and "clean" polarization signal and is surrounded by lines that instead of polarizing depolarize the weak continuum polarization. The Sr I line represents a $J = 0 \rightarrow 1 \rightarrow 0$ scattering transition. For this particular combination of J quantum numbers the scattering behaves like classical dipole-type scattering, while for other types of transitions the line polarizability dictated by quantum mechanics is generally smaller.

The scattering polarization is modified by magnetic fields via the *Hanle effect*, since the coherent superposition of the magnetic substates of the excited level leads to interference effects that depend on the relative frequency displacements of these sublevels due to the Zeeman effect. The Hanle effect manifests itself mainly in the form of a rotation of the plane of linear polarization, and as depolarization. While rotation can occur with both signs, depolarization has only one "sign". This means that for a tangled magnetic field with random orientations of the field vectors, the rotation effects (as well as the ordinary Zeeman effect polarization) cancel out, while all field vectors contribute in the same sense to the depolarization. A spatially unresolved turbulent magnetic field that does not have any other net polarization signatures should therefore reveal its presence by suppressing the amount of scattering polarization that is observed, in comparison with the amount that is expected from theory in the absence of magnetic fields.

The Hanle depolarization was first exploited by Stenflo (1982) in an attempt to interpret the observed polarization amplitudes, and led to an estimated value of 10–100 G for the strength of the turbulent field in the photosphere. This approach has later been much refined by Faurobert-Scholl (1993), who has carried out detailed radiative-transfer calculations to fit the center-to-limb variation of the polarization profiles of the Sr I line observed by Stenflo et al. (1980). This has allowed the admitted range of turbulent

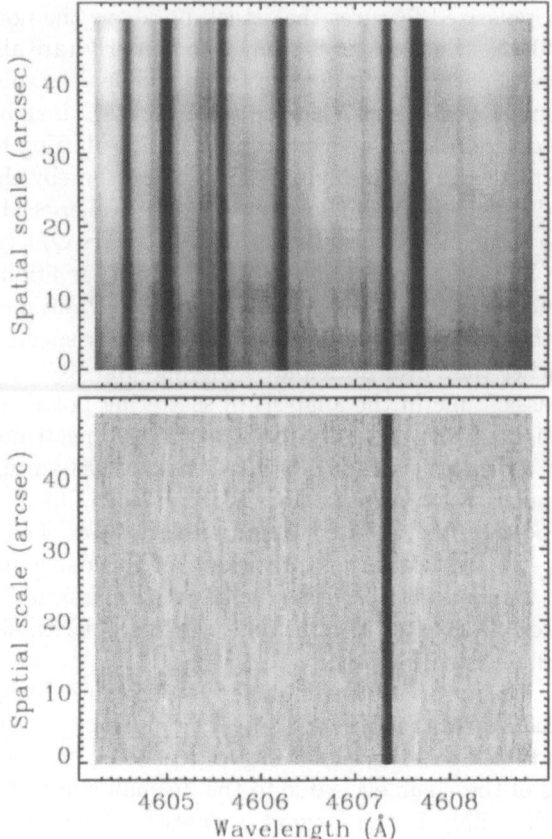

Fig. 1. 2-D Stokes I/I_c (top) and Q/I spectrum of the region around the Sr I 4607 Å line, obtained with ZIMPOL I in November 1994 at the NSO McMath telescope. The slit was positioned parallel to the solar limb near the heliographic north pole at a disk position of $\mu = 0.1$. While the Stokes I spectrum contains many spectral lines, only one of them, the Sr I line, stands out as a dark band in the Q/I image.

field strengths (assuming an isotropic distribution) to be narrowed down to 10–20 G (Faurobert-Scholl et al., 1995).

Another area where the Hanle effect has been much applied in the past is the diagnostics of magnetic fields in solar prominences (e.g. Leroy et al., 1977; Bommier, 1980; Querfeld et al., 1985). Most of the work has focused on the He I D₃ 5876 Å line, which, being optically thin, allows direct interpretations without involving radiative-transfer calculations. In this case not only the Hanle depolarization, but also the rotation of the plane of polarization, can be observed and studied in detail.

The example of Fig. 1 appears to suggest that, apart from a few isolated polarizing lines, the polarized Q/I spectrum is rather dull and devoid of interesting structures. A new survey performed with ZIMPOL I at the Mc-Math telescope in November 1994 and April 1995 (Stenflo and Keller, 1995) with a sensitivity of about 10^{-5} in the degree of linear polarization has however shown that the region around the Sr I 4607 Å line is not very typical for the rest of the spectrum. In other parts of the spectrum Q/I is more structured than the unpolarized Stokes I spectrum. An example is given in Fig. 2, which shows the region around the Mg I 5167 Å line. Many narrow polarization peaks occur at wavelengths, where there appear to be no counterparts in the Stokes I spectrum. A closer inspection reveals that the wavelengths of these "mysterious" polarization peaks generally coincide with the location of very weak molecular lines, in most cases MgH, sometimes C_2. These tentative identifications suggest new diagnostic opportunities that have yet to be exploited.

The partial polarization survey with ZIMPOL I has also revealed other hitherto unseen effects, like pronounced signatures of hyperfine structure in Ba II and Na I (an example of which will be shown later as Fig. 4 for the Na I D_1 and D_2 lines), various cases of interference between fine-structure components of a multiplet, as well as cases of not yet identified physics. These latter cases refer to observed strong polarization peaks in lines that should be intrinsically unpolarizable according to quantum mechanics for the particular combination of J quantum numbers. The "second solar spectrum" represents "virgin territory" that we only now, thanks to advances in observational techniques, are able to fully enter into and begin to explore. The first exploratory phase will be to identify and classify the physical processes responsible for the various structures observed. Then it will be possible to begin to exploit these structures for diagnostic purposes.

In particular one will soon want to make the observational step from center-to-limb studies to explore the effects of local inhomogeneities and magnetic fields on the scattering physics with full Stokes vector polarimetry. Such "Hanle observations" however demand large trade-offs with respect to polarimetric sensitivity, since the 10^{-5} accuracy could only be reached using extensive spatial and temporal averaging with the world's largest solar "light bucket" (the 1.5 m McMath entrance aperture).

With this background we will now try to outline our current understanding of scattering physics on the sun. We begin with a formulation of radiative scattering within the framework of classical physics and then describe how the picture is changed when going to quantum mechanics. After a discussion of the physics of interference between fine-structure components we turn to the general theory of partial redistribution, which we present in steps, starting with the unpolarized case and ending with the formulation for the general polarized case in the presence of magnetic fields of arbitrary strength and direction.

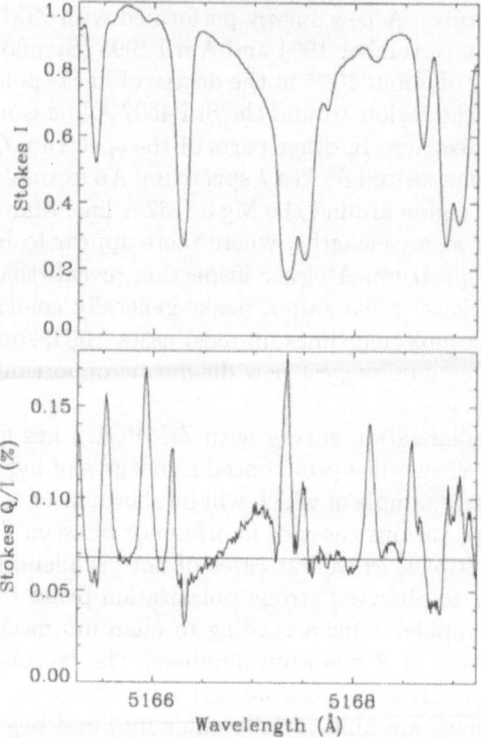

Fig. 2. Stokes I (top), normalized to its maximum value, and Q/I (bottom) of the region around the Mg I 5167 Å line, obtained with ZIMPOL I at the NSO McMath telescope in April 1995. The slit was positioned inside the limb near the heliographic north pole at a disk position of $\mu = 0.1$. Most of the sharp polarization peaks that surround the strong Mg I line coincide with the location of molecular MgH lines, which are so weak that they are almost "invisible" in the Stokes I spectrum. The level of the continuum polarization is marked by a horizontal line. The strong Stokes I line to the left of the Mg I line that depolarizes the continuum belongs to multiplet no. 1 of Fe I. (This multiplet also contains the well-known 5250/5247 line pair that has been extensively used for flux tube diagnostics.)

2. Classical View of Coherent Scattering in a Magnetic Field

The electric field $\mathbf{E}'$ of the exciting radiation drives a damped oscillator that represents the scattering system. The oscillations are described by the equation

$$\frac{d\mathbf{v}}{dt} + \frac{e}{m}(\mathbf{v} \times \mathbf{B}) + \gamma \mathbf{v} + \omega_0^2 \mathbf{r} = -\frac{e}{m}\mathbf{E}', \tag{1}$$

where $\mathbf{r}$ is the relative position vector. (Note that SI units are used throughout the present paper.)

It is impractical to solve this equation in Cartesian coordinates, since the $\mathbf{v} \times \mathbf{B}$ term makes the component equations coupled to each other. They however become decoupled if we instead of a Cartesian basis of unit vectors $\mathbf{e}_{x,y,z}$ use complex spherical vectors $\mathbf{e}_{0,\pm}$, defined by

$$
\begin{aligned}
\mathbf{e}_0 &= \mathbf{e}_z \, , \\
\mathbf{e}_\pm &= \mp (\mathbf{e}_x \pm i\mathbf{e}_y)/\sqrt{2} \, ,
\end{aligned}
\tag{2}
$$

The decoupled component equations of Eq. (1) then become

$$
\frac{dv_q}{dt} - (2qi\omega_L - \gamma)v_q + \omega_0^2 r_q = -\frac{e}{m} E_q' \, ,
\tag{3}
$$

where $q = 0, \pm 1$, and $\omega_L = eB/(2m)$ is the Larmor frequency.

The oscillatory solution of each of these three equations has the form

$$
r_q \sim (n_q - 1)E_q' \, ,
\tag{4}
$$

where the refractive index n_q is given by

$$
n_q - 1 \approx \frac{\omega_A^2}{4\omega_0} \frac{1}{\omega_0 - \omega - q\omega_L - i\gamma/2} \, ,
\tag{5}
$$

and

$$
\omega_A = \sqrt{\frac{e^2 \, N}{\epsilon_0 \, m}}
\tag{6}
$$

(Stenflo, 1994). N is the number density of classical oscillators.

Introducing Doppler broadening with Doppler width $\Delta\omega_D$, we can write

$$
n_q - 1 \sim iH(a, v_q) + 2F(a, v_q) \, ,
\tag{7}
$$

where H is the Voigt function and F the line dispersion function, and

$$
\begin{aligned}
v_q &= v - qv_H \, , \\
v_H &= \omega_L/\Delta\omega_D \, , \\
v &= (\omega_0 - \omega)/\Delta\omega_D \, , \\
a &= \gamma/(2\Delta\omega_D) \, .
\end{aligned}
\tag{8}
$$

We thus have three independent antennas radiating at three different frequencies, however with phase coherence between their overlapping frequency distributions.

The electric vector of the scattered radiation can now be obtained by simple geometric projection of the three r_q or E_q' on a plane perpendicular to the scattering direction that we consider. The E_q' components will be expanded in terms of linear electric vectors E_β', since it is convenient to describe both the incident and scattered radiation in terms of a linear polarization basis that can be easily related to the Stokes parameters.

Let $\mathbf{e}_\alpha$ be a linear electric vector of the scattered radiation. To simplify the notation we use the standard convention that repeated indices are summed over. Then, with the expansion

$$\mathbf{r} = r_q \mathbf{e}_q^* \tag{9}$$

(* means complex conjugation) and

$$\varepsilon_q^\alpha = \mathbf{e}_q \cdot \mathbf{e}_\alpha \tag{10}$$

being the notation for the scalar product between spherical and linear unit vectors, we get

$$\begin{aligned} E_\alpha &\sim \mathbf{r} \cdot \mathbf{e}_\alpha = \varepsilon_q^{\alpha^*} r_q, \\ E_q' &= \mathbf{E}_\beta' \cdot \mathbf{e}_q = \varepsilon_q^\beta E_\beta'. \end{aligned} \tag{11}$$

Combining these expansions with Eq. (4), we can express the relation between the linear electric vectors of the incident and scattered radiation as

$$E_\alpha \sim w_{\alpha\beta} E_\beta', \tag{12}$$

where

$$w_{\alpha\beta} = (n_q - 1)\varepsilon_q^{\alpha^*}\varepsilon_q^\beta \tag{13}$$

is the *Jones scattering matrix*.

The relation between the Jones matrix $\mathbf{w}$ and the *Mueller scattering matrix* $\mathbf{M}$, which determines the relation between the incident and scattered Stokes vector, is given by

$$\mathbf{M} \sim \mathbf{T}\,(\mathbf{w} \otimes \mathbf{w}^*)\,\mathbf{T}^{-1}, \tag{14}$$

The symbol $\otimes$ denotes a tensor product. Explicitly we have

$$\mathbf{w} \otimes \mathbf{w}^* = \begin{pmatrix} w_{11}w_{11}^* & w_{11}w_{12}^* & w_{12}w_{11}^* & w_{12}w_{12}^* \\ w_{11}w_{21}^* & w_{11}w_{22}^* & w_{12}w_{21}^* & w_{12}w_{22}^* \\ w_{21}w_{11}^* & w_{21}w_{12}^* & w_{22}w_{11}^* & w_{22}w_{12}^* \\ w_{21}w_{21}^* & w_{21}w_{22}^* & w_{22}w_{21}^* & w_{22}w_{22}^* \end{pmatrix}. \tag{15}$$

The transformation matrix $\mathbf{T}$ and its inverse are given in Stenflo (1994, p. 41). They are purely mathematical, without physical contents.

The frequency-dependent terms in $\mathbf{M}$ are of the form $(n_q - 1)(n_{q'}^* - 1)$. These distributions are smoothed by Doppler broadening. In the weak-field case, defined by the condition $\omega_L \ll \Delta\omega_D$, the frequency-averaged value can be written (cf. Stenflo, 1994, p. 87)

$$\langle (n_q - 1)(n_{q'}^* - 1) \rangle \sim \cos\alpha_{|q-q'|}\, e^{i\,\alpha_{|q-q'|}}. \tag{16}$$

Later we will give the full expression for the "Hanle angle" α_K (where $K = |q - q'|$) but here only note that the factor $\cos \alpha_K$ describes Hanle *depolarization*, while the factor $e^{i \alpha_K}$ describes *rotation* of the plane of linear polarization.

The scattering emission vector $\mathbf{j}$ can be written as

$$\mathbf{j} = \sigma \int \frac{d\Omega'}{4\pi} \int d\nu' \, \mathbf{R} \mathbf{I}_{\nu'} \,, \tag{17}$$

where $\mathbf{R}$ is the *redistribution matrix*. Our next task is to find the expression for $\mathbf{R}$.

As the Zeeman splitting is insignificant in comparison with the Doppler width in the weak-field case, it can be neglected for the frequency-redistribution problem. This has the great advantage that we can use the same frequency profile $f(\xi')$ for all the matrix components of $\mathbf{M}$. The frequency redistribution can then be factorized out of the polarization matrix as a scalar function. This means that one may write the redistribution matrix for resonant scattering in the rest frame in the absence of collisions as

$$\mathbf{R} = f(\xi') \, \mathbf{P} \, \delta(\xi' - \xi) \,, \tag{18}$$

where ξ', ξ are the rest frame frequencies of the incident and scattered beams (in the observer's frame denoted by ν', ν). $f(\xi')$ is given by the area-normalized $|n_0 - 1|^2$. The frequency-independent *phase matrix* $\mathbf{P}$ is proportional to $\mathbf{M}$ and normalized such that the first matrix component P_{11} becomes unity when averaging over all directions.

Note that although Eq. (18) is an exact expression for the rest frame in the absence of magnetic fields, it becomes useful for the weak-field case only *after* Doppler averaging has been applied to transform to the observer's frame, since the Zeeman splitting is compared with the Doppler, not the damping width. When the Zeeman splitting is small in comparison with the damping width as well, the Hanle effect vanishes, and the non-magnetic case is retrieved.

It is found convenient to express the normalized phase matrix $\mathbf{P}$ as a sum of three matrices (cf. Stenflo, 1994, p. 216):

$$\mathbf{P} = \sum_{K=0}^{2} \mathbf{P}_K \,. \tag{19}$$

The explicit expressions for these matrices have been given for the weak-field case ($\omega_L \ll \Delta\omega_D$) when the magnetic field vector is oriented along the polar axis in Stenflo (1994, pp. 88–89). The expressions for an arbitrary orientation of the field vector is then obtained simply by applying standard rotation matrices. $\mathbf{P}_0$ represents the isotropic part of $\mathbf{P}$ (with its first matrix element being unity, the remaining elements zero), $\mathbf{P}_1$ represents scattering of the linear polarization (expressed by the anisotropic part of the upper 3×3 portion of $\mathbf{P}$), while $\mathbf{P}_2$ describes scattering of the circular polarization (with only its last matrix element, P_{44}, being $\neq 0$). In the weak-field regime the circular and linear polarizations are uncoupled from each other in the scattering process.

3. Scattering in Quantum Mechanics

In a quantum-mechanical atomic system scattering can occur between different atomic states. When the n, L, and J quantum numbers of the initial and final states are the same, we speak of *Rayleigh* scattering, when they are different we have the more general case of *Raman* scattering. The special case of Rayleigh scattering when the frequency of the incident photon does not differ from the resonant frequency of the atomic transition by much more than the damping width is usually called *resonant* scattering. The corresponding special case of Raman scattering is called *fluorescent* scattering. From this classification hierarchy we see that Raman scattering is the most general concept, and that all the other types of scattering can be considered as special cases of Raman scattering.

Let us denote the set of quantum numbers for the initial, intermediate, and final states by a, b, and f, respectively. Time-dependent perturbation theory then gives the expression for the probability amplitude for scattering from an initial state characterized by the linear polarization unit vector $\mathbf{e}_\beta$ to the final polarization state characterized by $\mathbf{e}_\alpha$,

$$w_{\alpha\beta} \sim \sum_b \frac{\langle f | \hat{\mathbf{r}} \cdot \mathbf{e}_\alpha | b \rangle \langle b | \hat{\mathbf{r}} \cdot \mathbf{e}_\beta | a \rangle}{\omega_{bf} - \omega - i\gamma/2}, \tag{20}$$

where the resonant frequency $\omega_{bf} = (E_b - E_f)/\hbar$ includes the Zeeman displacements of the magnetic substates of b and f, similar to the inclusion of the Zeeman splitting in Eq. (5), which entered in the classical expression for $w_{\alpha\beta}$ in Eq. (13). The scalar products in Eq. (20) produce the same type of geometrical factors as in Eq. (13).

Interference between the substates occur as a result of the cross products in the tensor product $\mathbf{w} \otimes \mathbf{w}^*$ in the same way as in the classical case. The magnetic-field dependent effects that are due to interference between the partially overlapping magnetic substates of the excited state in a scattering process are usually called the *Hanle effect*.

The sum over the states b in Eq. (20) includes the sum over the magnetic quantum numbers M_u of the intermediate, upper (index u) state. To obtain the scattering phase matrix $\mathbf{P}$ we also need to sum over all the initial and final magnetic quantum numbers M_ℓ and M, weighting the initial

substates with their relative populations. This means that we should use for
the components of the tensor product $\mathbf{w} \otimes \mathbf{w}^*$ in Eq. (15)

$$\sum_{M_\ell} N_{J_\ell M_\ell} \sum_M w_{\alpha\beta} w_{\alpha'\beta'}^* \,, \tag{21}$$

where $N_{J_\ell M_\ell}$ is the population of the initial state with the total angular
momentum and magnetic quantum numbers J_ℓ and M_ℓ. In almost all cases
of practical interest it should be a very good approximation to assume complete redistribution between the initial magnetic substates. In this case we
may remove $N_{J_\ell M_\ell}$ from Eq. (21), since it is independent of M_ℓ and only a
common proportionality factor. .

The Zeeman displacement of an energy level is $gM\hbar\omega_L$, where g is the
Landé factor. For a transition between a lower (index ℓ) and upper (index u)
level the classical expression of $-q\omega_L$ for the Zeeman frequency displacement
in Eqs. (5) and (8) has to be replaced by

$$\Delta\omega_H = (g_\ell M_\ell - g_u M_u)\,\omega_L \,. \tag{22}$$

When using Eq. (20) to calculate the phase matrix $\mathbf{P}$ for electric dipole
transitions we find

$$\mathbf{P} = \sum_{K=0}^{2} W_K \mathbf{P}_K \,, \tag{23}$$

where $\mathbf{P}_K$ are the same matrices that we had in the classical expression of
Eq. (19), while the coefficients W_K represent the new feature introduced by
quantum mechanics. W_0 is always unity, while the values of $W_{1,2}$, which are
unity in the classical case, depend on the J quantum numbers of the levels
involved.

3.1. Negative polarizability in Raman scattering

The W_K coefficients have been given in terms of 6-j symbols and numerically listed by Landi Degl'Innocenti (1984), and in terms of simple algebraic
expressions in Stenflo (1994, pp. 188–189). We always have $0 \leq W_K \leq 1$ except in the Raman scattering case when the J quantum number for the final
state differs from that of the initial state by unity. This case is associated
with negative polarizability coefficients.

Since W_2 may be regarded as the fraction of scattering processes that
occur as dipole-type scattering it may seem odd that negative fractions may
occur. An example of a simple Raman scattering case with negative polarizability is the scattering transition $J = 0 \to 1 \to 1$. To understand how the
negative polarizability arises let us consider $90°$ scattering with a magnetic
field perpendicular to the scattering plane and the incoming radiation linearly polarized parallel to the magnetic field. With these assumptions there

will only be excitations with $\Delta M = 0$ (absorption of the π component), and as the initial state has only one magnetic substate, the excited state will only have the $M = 0$ substate populated. However, deexcitation into the $J = 1$, $M = 0$ final state is not allowed, so the emission can only occur in the sigma components, which means that the scattered radiation is linearly polarized perpendicular to the magnetic field. Now we invoke the principle of spectroscopic stability and let the magnetic field strength go to zero. The described property then remains, i.e., the scattered radiation is linearly polarized parallel to the scattering plane, perpendicular to what it would be for normal dipole-type scattering.

4. *J*-state Interference

Quantum-mechanical interference does not only occur between different magnetic substates (the Hanle effect) but also takes place between states of different total angular momentum quantum number (fine-structure components of a multiplet). A particularly striking observational example of this phenomenon is the interference between the scattering transitions of the Ca II H and K lines, which is illustrated in Fig. 3. As explained in Stenflo (1980), the interference between the $J = \frac{1}{2}$ and $\frac{3}{2}$ excited states leads to a polarizability

$$W_2 = \frac{(\nu_K - \nu)^{-2} + 2(\nu_H - \nu)^{-1}(\nu_K - \nu)^{-1}}{(\nu_H - \nu)^{-2} + 2(\nu_K - \nu)^{-2}}. \tag{24}$$

The asymptotic value of W_2 in the far dispersion wings, when $\Delta\nu \gg$ the fine-structure splitting, can easily be obtained for any complex atomic Raman transition with any combination of the J, L, and S quantum numbers by applying a generalized principle of spectroscopic stability. The asymptotic value must be the same as the value obtained by keeping the frequency fixed and instead letting the fine-structure splitting go to zero. Vanishing fine-structure splitting is however physically equivalent to vanishing electron spin, corresponding to $S = 0$, in which case $J = L$. Thus for a scattering transition with an arbitrary value of the spin S, the asymptotic values of its W_K coefficients are the same as the constant W_K for scattering with $S = 0, J = L$.

As a special example of this generalized principle of spectroscopic stability we consider the Ca II H and K case, for which $S = \frac{1}{2}$, $J = \frac{1}{2} \rightarrow \frac{1}{2} \rightarrow \frac{1}{2}$ and $\frac{1}{2} \rightarrow \frac{3}{2} \rightarrow \frac{1}{2}$. Although $W_2 = 0$ and $\frac{1}{2}$, respectively, for the H and K lines if they are considered separately, $W_2 \rightarrow 1$ in the far wings, since $L = 0 \rightarrow 1 \rightarrow 0$, which in the absence of electron spin is the quantum-mechanical analog to the classical case, for which we know that $W_2 = 1$.

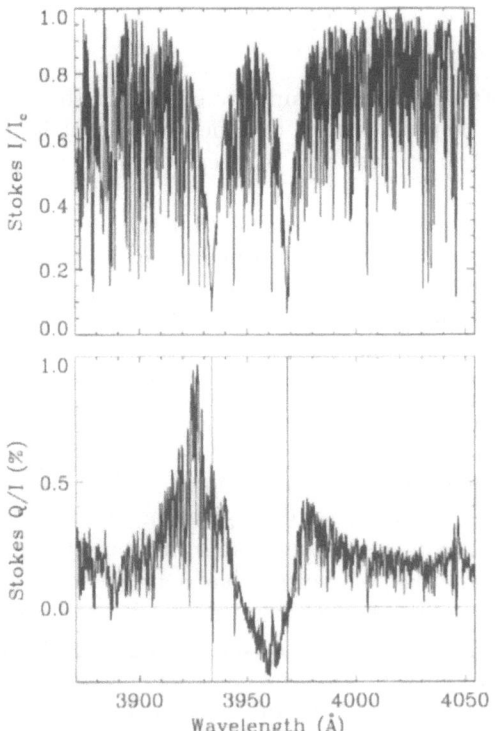

Fig. 3.   Stokes I/I_c (top) and Q/I spectrum of the region around the Ca II H and K lines,
obtained with the vertical spectrograph of the McMath telescope on October 6, 1978
(Stenflo, 1980). The slit was positioned inside the limb near the heliographic north pole at
a disk position of $\mu = 0.14$. The wavelength range that was repeatedly scanned by rotating
the grating covers 3870–4054 Å. The wavelength locations of the K and H resonances are
marked by the two vertical lines in the Q/I diagram. The blend lines generally depolarize
the Ca II polarization, but some lines polarize themselves. Thus the polarization bump
at the left end of the displayed wavelength range is due to the CN molecule, while the
polarizing feature near the right end is due to the 43 Fe I 4045.8 Å line.

If the interference terms did not have their proper signs and amplitudes,
spectroscopic stability would never be achieved. This principle is therefore
very useful for checking the correctness of the often very complicated algebraic expressions that occur in the general J-state interference case.

A background continuum adds another opacity source that competes with
the line scattering opacity. As the continuum polarization varies very slowly
with wavelength, we can represent the W_2 polarizability of the continuum
with a constant b. If all the continuum photons were due to Rayleigh or
Thomson scattering we would expect b to be unity, but since there are other
non-polarizing continuum sources b is smaller. We may define an effective
value of W_2 as

$$W_{2,\text{eff}} = W_2 \frac{\varphi_\nu}{\varphi_\nu + a} + b \frac{a}{\varphi_\nu + a}, \tag{25}$$

where φ_ν here stands for the composite profile for the line opacity, being a superposition of the profiles of all the lines of the contributing fine-structure components, while a represents a ν-independent continuum opacity contribution.

Although ignoring the circumstance that the anisotropy of the radiation field varies across the lines and is different in the continuum, the simple parametrization of Eq. (25) has been remarkably successful in explaining the observed polarization patterns of the Ca II H and K lines, the Na I D_1 and D_2 lines (Stenflo, 1980), and the Mg II h and k lines (Henze and Stenflo, 1987). A more quantitative analysis using the actual opacities in the solar atmosphere and detailed radiative transfer has been done by Auer et al. (1980).

With each new advance in observational techniques that allows us to enter into previously unexplored parameter domains, unexpected physical effects invariably surprise us. An example is given in Fig. 4, which shows a recent ZIMPOL I recording of the region around the Na I D_1 and D_2 lines. Similar to Ca II (cf. Fig. 3), the polarization has a broad, double S-shaped signature with sign reversals due to quantum-mechanical interference between the $J = \frac{1}{2}$ and $\frac{3}{2}$ upper states of the D_1 and D_2 lines. However, superposed on this broad feature we see a strong and narrow polarization peak in the D_1 line that was entirely unexpected but is most likely due to hyperfine structure. In addition, the shape of the D_2 profile is likely to be heavily influenced by partial redistribution effects. The scattering polarization thus appears to become a tool not only for exploring radiative-transfer physics and diagnosing the solar atmosphere, but also for exploring quantum-mechanical aspects of atomic structure. These effects are better exhibited in the "solar laboratory" than in man-made terrestrial laboratories.

5.   Partial Redistribution

5.1. UNPOLARIZED CASE

According to unpolarized standard redistribution theory (cf. Mihalas, 1978) the redistribution function R_0 in the rest frame (indicated by index 0) may be written as

$$R_0/g(\mathbf{n}, \mathbf{n}') = k_c R_{\text{II},0} + (1 - k_c) R_{\text{III},0}, \tag{26}$$

where

$$k_c = \frac{\gamma_N}{\gamma_N + \gamma_c}, \tag{27}$$

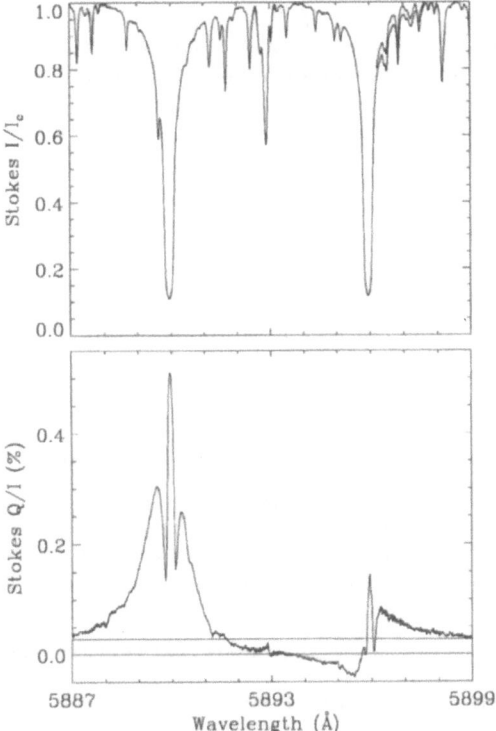

Fig. 4.   Stokes I/I_c (top) and Q/I spectrum of the region around the Na I D$_1$ and D$_2$ lines, obtained with ZIMPOL I at the McMath telescope in April 1995. The slit was positioned inside the limb near the heliographic north pole at a disk position of $\mu = 0.1$. The wavelength range shown has been covered by several partially overlapping scans. While the narrow polarization structures around the core of the D$_1$ line are likely to be due to hyperfine structure, the broad, double S-shaped feature with sign reversal near the wavelength of the D$_1$ resonance is due to quantum-mechanical interference between the $J = \frac{1}{2}$ and $\frac{3}{2}$ upper states of the D$_1$ and D$_2$ lines. The narrow polarization peak in the Doppler core of the D$_2$ line with surrounding wing maxima is likely to be due to partial redistribution effects, although hyperfine structure may also play a role. The estimated level of the continuum polarization is indicated by the horizontal line.

is the fraction of scattering events undisturbed by elastic collisions. Here γ_N is the natural or radiative damping constant, while γ_c is the collisional broadening rate. $g(\mathbf{n}, \mathbf{n}')$ is the *angular phase function* describing the directional probability distribution, and

$$
\begin{aligned}
R_{\mathrm{II},0} &= f(\xi')\,\delta(\xi' - \xi + \nu_{af})\,, \\
R_{\mathrm{III},0} &= f(\xi')\,f(\xi)\,.
\end{aligned}
\tag{28}
$$

$f(\xi')$ is the rest frame absorption profile, and $\nu_{af} = (E_a - E_f)/h$, where $E_{a,f}$ are the energies of the initial and final atomic states (which are different from each other in the general Raman scattering case).

While ξ' and ξ are the frequencies of the incident and scattered beams in the rest frame of the scattering particle, we let ν' and ν denote the corresponding frequencies in the observer's frame. The δ function in Eq. (28) expresses frequency coherence when the scattering process has not been disturbed by a collision, while the emission profile $f(\xi)$ in the second part of Eq. (28) appears as a result of complete frequency redistribution for the scattering events that have experienced elastic collisions.

The redistribution function in the observer's frame has the same form as Eq. (26) and is formally obtained simply by removing index 0. R_{II} and R_{III} are complicated but well defined expressions (cf. Stenflo, 1994, pp. 218–219) that are obtained from Eq. (28) if we replace ξ and ξ' by

$$
\begin{aligned}
\xi &= \nu - \nu_0(\mathbf{v} \cdot \mathbf{n})/c, \\
\xi' &= \nu' - \nu_0(\mathbf{v} \cdot \mathbf{n}')/c
\end{aligned}
\tag{29}
$$

and integrate over a Maxwellian velocity distribution P_{Mx}, where

$$
P_{\mathrm{Mx}}(v_x, v_y, v_z)\, \mathrm{d}v_x\, \mathrm{d}v_y\, \mathrm{d}v_z = \exp[-(v_x^2 + v_y^2 + v_z^2)/v_t^2]\, \mathrm{d}v_x\, \mathrm{d}v_y\, \mathrm{d}v_z.
\tag{30}
$$

Note that our definition of R_{II} and R_{III} does not contain the phase function g in contrast to much of previous literature (e.g. Mihalas, 1978). This definition allows us to retain the identical functions when we make the transition from unpolarized (scalar) to polarized (matrix) scattering.

5.2. POLARIZED CASE WITHOUT MAGNETIC FIELDS

The foundations for the theory of partial redistribution in the polarized case have been established by Omont et al. (1972, 1973), who expressed the redistribution matrix as the sum of contributions with well-defined branching ratios from three distinctly different cases with respect to collisions during the scattering event:

- Case (a): No collisions.
- Case (b): Elastic collisions that destroy the frequency coherence but not the atomic polarization (the $2K$-multipole).
- Case (c): Collisions that destroy both the frequency coherence and the atomic polarization.

For case (a) the branching ratio is k_c given by Eq. (27), while for case (b) it is

$$
\beta^{(K)} = k_c \, \frac{\gamma_c - \gamma_c^{(K)}}{\gamma_N + \gamma_c^{(K)}},
\tag{31}
$$

where upper index K refers to the $2K$-multipole ($K = 0, 1, 2$). $\gamma_c^{(K)}$ represents the destruction rate of the $2K$-multipole. When $K = 0$, $\gamma_c^{(0)}$ represents the rate of quenching collisions (inelastic collisions leading to collisional excitation, deexcitation, or ionization). Since $\gamma_c^{(0)} \ll \gamma_c$,

$$\beta^{(0)} \approx 1 - k_c. \tag{32}$$

Finally the remaining branching ratio, for case (c), is the difference between unity and the sum of the other branching ratios: $1 - k_c - \beta^{(K)}$.

Classical collision theory gives $\gamma_c^{(K)}/\gamma_c = 0.5$ for both $K = 1$ and 2 (Stenflo, 1994, p. 212). Earlier quantum-mechanical treatments using simple dipole-dipole interaction potentials have indicated somewhat different values (Berman and Lamb, 1969), but recent work based on more accurate interaction potentials (Spielfiedel et al., 1991; cf. also Faurobert-Scholl et al., 1995; Faurobert-Scholl, 1996) suggests that the classical value of 0.5 is in fact the best choice to make.

In the classical theory without collisions the redistribution matrix was given by Eq. (18), and the phase matrix $\mathbf{P}$ could be expanded in its multipole components $\mathbf{P}_K$ as in Eq. (19). This expansion also applies to the quantum-mechanical case if we include the W_K polarizability coefficients as in Eq. (23).

With our above-mentioned branching ratios we can now easily generalize Eqs. (18) and (23) to the collisional case as well as to general Raman scattering. It is convenient to write the resulting expression for the rest frame redistribution matrix $\mathbf{R}_0$ in the following form:

$$\mathbf{R}_0 = f(\xi') \sum_{K=0}^{2} p(\xi,\xi')^{(K)} W_K \mathbf{P}_K, \tag{33}$$

where

$$p(\xi,\xi')^{(K)} = k_c \delta(\xi' - \xi + \nu_{af}) + \beta^{(K)} f(\xi) \tag{34}$$

represents the probability of scattering from incoming frequency ξ' to emitted frequency ξ without collisional destruction of the $2K$-multipole (for $K = 1, 2$). Expression (33) is simpler, but equivalent, to the form for $\mathbf{R}_0$ given by Domke and Hubeny (1988).

We note that in the absence of collisions and when $\nu_{af} = 0$ (Rayleigh scattering), the classical redistribution matrix of Eq. (18) is retrieved. With collisions the unpolarized scalar redistribution function R_0 can be retrieved as the first term R_{11} of the $\mathbf{R}_0$ matrix, but it only takes the form of Eq. (26) when $W_{1,2} = 1$ (classical case) and $\gamma_c^{(1)} = \gamma_c^{(2)} = 0$ (which is unphysical). The general form given by Eq. (33) thus also better describes the special case of unpolarized scattering as compared with the "textbook" form of Eq. (26), which does not account for collisional destruction of the directional dependence of the angular phase function.

5.3. WEAK FIELD CASE

In the Hanle regime or weak-field case, when $\Delta\omega_H \ll \Delta\omega_D$, Eq. (33) remains valid, as explained in connection with Eq. (18), since the influence of the Zeeman splitting on the frequency profiles of the different matrix components can be neglected (after Doppler broadening has been applied). In this case the area-normalized absorption profile $f(\xi')$ can be factorized out as in Eqs. (33) and (18). In the observer's frame $R_{\mathrm{II},0}$ and $R_{\mathrm{III},0}$ become R_{II} and R_{III}. Thus we obtain from Eqs. (28), (33), and (34)

$$\mathbf{R}(\nu, \mathbf{n}; \nu', \mathbf{n}') = k_c\, R_{\mathrm{II}}\, \mathbf{P} + R_{\mathrm{III}} \sum_{K=0}^{2} \beta^{(K)}\, W_K \mathbf{P}_K \qquad (35)$$

(cf. Stenflo, 1994, p. 217).

The magnetic-field effects (Hanle effect) are contained in the phase matrices $\mathbf{P}_K$ and $\mathbf{P}$, but from Eq. (35) it cannot be seen how these effects vary from core to wings. The frequency dependence of the Hanle effect is hidden inside the phase matrices. It is however known that the weak-field Hanle effect takes place in the line core but is absent in the wings (cf. Stenflo, 1994, pp. 84 and 98–100). This behavior follows from the expression for the general redistribution matrix for arbitrary field strengths (see Sect. 5.4 below), when we from this expression go to the limit of weak fields. The weak-field Hanle phase matrix that is to be used in the line core has been given by Stenflo (1978) (cf. also Stenflo, 1994, pp. 88–89).

The "Hanle angle" that was introduced in Eq. (16) is in the presence of collisions given by

$$\tan \alpha_K = \frac{K g_u \omega_L}{\gamma_N + \gamma_c^{(K)}}, \qquad (36)$$

where g_u is the Landé factor for the upper (index u), excited state.

If we define

$$k_c^{(K)} = \frac{\gamma_N}{\gamma_N + \gamma_c^{(K)}}, \qquad (37)$$

we get

$$\tan \alpha_K = K\, k_c^{(K)} B/B_0, \qquad (38)$$

where the "characteristic field strength" B_0 for which the Hanle effect is most sensitive is

$$B_0 = \frac{2m\, \gamma_N}{e\, g_e}. \qquad (39)$$

5.4. GENERAL CASE: ARBITRARY FIELD STRENGTH

In the general case, when the Zeeman splitting cannot be assumed to be small as compared with the Doppler broadening, the different components of the polarization matrix $\mathbf{M}$ or $\mathbf{w} \otimes \mathbf{w}^*$ of Eqs. (14) and (15) may have significantly different frequency profiles, even after Doppler averaging through transformation to the observer's frame has been performed. The frequency redistribution and the directional properties of the polarization matrix then become intertwined and cannot be factorized out of each other.

In the absence of collisions the rest frame redistribution matrix then becomes

$$\mathbf{R}_{0,\,\text{no coll.}} = \mathbf{M}\,\delta(\xi' - \xi + \nu_{af}) \tag{40}$$

instead of Eq. (18), since we have frequency coherence, and the absorption profiles are now contained inside $\mathbf{M}$ instead of having been factorized out as in Eq. (18). We implicitly assume in Eq. (40) that $\mathbf{M}$ has been normalized such that its first component M_{11}, when integrated over all frequencies and averaged over all directions, equals unity.

With this normalization $\mathbf{M}$ becomes $f(\xi')\mathbf{P}$ in the zero-field limit. Similar to $\mathbf{P}$ in Eq. (23), we may express $\mathbf{M}$ as the sum of three multipolar components $\mathbf{M}_K$ $(K = 0, 1, 2)$. The frequency dependence of the various matrix terms are determined by $(n_q - 1)(n_{q'}^* - 1)$, where $n_q - 1$ is given by Eq. (5).

In the presence of collisions Eq. (40) can be generalized in the same manner as for Eq. (33), which gives us the general rest frame redistribution matrix

$$\mathbf{R}_0 = \sum_{K=0}^{2} p(\xi, \xi')^{(K)}\, W_K\, \mathbf{M}_K , \tag{41}$$

or, with Eq. (34),

$$\mathbf{R}_0 = k_c\,\delta(\xi' - \xi + \nu_{af})\,\mathbf{M} + f(\xi) \sum_{K=0}^{2} \beta^{(K)}\, W_K \mathbf{M}_K . \tag{42}$$

What remains to be clarified is what the form of the emission profile $f(\xi)$ should be when the Zeeman splitting is not small. The second term in Eq. (42) that contains $f(\xi)$ physically represents those scattering processes, for which the frequency coherence is destroyed by collisions. The resulting emission profile therefore becomes uncorrelated with the excitation process. The intensity profile of such uncorrelated emission is obtained by integrating the first component R_{11} of $\mathbf{R}_0$ over all frequencies and averaging over all directions of the *incident* radiation. The emission profile $f(\xi)$ that is thus obtained represents the probability that an emitted photon appears at frequency ξ (regardless of its polarization state). The polarization properties of that photon are determined by the factor $\sum_{K=0}^{2} \beta^{(K)}\, W_K \mathbf{M}_K$ in Eq. (42).

The averaging of R_{11} gives the result

$$f(\xi) = \phi_I,$$ (43)

where ϕ_I is the first component of the area-normalized Mueller absorption matrix. Explicitly

$$\phi_I = \phi_\Delta \sin^2 \gamma + \tfrac{1}{2}(\phi_+ + \phi_-),$$ (44)

where

$$\phi_\Delta = \tfrac{1}{2}[\phi_0 - \tfrac{1}{2}(\phi_+ + \phi_-)].$$ (45)

ϕ_q, $q = 0, \pm1$, are the area-normalized versions of the Voigt functions $H(a, v_q)$ of Eq. (7), and γ is the angle between the magnetic field and the line of sight.

If we integrate Eq. (42) over a Maxwellian velocity distribution (cf. Eqs. (29) and (30)), we obtain the general redistribution matrix **R** in the observer's frame, which describes how the magnetic-field effects vary from core to wings, including the detailed behavior at all intermediate frequencies. When we go to the limit of weak magnetic fields we get a full description of the transition at intermediate frequencies between the Hanle effect in the line core and the non-magnetic scattering in the line wings.

Acknowledgements

I thank Christoph Keller for collaborating on the observations of the scattering polarization with ZIMPOL at NSO/Kitt Peak.

References

Auer, L.H., Rees, D., Stenflo, J.O.: 1980, *Astron. Astrophys.* **88**, 302
Berman, P.R., Lamb, W.E. Jr.: 1969, *Phys. Rev.* **187**, 221
Bommier, V.: 1980, *Astron. Astrophys.* **87**, 109
Domke, H., Hubeny, I.: 1988, *Astrophys. J.* **334**, 527
Faurobert-Scholl, M.: 1993, *Astron. Astrophys.* **268**, 765
Faurobert-Scholl, M.: 1996, these proceedings
Faurobert-Scholl, M., Feautrier, N., Machefert, F., Petrovay, K., Spielfiedel, A.: 1995, *Astron. Astrophys.* **298**, 289
Henze, W. Jr., Stenflo, J.O.: 1987, *Solar Phys.* **111**, 243
Keller, C.U., Aebersold, F., Egger, U., Povel, H.P., Steiner, P., Stenflo, J.O.: 1992, *LEST Foundation Technical Report* No. 53
Landi Degl'Innocenti, E.: 1984, *Solar Phys.* **91**, 1
Leroy, J.-L., Ratier, G., Bommier, V.: 1977, *Astron. Astrophys.* **54**, 811
Mihalas, D.: 1978, *Stellar Atmospheres*, Freeman and Company, San Francisco
Omont, A., Smith, E.W., Cooper, J.: 1972, *Astrophys. J.* **175**, 185
Omont, A., Smith, E.W., Cooper, J.: 1973, *Astrophys. J.* **182**, 283
Povel, H.P., Keller, C.U., Stenflo, J.O.: 1991, in L.J. November (ed.), *Solar Polarimetry*, NSO/SP Summer Workshop Series No. 11, p. 102

Querfeld, C.W., Smartt, R.N., Bommier, V., Landi Degl'Innocenti, E., House, L.L.: 1985, *Solar Phys.* **96**, 277

Spielfiedel, A., Feautrier, N., Chambaud, G., Lévy, B.: 1991, *J. Phys. B: At. Mol. Opt. Phys.* **24**, 4711

Stenflo, J.O.: 1978, *Astron. Astrophys.* **66**, 241

Stenflo, J.O.: 1980, *Astron. Astrophys.* **84**, 68

Stenflo, J.O.: 1982, *Solar Phys.* **80**, 209

Stenflo, J.O.: 1994, *Solar Magnetic Fields — Polarized Radiation Diagnostics*, Kluwer, Dordrecht

Stenflo, J.O., Keller, C.U.: 1996, to be published

Stenflo, J.O., Baur, T.G., Elmore, D.F.: 1980, *Astron. Astrophys.* **84**, 60

Stenflo, J.O., Twerenbold, D., Harvey, J.W.: 1983a, *Astron. Astrophys. Suppl. Ser.* **52**, 161

Stenflo, J.O., Twerenbold, D., Harvey, J.W., Brault, J.W.: 1983b, *Astron. Astrophys. Suppl. Ser.* **54**, 505

Stenflo, J.O., Keller, C.U., Povel, H.P.: 1992, *LEST Foundation Technical Report* No. **54**

THE DENSITY MATRIX APPROACH TO POLARIZED
RADIATIVE TRANSFER

E. LANDI DEGL'INNOCENTI*

Instituto de Astrofísica de Canarias, E-38200 La Laguna, Tenerife, España

Abstract. The density matrix approach to polarized radiative transfer is reviewed, with particular emphasis on the physical assumptions that are at the basis of the recent developments achieved by means of this formalism. In particular it is shown that two of the basic hypotheses (the hypothesis of neglecting correlation effects between thermal velocity and density matrix, and the hypothesis of neglecting atomic polarization in the atomic ground level) are highly questionable for the description of resonance polarization –and its modifications due to the presence of a magnetic field– in spectral lines formed in the solar atmosphere.

Key words: Spectropolarimetry – Radiative transfer – Line Formation

1. Introduction

Density Matrix is a fundamental concept in atomic physics theory. Introduced in the literature almost 70 years ago (von Neumann, 1927) this concept has become quite popular among atomic physicists since the late forties and has proved to be of fundamental importance for the description of optical pumping experiments. More particularly, the formalism of the irreducible tensor components of the density matrix (sometimes called the *statistical tensors* –see Omont, 1977, for a detailed review of the subject) has shown to be particularly suitable also for astrophysical applications.

The first of such applications were developed for the interpretation of the linear polarization observed in the He I D_3 line in the spectrum of prominences at the solar limb (Bommier, 1977; Bommier and Sahal-Bréchot, 1978; Landi Degl'Innocenti, 1982). In these previous works, only the equations of statistical equilibrium for the statistical tensors were solved, as the radiation field that is illuminating the He atoms of the solar prominence was supposed to be known (being the photospheric radiation field, not modified by the prominence plasma, assumed as optically thin in the He lines).

Following some advancements in the theory of the generation and transfer of polarized radiation (Landi Degl'Innocenti, 1983, 1984, 1985), a self-consistent formalism has been recently developed. This formalism is based on the use of statistical tensors, and is capable of describing the physics

* On leave from the *Dipartimento di Astronomia e Scienza dello Spazio, Università di Firenze, Largo E. Fermi 5, I-50125 Firenze, Italia*

of resonance polarization in an optically thick, magnetized medium. The approach is the following.

The values of the statistical tensors at any assigned point of the medium are coupled to the polarized radiation field, and, through the effect of collisions, to the local thermodynamical properties of the medium (like temperature and pressure). This coupling is described by the statistical equilibrium equations for the statistical tensors, which also contain the effect of the magnetic field. On the other hand, the polarized radiation field propagating at any assigned point and along any given direction, is coupled, through the radiative transfer equations, to the local value of the statistical tensors.

The formalism thus developed is nothing but the generalization (to the "polarized case") of the well-known non-LTE problem for a multi-level atom, with the only difference that the atom is described by its statistical tensors (rather than by level-populations) while the radiation field is described by the four Stokes parameters (and not simply by its intensity). For this more general non-LTE problem the name of *non-LTE of the 2nd kind* has been suggested (Landi Degl'Innocenti, 1987).

In the framework of this theory, some results have been obtained for a simplified two-level atomic model with unpolarized ground level and with stimulated emission neglected. The case of the Hanle effect regime, when the Zeeman splitting is of the order of the natural broadening of the upper level, but completely negligible with respect to the Doppler broadening of the line, has been developed in Landi Degl'Innocenti, Bommier and Sahal-Bréchot (1990) and in Bommier, Landi Degl'Innocenti and Sahal-Bréchot (1991); the first paper deals with the general formulation for a medium of arbitrary shape, while the second paper considers the case of a plane-parallel atmosphere. The case of the *strong field* regime, when the Zeeman splitting is much larger than the natural broadening of the upper level (so that coherences can be neglected), has been developed in Landi Degl'Innocenti, Bommier and Sahal-Bréchot (1991a,b). Again, the first paper deals with the general formulation, while the second with the case of a plane-parallel atmosphere.

It is the aim of the present contribution to re-examine in detail the physical assumptions that are at the basis of the results obtained in the papers quoted above.

2. Physical Assumptions

Consider a medium of arbitrary shape composed of atoms of a given species, plus a collection of "perturbers", such as electrons, Hydrogen and Helium atoms, and atoms (or molecules) of different species. The medium is permeated by a magnetic field $\mathbf{B}(P)$ which is, in general, a function of the point P.

All atoms have maxwellian distributions of velocities locally characterized by a unique temperature $T(P)$. The medium is, in general, non-static in the sense that there are macroscopic velocity fields described by the function $\mathbf{V}(P)$. The problem is the one of finding, for any assigned direction and frequency, the radiation emerging from such a medium. Specifically, one is interested in obtaining the spectropolarimetric profiles of the lines belonging to the atomic species considered.

This problem, which is the obvious generalization to the "polarized" case of the standard non-LTE problem, is, in its generality, enormously complex. For the time being, the attempts to find its solution have been based on rather drastic assumptions. These are discussed in the following.

2.1. FACTORIZATION OF THE DENSITY MATRIX

At any point P of the medium, the atoms of the species considered can be described, in full generality, by an overall density matrix of the form $R(P; \mathbf{v}, n, m)$. The diagonal terms $(m = n)$ represent the joint probability of finding the atom with velocity $\mathbf{v}$ and "sitting" in the atomic level specified by the (collection of) quantum number(s) n. The non-diagonal terms $(m \neq n)$ represent the analogous probability of finding the atom with velocity $\mathbf{v}$ and showing a *coherence* between the atomic levels n and m.

The assumption that is currently made is that of considering the joint probability as the product of two separate probabilities, so that the overall density matrix can be written in the form

$$R(P; \mathbf{v}, n, m) = F(P; \mathbf{v} - \mathbf{V}) \, \rho(P; n, m) \tag{1}$$

where $F(P; \mathbf{v} - \mathbf{V})$ is the standard maxwellian distribution of velocities (defined in the local frame), while $\rho(P; n, m)$ is the usual atomic density matrix.

The approximation described by this equation is rather rough, because it completely neglects correlation effects between velocity and atomic density matrix. It can be expected that such correlations may indeed exist, and they may turn out to be particularly important for polarization phenomena.

To understand, in a qualitative way, how such correlations can originate, let us consider the particular case of a static atmosphere, with a vertical magnetic field. Let us consider, for the sake of simplicity, two-level atoms having a lower level with $J_l = 0$ and an upper level with $J_u = 1$ (J_l and J_u are the angular momentum quantum numbers). Among such atoms, consider one that is sitting close to the top of the atmosphere and which has an upward velocity such that the Doppler shift is comparable with the Zeeman splitting due to the magnetic field. As this atom "sees" a redshifted radiation field with a well formed absorption line, its Zeeman sublevel $M_u = 1$ (M_u is the magnetic quantum number) will be overpopulated with respect to the sublevel $M_u = 0$ and, even more conspicuously, with respect to the sublevel $M_u = -1$. The opposite will be true for an atom that is moving downwards, instead of upwards. The net result of such a physical mechanism is that the atomic density matrix depends on the velocity of the atom.

In principle, one may think that collisions could destroy such correlation effects. However, it is easy to show that, when velocity-density matrix correlations (VDMC) are negligible, the atomic density matrix becomes diagonal, which implies that there are no more polarization effects due to the anisotropy of the radiation field (such as resonance polarization or the Hanle effect). The reason for this fact is that there are three typical relaxation times involved in the problem: τ_{rad}, the radiative mean-free-time of the atom in the upper level (the inverse of the Einstein coefficient for spontaneous emission), τ_{dep}, the mean-free-time between two successive depolarizing collisions, and τ_{col}, the mean-free-time between two successive, velocity-changing collisions. As the cross-sections for depolarizing collisions are always larger than the cross-sections for velocity-changing collisions (because only collisions with a very small impact parameter are capable of changing the velocity of the atom, whereas larger impact parameters contribute to its depolarization by the Van der Waals interaction potential), it always results:

$$\tau_{dep} \ll \tau_{col} \; , \tag{2}$$

the ratio between the two mean-free-times being estimated, typically, in a value of the order of 10^{-2}. On the other hand, to avoid VDMC it would be necessary that, while the atom is excited, it suffers a large number of velocity-changing collisions, or, in other words, it would be necessary that:

$$\tau_{col} \ll \tau_{rad} \; . \tag{3}$$

However, from inequalities (2) and (3) it follows:

$$\tau_{dep} \ll \tau_{rad} \; , \tag{4}$$

which is just the condition implying that the atomic density matrix is diagonal (in the Zeeman sublevels' subspace of a given level).

Relaxing the approximation described by Equation (1) is, however, a rather complicated matter because it would entail the introduction into the problem of a substantially larger number of unknowns (the statistical tensors at any point P *and for any velocity* v). The approximation described by Equation (1) can then be considered as an *Ansatz* to reduce to its simpler form a problem that –although solvable with modern machines– would require a conspicuous increase in computing resources.

It has also to be mentioned that in a different context, namely in the framework of the redistribution theory of scattering, neglecting correlation effects related to the velocity of the atoms may well result in being a good approximation. For instance, quoting from Mihalas (1978, pag. 430) *the case of noncoherence in the atom's frame may be treated as noncoherence in the laboratory frame also, without serious errors.* However, this statement has been proved only for the non-polarized case and detailed calculations are necessary in order to be confident with the approximation described by Equation (1) when polarization is concerned.

2.2. Two-level Atom with Unpolarized Ground Level

This further approximation has mainly to do with the overall number of unknowns that are needed to fully describe (from the density matrix point of view) the excitation state of the atom in a particular point P of the medium. Considering, for instance, a multilevel atom composed of N levels, each having angular momentum J_i $(i = 1, ..., N)$, the overall number of density matrix elements to be considered is given by

$$N_{\text{elements}} = \sum_{i=1}^{N} (2J_i + 1)^2 , \qquad (5)$$

a number that can attain pretty large values even when N is relatively small. Refer for instance to the case of a model atom with $N = 5$. Assuming 2 as a typical average value for J_i, one gets $N_{\text{elements}} = 125$, which starts being a quite conspicuous number of unknowns (remember that this has to be multiplied by the overall number of grid-points).

The easiest way of reducing the number of unknowns, still keeping some degree of significance in the physical problem, is the one of restricting the analysis to a two-level atom and, moreover, to assume that the ground level does not have any atomic polarization. By so doing, one gets for the number of unknowns the more tractable value $(2J_u + 1)^2$, where J_u is the angular momentum of the upper level.

It has also to be remarked that these assumptions (together with the further assumption of neglecting stimulated emission, see next subsection) are also fundamental to keep the problem simpler from the mathematical point of view. Indeed, with these assumptions, the problem is *linear* because the unknowns (the statistical tensors of the upper level at the different grid-points) do not enter explicitly into the coupling coefficients relating the statistical tensor at point P with the statistical tensors at each of the other grid-points P'.

The approximation of neglecting atomic polarization in the lower level is, however, a rather questionable one, apart from the particular case $J_l = 0$ (J_l is the angular momentum of the lower level) where, obviously, it is strictly satisfied. For this approximation to be valid one has to invoke depolarizing collisions, or, more precisely, one has to suppose that the typical mean-free-time $\tau_{\text{dep}}^{(l)}$ for depolarizing collisions of the lower level is much smaller than the mean-free-time τ_{abs} for absorption from the lower level:

$$\tau_{\text{dep}}^{(l)} \ll \tau_{\text{abs}} \ . \tag{6}$$

On the other hand, as the mean-free-times for depolarizing collisions in the lower and upper levels are of the same order of magnitude, a physical situation where the lower level is unpolarized and, at the same time, the upper level is polarized, can be obtained only if:

$$\tau_{\text{rad}} < \tau_{\text{dep}}^{(u)} \approx \tau_{\text{dep}}^{(l)} \ll \tau_{\text{abs}} \ , \tag{7}$$

where τ_{rad} is the quantity introduced in the former subsection. This chain of inequalities leads to the following condition:

$$\tau_{\text{rad}} \ll \tau_{\text{abs}} \ , \tag{8}$$

which is only partially satisfied for optical lines in the solar atmosphere (where, typically, τ_{rad} is only 10 times smaller than τ_{abs}).

These remarks point to the fact that the hypothesis of neglecting atomic polarization in the lower level may be a poor one for describing the physics of resonance scattering in the solar atmosphere. As the inclusion of such physical phenomenon dramatically changes the signature of resonance polarization (Landolfi and Landi Degl'Innocenti, 1986), it can be suggested that the rather peculiar polarization signals recently observed in some resonance lines (Stenflo, 1996) may be due to the presence of atomic polarization in the lower level.

2.3. NEGLECTING STIMULATED EMISSION

Neglecting stimulated emission simplifies the statistical equilibrium equations for the statistical tensors of the upper level, and, moreover, avoids the introduction of non-linearities into the mathematical problem (see previous subsection). For this reason, the physical phenomena caused by the stimulation of polarized radiation have been systematically neglected in the papers quoted in the introduction.

The importance of such phenomena can be estimated by comparing the characteristic mean-free-times involved. In this case one has to compare τ_{rad} with τ_{stim}, the mean-free-time for stimulated de-excitation of the upper level. It turns out that the ratio of these two quantities is, typically, of the order of 10% for optical lines in the solar atmosphere (indeed $\tau_{\text{stim}} \approx \tau_{\text{abs}}$, see previous subsection).

It can be concluded that resonance polarization evaluated by neglecting stimulated emission can be in error by a factor of the order of 10%. The situation changes dramatically, however, for infrared lines, where the ratio between τ_{rad} and τ_{stim}, can approach unity. Any diagnostic of resonance polarization in infrared lines cannot be reasonably performed without taking stimulated emission properly into account.

3. Discussion and Conclusions

The physical assumptions described in the former section, and, more particularly, those itemized in Sections 2.1 and 2.2, are indeed rather questionable. For the time being, they can just be considered as working hypotheses which reduce to its simplest form a physical problem of extreme complexity.

Once these hypotheses are incorporated in the relevant equations (radiative transfer equations for polarized radiation, plus equilibrium equations for the statistical tensors), one finds a coupled system of integral, linear equations where the unknowns are the statistical tensors of the upper level at each point P of the medium. So far, these system of equations has been solved numerically only for the case of a semi-infinite, plane-parallel, static atmosphere, in the *Hanle effect* regime and in the *strong field* regime. For an isothermal atmosphere an important analytical result has also been found (Landi Degl'Innocenti and Bommier, 1994) which generalizes to the "polarized case" the well known $\sqrt{\epsilon}$-law for the source function at the surface of the atmosphere. Such a generalization, in its simplest form, can be expressed through the equation

$$\sqrt{\sum_{K,Q} [S_Q^K(0)]^2} = \sqrt{\epsilon} \quad , \tag{9}$$

where $S_Q^K(0)$ is the *statistical source function* (proportional to the statistical tensor of the upper level) evaluated at the top of the atmosphere and where ϵ is the standard parameter of the non-LTE theory which describes the coupling with the thermal properties of the atmosphere.

The density matrix approach to polarized radiative transfer describes the physics of resonance scattering (and its modifications due to the presence of a magnetic field) in the so-called approximation of *complete redistribution in frequency*. Other approaches are currently used to treat the same physical phenomena (see Faurobert-Scholl, 1996, for a review). These alternative approaches are based on the theory of *partial redistribution in frequency* developed by Omont, Smith and Cooper (1972, 1973).

Each of the two approaches has its own advantages and disadvantages. Although with the density matrix formalism (as developed up to the present time) it is not possible to have a realistic description of the strongest resonance lines that are formed in the solar atmosphere, it is possible to describe with such a formalism a number of physical phenomena (like, for instance, the presence of atomic polarization in the lower level) that cannot be described in the framework of the present theory of partial redistribution. It is the impression of the author that a definite progress in this field can only be achieved with the development of a new theoretical scheme capable of unifying the two theoretical approaches mentioned above, or, in other words, capable of incorporating the effects of partial redistribution into the framework of the density matrix formalism. The contribution presented by Bommier in these proceedings is probably a first step forward in this direction.

Acknowledgements

The author acknowledges the support of the spanish DGYCIT for a sabbatical leave at the Instituto de Astrofísica de Canarias. Thanks are due to J.C. del Toro Iniesta, J. Trujillo Bueno, and V. Bommier for helping in the presentation of the paper.

References

Bommier, V.: 1977, *Thèse de 3ème cycle*, Université de Paris VI
Bommier, V., Landi Degl'Innocenti, E., and Sahal-Bréchot, S.: 1991, *Astron. Astrophys.* **244**, 383
Bommier, V. and Sahal-Bréchot, S.: 1978, *Astron. Astrophys.* **69**, 57
Faurobert-Scholl, M.: 1996, *in these Proceedings*
Landi Degl'Innocenti, E.: 1982, *Solar Phys.* **79**, 291
Landi Degl'Innocenti, E.: 1983, *Solar Phys.* **85**, 3
Landi Degl'Innocenti, E.: 1984, *Solar Phys.* **91**, 1
Landi Degl'Innocenti, E.: 1985, *Solar Phys.* **102**, 1
Landi Degl'Innocenti, E.: 1987, in W. Kalkofen (ed.), *Numerical Radiative Transfer*, Cambridge University Press, 265
Landi Degl'Innocenti, E. and Bommier, V.: 1994, *Astron. Astrophys.* **284**, 865
Landi Degl'Innocenti, E., Bommier, V., and Sahal-Bréchot, S.: 1990, *Astron. Astrophys.* **235**, 459
Landi Degl'Innocenti, E., Bommier, V., and Sahal-Bréchot, S.: 1991a, *Astron. Astrophys.* **244**, 391
Landi Degl'Innocenti, E., Bommier, V., and Sahal-Bréchot, S.: 1991b, *Astron. Astrophys.* **244**, 401
Landolfi, M. and Landi Degl'Innocenti, E.: 1986, *Astron. Astrophys.* **167**, 200
Mihalas, D.: 1978, *Stellar Atmospheres, 2nd Ed.*, W.H. Freeman and Co., New York
Omont, A.: 1977, *Prog. Quantum Electron.* **5**, 69
Omont, A., Smith, E.W., and Cooper, J.: 1972, *Astrophys. Journ.* **175**, 185
Omont, A., Smith, E.W., and Cooper, J.: 1973, *Astrophys. Journ.* **182**, 283
Stenflo, J.O.: 1996, *in these Proceedings*
von Neumann, J.: 1927, Göttingen Nachrichten **245**

ATOMIC COHERENCES
AND LEVEL-CROSSINGS PHYSICS

VÉRONIQUE BOMMIER

Laboratoire 'Atomes et Molécules en Astrophysique', CNRS URA 812 – DAMAp,
Observatoire de Paris, Section de Meudon, F-92195 Meudon Cedex, France

Abstract. A synthesis work about the interaction of matter with polarized radiation, applied to solar magnetic field diagnostics, has recently been done by Stenflo (1994). This synthesis uses the classical theory of matter-radiation interaction – supplemented by the theory of partial redistribution of Omont, Smith, and Cooper (1972), on the one hand, and full quantum matter-radiation interaction theory, unable to take into account the partial frequency redistribution effects, on the other hand. The need of a full quantum approach taking into account the partial frequency redistribution effects appears as a unifying purpose; the present work, using the density matrix formalism, is a first attempt in this direction.

Key words: Density Matrix – Radiative Transfer – Partial Redistribution

1. Introduction

Scattering physics is at the basis of the interpretation of line profiles emitted from optically thick (or thin) atmospheres. Although it has been widely studied for a long time through classical and semi-classical approaches (see a synthesis in Stenflo, 1994), a full quantum approach has been developed in the two last decades, taking advantage of the density matrix formalism. The main features of this formalism, together with its main advantages for astrophysical purposes, are summarized in Section 2 of this paper. This formalism is particularly fruitful for polarization studies: the signature of the emitted polarization is contained in the atomic density matrix, including its off-diagonal elements called *coherences*, which can be modified by the magnetic field in the vicinity of level-crossings, leading to the Hanle effect. The development of the formalism has induced the possibility of magnetic field diagnostics via the Hanle effect in solar prominences (see Section 1 of Bommier *et al.*, 1994, and references therein).

For optically thin lines, the polarization of the emitted line can be computed by solving the statistical equilibrium equation for the atomic density matrix. However, as optically thick lines can also be found in astrophysics, the statistical equilibrium equation for the atomic density matrix has to be supplemented by the transfer equation for polarized radiation. Both equations can be derived from the Schrödinger equation through the Master Equation theory, which we briefly describe in Section 3 of the present paper (the main references are given there), putting the emphasis on the main physical assumptions that are at the basis of this theory.

When both equations are obtained, they have to be solved in a consistent manner to get a solution of the non-LTE problem, taking polarization into account. A new scheme for the solution, in the density matrix formalism, has recently been proposed by E. Landi Degl'Innocenti (see also Landi Degl'Innocenti, 1996 – these proceedings). It is briefly described in Section 4 of this paper, where it is also shown how the usual transfer equation for the Stokes parameters can be recovered from the density matrix approach.

However, the density matrix approach mentioned above is unable to take into account partial frequency redistribution – the main physical assumptions, which are at the basis of the method and which are summarized in Subsection 3.1, induce complete frequency redistribution in the scattering process. It is well known, since the fundamental work of Omont, Smith, and Cooper (1972), that the frequency redistribution function in the atomic rest frame is not only the one resulting from complete redistribution: coherent scattering has also to be taken into account. It is also well known that partial frequency redistribution plays an important role on the shape of thick resonance lines emitted from the solar atmosphere. Nevertheless, the density matrix approach mentioned above does not use frequency redistribution functions as basic tools – although it is able to derive them as results. The result of the theoretical work of Omont, Smith, and Cooper (1972) cannot then be simply added to the density matrix approach, in order to make it able to describe coherent scattering phenomena.

The author has obtained that coherent scattering phenomena can be described in the density matrix approach if the perturbation development of the matter-radiation interaction is driven up to orders higher than the lowest non-zero one, namely the order 2. If order 4 terms are taken into account, as briefly described in Subsection 3.2, the previous results of Omont, Smith, and Cooper (1972), generalized to the polarized case by Domke and Hubeny (1988), are fully recovered (see Subsection 5.2.1). New results can be inferred in the presence of a weak magnetic field, responsible of the Hanle effect (see Subsection 5.2.2). If orders higher than 4 are introduced, non-negligible contributions appear and new results are derived, which are given in Subsection 5.3 in the usual form of the frequency redistribution functions. However, the detailed calculation of this perturbation development cannot be included in the present paper, which has to be considered only as a preliminary one: the method will be published in a following paper devoted to this particular aim. Only the main physical assumptions that are at the basis of the development are given here in Section 3, and the results on the partial frequency redistribution are provided in the usual form of the frequency redistribution functions in Section 5.

2. Atomic Density Matrix definition and properties

2.1. DEFINITION AND INTEREST OF THE ATOMIC DENSITY MATRIX

For the description of an atomic system, the density matrix has been first introduced by von Neumann (1927). Application to the theory of optical pumping by polarized radiation has then be developed by Fano (1957), using an operator formalism. Well-written introductory lecture notes on the density matrix can be found in Messiah (1969, Chapter VIII, §21) and Cohen-Tannoudji, Diu, and Laloë (1977, Complément E_{III}); see also Blum (1981). Let us only summarize below the main interest of this formalism, for astrophysical purposes.

For one single (isolated) atom, the wavefunction $|\psi\rangle$ and the density matrix defined by

$$\rho = |\psi\rangle\langle\psi| \tag{1}$$

are two exactly equivalent descriptions of the considered physical system. The particular interest of the density matrix, with respect to the wavefunction, appears when one is interested in the description of an *ensemble* of N identical atoms, because it is then possible to define a density matrix for the total system of N atoms by averaging the density matrices ρ_i of individual atoms,

$$\rho = \frac{1}{N}\sum_{i=1}^{N}\rho_i \quad , \tag{2}$$

whereas it is not possible to define an average wavefunction (the wavefunction of the total system being the symmetrized – or anti-symmetrized – product of individual wavefunctions). The density matrix ρ of the total system can be obtained by solving the Schrödinger equation

$$i\hbar\frac{d}{dt}\rho = [H,\rho] \quad , \tag{3}$$

where H is the Hamiltonian of one single atom (here and in the following the atoms are also assumed to be independent), which can be solved even if the individual density matrices ρ_i are not evaluated. Then, an average information on the system can be obtained through ρ, even if a detailed information (*i.e.*, on each particle) cannot be obtained, or is not desirable. In that sense, the density matrix ρ describes a 'mean atom'.

Therefore, the density matrix is able to describe a system, in which the atoms are not all in the same state (*i.e.*, the system state is a 'statistical mixture' of states); in particular, it is able to describe quantitatively partial anisotropies. It is then well suited to the quantitative modelisation of natural systems (considered as opposite to laboratory systems, which may have been prepared in a given 'pure' state), such as those encountered in astrophysics.

2.2. ATOMIC DENSITY MATRIX ELEMENTS: POPULATIONS AND COHERENCES

The density matrix elements have to be distinguished, as follows:

i) the *diagonal* elements of the atomic density matrix $\rho_{kk} = \langle k|\rho|k \rangle$ give the average probability of finding one atom of the system in a given state $|k\rangle$; this is the well-known so-called k-state *population*[1];

ii) the *off-diagonal* elements of the atomic density matrix $\rho_{kl} = \langle k|\rho|l \rangle$ give the average interference between the probability amplitudes associated with the states $|k\rangle$ and $|l\rangle$; this is the so-called *coherence* between the k-state and the l-state.

At this stage, it must be noticed that there is no physical difference between populations and coherences; the separation between populations and coherences depends on the chosen basis $\{|k\rangle\}$: in a basis (or axes) change, populations can be transformed into coherences and viceversa.

The density matrix is an hermitian positive operator, which implies that a population is necessarily given by a real positive quantity, whereas the coherences, which are given by complex numbers, obey the Schwartz inequality

$$|\rho_{kl}|^2 \leq \rho_{kk}\rho_{ll} , \tag{4}$$

which implies in particular that a coherence can be non-zero only if the two states that it connects are populated.

For an isolated system of N identical independent atoms, and if $\{|k\rangle\}$ is the basis of eigenstates of the Hamiltonian H (eigenvalues E_k = level energies), the solution of the Schrödinger equation (3) is obvious and implies that:

i) the populations are constant (time-independent)

ii) the coherences are oscillating quantities at the Bohr frequencies of the atomic system (in pulsation units)

$$\omega_{kl} = \frac{E_k - E_l}{\hbar} . \tag{5}$$

One can then distinguish between *Zeeman coherences*, connecting Zeeman substates, which typically oscillate at frequencies of a few MHz/Gauss, and *optical coherences*, connecting the upper and lower level of a line, which oscillate at the optical frequencies $10^8 - 10^9$ MHz.

---

[1]The normalization condition of the density matrix is given by $\mathrm{Tr}\rho = \sum_k \langle k|\rho|k \rangle = 1$

If the system is not isolated, one has to take into account the interaction between the system and its surroundings; this can be qualitatively achieved by adding a *non-hermitian* damping term[2] $-i\hbar\Gamma/2$ to the energies of the excited sublevels[3] (see Cohen-Tannoudji, 1977); the Zeeman coherence evolution (in the excited level) is then given by

$$\rho_{kl}(t) = e^{-[\Gamma + i\omega_{kl}]t}\rho_{kl}(0) \qquad , \tag{6}$$

which means that the coherence is both oscillating (due to the presence of ω_{kl}), or rotating in the complex plane, and damped (due to the presence of Γ), which leads to the consideration of two regimes:

i) $\omega_{kl} \approx \Gamma \quad \Leftrightarrow \quad \omega_{kl}\tau \approx 1$ (where τ is the lifetime of the upper level $\tau = 1/\Gamma$): the characteristic times of the oscillation and of the damping are of the same order of magnitude. Then the time-averaged coherence is partially, but not totally destroyed; if the Bohr frequency is non-zero, the time-average leads moreover to a non-zero average rotation in the complex plane;

ii) $\omega_{kl} \gg \Gamma \quad \Leftrightarrow \quad \omega_{kl}\tau \gg 1$: the coherence performs many oscillations before being damped; then, the time-averaged coherence is totally destroyed.

The first of these two regimes is that of the Hanle effect, the damping being due to the spontaneous emission: the magnetic field acts upon the Bohr frequencies of the upper level, and then modifies the time-averaged coherences; if the magnetic field strength corresponds to the first regime (which gives the sensitivity domain of the Hanle effect, typically $B \approx 10$ Gauss for $\Lambda \approx 10^8$ s^{-1}), the coherences, and then the emitted line polarization, are modified but not destroyed by the magnetic field, leading to the Hanle effect (depolarization and rotation of the linear polarization direction due to the magnetic field).

If the magnetic field is large enough to be in the second regime, the Zeeman coherences are totally destroyed.

More generally, coherences can appear when level-crossings occur, leading to vanishing Bohr frequencies in their vicinity: this is obviously the case of the zero-field level-crossings due to the Zeeman effect, where the usual Hanle effect takes place; this can be also the case when non-zero-field level-crossings occur, such as those encountered with the Helium He I D$_3$ line $3d^3D_{3,2,1} \rightarrow 2p^3P_{2,1,0}$, owing to its non-zero fine structure, combined with the Zeeman effect on sublevels energies; for values of the magnetic field close to those producing a level-crossing, a peculiar form of the Hanle effect can take place (see Bommier, 1980).

---

[2]The derivation of this behavior requires in fact the theory of the Master equation, together with a quantum description of matter-radiation interaction; this is further described in the present paper.

[3]Presently, one considers, for simplicity purposes, 2-(degenerate) level atoms.

The Zeeman splitting has then to be compared to the natural width, to determine whether the Zeeman coherences have to be considered, and to the total ($\approx$ Doppler) width of the line, to determine whether the Zeeman splitting has to be taken into account in the line profile polarization. In physical conditions typical of the solar atmosphere, the Doppler width of spectral lines is much larger (≈ 500 times) than their natural width. Taking this fact into account, Landi Degl'Innocenti, Bommier and Sahal-Bréchot have been able to give a solution of the non-LTE problem for polarized transfer in two field regimes:

i) the weak field regime ($B \approx 10$ Gauss), where the Zeeman coherences are non-zero and the Zeeman splitting is negligible in the line profile: this is the Hanle effect regime (Landi Degl'Innocenti, Bommier, and Sahal-Bréchot, 1990; Bommier, Landi Degl'Innocenti, and Sahal-Bréchot, 1991)

ii) the intermediate field regime, ($B \approx 100 - 5000$ Gauss), where the Zeeman coherences can be neglected, whereas the Zeeman splitting is non-negligible in the line profile (Landi Degl'Innocenti, Bommier, and Sahal-Bréchot, 1991a,b; Landi Degl'Innocenti, and Bommier, 1994).

2.3. IRREDUCIBLE TENSORIAL OPERATORS BASIS

In many cases it is preferable to expand the density matrix in terms of the basis of irreducible tensorial operators $^{JJ}T_Q^K$ (which form an orthonormal basis of the Hilbert space of operators, irreducible with respect to the rotation group), rather than in terms of the basis of dyadic operators $|l\rangle\langle k|$ associated with the Hamiltonian eigenstates

$$\rho = \sum_{\alpha J} \sum_{KQ} \left(^{\alpha J}\rho_Q^K \right)\left(^{JJ}T_Q^K \right) \quad , \tag{7}$$

where the summation runs over the atomic states denoted by αJ. The reason for this lies in the linear relations between ITOs[4] in a frame rotation, which make such an operation easier to achieve; in astrophysical situations, frame rotations are frequent, for instance from the quantization reference frame (whose OZ axis is given by the magnetic field direction), to the observer's reference frame (whose Oz axis is given by the line-of-sight).

Definition of the ITOs can be found for instance in Brink and Satchler (1968), or in other books devoted to the angular momentum theory (see also Messiah, 1969). Let us recall that $^{JJ}T_Q^K$ is associated with the level angular momentum J, that K is a positive or null integer ranging from 0 to $2J$, and Q is a relative integer ranging from $-K$ to K.

In a frame rotation, K is conserved and the linear expansion runs over Q only. This is why, if symmetries can be found in the physical problem under

---

[4]Abbreviation for 'Irreducible Tensorial Operators'

study, they cause simplifications (vanishing terms) in the various operators or relationships. For instance, if the physical system is isotropic, only the $K = Q = 0$ component of its density matrix is non-vanishing; if the physical system is axially symmetric about OZ (which is the case if the magnetic field lies along the preferred direction of the incident radiation, for instance), all the $Q \neq 0$ elements of the density matrix vanish. Because $Q = M - M'$, where M and M' are magnetic quantum numbers associated with two Zeeman sublevels, all the coherences connecting different Zeeman sublevels vanish: there is then no Hanle effect in that case (vertical magnetic field in the solar atmosphere geometry).

The various components of the density matrix expanded in terms of the ITOs basis are denoted as follows:

i) the $K = Q = 0$ element is called *population*, because it is simply proportional to the overall population of the αJ level

$$^{\alpha J}\rho_0^0 = \frac{1}{\sqrt{2J+1}} \sum_M {}^{\alpha J}\rho_{MM} \quad . \tag{8}$$

ii) the three $K = 1$, $Q = 0, \pm 1$ elements are called *orientation* components, because they are proportional to the average values of the 3 components of the angular momentum J; for instance

$$^{\alpha J}\rho_0^1 \propto \sum_M M \, {}^{\alpha J}\rho_{MM} = \langle J_Z \rangle \quad . \tag{9}$$

If at least one of them is non-zero, this means that the average angular momentum of the system is non-zero; due to the angular momentum conservation law, this angular momentum has, in most cases[5], to be brought from outside; in most of the systems we have studied, this is not the case, and the orientation vanishes: the medium is said to be 'non-oriented'. This implies that the sublevels M and $-M$ are equally populated. As for the photons states, this implies also that the atom is not able to emit non-zero integrated circular polarization.

iii) the five $K = 2$, $Q = 0, \pm 1, \pm 2$ components are called *alignment* components; they describe the anisotropy of the medium, which can be seen for instance on the $K = 2$, $Q = 0$ component, which is proportional to

$$^{\alpha J}\rho_0^2 \propto \sum_M \left[3M^2 - J(J+1) \right] {}^{\alpha J}\rho_{MM}$$

$$= 3\langle J_Z^2 \rangle - \langle J^2 \rangle = 2\langle J_Z^2 \rangle - \left[\langle J_X^2 \rangle + \langle J_Y^2 \rangle \right] \quad ; \tag{10}$$

as for the photons states, non-zero atomic alignment implies emission of non-zero integrated linear polarization.

The use of ITOs can also be simplifying for the various physical interactions, which appear as super-operators (operators acting on the

---

[5]In the absence of an internal source of angular momentum

density matrix operator), if they obey any symmetry. For instance, let us consider the magnetic field interaction. If the quantization axis OZ is taken along the magnetic field direction, which implies that the magnetic field interaction is axially symmetric about OZ, the super-operator is diagonal in K and Q, which means that the $^{\alpha J}\rho_Q^K$ component is self-coupled only. Another example is given by the depolarizing collisions with the surrounding particles: if they are isotropic, the corresponding term depends only on K and is Q-independent[6].

The hermiticity of the density matrix implies moreover the following conjugation law

$$\left[^{\alpha J}\rho_Q^K\right]^* = (-1)^Q \left[^{\alpha J}\rho_{-Q}^K\right] \quad , \tag{11}$$

which implies that, if the medium is non-oriented and if one is interested in the $K = 0, 2$ components only, the density matrix of a 2-level atom, with unpolarized ground level, is completely described by 6 real numbers.

3. Overview of the Master equation theory

One considers here two interacting systems A and B, A being a 'small system' interacting with a 'bath' B. The density matrix ρ of the total system $A + B$ (which is an isolated system) obeys the Schrödinger equation (3) with the Hamiltonian H given, with evident notations, by

$$H = H_A + H_B + V \quad , \tag{12}$$

where V is the interaction Hamiltonian. However, if information on one of the two systems only is desired, one can average over the other system by means of the *partial trace* operation, thus obtaining a *reduced density matrix* for one of the two systems, for instance the small system A

$$\sigma_A = \text{Tr}_B \rho \quad \Leftrightarrow \quad \langle k|\sigma_A|l\rangle = \sum_\alpha \langle k\alpha|\rho|l\alpha\rangle \quad , \tag{13}$$

$k, l, \ldots$ being states of the system A and $\alpha, \beta, \ldots$ states of the system B. The reduced density matrix σ_A does not obey a Schrödinger equation, due to the interaction with the bath B, which makes the small system A a non-isolated system. Instead it obeys a 'Master equation'.

The Master equation theory is able to provide, in symmetrical derivations, both the statistical equilibrium equation for the atomic density matrix (the small system is then the atom interacting with the bath of photons and/or perturbers), and the transfer equation of polarized radiation (the small system is then the beam of radiation, interacting with the bath of atoms along the line-of-sight), through a quantum theory of matter-radiation

---

[6]Operator $D^{(K)}$, see Section 4

interaction: see Cohen-Tannoudji (1977)[*] , Bommier (1977)[*] , Bommier and Sahal-Bréchot (1978)[*] , Landi Degl'Innocenti (1983, 1984)[**] , Bommier and Sahal-Bréchot (1991)[*] , Bommier (1991)[***] .

Until recently, such derivations have been achieved up to the second order of the perturbation theory for the matter-radiation interaction, which implies, as we will see below, that one considers only absorption followed by emission processes (Rayleigh scattering is then discarded), which induces complete frequency redistribution in the rest frame. Recent work by Bommier (to be published) has achieved development to higher orders, which makes the formalism able to take into account coherent scattering processes, leading to partial frequency redistribution.

3.1. THE MASTER EQUATION TO SECOND ORDER OF THE PERTURBATION THEORY

Let us give here an overview of the derivation of the Master equation for the 'small system' A, up to the second order of the perturbation theory. To do so, let us introduce the *interaction representation*

$$\tilde{\rho}(t) = e^{\frac{i}{\hbar}H_0 t}\, \rho(t)\, e^{-\frac{i}{\hbar}H_0 t} \quad , \quad \tilde{V}(t) = e^{\frac{i}{\hbar}H_0 t}\, V\, e^{-\frac{i}{\hbar}H_0 t} \quad , \tag{14}$$

where $H_0 = H_A + H_B$ is the unperturbed total Hamiltonian. The interaction representation then 'follows' the unperturbed evolution. The Schrödinger equation (3) for the total system in this representation depends only on $\tilde{V}$ and is therefore

$$i\hbar \frac{d}{dt}\tilde{\rho}(t) = \left[\tilde{V}(t), \tilde{\rho}(t)\right] \quad , \tag{15}$$

which can be formally integrated

$$\tilde{\rho}(t) = \tilde{\rho}(0) + \frac{1}{i\hbar}\int_0^t \left[\tilde{V}(t-\tau), \tilde{\rho}(t-\tau)\right]d\tau \quad . \tag{16}$$

This formal solution can be inserted in the differential equation (15), resulting in

$$\frac{d}{dt}\tilde{\rho}(t) = \frac{1}{i\hbar}\left[\tilde{V}(t), \tilde{\rho}(0)\right] - \frac{1}{\hbar^2}\int_0^t \left[\tilde{V}(t), \left[\tilde{V}(t-\tau), \tilde{\rho}(t-\tau)\right]\right]d\tau \quad , \tag{17}$$

which is an exact equation, without, for the moment, any approximation.

To derive the Master equation, one has to introduce the reduced density operators

---

[*] Statistical equilibrium equation only (no radiative transfer equation)
[**] Both statistical equilibrium and radiative transfer equations
[***] Radiative transfer equation only

$$\begin{cases} \tilde{\sigma}_A(t) = \text{Tr}_B[\tilde{\rho}(t)] \\ \tilde{\sigma}_B(t) = \text{Tr}_A[\tilde{\rho}(t)] \end{cases} . \tag{18}$$

The Master equation is obtained by taking the partial trace of Eq. (17), which is possible after having introduced some assumptions.

The first assumption is that the total density matrix ρ can be factorized at each time t

$$\tilde{\rho}(t) = \tilde{\sigma}_A(t) \otimes \tilde{\sigma}_B(t) \quad , \tag{19}$$

which means that the correlations between the two systems A and B, induced by the interaction, are negligible, owing to the fact that the bath B has so many degrees of freedom that the effects of the interaction with A dissipate away quickly, do not react back onto A to any significant extent ('no back-reaction approximation'), and do not modify the mean state of the system B (Fano, 1957; Blum, 1981; Cohen-Tannoudji, 1977)[7].

Besides, it is assumed about the interaction potential V that

$$\text{Tr}_B[\tilde{\sigma}_B(t)\tilde{V}(t')] = 0 \quad . \tag{20}$$

This hypothesis is always satisfied when one considers electric dipole interactions and incident radiation containing only polarization coherences, as typical of stellar atmospheres[8]. It implies that the first term of Eq. (17) vanishes in the partial trace operation. One then obtains the Master equation

$$\frac{d}{dt}\tilde{\sigma}_A(t) = -\frac{1}{\hbar^2} \int_0^t \text{Tr}_B[\tilde{V}(t),[\tilde{V}(t-\tau),\tilde{\sigma}_A(t-\tau)\otimes\tilde{\sigma}_B(t-\tau)]]d\tau \quad . \tag{21}$$

The treatment of $\int_0^t$ now requires an approximation, which has several aspects:

i) the interactions between the two systems are assumed to be time sequenced, in such a way that the duration of one interaction τ_c is very small compared to the mean duration between two interactions, which is given by the lifetime of the level $\bar{T} = \Gamma^{-1}$ (the lifetime is the inverse of the number of interactions per time unit); one can then define a *time of interest s* such that

$$\tau_c \ll s \ll \bar{T} \quad . \tag{22}$$

It is then possible to extend the limit t of the integral in Eq. (21) to infinity, although the integral actually covers the time interval s containing one single interaction. Then multiple time-correlated interactions (such as Rayleigh or coherent scattering) are neglected. One considers only as the scattering process the absorption of a photon, followed by emission of another photon. As a result, one obtains complete frequency redistribution in the scattering process.

---

[7]By doing so, one introduces some irreversibility in the evolution equation, which was previously (at the Schrödinger equation stage) time reversible.
[8]This could be invalid for coherent states of radiation (laser radiation for instance)

ii) The reduced density matrix, in the interaction representation that follows the unperturbed evolution, is assumed to evolve slowly enough to allow the 'short memory' (or Markov) approximation, which consists in substituting, in Eq. (21)

$$\tilde{\sigma}_{A,B}(t-\tau) \quad \rightarrow \quad \tilde{\sigma}_{A,B}(t) \quad . \tag{23}$$

Then, the integration over time τ can be performed in Eq. (21). These approximations, which are discussed in detail in Cohen-Tannoudji (1977), are in fact not independent of each other but are two aspects of the same approximation, namely the perturbation theory of second order. The Master equation is then

$$\frac{d}{dt}\tilde{\sigma}_A(t) = -\frac{1}{\hbar^2}\int_0^\infty \text{Tr}_B\Big[\tilde{V}(t),\big[\tilde{V}(t-\tau),\tilde{\sigma}_A(t)\otimes\tilde{\sigma}_B(t)\big]\Big]d\tau \quad . \tag{24}$$

At this stage, the main physical assumptions that are at the basis of the Master Equation theory have been discussed. This is the first step in the matter-radiation interaction description. The following steps – expansion over atomic and photons states in a full quantum scheme, particular case of the dipolar electric interaction, time and frequency integrations – can be found for instance in Cohen-Tannoudji (1977), Bommier (1977), Landi Degl'Innocenti (1983, 1984), Bommier and Sahal-Bréchot (1991), Bommier, (1991). The result is the statistical equilibrium equation for the atomic density matrix on the one hand, and the transfer equation of polarized radiation on the other hand. Other physical assumptions have nevertheless to be introduced throughout the development. They are discussed in the references quoted above – see also the contribution of Landi Degl'Innocenti in these proceedings.

3.2. THE MASTER EQUATION TO HIGHER ORDERS OF THE PERTURBATION THEORY

To derive higher orders, one has to repeat as many times as necessary the formal solution of the Schrödinger equation, given by Eq. (16), inside the differential Schrödinger equation Eq. (15, 17, etc.). One then has to make the basic approximation described above, namely:
i) to extend to infinity the limits of the time integrals, which means that at each order one considers the interactions two by two, three by three, etc.
ii) to apply the Markov approximation to the density matrix.
If the interactions are now considered two by two, three by three, etc., it is obvious that, for instance at order 4, two cases have to be considered:
a) the two interactions are separated in time: they have already been considered at order 2 and do not have to be taken into account at order 4;

b) the two interactions are time-interlaced. They have not been taken into account at order 2 and appear as new terms at order 4.

One then obtains at order 4 a term that has to be added to the right hand side of the Master equation Eq. (24)

$$+\frac{1}{\hbar^4}\mathcal{E}\int_0^\infty d\tau_1\int_0^\infty d\tau_2\int_0^\infty d\tau_3$$

$$\mathrm{Tr}_B\left[\tilde{V}(t),\left[\tilde{V}(t-\tau_1),\left[\tilde{V}(t-\tau_1-\tau_2),\left[\tilde{V}(t-\tau_1-\tau_2-\tau_3),\tilde{\sigma}_A(t)\otimes\tilde{\sigma}_B(t)\right]\right]\right]\right]$$ (25)

where the symbol $\mathcal{E}$ means that the non-new terms, with respect to order 2, have to be discarded. This cannot be done 'a priori' (i.e. at this stage). This can only be achieved at the end of the calculation, when expansion over atomic and photons states has been performed. At that moment, an order-4 term can be identified as non-new if it exactly counterbalances another order-2 term. This identification can be confirmed by the fact that, if only non-new terms are introduced, the solution of the statistical equilibrium and transfer equation is unchanged (although the statistical equilibrium equation becomes then undetermined). Finally, non-new terms can be recognized by their characteristic expression (which will not be written here) involving expected branching ratios.

The different order-4 terms can be represented as in Fig. 1, which has been plotted for a 2-level atom, assuming that the atom is in the lower state a at the beginning of the interaction, which is represented by a dot. The end of the interaction is represented by another dot. The initial dot and the final dot are linked by two transition amplitudes represented by two lines, each amplitude containing two interactions V, which are represented by vertical lines. It appears in this figure that the order-4 terms can be classified in two groups:

i) Terms (1) and (2): During the intermediate time τ_2 the two transition amplitudes are both in the upper state b, which is then *populated* during the process. There is thus one photon absorption (duration τ_3), followed by one photon emission (duration τ_1), with population of the excited state during the process. This process consists of two order-2 processes (absorption, spontaneous emission), and these terms have to be discarded from the Master equation at order 4, because they have been taken into account at order 2.

ii) Terms (3) and (4): The two transition amplitudes are never at the same time in the upper state b, which is then never populated. In term (3) the duration of the photon absorption is $\tau_2+\tau_3$, the duration of the photon emission is $\tau_1+\tau_2$. The emission begins before the end of the absorption; both are then time-interlaced. In term (4) the emission begins and ends before the end of the absorption. As the final dot is not at the final time, this term does not appear in previous work like Omont, Smith, and Cooper

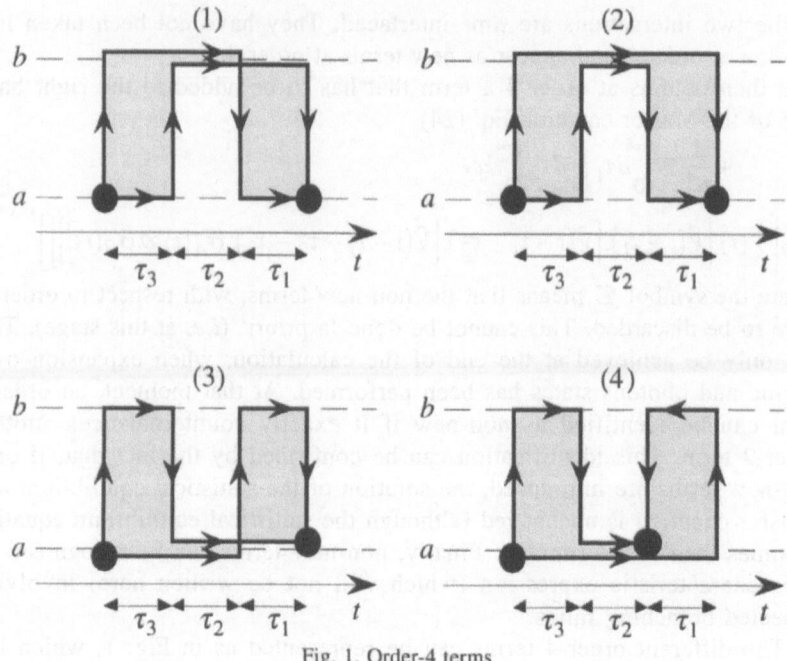

Fig. 1. Order-4 terms

(1972) (see also Nienhuis and Schuller, 1977), since such earlier work only considered transition probabilities, for which the final dot *has to be* at the end of the process. However, term (4) plays a role in the statistical equilibrium and radiative transfer equations and has to be considered in the present work. It also allows new results to be derived. Terms (3) and (4) represent reemission of the absorbed photon *without population* of the excited state during the process. These are new terms, with respect to order 2, which have to be considered at order 4.

These higher orders terms include the effect of possible reemission of the absorbed photon, or possible reabsorption of the emitted photon (the inverse process has to be taken into account). They can lead to partial frequency redistribution effects. New terms appear at each higher order.

4. Solution of the non-LTE problem

Using the Master equation theory outlined above, one obtains through quantum theory of atom and radiation, for instance as in Landi Degl'Innocenti, Bommier, and Sahal-Bréchot (1991a):
i) the statistical equilibrium for the atomic density matrix
– in the weak magnetic field regime, see Eq. (42) of that paper

– in the intermediate magnetic field regime, see Eq. (37) of that paper
ii) the transfer equation of polarized radiation
– in the weak magnetic field regime, see Eq. (43) of that paper
– in the intermediate magnetic field regime, see Eqs. (39)-(41) of that paper
under the hypothesis of a 2-level atom with unpolarized ground level,
negligible induced emission, and factorization of the atomic density matrix
between internal and external (atomic velocity) states (see Landi
Degl'Innocenti, 1996, for a discussion of these various approximations). The
equations listed above are derived using the second-order perturbation
theory for matter-radiation interaction. The equations obtained to higher
(infinite) orders will be published elsewhere (Bommier, to be published).
Only some consequences will be given below.

The solution of the non-LTE problem requires the solution of these two
coupled equations. To do this, one can take advantage of the fact that both
equations admit, under some hypothesis, a formal solution:
i) For the statistical equilibrium equation, the formal solution is obvious in
the case of a 2-level atom with unpolarized ground level and negligible
induced emission;
ii) For the radiative transfer equation, the formal solution requires a depth-
independent absorption profile width, and can be derived (see Eq. (45) of
Landi Degl'Innocenti, Bommier, and Sahal-Bréchot, 1991a, or Eq. (15) of
Landi Degl'Innocenti, Bommier and Sahal-Bréchot, 1990).
The solution of the non-LTE problem can then be obtained in two ways[9]:
i) Either one inserts the formal solution of the statistical equilibrium
equation inside the transfer equation, eliminating those unknowns that are
the atomic density matrix elements. One obtains in this way the usual
transfer equation for the Stokes parameters. This approach could be called
the 'scattering method', because the obtained equation describes the
scattering of the Stokes parameters. By so doing, with the equations listed
above, one recovers every element of the Stokes parameters transfer
equation: for weak magnetic fields, the decoupling of the frequency and
directional redistributions in the rest frame[10] is derived. In particular, the
polarization phase matrix is recovered (see Landi Degl'Innocenti, 1984). As
regards the frequency redistribution function, see next section.
ii) Or one inserts the formal solution of the radiative transfer equation inside
the statistical equilibrium equation, eliminating those unknowns that are the
Stokes parameters of the radiation at each point, direction and frequency,
obtaining a system of integral equations (to be solved) in which the density
matrix elements at all points of the medium are coupled. This approach, due

---

[9]Full numerical solutions of the coupled problem of statistical equilibrium and radiative
transfer also exist. Such solutions are disregarded in the present discussion.
[10]This is no more true for intermediate magnetic fields

to an original idea of E. Landi Degl'Innocenti, could be called the 'density matrix method' and is described in Landi Degl'Innocenti, Bommier, and Sahal-Bréchot (1990), Bommier, Landi Degl'Innocenti, and Sahal-Bréchot (1991) for weak magnetic fields, in Landi Degl'Innocenti, Bommier and Sahal-Bréchot (1991a,b), Landi Degl'Innocenti and Bommier (1994) for arbitrary magnetic fields.

5. Partial frequency redistribution

The notations used throughout this section will be those of Domke and Hubeny (1988, consider in particular their Equations (44)-(51)), which are not very different from those of the basic paper of Omont, Smith, and Cooper (1972). In fact, Domke and Hubeny (1988) express in the polarized case the results implicitly already contained in Omont, Smith and Cooper (1972). In this section we examine the frequency redistribution function that can be derived from the theory outlined in the preceding sections, and we derive, in particular, its expression in the presence of a weak magnetic field[11] (Hanle effect).

5.1. SECOND-ORDER PERTURBATION THEORY: COMPLETE FREQUENCY REDISTRIBUTION

Using the second-order perturbation theory outlined above, one obtains the complete frequency redistribution function r_{III} in the rest frame.

5.2. FOURTH-ORDER PERTURBATION THEORY: USUAL PARTIAL FREQUENCY REDISTRIBUTION FUNCTION

5.2.1. *Zero magnetic field*

At order 4 of the perturbation development (and, in fact, considering only terms (1), (2) and (3) of Fig. 1, as Omont, Smith, and Cooper, 1972), one recovers, through the theory outlined above, the usual partial frequency redistribution function, given in Eq. (63) of Omont, Smith, and Cooper (1972) for the unpolarized case, and Eq. (49) of Domke and Hubeny (1988) for the polarized case in zero magnetic field. Let us give here this

---

[11]These results should in fact be used with some care, because for their derivation we have neglected the Zeeman splitting in the line profile. Though this can be considered as correct if the profile refers to the laboratory reference frame, this is not correct, even in the Hanle effect regime, for profiles in the rest frame, which is actually the case for the profiles introduced throughout this section. So the present results have to be considered as only the best that we can do for the moment.

last equation, written in a slightly different manner, and in the case of a 'non-oriented' medium (no circular polarization)[12]

$$F/\zeta = p_R W_2 \left[\alpha r_{II} + \beta^{(2)} r_{III} \right]$$
$$+ p_{is} \left\{ \left(1 - W_2 \right) \left[\alpha r_{II} + \beta^{(2)} r_{III} \right] + \left[\beta^{(0)} - \beta^{(2)} \right] r_{III} \right\} \qquad (26)$$

where p_R is the Rayleigh phase matrix, p_{is} is the isotropic phase matrix, W_2 depends on the line quantum numbers, α and $\beta^{(J)}$ are given, in the case of a resonance line (infinite lower-level lifetime), by

$$\alpha = \frac{\Gamma_R}{\Gamma_R + \Gamma_I + \Gamma_E} \quad , \quad \beta^{(J)} = \alpha \frac{\Gamma_E - D^{(J)}}{\Gamma_R + \Gamma_I + D^{(J)}} \qquad , \qquad (27)$$

$$\Rightarrow \quad \alpha + \beta^{(J)} = \frac{\Gamma_R}{\Gamma_R + \Gamma_I + D^{(J)}}$$

where Γ_R is the radiative deexcitation rate, Γ_I is the inelastic deexcitation rate, Γ_E is the elastic collision rate for the excited level, $D^{(0)} = 0$, $D^{(2)}$ is the depolarizing elastic collision rate (alignment destruction rate). The frequency redistribution functions r_{II} and r_{III} are given by

$$r_{II}(\omega_1, \omega_2) = L(\omega_1) \delta(\omega_1 - \omega_2) \quad , \quad r_{III}(\omega_1, \omega_2) = L(\omega_1) L(\omega_2) \quad , \quad (28)$$

where $L(\omega)$ is the rest frame (lorentzian) absorption profile given by

$$L(\omega) = \frac{1}{\pi} \frac{\Gamma/2}{\left(\omega - \omega_0 \right)^2 + \left(\Gamma/2 \right)^2} \quad , \quad \Gamma = \Gamma_R + \Gamma_I + \Gamma_E \quad . \qquad (29)$$

The quantity F/ζ is the redistribution matrix, which provides the joint probability that a photon of frequency $\omega_1/2\pi$ and direction $\bar{n}_1$ is absorbed and reemitted at frequency $\omega_2/2\pi$ and in direction $\bar{n}_2$, in the rest frame. The angular redistribution is contained in the phase matrix, whereas the frequency redistribution is described by the functions r_{II} and r_{III}. The redistribution matrix is not normalized to unity in the presence of inelastic collisions.

In the absence of collisions the frequency redistribution function is simply the coherent r_{II}, opposite to what has been obtained at second order of the perturbation theory.

5.2.2. Weak magnetic field: Hanle effect

To derive the expressions in the presence of a weak magnetic field we first note that the frequency redistribution functions are derived from two

---

[12]The unpolarized result is recovered by ignoring polarization, namely $p_R = p_{is}$, which results in:
$$F/\zeta = \alpha r_{II} + \beta^{(0)} r_{III}$$

functions (see Omont, Smith, and Cooper, 1972, Domke and Hubeny, 1988), namely

$$f_1 = \frac{2\pi^2}{\Gamma_R + \Gamma_I + \Gamma_E} [\underbrace{\mathcal{L}(\omega_1)\delta(\omega_1 - \omega_2)}_{r_{II}(\omega_1,\omega_2)} - \underbrace{\mathcal{L}(\omega_1)\mathcal{L}(\omega_2)]}_{r_{III}(\omega_1,\omega_2)}$$

$$f_{23}^{(J)} = \frac{2\pi^2}{\Gamma_R + \Gamma_I + D^{(J)}} \underbrace{\mathcal{L}(\omega_1)\mathcal{L}(\omega_2)}_{r_{III}(\omega_1,\omega_2)}$$

(30)

The function f_1 is derived from processes corresponding to term *(i)* of Fig. 2 of Omont, Smith, and Cooper (1972), which is also term (3) of Fig. 1 of the present paper. In this term there is no population of the excited state, in the coherences of which the Hanle effect takes place. There is thus no Hanle effect associated with this function. As regards the function $f_{23}^{(J)}$, it is derived from processes corresponding to terms *(ii)* and *(iii)* of Fig. 2 of Omont, Smith, and Cooper (1972), which are also terms (1) and (2) of Fig. 1 of the present paper. In this case there is population of the excited level and therefore the possibility of the Hanle effect.

It can be suggested from this (see Frisch, 1996 – these proceedings), as has also been demonstrated through the quantum calculations presented in the present paper (considering only terms (1), (2) and (3) of Fig. 1, as Omont, Smith, and Cooper, 1972), that the presence of a weak magnetic field (Hanle effect) can be taken into account by replacing the Rayleigh phase matrix p_R by the Hanle phase matrix p_H (Landi Degl'Innocenti and Landi Degl'Innocenti, 1988) in front of the terms proportional to $f_{23}^{(J)}$, but not in front of the terms proportional to f_1, in Eq. (26). One then obtains

$$F/\zeta = p_R W_2 [\alpha(r_{II} - r_{III})] + p_H W_2 [\alpha + \beta^{(2)}] r_{III}$$

$$+ p_{is} \left\{ (1 - W_2) [\alpha r_{II} + \beta^{(2)} r_{III}] + [\beta^{(0)} - \beta^{(2)}] r_{III} \right\}$$

(31)

In Landi Degl'Innocenti and Landi Degl'Innocenti (1988), the Hanle phase matrix has however been derived in the absence of collisions. In the presence of collisions, one has only to replace the γ factor, which contains the dependence on the magnetic field strength and which is given in Eq. (5) of that paper, by

$$\gamma = g_{J'} \frac{2\pi v_L}{\Gamma_R + \Gamma_I + D^{(2)}}$$

(32)

where $g_{J'}$ is the Landé factor of the upper level and v_L is the Larmor frequency.

It appears from this equation that the Hanle effect vanishes in the far wings, as outlined by Stenflo (1994), and in accordance with the results of

Omont, Smith, and Cooper (1973), as analyzed by Faurobert-Scholl *et al.* (1995).

5.3. HIGHER (INFINITE) ORDERS PERTURBATION THEORY

Using the theory briefly outlined in the present paper, we have been able to take into account all the order 4 terms given in Fig. 1, and to perform calculations up to orders higher than 4, and even to sum over infinite orders, in the presence of a weak magnetic field. The only difference, with respect to the results given in the preceding sections, lies in the fact that Γ_R has now to be replaced by $2\Gamma_R$ in the denominator of the α coefficient ($\beta^{(J)}$ is accordingly modified)

$$\alpha = \frac{\Gamma_R}{2\Gamma_R + \Gamma_I + \Gamma_E} \quad , \quad \beta^{(J)} = \alpha \frac{\Gamma_R + \Gamma_E - D^{(J)}}{\Gamma_R + \Gamma_I + D^{(J)}} \quad , \tag{33}$$

$$\Rightarrow \quad \alpha + \beta^{(J)} = \frac{\Gamma_R}{\Gamma_R + \Gamma_I + D^{(J)}}$$

and in the absorption profile width

$$\mathcal{L}(\omega) = \frac{1}{\pi} \frac{\Gamma/2}{(\omega - \omega_0)^2 + (\Gamma/2)^2} \quad , \quad \Gamma = 2\Gamma_R + \Gamma_I + \Gamma_E \quad . \tag{34}$$

The redistribution matrix is again given by Eqs. (26) and (31), using these new coefficients.

This factor of 2 for the natural width is due to the fact that, if all orders are taken into account, an absorbed photon can either lead to upper level excitation, or be immediately – and coherently – reemitted (an emitted photon can either lead to deexcitation of the upper level, or be immediately – and coherently – reabsorbed). The mean duration of the excited state is the average between the natural lifetime $1/\Gamma_R$ and 0, namely $1/2\Gamma_R$. The natural width Γ_R has accordingly to be multiplied by 2.

The effect of the order-higher-than-4 terms should be important in the line core only, if partial frequency redistribution effects are visible there. The far wings should not be modified much.

It is interesting to consider the case of the absence of collisions. Then the frequency redistribution function is the half-sum of r_{II} and r_{III}, which is exactly intermediate between what has been obtained at orders 2 and 4 in this particular case. It shows that spontaneous emission can also be responsible, besides collisions, of complete frequency redistribution (in the rest frame). This is in fact to be expected from such a random process.

Acknowledgments

The author is indebted to E. Landi Degl'Innocenti, N. Feautrier, E. Roueff, S. Sahal-Bréchot, H. Frisch, K.N. Nagendra and J.O. Stenflo for helpful discussions and suggestions during this work and the preparation of this paper.

References

Blum, K.: 1981, *Density Matrix Theory and Applications*, Plenum Press, New-York
Bommier, V.: 1977, Thèse de 3ème cycle, Paris VI University
Bommier, V.: 1980, *Astron. Astrophys.* **87**, 109
Bommier, V.: 1991, *Ann. Phys. Fr.* **16**, 599
Bommier, V., and Sahal-Bréchot, S.: 1978, *Astron. Astrophys.* **69**, 57
Bommier, V., and Sahal-Bréchot, S.: 1991, *Ann. Phys. Fr.* **16**, 555
Bommier, V., Landi Degl'Innocenti, E., and Sahal-Bréchot, S.: 1991, *Astron. Astrophys.* **244**, 383
Bommier, V., Landi Degl'Innocenti, E., Leroy, J.L., and Sahal-Bréchot, S.: 1994, *Solar Physics* **154**, 231
Brink, D.M., and Satchler, G.R.: 1968, *Angular Momentum*, 2nd edition, Clarendon Press, Oxford
Cohen-Tannoudji, C.: 1977, in Balian, R., Haroche, S. and Liberman, S. (eds), *Frontiers in Laser Spectroscopy*, Les Houches, Session XXVII, 1975, North-Holland, Amsterdam (Part I, Course 1)
Cohen-Tannoudji, C., Diu, B., and Laloë, F.: 1977, *Quantum Mechanics* (2nd edition), Hermann, Paris
Domke, H., and Hubeny, I.: 1988, *Ap. J.* **334**, 527
Fano, U.: 1957, *Rev. Mod. Phys.* **29**, 74
Faurobert-Scholl, M., Feautrier, N., Machefert, F., Petrovay, K., and Spielfiedel, A.: 1995, *Astron. Astrophys.* **298**, 289
Frisch, H.: 1996 (these proceedings)
Landi Degl'Innocenti, E.: 1983, *Solar Physics* **85**, 3
Landi Degl'Innocenti, E.: 1984, *Solar Physics* **91**, 1
Landi Degl'Innocenti, E.: 1996 (these proceedings)
Landi Degl'Innocenti, M., and Landi Degl'Innocenti, E.: 1988, *Astron. Astrophys.* **192**, 374
Landi Degl'Innocenti, E., Bommier, V., and Sahal-Bréchot, S.: 1990, *Astron. Astrophys.* **235**, 459
Landi Degl'Innocenti, E., Bommier, V., and Sahal-Bréchot, S.: 1991a, *Astron. Astrophys.* **244**, 391
Landi Degl'Innocenti, E., Bommier, V., and Sahal-Bréchot, S.: 1991b, *Astron. Astrophys.* **244**, 401
Landi Degl'Innocenti, E., and Bommier, V.: 1994, *Astron. Astrophys.* **284**, 865
Messiah, A.: 1969, *Quantum Mechanics*, Dunod, Paris
Nienhuis, G., and Schuller, F.: 1977, *Physica* **92C**, 397
Omont, A., Smith, E.W., and Cooper, J.: 1972, *Ap. J.* **175**, 185
Omont, A., Smith, E.W., and Cooper, J.: 1973, *Ap. J.* **182**, 283
Stenflo, J.O.: 1994, *Solar Magnetic Fields – Polarized Radiation Diagnostics*, Kluwer, Dordrecht
von Neumann, J.: 1927, *Göttingen Nachrichten* **245**

PARTIAL FREQUENCY REDISTRIBUTION OF POLARIZED RADIATION

H. FRISCH

C.N.R.S., Observatoire de Nice, B.P. 229, 06304 Nice Cedex 4, France

Abstract. Resonance polarization, which is created by the scattering of an anisotropic radiation field in regions of zero or weak magnetic fields, is strongly dependent on the frequency redistribution taking place during the scatterings. Here we discuss the frequency redistribution matrix relevant to resonance lines, concentrating on linear polarization. First we analyze in detail the redistribution matrix in a zero magnetic field given by the theory of Omont, Smith and Cooper (1972), revisited by Domke and Hubeny (1988). We explain that the linear polarization maxima which may appear in the wings of the Stokes Q profiles of strong resonances lines such as the Ca I 4227 Å line are coherent frequency redistribution effects. Various approximate forms of the frequency redistribution matrix are also examined. For resonance polarization in a weak magnetic field, we suggest a new expression for the redistribution matrix which can be used at all line frequencies, and is consistent with the condition that the Hanle effect acts only in the line core.

Key words: Spectral lines – Polarization – Frequency redistribution

1. Introduction

Imagine that one can follow a photon which is scattered by an atom. One says that there is *partial frequency redistribution* when the frequency of the reemitted (scattered) photon is correlated to the frequency of the absorbed (incident) photon. When there is no correlation between the frequencies of the absorbed and emitted photons, one says that there is *complete frequency redistribution*. Transfer problems with complete frequency redistribution are much easier to solve than transfer problems with partial frequency redistribution. Subordinate lines and weak resonance lines are adequately described by complete frequency redistribution. However, it is absolutely necessary to take partial frequency redistribution into account for strong resonance lines such as the resonance line of Ca I at 4227 Å, in particular when one considers resonance polarization in regions of zero or weak magnetic fields. We recall that resonance polarization is created by the scattering of an anisotropic radiation field. The polarization mechanism is the quantum counterpart of Rayleigh scattering (see the paper by Faurobert–Scholl, same volume, for details). A weak magnetic field produces the so-called Hanle effect which induces a rotation of the plane of polarization and, in general, a decrease in the degree of polarization. The Hanle effect acts only in the line core.

Partial frequency redistribution effects are observable in the wings of strong resonance lines and to a lesser extent in the line cores. They affect

Solar Physics **164**: 49–66, 1996.

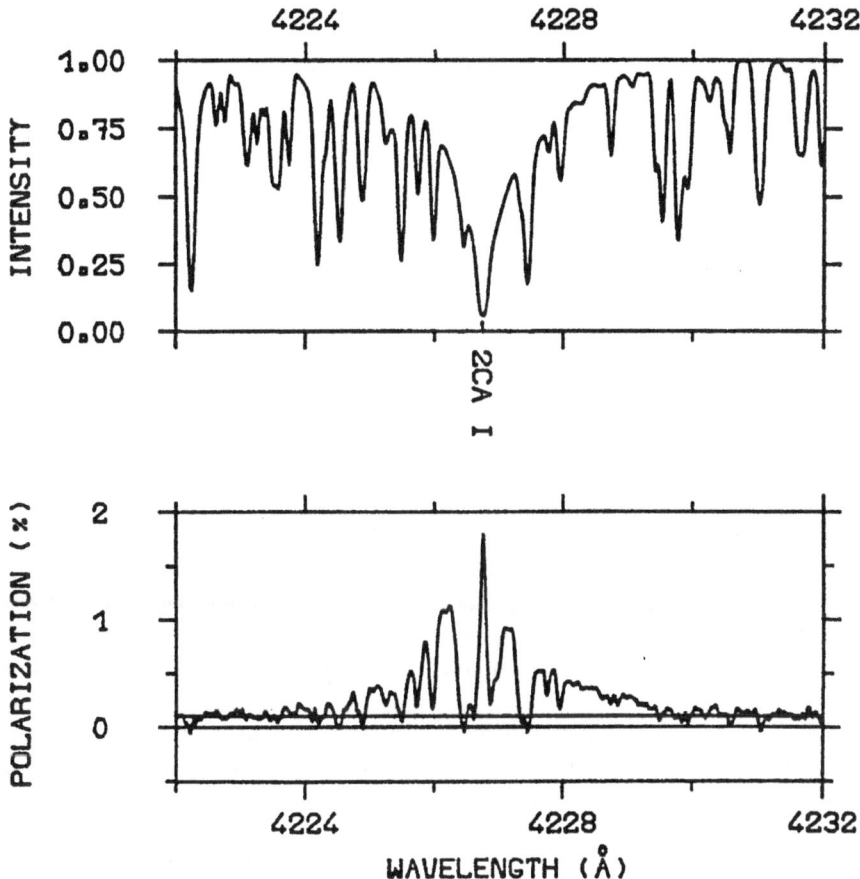

Fig. 1.   Intensity and polarization profile of the Ca I 4227 Å line. From Stenflo *et al.*
(1980) (see also Stenflo, 1994).

both the intensity spectrum and the polarization spectrum and are particularly conspicuous in the latter.

Figure 1 shows the intensity and the linear polarization profile of the
Ca I 4227 Å line, obtained inside the south polar limb in a non-magnetic
region (Stenflo *et al.*, 1980; Stenflo, 1994, p. 99). This line is formed in the
low chromosphere. The intensity spectrum shows a strong absorption line
with many blends. The linear polarization spectrum shows a peak in the line
core and two rather broad maxima in the line wings. These maxima are a
signature of partial redistribution effects as will be shown below.

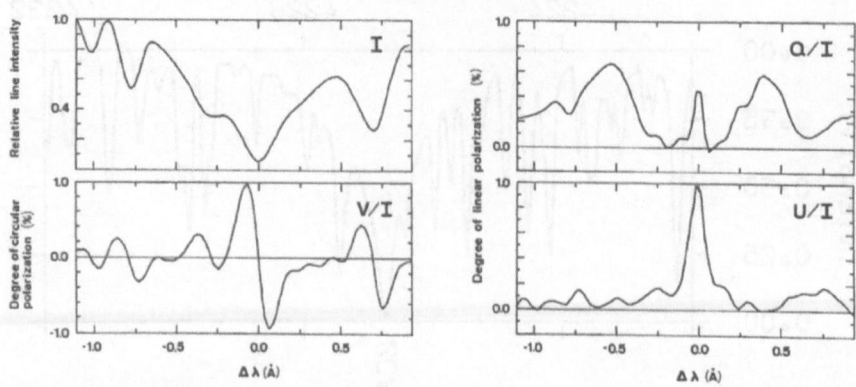

Fig. 2. Intensity and polarization profiles of the Ca I 4227 Å line in a magnetic region. From Stenflo (1982)(see also Stenflo, 1994).

The Hanle effect is illustrated in Figure 2 which shows the intensity and polarization profile of the same line recorded in a magnetic field region (Stenflo, 1982; Stenflo, 1994, p. 100). Because of the rotation of the plane of polarization the core signal is larger in U/I than in Q/I. The strong wing polarization in Q/I is however unaffected by the Hanle effect in agreement wih the theory.

Partial frequency redistribution effects are described by the redistribution matrix $\hat{R}(\nu, \mathbf{n}, \nu', \mathbf{n}')$ which gives the joint probability of absorbing a photon with frequency ν' and direction $\mathbf{n}'$ and reemitting a photon with frequency ν and direction $\mathbf{n}$. The redistribution matrix depends also on the local properties of the atmosphere, such as the value and direction of the magnetic field. If we represent the polarized radiation field at a point $\mathbf{r}$ by the Stokes vector $\mathbf{I}(\mathbf{r}, \nu, \mathbf{n})$, the scattered radiation at $\mathbf{r}$ is given by

$$\mathcal{E}(\mathbf{r}, \nu, \mathbf{n}) = \int d\nu' \int \frac{d\Omega'}{4\pi} \hat{R}(\nu, \mathbf{n}, \nu', \mathbf{n}') \mathbf{I}(\mathbf{r}, \nu', \mathbf{n}'), \qquad (1)$$

where the integration is over all frequencies and directions of the absorbed (incident) radiation field.

Knowing the redistribution matrix, one can then calculate polarization profiles by solving a transfer equation. For resonance polarization, including the case of the Hanle effect, the transfer equation is of the form

$$\mathbf{n} . \nabla \mathbf{I}(\mathbf{r}, \nu, \mathbf{n}) = -\phi(\mathbf{r}, \nu) \mathbf{I} + (1 - \epsilon)\mathcal{E} + \text{primary sources.} \qquad (2)$$

For zero or weak magnetic fields, the absorption coefficient is simply the absorption profile $\phi(\mathbf{r}, \nu)$. It is a scalar. The last term is the source term for primary photons. They can be created inside the medium by thermal emission or be incident upon the medium. After absorption, a photon can be destroyed by a collisional deexcitation of the excited level of the transition. The scattering term is thus multiplied by a scattering probability

$$1 - \epsilon = \frac{\Gamma_R}{\Gamma_R + \Gamma_I} \, . \tag{3}$$

For resonance lines, the lower level is infinitely sharp. The $\Gamma's$ refer to the upper level of the transition: Γ_R is the radiative deexcitation rate (inverse life-time) and Γ_I the inelastic collisions deexcitation rate. Inelastic collisions are mainly due to electronic collisions. For the Ca I 4227 Å line, $\Gamma_R = 2 \times 10^8 \, s^{-1}$. In the chromosphere, $\Gamma_I \simeq 2 \times 10^3 \, s^{-1}$, thus $\epsilon \simeq 10^{-5}$ (Faurobert–Scholl, 1992). Because of the very small value of ϵ, the line formation is controled by multiple scatterings and one needs a reliable redistribution matrix.

The form of the redistribution matrix for resonance polarization is still a subject of research (see the paper by Bommier, same volume) which began with Zanstra (1941a). A few "historical" references concerning the redistribution matrix and polarized transfer equations with partial frequency redistribution are presented here. References to recent work are to be found in the following Sections.

In Volume **101** (1941) of the Monthly Notices of the Royal Astronomical Society, three papers are devoted to resonance polarization. A first paper by Zanstra (1941a) presents a "Theory of a polarisation effect in Fraunhofer lines due to oscillator scattering" and a prediction concerning the linear polarization of the Ca I resonance line near the solar limb. The second paper by Redman (1941) describes an observational search for this polarization and reports values which are ten times smaller than the prediction of Zanstra. In a third paper, Zanstra (1941b) suggests that elastic collisions are responsible for the weakening of the polarization, because they "convert much of the light absorbed in the line wing to light of a frequency close to that of the line center". It is amusing to note that partial frequency redistribution was first introduced to interpret the linear polarization of resonance lines.

The first quantum mechanical calculation of the redistribution matrix for resonance polarization, taking into account the effects of elastic collisions, was performed by Omont et al. (1972), hereafter refered to as OSC. The main emphasis of this paper is on frequency redistribution. A year later, the same authors (Omont et al., 1973) considered the effect of a magnetic field. Not quite as famous as the first paper, it still contains several interesting comments and remarks, in particular that the Hanle effect may act in the core of a line but not in its wings because of the very short life-time of the excited level for wing frequencies. Starting from the work of OSC, Domke and Hubeny (1988) calculated explicitly the angular redistribution phase matrix for the resonance scattering of an arbitrary polarized light in the case of a zero magnetic field. Their expression for $\hat{R}(\nu, \mathbf{n}, \nu', \mathbf{n}')$ is now a standard reference.

Solution of the transfer problem for polarized radiation with partial frequency redistribution was first attempted by Stenflo (1976, 1978). He suggested a compact way of writing the transfer equations incorporating the results of OSC and Omont et al. (1973). Linear polarization profiles showing an emission peak in the line core with two maxima in the wings were first calculated by Rees and Saliba (1982).

In the next two Sections, I discuss the redistribution matrix for resonance scattering and then the redistribution matrix for the Hanle effect. My discussion is limited to linear polarization. For resonance scattering, circular polarization is totally decoupled from linear polarization. Only an already circularly polarized radiation can produce circular polarization in the scattered radiation.

2. Resonance polarization

After an analysis of the physical meaning of the redistribution matrix in the atomic rest frame (Section 2.1), I present various approximate forms of the redistribution matrix in the laboratory frame (Section 2.2) and then discuss the influence of partial frequency redistribution on linear polarization profiles (Section 2.3).

2.1. ATOMIC FRAME REDISTRIBUTION MATRIX

For a resonance line, when one takes into account the various mechanisms which contribute to the broadening and population of the atomic levels, elastic and inelastic collisions, natural broadening, radiative transitions, one finds (Domke and Hubeny, 1988) that the (3×3) redistribution matrix describing the scattering of a linearly polarized radiation may be written as

$$r(\xi, \mathbf{n}, \xi', \mathbf{n}') = \gamma r_{II}(\xi, \xi')\hat{P}(\mathbf{n}, \mathbf{n}') + r_{III}(\xi, \xi')[c\hat{P}_{is} + (1 - c - \gamma)\hat{P}(\mathbf{n}, \mathbf{n}')], (4)$$

where ξ and ξ' are the frequency of scattered and incident radiation in the atomic rest frame. The impact and isolated line approximations, and the assumption that the lower level is non–polarized are the main physical limitations of this expression. We note that in Domke and Hubeny (1988) the redistribution matrix contains the factor $(1 - \epsilon)$. The first term in the r.h.s. describes frequency coherent scattering and the second term frequency incoherent scattering. They are discussed separately below. Each term is the product of a scalar function describing the frequency redistribution and of a phase matrix describing the angular redistribution. In the rest frame there is no correlation between the frequency and the angular redistribution. It will be seen in Section 2.3 that the incoherent term gives zero polarization in the wings. It is the frequency coherent term which is responsible for the polarization maxima which may be observed in the wings of resonance lines.

The frequency coherent redistribution function $r_{II}(\xi, \xi')$, which was introduced by Hummer (1962), describes the absorption of radiation with frequency ξ' followed by a reemission at the same frequency. It is defined by

$$r_{II}(\xi, \xi') = \mathcal{L}(\xi)\delta(\xi - \xi'),\tag{5}$$

where δ is the Dirac distribution and $\mathcal{L}(\xi)$ the rest frame absorption profile. It is a Lorentzian

$$\mathcal{L}(\xi) = \frac{1}{\pi}\frac{a}{(\xi - \xi_o)^2 + a^2},\tag{6}$$

where ξ_o is the rest frame frequency of the transition and

$$a = (\Gamma_R + \Gamma_I + \Gamma_c)/4\pi.\tag{7}$$

We have seen that Γ_R and Γ_I are the radiative and inelastic collisional rates of the upper level, respectively. The parameter Γ_c is the elastic collision rate. In the solar atmosphere, elastic collisions are due to collisions with hydrogen atoms.

The angular redistribution, in the frequency coherent term, is given by the phase matrix

$$\hat{P}(\mathbf{n}, \mathbf{n}') = W_2\hat{P}_R(\mathbf{n}, \mathbf{n}') + (1 - W_2)\hat{P}_{is}.\tag{8}$$

The matrix $\hat{P}_R$ is the general (4×4) Rayleigh matrix limited to its first three rows and columns. All the entries of the isotropic matrix $\hat{P}_{is}$ are zero except $P_{11} = 1$. The factor $1 - W_2$ measures the effect of intrisic depolarization, which occurs even in the absence of collisions and which depends only on the angular momenta of the atomic levels of the transition. The Rayleigh matrix and the analytic expression of W_2, which was first established by Hamilton (1947), can be found, e.g., in Chandrasekhar (1950), van de Hulst (1980), Domke and Hubeny (1988) and Stenflo (1994). For normal Zeeman triplets $W_2 = 1$. These transitions thus show the largest linear polarization signals.

The branching ratio γ is given by

$$\gamma = \frac{\Gamma_R + \Gamma_I}{\Gamma_R + \Gamma_I + \Gamma_c}.\tag{9}$$

It represents the probability that reemission of radiation occurs before a collision, elastic or inelastic, destroys the frequency correlation between the frequencies of the absorbed and reemitted photons. Domke and Hubeny (1988) introduce the branching ratio $\alpha = \gamma(1 - \epsilon)$.

The frequency incoherent redistribution function r_{III} introduced by Hummer (1962), is defined by

$$r_{III}(\xi, \xi') = \mathcal{L}(\xi)\mathcal{L}(\xi'). \tag{10}$$

Since $r_{III}(\xi, \xi')$ is factorized into the product of a function of ξ and a function of ξ', there are no correlations between the frequencies of the absorbed and reemitted photons. The correlations are destroyed by collisions occurring between the absorption and the reemission. As noted in Section 1, the absence of correlation is known as complete frequency redistribution.

For the frequency incoherent term, the angular redistribution is a linear combination of isotropic scattering and Rayleigh scattering (see Equation (4)). The coefficient c is defined by

$$c = \frac{\mathcal{D}^{(2)}}{\Gamma_R + \Gamma_I + \mathcal{D}^{(2)}}, \tag{11}$$

where $\mathcal{D}^{(2)}$ is the rate of elastic collisions which are destroying the alignment of the atomic dipole, i.e. the phase relationships between the Zeeman sub–levels (Faurobert–Scholl et al., 1995). The corresponding coefficient in Domke and Hubeny (1988) is $\beta^{(0)} - \beta^{(2)} = c(1 - \epsilon)$. Note that $\mathcal{D}^{(2)} \leq \Gamma_c$. The second term in the square bracket of Equation (4) corresponds to collisions which change the phases of the atomic dipole but not its alignment, thus preserving an angular redistribution which is of the Rayleigh type.

We note that in the absence of elastic collisions, $\Gamma_c = 0$ and $\mathcal{D}^{(2)} = 0$, hence $c = 0$ and $\gamma = 1$. The redistribution matrix reduces to the frequency coherent term. This limiting case is also discussed in Bommier (same volume).

Whether the redistribution matrix is dominated by coherent or by incoherent scattering depends on the value of the branching ratio γ. In general the elastic collision rate Γ_c is much larger than the inelastic collision rate Γ_I so that

$$\gamma \simeq \frac{\Gamma_R}{\Gamma_R + \Gamma_c}. \tag{12}$$

The value of γ is thus very sensitive to the value of elastic collision rate Γ_c, which is not very well known, in general. In a first approximation Γ_c can be evaluated with the van der Waals approximation (Allen, 1964):

$$\Gamma_c = \gamma_w n_H (T_e/5000)^{0.3}, \tag{13}$$

where n_H is the density of hydrogen atoms and T_e the electronic temperature. For the Ca I resonance line, the van der Waals approximation yields
$\gamma_w = 1.7 \times 10^{-8}\,cm^{-3}s^{-1}$. Empirical determinations of Γ_c relying on observations of the wing intensity and the wing polarization, which are domi-
nated by the coherent scattering term, have been attempted by Ayres (1977)
and Faurobert–Scholl (1992). Nagendra (1994) has shown that the line core
polarization is sensitive to the ratio $\mathcal{D}^{(2)}/\Gamma_c$, but it is not clear whether
observations can be used for empirical determinations of $\mathcal{D}^{(2)}$. Direct quantum mechanical calculation is in principle the best approach for obtaining
reliable values of Γ_c and $\mathcal{D}^{(2)}$ but requires sophisticated calculations. Such
calculations were performed for the resonance lines of Ca I, at 4227 Å,
and of Sr I, at 4607 Å, using accurate interatomic potentials (Spielfiedel et
$al.$, 1991; Faurobert–Scholl et $al.$, 1995). They have an accuracy of approximately 20%. For the Ca I line, these authors find $\gamma_w = 3 \times 10^{-8}\,cm^{-3}s^{-1}$, a
value which is significantly different from the van der Waals approximation,
and a slightly different temperature dependence. The same set of calculations
yields $\mathcal{D}^{(2)}/\Gamma_c = 0.6$. These values of γ_w and $\mathcal{D}^{(2)}$ lead to $\Gamma_c \simeq 2 \times 10^6\,s^{-1}$,
$c \simeq 5 \times 10^{-3}$ and $1 - \gamma \simeq 10^{-2}$ in the chromosphere (Faurobert–Scholl,
1992). Note that the elastic collision rate Γ_c is indeed much larger than the
inelastic collision rate $\Gamma_I \simeq 2 \times 10^3\,s^{-1}$.

2.2. REDISTRIBUTION MATRIX IN THE OBSERVER'S FRAME

To determine the redistribution matrix in the observer's frame, it is necessary to take into account the thermal movements of the atoms. Two assumptions are currently made, first that the atomic velocity is unchanged
during the scattering process, second that the velocity distribution for the
lower level of the transition is a Maxwellian. The former approximation is
justified by the short duration of the scattering process. The latter approximation certainly holds for resonance lines because the lower level of the
transition can be considered as having an infinite life–time. It use for subordinate lines is more questionable (see, e.g., the discussion in Ivanov, 1973,
p. 10)

Making in Equation (4) the substitutions

$$\xi \to \nu - \nu_o \frac{\mathbf{n}.\mathbf{v}}{c}, \quad \xi' \to \nu' - \nu_o \frac{\mathbf{n'}.\mathbf{v}}{c}, \tag{14}$$

where $\mathbf{v}$ is the thermal atomic velocity and c the velocity of light, and
performing a convolution with a Maxwellian distribution, one obtains a lab-
oratory frame redistribution matrix $\hat{R}(\nu, \mathbf{n}, \nu', \mathbf{n'})$ which has an expression
similar to Equation (4) with $r_m(\xi, \xi')$, $m = II,III$, replaced by some functions
$R_m(\nu, \mathbf{n}, \nu', \mathbf{n'})$ first given by Hummer (1962) (see also Mihalas, 1978). In the
observer's frame the frequency and the angular redistribution are coupled.

As suggested in Domke and Hubeny (1988), a separation between the angular and frequency variables may be achieved by an expansion of $\hat{R}(\nu, \mathbf{n}, \nu', \mathbf{n}')$ in Legendre polynomials. According to these authors this expansion does not have good numerical properties. In the case of an azimuthally independent radiation field, the problem is fully described by the azimuthally averaged redistribution matrix. It involves six different redistribution functions (three for each redistribution mechanism) which depend on the frequencies and polar angles of the incident and scattered radiation. They are defined in Faurobert-Scholl (1987) and in Domke and Hubeny (1988). The use of the exact angle and frequency dependent redistribution matrix is fairly cumbersome, therefore several approximations inspired from the scalar case have been introduced.

In the scalar case, and for optically thick lines, because the radiation field is almost isotropic, it is justified to replace the angle dependent redistribution functions $R_m(\nu, \mathbf{n}, \nu', \mathbf{n}')$, $m = II,III$, by their angle averages

$$R_m(\nu, \nu') = \int \frac{d\Omega'}{4\pi} R_m(\nu, \mathbf{n}, \nu', \mathbf{n}'). \tag{15}$$

The functions $R_{II}(\nu, \nu')$ and $R_{III}(\nu, \nu')$, corresponding to coherent and incoherent scattering, respectively, can be found in Hummer (1962) (see also Mihalas, 1978). This angle averaging greatly simplifies numerical solutions of transfer equations and saves a substantial amount of memory and computing time. It is therefore tempting to use it also for polarized transfer. However, since resonance scattering is very sensitive to the anisotropy of the radiation field, the use of angle averaged functions may not be as good for polarized as for non polarized transfer. Tests of this approximation performed by Faurobert (1988), for purely coherent scattering, show that the errors on the linear polarization stay very small (see Section 2.3). No tests have yet been made with a general redistribution matrix combining coherent and incoherent scattering.

In the case of resonance lines, the most frequently used approximation is a hybrid model which retains the angular dependence in the phase matrix $\hat{P}(\mathbf{n}, \mathbf{n}')$ but replaces the angle and frequency dependent redistribution functions $R_m(\nu, \mathbf{n}, \nu', \mathbf{n}')$, $m = II,III$, by their angle averaged $R_m(\nu, \nu')$ (Stenflo and Stenholm, 1976; Rees and Saliba, 1982; Faurobert-Scholl, 1987).

Even with the angle averaged approximation, the numerical work for solving transfer problems with partial frequency redistribution is much larger than with complete redistribution. So many efforts have been made for finding approximations for $R_{II}(\nu, \nu')$ and $R_{III}(\nu, \nu')$. The main criteria for deciding whether an approximation is good or not, at least for optically thick lines, is that it preserves the large scale behaviour of the radiation field (Ivanov, 1973; Frisch, 1980). For R_{II} it should preserve the large scale diffusive behaviour.

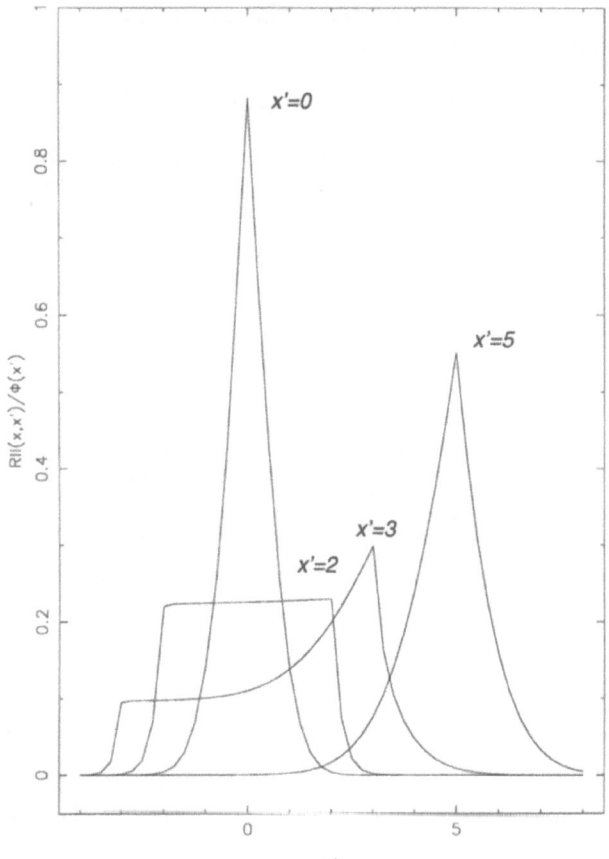

Fig. 3. Angle averaged R_{II} redistribution function with a Lorentz parameter $a = 10^{-3}$. Ordinate $R_{II}(x, x')/\phi(x')$, with x and x' the emission and absorption frequencies measured from line center in Doppler width units. Curves are labeled with frequency x'.

For the incoherent frequency term one may safely assume complete frequency redistribution in the laboratory frame, i.e. use the approximation

$$R_{III}(\nu, \nu') \simeq \phi(\nu)\phi(\nu'), \tag{16}$$

which is already suggested in Hummer (1962) and has been tested for scalar problems by Finn (1967).

For the coherent term, a very popular approximation, first introduced by Jefferies and White (1960) and revised by Kneer (1975), is to assume that there is complete frequency redistribution in the line core and purely coherent scattering in the line wings. This approximation, called here the CS approximation, may be written as

$$R_{II}(\nu, \nu') \simeq [1 - a(\nu)]\phi(\nu)\phi(\nu') + a(\nu)\phi(\nu)\delta(\nu - \nu'), \tag{17}$$

where $a(\nu)$ is an ad-hoc function which is zero in the line core and goes to unity in the line wings, the transition taking place at a few Doppler widths from line center. It is clear that the first term in Equation (17) yields the core redistribution function and the second term the wing redistribution. The origin of this approximation can be understood by looking at Figure 3 which shows the angle averaged R_{II} function normalized by the absorption profile. Photons with a frequency close to that of the line center are reemitted essentially according to complete frequency redistribution. In contrast, the rest frame frequency coherence is largely preserved in the wings. The reason is that the absorption profile has Lorentzian (algebraic) wings, whereas the velocity distribution function falls off exponentially. Thus photons with large frequencies are more likely to be absorbed in the wings by atoms which are almost at rest than near line center by atoms having a large velocity (Jefferies, 1968). This CS approximation does not preserve the large scale diffusive behaviour of the radiation field (Frisch, 1980).

The redistribution matrix which is commonly used for resonance polarization with partial frequency redistribution is thus

$$\hat{R}(\nu, \mathbf{n}, \nu', \mathbf{n}') = \gamma R_{II}(\nu, \nu')\hat{P}(\mathbf{n}, \mathbf{n}') + \phi(\nu)\phi(\nu')[c\hat{P}_{is} + (1-c-\gamma)\hat{P}(\mathbf{n}, \mathbf{n}')]. \tag{18}$$

The decoupling between frequency and angular redistribution allows one to use efficient numerical methods for the solution of the transfer equation, such as the azimuthal Fourier decomposition of the radiation field (Faurobert–Scholl, 1991).

2.3. POLARIZATION PROFILES

Figure 4 shows the emergent intensity and the linear polarization profiles calculated by Faurobert (1988) for a normal Zeeman triplet ($W_2 = 1$), assuming a semi–infinite and isothermal plane–parallel atmosphere, purely coherent scattering, a destruction probability $\epsilon = 10^{-4}$ and a Voigt (Lorentz) parameter $a = 10^{-3}$. For this problem the exact redistribution matrix is

$$\hat{R}(\nu, \mathbf{n}, \nu', \mathbf{n}') = R_{II}(\nu, \mathbf{n}, \nu', \mathbf{n}')\hat{P}_R(\mathbf{n}, \mathbf{n}'). \tag{19}$$

The calculations have been performed for several choices of the first factor which describes the redistribution in frequency : frequency and angle dependent R_{II}, angle averaged R_{II}, CS approximation of Equation (17), and also complete frequency redistribution. Because of the symmetry of the profiles with respect to line center only one half of the profiles are shown. They

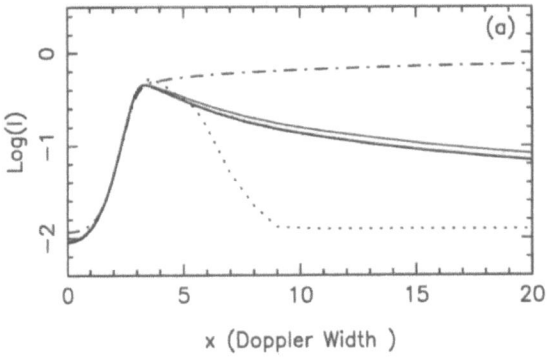

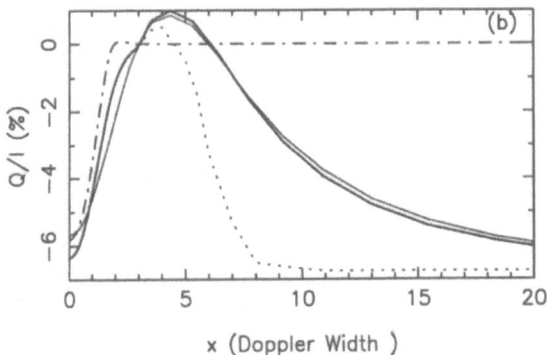

Fig. 4.   Resonance polarization in a semi-infinite medium. Emergent line intensity and linear polarization; angle–dependent R_{II}: thick solid line; angle–averaged R_{II}: thin solid line; CS approximation : dotted line; complete frequency redistribution : dash–dotted line.

correspond to an heliocentric angle $\theta = \arccos(0.11)$ inside the solar limb. Note that Figure 4 shows Q/I whereas Figures 1 and 2 show $-Q/I$. A few comments are in order regarding these results:

— With complete frequency redistribution, the polarization in the wings is zero. The reason is that the radiation field becomes essentially isotropic at the optical depths where the wings are formed.

— The use of the angle averaged R_{II} redistribution function leads to errors in the linear polarization which are very small in the wings. In the line core, the polarization peak obtained with the angle–dependent R_{II} redistribution function is slightly sharper than with the angle averaged R_{II}. However, this difference will likely be smeared out in observed

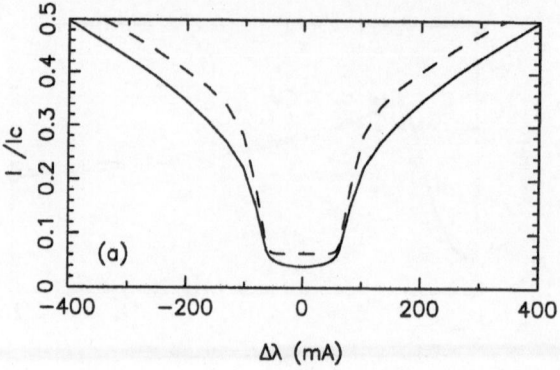

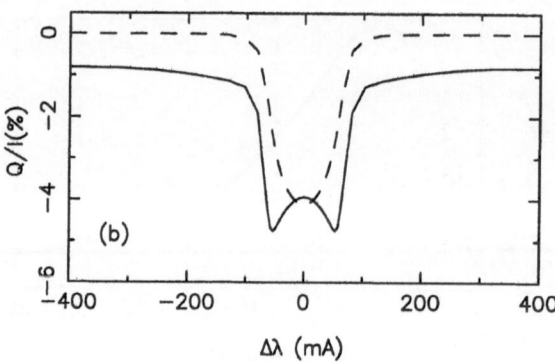

Fig. 5.   Intensity and linear polarization profiles in the core of the Ca I 4227 Å line;
partial redistribution : solid line; complete redistribution : dashed line.

profiles due to the finite spectral resolution and macroturbulent motions
in the atmosphere.

– The CS approximation underestimates both the intensity and the linear
  polarization in the line wings. The reason is that it prevents photons
  from the line core to be fed into the line wings.

– For R_{II} redistribution, the wing polarization is of the same order as the
  line core polarization because the anisotropy of the radiation field in
  these two frequency ranges follows the Chandrasekhar $H(\mu)$ function
  (Faurobert, 1988).

– With the addition of a background continuum, the wing polarization
  would finally go to zero and one would observe a maxima (in absolute
  value) as in Rees and Saliba (1982).

Tests using isothermal models, which in particular imply a constant Doppler width, suggest that partial redistribution effects are negligible in the line core. For strong resonance lines and realistic atmospheric models this conclusion is not valid. Figure 5 shows the intensity profile and the linear polarization of the Ca I 4227 Å resonance line near line center, calculated with the atmospheric model C of Fontenla *et al.* (1993) and the elastic cross–sections derived by Spielfiedel *et al.* (1991) (see also Faurobert *et al.*, 1995). First we note that partial redistribution effects are significant, especially on the linear polarization. In the wings, the polarization is smaller by a factor 2 than in Faurobert–Scholl (1994) because the elastic collision cross–sections are twice larger than the van der Waals approximate values used in this latter work. The line core is not sensitive to the values of the elastic cross–sections because it is formed in higher layers where elastic collisions are negligible.

We stress here that partial redistribution effects are noticeable only for strong lines with well developed wings where frequency coherent scattering can play a role. In that case they are observable both in the line wings and in the line core. The CaI resonance line which has a total optical depth at line center of the order of 10^5 is such a line. A weak resonance line, such as the resonance line of SrI at 4607 Å, which has a total optical depth around 10, shows no partial redistribution effects although the branching ratio γ is very close to unity. For these lines one can replace R_{II} by R_{III} and use

$$\hat{R}(\nu, \mathbf{n}, \nu', \mathbf{n}') \simeq \phi(\nu)\phi(\nu')[c\hat{P}_{is} + (1 - c)\hat{P}(\mathbf{n}, \mathbf{n}')]. \tag{20}$$

3. Resonance polarization with Hanle effect

The form of the redistribution matrix for the Hanle effect (resonance polarization in the presence of a weak magnetic field) may still be considered as an open problem and work by Bommier, presently in progress, is presented in this volume. Stenflo (1978) and Faurobert–Scholl (1992) have proposed more or less empirical formulas for the core and the wings of a line (see below). Here we propose a new "empirical formula" based on the work of OSC and comments by Faurobert *et al.* (1995).

In the theory of OSC, an essential ingredient entering into the determination of the rest frame redistribution matrix is the matrix element

$$S_{if} = \int_0^T dt \int_0^t dt' e^{i(\omega t - \omega t')} < f|\mathcal{O}(T, t, t', 0)|i > . \tag{21}$$

Here $|i>$ and $|f>$ are the initial and final states of the transition, ω' and ω are the angular frequencies of the absorbed and reemitted radiation and $\mathcal{O}$ is an operator describing the evolution of the atom under collisions as well

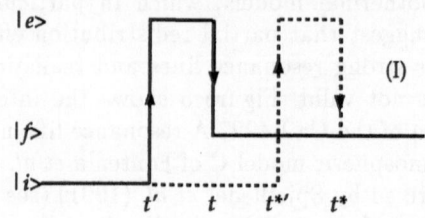

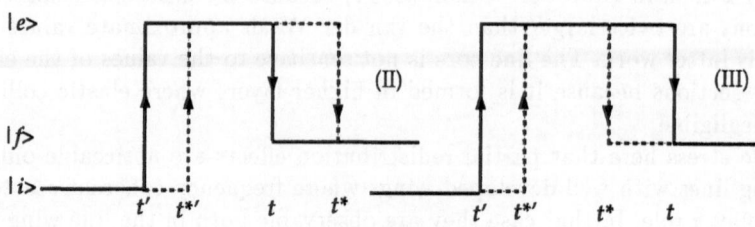

Fig. 6. Diagrams of the different orderings of the instants of absorption $t', t^{*'}$ and of emission t, t^*. Only half the diagrams are shown. The three others are derived by exchanging the positions of t and t^* and of t' and $t^{*'}$.

as the absorption of a photon with frequency ω' at time t' followed by the emission of a photon with frequency ω at time t. During the absorption-emission process, the atom initially in a state $|i>$ is excited to a state $|e>$ and reaches a final state $|f>$. The quantity which appears in the expression of the redistribution matrix is not S_{if} itself but its amplitude $|S_{if}|^2$. The product of S_{if} by its complex conjugate S_{if}^* involves a set of time (t', t) associated with S_{if} and another set, say, $(t^{*'}, t^*)$ associated with S_{if}^*. The value of $|S_{if}|^2$ depends on the ordering of these two sets of time. Three types of configurations are possible. They are shown in Figure 6 (see also Figure 2 in OSC).

In the problem at hand, $|i>$ and $|f>$ are identical and correspond to the ground state which we assume infinitely sharp (infinite radiative life time and negligible collisional broadening). The frequency redistribution function corresponding to the first diagram is then of the form

$$\mathcal{L}(\omega)\delta(\omega - \omega') - \mathcal{L}(\omega)\mathcal{L}(\omega'), \tag{22}$$

where $\mathcal{L}(\omega)$ is the Lorentzian defined in Equations (6) and (7) with 4π replaced by 2. This diagram yields the frequency coherent part of the redistribution matrix. It is difficult to give it a physical interpretation because "the incident and scattered polarization are mixed and ones does not obtain a simple product of absorption and emission operators" (OSC, p. 194). The frequency redistribution corresponding to the second and third diagram is of the form

$$\mathcal{L}(\omega)\mathcal{L}(\omega'). \tag{23}$$

These two diagrams contribute thus only to the incoherent part of the frequency redistribution function. The full calculation leads to the rest frame redistribution written in Equation (4).

In the presence of a weak magnetic field $\mathcal{H}$, the redistribution matrix is much more complicated and even in the rest frame the frequency redistribution is coupled to the angular redistribution. Now there is an important remark to be made. The first diagram, which corresponds to absorption immediately followed by emission, induces no population of the atom in the excited state. Therefore this term is not affected by the Hanle effect. This suggests a heuristic method for finding the redistribution matrix.

Let us first rewrite the laboratory frame redistribution matrix for resonance polarization with zero magnetic field in the form

$$\hat{R}(\nu, \mathbf{n}, \nu', \mathbf{n}') = \gamma[R_{II} - R_{III}]\hat{P}(\mathbf{n}, \mathbf{n}') + R_{III}[c\hat{P}_{is} + (1 - c)\hat{P}(\mathbf{n}, \mathbf{n}')], \tag{24}$$

where $\hat{P}(\mathbf{n}, \mathbf{n}')$ is the linear combination of Rayleigh and isotropic scattering written in Equation (8). The first term in the r.h.s. is the contribution from the first diagram and the second one the contribution from the second and third diagram. Since the first term is not affected by the magnetic field, the idea is to replace the Rayleigh phase matrix by the Hanle phase matrix only in the second term. Bommier (this volume) proposes a similar expression based on a detailed quantum mechanical calculation.

For a normal Zeeman triplet ($W_2 = 1$), our suggestion leads to

$$\hat{R}(\nu, \mathbf{n}, \nu', \mathbf{n}') = \gamma[R_{II} - R_{III}]\hat{P}_R(\mathbf{n}, \mathbf{n}') + R_{III}[c\hat{P}_{is} + (1 - c)\hat{P}_H(\mathbf{n}, \mathbf{n}')], \tag{25}$$

where $\hat{P}_H(\mathbf{n}, \mathbf{n}')$ is the Hanle phase matrix (see, e.g., Landi degl'Innocenti and Landi degl'Innocenti, 1988; Faurobert, 1991; Stenflo, 1994). Equation (25) has not yet been used for numerical applications. Faurobert–Scholl (1992) has employed

$$\hat{R}(\nu, \mathbf{n}, \nu', \mathbf{n}') = \gamma[R_{II} - R_{III}]\hat{P}_R(\mathbf{n}, \mathbf{n}') + R_{III}[c\hat{P}_{is} + (1 - c)\hat{P}_R(\mathbf{n}, \mathbf{n}')], \tag{26}$$

in the line wings and

$$\hat{R}(\nu, \mathbf{n}, \nu', \mathbf{n}') = \gamma[R_{II} - R_{III}]\hat{P}_H(\mathbf{n}, \mathbf{n}') + R_{III}[c\hat{P}_{is} + (1 - c)\hat{P}_H(\mathbf{n}, \mathbf{n}')], \tag{27}$$

in the line core. The expression (25) avoids a change of redistribution matrices at a somewhat arbitrary frequency. For the wings it will give the same results as Equation (26) since the complete frequency redistribution term proportional to R_{III} does not contribute to the wing polarization. For the core, whether (25) and (27) give the same polarization is harder to predict. An educated guess is that differences, if any, will be noticeable only for lines showing strong partial redistribution effects such as the resonance line of Ca I.

4. Conclusion

In this paper we have discussed the frequency and angle redistribution matrix for resonance scattering in regions of zero or weak magnetic fields. For strong resonance lines in zero magnetic fields, a reasonable approximation is the expression given in Equation (18) and for subordinate lines and weak resonance lines the expression given in Equation (20). For regions with weak magnetic fields, we propose a new expression given in Equation (25). Its consequences on line core polarization remain to be examined. The formulae we have presented here are based on the theory of Omont *et al.* (1972, 1973) and its subsequent developements by Domke and Hubeny (1988). Work currently in progress concerning partial redistribution of polarized radiation is to be found in articles by Bommier and Stenflo in these Proceedings.

Acknowledgements

I gratefully acknowledge stimulating discussions with Drs. V. Bommier, M. Faurobert–Scholl and N. Feautrier during the preparation of my conference and the kind help of Drs. M. Faurobert–Scholl and J. Stenflo in the production of some of the figures.

References

Allen, C.W.: 1964, *Astrophysical Quantities*, University of London, The Athlone Press, London
Ayres, T.R.: 1977, Astrophys. J. **213**, 296
Bommier, V.: 1996, Solar Physics (this volume)
Chandrasekhar, S.: 1950, *Radiative Transfer*, Oxford University Press, (Dover, New York, 1960)
Domke, H., and Hubeny, I.: 1988, Astrophys. J. **334**, 527
Faurobert, M.: 1987, Astron. Astrophys. **178**, 269
Faurobert, M.: 1988, Astron. Astrophys. **194**, 268
Faurobert–Scholl, M.: 1991, Astron. Astrophys. **246**, 469
Faurobert–Scholl, M.: 1992, Astron. Astrophys. **258**, 521
Faurobert–Scholl, M.: 1994, Astron. Astrophys. **285**, 655
Faurobert–Scholl, M.: 1996, Solar Physics (this volume)
Faurobert–Scholl, M., Feautrier, N., Machefert, F., Petrovay, K., and Spielfiedel, A.: 1995, Astron. Astrophys. **298**, 289

Finn, G.D., 1967, Astrophys. J. **147**, 1085

Fontenla, J.M., Avrett, E.H., and Loeser, R. : 1993, Astrophys. J. **406**, 319

Frisch, H.: 1980, Astron. Astrophys. **83**, 166

Hamilton, D.R.: 1947, Astrophys. J. **105**, 424

Hummer, D.G.: 1962, Mon. Not. R. astr. Soc. **125**, 21

Landi degl'Innocenti, M., and Landi degl'Innocenti, E.: 1988, Astron. Astrophys. **192**, 374

Ivanov, V.V.: 1973, *Transfer of Radiation in Spectral Lines*, NBS Spec. Publ. 385, Washington Governement Printing Office

Jefferies, J.T.: 1968 *Spectral Line Formation*, Blaisdell Publishing Company, Waltham, Massachusetts

Jefferies, J.T., and White, O.: 1960, Astrophys. J. **132**, 767

Kneer, F.: 1975, Astrophys. J. **200**, 367

Nagendra, K.N.: 1994, Astrophys. J. **432**, 274

Mihalas, D.: 1978, *Stellar Atmospheres*, Freeman and Company, San Francisco

Omont, A., Smith, E.W., and Cooper, J: 1972, Astrophys. J. **175**, 185

Omont, A., Smith, E.W., and Cooper, J, 1973, Astrophys. J. **182**, 283

Redman, R.O.: 1941, Mon. Not. R. astr. Soc. **101**, 266

Rees, D., and Saliba, G.J.: 1982, Astron. Astrophys. **115**, 1

Spielfiedel A., Feautrier N., Chambaud G., and Lévy B.: 1991, J. Phys. B: At. Mol. Opt. Phys. **24**, 4711

Stenflo, J.: 1976, Astron. Astrophys. **46**, 61

Stenflo, J.: 1978, Astron. Astrophys. **66**, 241

Stenflo, J.: 1982, Solar Phys. **80**, 209

Stenflo, J.: 1994, *Solar Magnetic Fields*, Kluwer Academic Publishers, Dordrecht

Stenflo, J.: 1996, Solar Physics (this volume)

Stenflo, J., Baur, T.G. and Elmore D.F.: 1980, Astron. Astrophys. **84**, 60

Stenflo, J., and Stenholm, L.: Astron. Astrophys. **46**, 69

Van de Hulst H.C.: 1980, *Multiple Light Scattering, Tables, Formulas and Applications*, Academic Press, New York

Zanstra, H.: 1941a, Mon. Not. R. astr. Soc. **101**, 250

Zanstra, H.: 1941b, Mon. Not. R. astr. Soc. **101**, 273

PARTIAL REDISTRIBUTION EFFECTS ON LINE POLARIZATION IN THE PRESENCE OF VELOCITY FIELDS

K.N. NAGENDRA

Indian Institute of Astrophysics, Bangalore 560 034, India

Abstract. Velocity fields in line formation regions strongly affect the line polarization. The conventionally used observer's frame method of solving the polarized transfer equation becomes expensive and inaccurate for partial redistribution problems, when large amplitude velocity fields have to be considered in the observer's frame. An alternative method of solution is the comoving frame method. Partial redistribution problems are solved using comoving frame formalism for line polarization caused by resonance scattering.

Key words: Line Polarization – Radiative transfer – Comoving frame method

1. Introduction

The systematic velocity fields in stellar and solar atmospheres arise due to dynamical state of the medium. A discussion of line formation problems in moving media is given in Mihalas (1978). The numerical methods required for solving this general problem are discussed in greater detail in the review articles by Mihalas and Kunasz (1986), and by Grinin (1984). In most of the theoretical work in this area, the polarization state of the radiation is not considered. Since polarization measurements bring invaluable informations on vectorial quantities such as systematic velocity fields, it appears important to develop our knowledge on the effects of such fields on line polarization. Not much work has been done up to now. In a recent article, Sengupta (1993) has studied the resonance polarization of optically thick lines formed in a moving medium, under the approximation of complete frequency redistribution. This approximation however does not hold for strong resonance lines which are often the most conspicuous lines in stellar spectra.

In the solar atmosphere, rather small velocity fields are observed. Hence the usual observer's frame method of solving the radiative transfer problem is appropriate (see Mihalas, 1978). Velocity fields have thus been included in a realistic manner in polarized Zeeman line formation theory, neglecting however the partial frequency redistribution effects (see Stenflo, 1994). For stellar atmospheres we often encounter high velocity flows. In such cases, the comoving frame method (abbreviated hereafter as CMF) is normally employed in the solution of line transfer problems in moving stellar atmospheres (see Mihalas and Kunasz, 1986). This method can be used even in small amplitude velocity fields, as encountered in solar atmospheric features.

Solar Physics **164**: 67–78, 1996.

The purpose of this paper is to show the advantages as well as difficulties in using the CMF method for partial redistribution problems, including polarization.

2. CMF polarized transfer equation for moving media

For axisymmetric problems in planar geometry, it is sufficient to use the radiative transfer equation for the Stokes vector $\mathbf{I} = (I \ Q)^T$, written in the conventional notation (see Nagendra, 1994) as

$$\mu \frac{\partial \mathbf{I}}{\partial z} = k^L (\beta + \phi) [\mathbf{S} - \mathbf{I}] + \left[\mu^2 \frac{dv(z)}{dz} \right] \frac{\partial \mathbf{I}}{\partial x}, \tag{1}$$

where $v(z)$ is the macroscopic velocity of the gas at a geometrical height z. The frequency grid is expressed in terms of the reduced frequency (x) units. Similarly, the velocities are expressed in terms of mean thermal unit $v_{th} = (2kT/M)^{1/2}$. The quantity μ is the angle between the ray and the symmetry axis of the slab. The optical depth scale is defined as $\partial \tau = -k^L \partial z$. The last term in the above equation appears because the transfer equation is written in the comoving frame of the fluid, in which there is no relative motion between the observer and the moving layers of gas. Alternatively, the transfer equation may also be written in the laboratory (observer's) frame, which is called as the rest frame in literature (see Mihalas, 1978, p. 449). The advantages of both the methods are discussed by Mihalas and Kunasz (1986). The most important practical advantage in working with CMF is that, it is equivalent to solving the transfer equation in a static media, which is considerably easier. The polarized source vector $\mathbf{S}$ in the above equation is given by

$$\mathbf{S} = \frac{\phi \mathbf{S}^L + \beta \mathbf{S}^C}{\phi + \beta}, \tag{2}$$

where $\mathbf{S}^C = \mathbf{B} = (B \ 0)^T$ is the continuum source vector, which is assumed as unpolarized. The quantities ϕ, β, and a are the profile function, continuum absorption parameter, and the damping parameter respectively. The CMF line source vector $\mathbf{S}^L$ is written as

$$\mathbf{S}^L = \frac{(1 - \epsilon)}{\phi} \int_{-\infty}^{+\infty} dx' \frac{1}{2} \int_{-1}^{+1} d\mu' \, \mathbf{R}(x, \mu, x', \mu') \mathbf{I}(x', \mu') + \epsilon \mathbf{B}, \tag{3}$$

where ϵ is the thermalization parameter, and B is the Planck function. The line source vector has the same form as in the static case, which allows us to use approximations for $\mathbf{R}(x, \mu, x', \mu')$ employed in the static case. Since the absorption and emission coefficients in a moving media, are sufficiently isotropic in CMF, as compared to those in the observer's frame, the angle and frequency decoupling:

$$\mathbf{R}(x,\mu,x',\mu') = R(x,x')\ \mathbf{P}(\mu,\mu'),\tag{4}$$

remains a good approximation (see Rees and Saliba, 1982; Faurobert, 1987). This approximation greatly simplifies the evaluation of angle and frequency integrals in the line source vector. The expressions for $R(x,x')$ and $\mathbf{P}(\mu,\mu')$ can be found in Faurobert (1987). The difference in polarization profiles obtained with angle averaged and angle dependent partial redistribution functions is discussed in the static case by Faurobert (1987) and Frisch (1996). Such a comparison for the case of moving atmospheres, in the CMF formalism is given by Mihalas (1980). It is found by these authors that the use of angle averaged functions in stationary frames is a reasonable approximation. We have found this conclusion to be true even in the present situation of polarized line transfer in the CMF. For computations presented in this paper we employ angle averaged functions. Whereas, it is consistent to employ an angle averaged version of a partial frequency redistribution (PRD) function in CMF, one should employ explicitly angle dependent redistribution function (see Equation (3) above) when solving the transfer equation in the observer's frame (see Vardavas, 1976). The two-level atom frequency redistribution models used for this paper are complete frequency redistribution

$$\text{CRD}:\quad R(x,x') = \phi(x)\,\phi(x'),\tag{5}$$

with normalised Voigt profile functions, and partial frequency redistribution

$$\text{PRD}:\quad R(x,x') = R_{II-A}(x,x'),\tag{6}$$

in Hummer's standard notation. A more general PRD mechanism, including collisional redistribution, can be easily included in the computations.

3. Method of Solution

The numerical method used for solving the radiative transfer problem is the vector analogue of the scalar (unpolarized) comoving frame method described in Peraiah (1984). The applications of this numerical method to polarized PRD line scattering in static planar and spherical geometries are described in Nagendra (1988, 1994), where the model parametrization etc. are also given. For the computations in this paper, homogeneous, isothermal slab models are employed, to serve the purpose as benchmark problems. In CMF formalism, we also have to specify the initial conditions $\partial\mathbf{I}/\partial x$, at the blue and red frequency edges of the line profile. We employ $\partial\mathbf{I}/\partial x = 0$, which is accurate enough if an unpolarized continuum radiation ($\beta \neq 0$) is present in the medium, or when a thermal emission is imposed as a boundary condition at the lower boundary $\tau = T^L$ (Schuster problem). A careful

selection of these initial conditions is required for the case of optically thin
self emitting slabs, since the far wings are not fully saturated to the level
of thermal continuum (see Noerdlinger 1981; and Mihalas et al., 1976 for
details in the scalar case).

The full computation consists of two steps: In the step-1, the radiative
transfer problem is solved in the CMF to obtain the frequency and angle dependent specific intensities $I(x, \mu, \tau)$ and the source vectors $S(x, \mu, \tau)$ from
the defining Equation (3). In the step-2, a formal integration of the transfer equation has to be carried out in the observer's frame, in order to get
emergent specific intensity profiles at required angles (μ). These angle points
form a much finer grid, than the 3-point Gaussian μ-grid used in the CMF
radiative transfer solution. The same symbol μ is retained, but unlikely now
to cause confusion. The profiles in Figure 1 for instance, are shown for $\mu_* =$
0.83, which is one of the points in this 19-point Simpson μ-grid. For a short
optical path characterised by the depth points $(d-1, d)$, along a given line
of sight, the formal integral for a moving slab atmosphere is written as

$$I(x, \mu, \tau_{x,\mu,d}) = I(x, \mu, \tau_{x,\mu,d-1})\, exp\left(-\mid \tau_{x,\mu,d} - \tau_{x,\mu,d-1}\mid\right)$$
$$+ \int_{\tau_{x,\mu,d-1}}^{\tau_{x,\mu,d}} S(x, \mu, t_{x,\mu,\bar{d}})\, exp\left(-\mid t_{x,\mu,\bar{d}} - \tau_{x,\mu,d-1}\mid\right)\, dt_{x,\mu,\bar{d}}, \qquad (7)$$

where $d = 2, 3, 4, ...N$, is the optical depth grid index, with $\tau_1 = 0$, and
$\tau_N = T^L$. The observer's frame depth scale is represented for any subinterval
dt as

$$dt_{x,\mu,\bar{d}} - dt\, \phi(x - \mu v_{\bar{d}})/\mu, \qquad (8)$$

where $\bar{d}$ indicates an average of the physical quantities over the optical depth
region $(d-1, d)$, or a smooth functional form of the concerned variable in this
region. Notice that the source vector on a transformed (observer's frame)
frequency grid, and in general, an arbitrary angle and depth grids as well,
are required to perform the formal integral. But, for an arbitrary frequency,
angle and depth point, the CMF source vector may not generally be available from step-1 (see Equations (1)-(6)). If, for instance, the CMF source
vectors are computed on a very wide frequency grid, the problem in step-2
involves performing interpolations using the original $S(x, \mu, \tau)$. However, the
basic idea of using CMF formalism is, to be able to obtain as accurate solutions as the observer's frame solution itself, but only using a small frequency
bandwidth, normally used in a corresponding static case. In this paper, we
are concerned only with small velocities. But in practical cases involving
high velocity flows ($v(\tau) = 50$ mean thermal units (mtu) or more in early
type stellar atmospheres), it is computationally advantageous to extrapolate
the local $S(x, \mu, \tau)$ for large x, using asymptotic expansions for the source
vector (see Hamman and Kudritzki, 1977, where such expansion for the case

of $R_{II-A}(x, x')$ is used in the scalar case). It is important to develop analytic approximations to far wing polarized source vector (see Frisch, 1988; Faurobert, 1988; Faurobert-Scholl and Frisch, 1989 for such analyses in polarized transfer problems, and Hubeny, 1985; Grachev, 1988; and Gayley, 1992 for the scalar transfer problems), in order to economically solve the polarized line transfer problems in high velocity stellar winds. However, for polarized PRD problems, care is required in using this procedure of extrapolations. We emphasise that, to produce the benchmark cases presented in this paper, we employed a fine as well as wide enough CMF frequency grid, so as to contain within it the transformed frequency grid. As a result, only interpolations on original $\mathbf{S}(x, \mu, \tau)$ were necessary.

Profiles are computed for the input parameters: $[T^L, a, \epsilon, \beta, B(\tau), v(\tau)]$. The boundary conditions $\mathbf{I}(x, \mu, \tau = T^L)$ and $\mathbf{I}(x, \mu, \tau = 0)$ are imposed at the lower and upper free boundaries. The initial conditions $\partial \mathbf{I}/\partial x = 0$ are used at the blue and red edges of the line profile. A three-point Gaussian angle quadrature ($\mu = 0.11$, 0.50, 0.89), a non-uniform trapezoidal frequency grid with 41 points, and a logarithmic depth grid with 5-8 points per decade of optical depth scale are employed in CMF radiative transfer computations. A finer τ - grid is used near the boundaries. The emergent flux profile $\mathrm{F}(x)$, and polarization profile $\mathrm{P}(x)$ are computed using a 19 point Simpson angle quadrature ($0 < \mu < 1$) in the line of sight computations (transformation from CMF, and then formal integration). The specific intensities $\mathbf{I}(x, \mu, \tau)$ are computed on this fine angle grid using the formal integral. In the solar atmosphere, the intensity $\mathrm{I}(x, \mu_*)$ and polarization $\mathrm{p}(x, \mu_*)$ can be observed on the resolved disk. In the case of stellar atmospheres, only the flux profiles are observed, and the $\mathrm{F}(x)$ and $\mathrm{P}(x)$ profiles are necessary for modelling the observed data. Two types of velocity profiles are considered in this paper namely, $v(\tau) = constant$, or $v(\tau) = v_g \tau$, where $v_g = dv(\tau)/d\tau$ is the velocity gradient, which is inputed as a free parameter. We have selected $v_g = (0, 1/T, 3/T, 5/T)$ in the models presented here. Details of our method of solution for polarized transfer can be found in Nagendra (1988, 1994) and Nagendra and Peraiah (1985).

4. Results and Discussion

The use of CMF technique is well established in stellar line formation theory (see Mihalas and Kunasz, 1986; and Noerdlinger and Rybicki, 1974). We find that the inclusion of polarization, particularly in line transfer with PRD demands higher accuracy in all aspects of the solution. These aspects include the frequency, angle and spatial resolution of the radiation field, numerical discretizations and quadratures etc. Thus we find it useful to generate a set

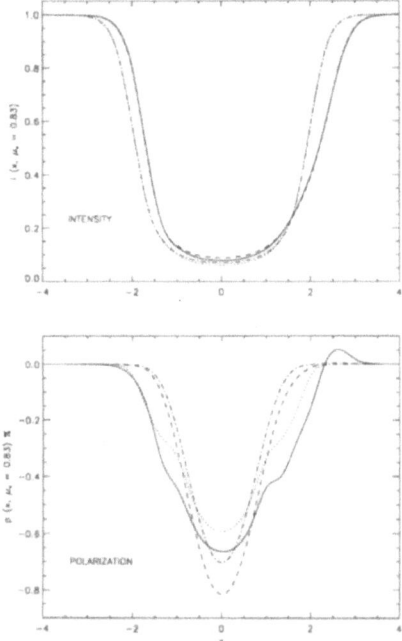

Fig. 1.   Emergent intensity (I) and the linear polarization (p) profiles at
$\mu_* = 0.83$ for an absorption line formed in a model slab with following parameters:
$\left[T^L, a, \epsilon, \beta, B(\tau)\right] = \left[100, 10^{-3}, 10^{-6}, 10^{-8}, 1\right]$. A linear velocity law $v(\tau) = v_g\tau$, with
a gradient $v_g = 10^{-2}$ is used. x is the frequency in dimensionless units. The boundary
condition $\mathbf{I}(x, \mu, T^L) = (1 \quad 0)^T$ is used for this model. Dotted and Dash-dotted curves
represent PRD and CRD cases for a static slab. The Solid and Dashed curves represent
the respective profiles for a moving slab. Notice a differential blue shift of I and p profiles.

of benchmark solutions, which also show the characteristics of polarized line
profiles formed in moving planar media.

4.1. ABSORPTION LINE POLARIZATION IN A SLOWLY EXPANDING
SLAB

In Figure 1 we show emergent intensity and polarization profiles for $\mu_* = 0.83$. Since the velocity vector is directed towards the observer at all the
depth points, the intensity (I) and polarization (p) profiles undergo a fre-
quency dependent blueshift, due to a linear increase of velocity with optical
depth τ. There is significant difference between emergent polarization for
static $(v(\tau) = 0)$ and moving media, at the line center and the near wings.
The intensity profile is much less sensitive. Velocity of expansion increases
the line center polarization, and the width of polarization profile, for both
CRD and PRD mechanisms. Due to a relative decrease in the red wing opti-

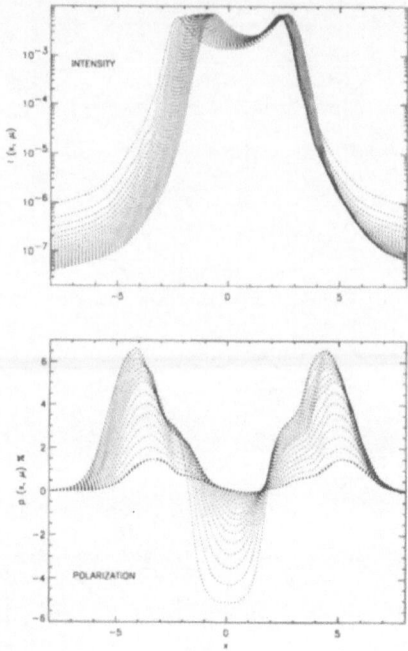

Fig. 2a. Limb-to-center $(0.16 < \mu < 0.99)$ variation of I and p profiles for PRD. These emission lines are formed in a self-emitting slab moving with a constant velocity of $v(\tau) = 1$ mtu. The model parameters are: $\left[T^L, a, \epsilon, \beta, B(\tau)\right] = \left[100, 10^{-3}, 10^{-4}, 0, 1\right]$.

cal depths, the photons in red wing escape easily from the deeper layers, and undergo a number of scatterings at the line center, in the outermost layers, where the polarization is created. The overall effect of the velocity fields is caused by the manner in which the macroscopic motion of the gas layers makes them transparent to the radiation from deeper layers. This occurs through a relative displacement of resonantly absorbing atoms in each layer with respect to other layers. The PRD polarization profile is more sensitive to the velocity field compared to the corresponding CRD profile.

4.2. Emission Line Polarization in a Slowly moving Slab

In this Section we show the emission lines formed in a self-emitting slab, moving as a whole with a constant velocity of 1 mtu. Since the velocity gradient is zero, the radiative transfer solutions are equivalent to static solutions. As a constant Doppler velocity shift is imparted at all the optical depths in the domain of formal integration, the line core and the wings are blue shifted to the same extent. Their shapes are same as in the static case (see Faurobert, 1987 for a comparison of CRD and PRD emission profiles).

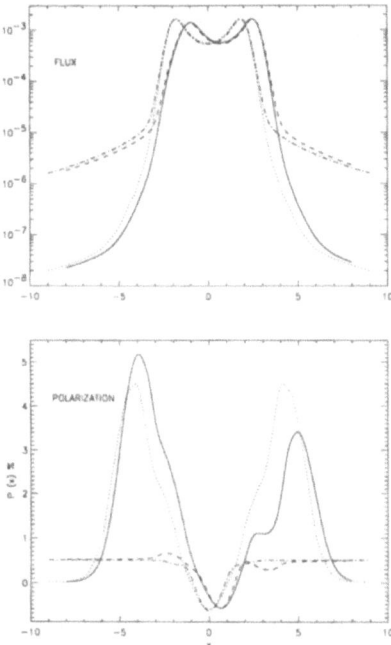

Fig. 2b.   PRD and CRD flux (F) and polarization (P) profiles. Dotted and Dash-dotted curves represent PRD and CRD cases for a static slab; the Solid and Dashed curves, the corresponding cases for a moving slab. Notice diminished blue wing polarization peak, compared to the red wing peak.

In Figure 2a we show the limb to center variation of I and p profiles only for the PRD mechanism. As the line of sight changes from limb to the center, line center and near wing polarizations decrease in magnitude at all frequencies. In Figure 2b we show the emergent flux F and polarization P profiles. An asymmetry of the I and p profiles about the line center, in addition to their strong frequency dependence, causes the integrated polarization profile P to exhibit a difference between the blue and red wings. A similar behaviour is seen in the CRD case also, but less significant.

A remark here is necessary, regarding the initial conditions $\partial I/\partial x=0$ imposed at the end points of the frequency grid. If the frequency bandwidth is not wide enough to ensure that the emission profile becomes indeed flat in the wings, the initial conditions cause slight numerical oscillations in the CMF far-wing source functions. This difficulty can be avoided by using a finite value of the continuous absorption parameter β.

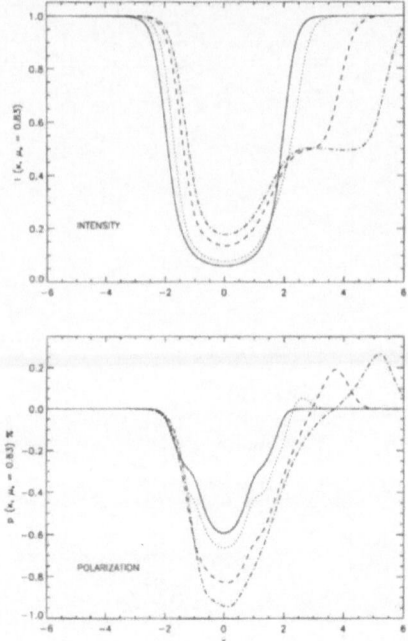

Fig. 3. I and p profiles for PRD absorption lines formed under large differential velocities $v(\tau) = v_g\tau$, with $v_g = 10^{-2}$ (Dotted curves); 3×10^{-2} (Dashed curves); and 5×10^{-2} (Dash-dotted curves). The static profiles $v_g = 0$ (Full curves) are shown for comparison. The model employed is same as that given in Fig. 1.

4.3. Absorption Line polarization under high velocity gradients

In the present Section we explore the PRD line transfer for planar layers expanding with fixed velocity gradients, and large velocities (see Figure 3). As v_g is increased from 10^{-2} to $5\ 10^{-2}$, the I profile becomes highly asymmetric due to enhanced blue wing opacity, further resulting in an increase in the number of coherent scatterings. There is also an increased escape of red wing and line center photons from the deeper layers. These processes together increase the anisotropy throughout the atmosphere. Thus we have enhanced polarization peaks at the line center and the blue wings.

4.4. Absorption Line Polarization in a Uniformly Moving Slab

In this Section, we discusss the PRD line polarization for $v(\tau) = constant$. The intensity and polarization profiles are Doppler shifted in frequency uniformly, without any changes in their shape. For the case of a slab moving

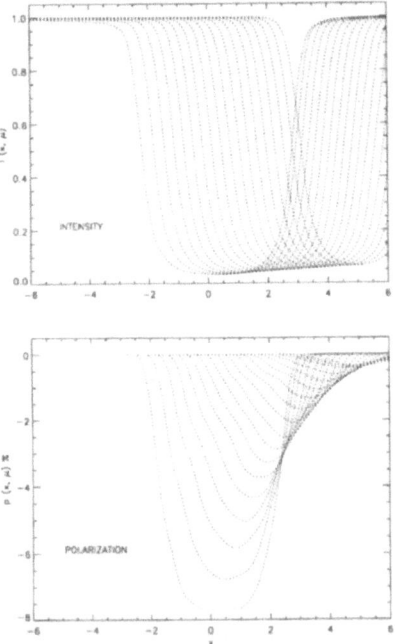

Fig. 4a.   Limb-to-center ($0.16 < \mu < 0.99$) variation of I and p profiles for PRD. These absorption lines are computed for the case of a constant velocity of $v(\tau) = 5$ mtu. The model employed is same as that given in Fig. 1.

with high velocity with respect to the central star, there is a sharp increase of projected Doppler shifts from limb to the center, leading to a large polarization asymmetry about the line center (see Figure 4a). As a result, the disk integrated polarization profiles P in Figure 4b show a shift and weakening of the line core polarization, as the velocity of the slab increases from 0 to 5 mtu. It is a consequence of the resonance absorption by atoms, which is now strongly angle dependent due to variation of projected Doppler shifts, but depth independent in the moving slab since the slab moves as a whole.

5. Conclusions

In this paper, we have applied the comoving frame formulation of the polarized transfer equation in planar geometry to a study of line polarization by CRD as well as PRD mechanisms. Few test cases are discussed which can serve as benchmark problems. Whenever monotonic velocity fields are encountered in stellar and solar atmospheres (absence of shocks etc.), it is advantageous to use CMF formalism for a fast and accurate solution of the

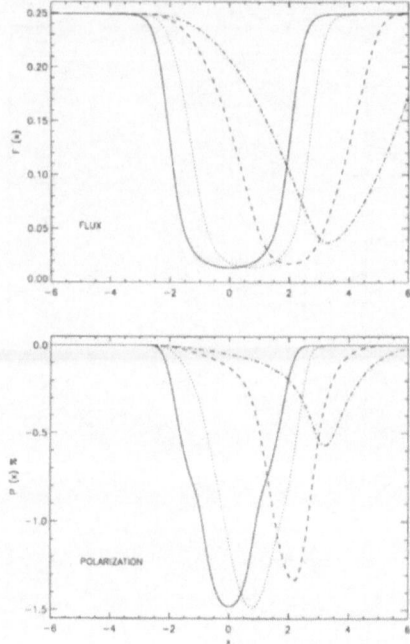

Fig. 4b.   The flux (F) and polarization (P) profiles for PRD. The cases shown are: $v(\tau) = 1$ mtu (Dotted curves; $v(\tau) = 3$ mtu (Dashed curves; and $v(\tau) = 5$ mtu (Dash-dotted curves). The static profiles $v(\tau) = 0$ mtu (Solid curves) are also shown. Compare with PRD polarized flux profiles of Fig. 2b. Notice the shift and weakening of P profiles.

transfer equation. The main advantages of CMF formalism is that it allows, through its formal similarity to a static solution, low resolution angle and frequency grids, and smallar frequency bandwidths (like those employed in this paper). To perform the same computations using the observer's frame method, we would require at least twice as many number of frequency, angle and depth points. It can be observed from the figures, that polarization profiles have more structure than the intensity profiles. Hence the treatment of polarization requires some additional care in the selection of angle and frequency grids, particularly for high velocities, or optically thin emission lines formed in self-excited media. The explicit use of angle and frequency dependent $S(x, \mu, \tau)$, instead of angle and frequency averaged source vector, is necessary in the polarized formal integration code, to obtain accurate results. For polarized line transfer, the computational advantages of CMF method still remain, compared to observer's frame method, where we should employ fully angle dependent redistribution matrices, computed on a much wider frequency grid, compared to the CMF case. Our solution of this polarization problem is a modest attempt, to fill a little gap in the 'Physics Tree' of Ivanov (1991, p.83).

Acknowledgements

I am grateful to Prof. V.V. Ivanov and Prof. Yu.N. Gnedin for providing local hospitality during my visit to Pulkovo Observatory. I am thankful to Prof. H. Frisch for useful discussions and suggestions which helped in improving this paper a great deal. I am grateful to Mrs. S. Vimala for her encouragement during the course of this work.

References

Faurobert, M.: 1987, *Astron. Astrophys.* **178**, 269
Faurobert, M.: 1988, *Astron. Astrophys.* **194**, 268
Faurobert-Scholl, M., and Frisch, H.: 1989, *Astron. Astrophys.* **219**, 338
Frisch, H.: 1988, in Y. Chmielewski and T. Lanz, (eds.), *Radiation in Moving Gaseous Media*, Saas-Fee, Geneva Observatory, p. 337
Frisch, H.: 1996 (in this procedings)
Gayley, K.G.: 1992, *Astrophys. J.* **390**, 583
Grachev, S.I.: 1988, *Astrophysics* **28**, 205
Grinin, V.P.: 1984, *Astrophysics* **20**, 190
Hamman, W.R., and Kudritzki, R.P.: 1977, *Astron. Astrophys.* **54**, 525
Hubeny, I.: 1985, *Astron. Astrophys.* **145**, 461
Ivanov, V.V.: 1991, in L.Crivellari, I. Hubeny, and D.G. Hummer, (eds.), *Stellar Atmospheres: Beyond Classical Models*, Kluwer Academic Publishers, p. 81
Mihalas, D.: 1978, *Stellar Atmospheres*, Freeman, San Francisco
Mihalas, D.: 1980, *Astrophys. J.* **238**, 1034
Mihalas, D., and Kunasz, P.B.: 1986, *J. Comp. Phys.* **64**, 1
Mihalas, D., Shine, R.A., Kunasz, P.B., and Hummer, D.G.: 1976, *Astrophys. J.* **205**, 492
Nagendra, K.N.: 1988, *Astrophys. J.* **335**, 269
Nagendra, K.N.: 1994, *Astrophys. J.* **432**, 274
Nagendra, K.N., and Peraiah, A.: 1985, *Monthly Notices Roy. Astron. Soc.* **214**, 203
Noerdlinger, P.D.: 1981, *Astrophys. J.* **245**, 682
Noerdlinger, P.D., and Rybicki, G.B.: 1974, *Astrophys. J.* **193**, 651
Peraiah, A.: 1984, in W. Kalkofen, (ed.), *Methods in Radiative Transfer*, Cambridge Univ. Press, p. 281
Rees, D.E., and Saliba, G.J.: 1982, *Astron. Astrophys.* **115**, 1
Sengupta, S.: 1993, *Monthly Notices Roy. Astron. Soc.* **265**, 513
Stenflo, J.O.: 1994, *Solar Magnetic Fields – Polarized Radiation Diagnostics*, Kluwer, Dordrecht
Vardavas, I.M.: 1976, *J. Quant. Spectrosc. Radiat. Transfer* **16**, 781

DIAGNOSTICS WITH THE HANLE EFFECT

M. FAUROBERT-SCHOLL

Observatory of Nice, B.P. 229, F- 06304 Nice Cedex 4, France

Abstract. The Hanle effect has been extensively used for the determination of the magnetic field strength and direction in solar prominences. Here we address the problem of the diagnostics of weak magnetic fields in the solar photosphere and chromosphere by means of their Hanle effect in some selected absorption lines. As this is a relatively new area we will focus on the diagnostic methods and summarize some results that relate to the presence of a weak, turbulent magnetic field in the photosphere and to the chromospheric magnetic canopy. Finally we will outline some directions for future work.

Key words: Polarization – Magnetic fields – Solar spectrum

1. Introduction

Stenflo et al. (1983a, b) have made an extensive survey of linear polarization in solar lines covering the wavelength range 3165–9950 Å. These observations carried out near the solar limb and outside active regions show evidence of a wide variety of resonant and fluorescent scattering effects. Even more features are discovered when the polarization sensitivity is increased (see Stenflo, these proceedings). The diagnostic contents of this polarization spectrum is still far from being fully investigated.

The observed resonance polarization is due to the coherent scattering of the anisotropic photospheric radiation field by the atoms. In the presence of a weak magnetic field, when the Zeeman splitting is on the order of the natural line width, the resonance polarization is reduced, and the polarization plane is rotated. This is the so-called Hanle effect, which is due to the relaxation of the coherences between the Zeeman sublevels. In the presence of a weak magnetic field with mixed polarities over the resolution element of the telescope there is no prefered direction for the rotation of the polarization plane, but the depolarization is not cancelled out. This provides a diagnostic tool for weak magnetic fields with mixed polarities on small scales, which do not give rise to detectable Zeeman polarization. This was first pointed out by Stenflo (1982) who suggested that such weak magnetic fields, although hidden from magnetograms, could carry a significant part of the solar magnetic flux.

The indirect diagnostic of these fields, based on the Hanle effect, requires detailed non-LTE radiative transfer calculations including polarization. Until now this has been done essentially for two resonance lines, namely the Ca I 4227 Å and the Sr I 4607 Å lines (Faurobert-Scholl, 1992, 1993, 1994;

Solar Physics **164**: 79–90, 1996.

Faurobert-Scholl et al., 1995). Both are normal triplets which show relatively
high linear polarization rates outside active regions, typically on the order
of a few percent at 10 arcsec inside the solar limb. We have used center-
to-limb observations performed by Stenflo et al. (1980) with the Stokes I
spectro-polarimeter at Sac Peak. New observations have been made recently
by Keller and Stenflo with ZIMPOL, and by Arnaud and Penn with ASP.

The sensitivity of a line to the Hanle effect is in that part of the magnetic
field strength domain, where the shift of the Zeeman sub-levels is on the
order of the natural line width (Bommier, these proceedings). This gives

$$B \simeq \Gamma_R/0.88 g_J, \tag{1}$$

where B is in G and Γ_R in units of 10^7 s^{-1}, while g_J is the Landé factor of
the upper level. For the Sr I and Ca I lines $g_J = 1$ and $\Gamma_R = 2.01 \ 10^8$ s^{-1} and
$2.18 \ 10^8$ s^{-1}, respectively. These lines are sensitive to the Hanle effect for
magnetic fields of about 20 G, the useful range being approximately 5–100
G.

2. Diagnostic method

The first step of the diagnostic method is to compute as accurately as possible the resonance polarization which would be observed in the absence of
a magnetic field. Resonance polarization is very sensitive to the non-LTE
processes which contribute to the line formation, such as frequency redistribution, depolarizing collisions, velocity fields and multiple scattering. It is
obviously highly sensitive to the anisotropy of the line radiation field, which
depends on the atmospheric structure. The results are then compared with
the observed polarization, and the discrepancy is interpreted in terms of the
Hanle effect. Let us note that the Hanle effect does not affect the polarization in the line wings (Frisch, these proceedings) and does not change the
intensity profiles, since very weak magnetic fields do not significantly modify
the line absorption profile. This allows us to constrain, at least partly, the
first step of the calculations by requiring that the center-to-limb variations
of the line intensity profile should be well reproduced. This is a necessary
requirement, but it does not completely ensure that we are computing the
line polarization correctly . This will be discussed further in the following.

2.1. MULTI-LEVEL TRANSFER OF POLARIZED RADIATION

As we are dealing with low polarization rates (a few % at most), we assume that the polarization in the lines does not affect the populations of the
atomic levels. We first solve a "standard" non-LTE multi-level transfer problem without polarization and then compute the resonance line polarization

using an equivalent two-level atom formalism for the line source function. The non-polarized calculation provides the optical thickness in the line of interest, together with the pseudo creation and destruction terms, due to level coupling, which appear in the equivalent two-level form of the source function. This procedure assumes that level coupling does not affect the polarization of the line. This would not apply to the computation of Ca II H and K resonance polarization because, interference takes place between the $J = \frac{1}{2}$ and $\frac{3}{2}$ upper levels (Stenflo, 1980). It is however a reasonable assumption for the Ca I and Sr I lines.

The Ca I and Sr I model atoms have been represented by two bound levels and a continuum. As both lines have large oscillator strengths, level coupling with other atomic levels has been neglected. The non-polarized transfer equation, coupled with the statistical equilibrium equations for the atomic levels, has been solved with an iterative method based on the equivalent two-level atom approach. Partial frequency redistribution in the line has been accounted for in these calculations.

2.2. CALCULATIONS OF RESONANCE POLARIZATION

Let us first say a few words about the numerical method that is used to compute the line polarization.

2.2.1. *Numerical method*
We have to solve the following vector transfer equation

$$\mu \frac{\partial \mathbf{I}(\tau, x, \mathbf{\Omega})}{\partial \tau} = (\beta + \phi(x))(\mathbf{I}(\tau, x, \mathbf{\Omega}) - \mathbf{S}(\tau, x, \mathbf{\Omega})). \tag{2}$$

Standard notations are used. $\mu = \cos\theta$, where θ is the heliocentric angle, τ is the line average optical depth and x is the frequency. $\mathbf{I}$ is the Stokes vector. The vector source function $\mathbf{S}$ is related to the line source function $\mathbf{S}_L$ by

$$\mathbf{S}(\tau, x, \mathbf{\Omega}) = \frac{\phi(x)\mathbf{S}_L + \beta \mathcal{B}\mathbf{U}}{\phi(x) + \beta}, \tag{3}$$

where $\mathcal{B}$ is the Planck function and $\mathbf{U}$ is the unpolarized unit vector. We stress that, as we are dealing with a realistic solar model, the Voigt absorption profile $\phi(x)$ and the ratio of the continuum opacity to line integrated opacity, β, are both depth-dependent. The line source function $\mathbf{S}_L$ is written

$$\mathbf{S}_L(\tau, x, \mathbf{\Omega}) = \{\mathbf{sc}(\tau, x, \mathbf{\Omega}) + (\varepsilon'\mathcal{B} + \eta B^*)\mathbf{U}\}/(1 + \varepsilon' + \eta), \tag{4}$$

where sc denotes the scattering term. This is the generalization to the vectorial case of the equivalent two-level atom formulation given by Mihalas

(1978, pp. 359, 360), for a two-level atom with continuum. The extra creation term ηB^* and sink term η are due to the coupling with the continuum. The scattering term is discussed in detail in Frisch's paper (these proceedings).

In the absence of a magnetic field the radiation field is axially symmetric, and we can apply the Feautrier method to the vector transfer equation in order to compute both the intensity and the polarization profiles (Faurobert, 1987). This also holds in the presence of a weak, isotropic turbulent magnetic field that we will consider in the following. The phase matrix is only modified in the line core by magnetic fields.

In the presence of an anisotropic magnetic field, the radiation field is no more axially symmetric. An iterative procedure, based on the azimuthal Fourier expansion of the radiation field, can be used to compute the line polarization in the presence of the Hanle effect (Faurobert-Scholl, 1991).

Let us now examine the effect of various physical mechanisms on the resonance polarization in the CaI and SrI lines. As part of this study we need to find out whether it is possible to constrain these mechanisms by fitting the intensity profiles, which are independent of the Hanle effect. When this is not possible we have to be confident enough about the way that we model these mechanisms, because they affect the diagnostics of the weak magnetic fields.

2.2.2. *Partial frequency redistribution*

Partial frequency redistribution affects more strongly the polarization profiles than the intensity profiles. This is illustrated for the CaI line in Fig. 5 of Frisch's paper (these proceedings). A correct form for the redistribution function must be used, as discussed in Frisch's paper. Let us just recall here that angle-averaged forms of the redistribution function may be safely used, even for strong resonance lines.

In contrast to the CaI line, the SrI line core polarization is almost insensitive to partial frequency redistribution effects. The reason is that, because of the very low abundance of strontium in the solar atmosphere, the SrI line is not a strong line.

2.2.3. *Elastic collisions*

Elastic collisions may have much stronger effects on the polarization profiles than on the intensity profiles. This is illustrated, in the case of the SrI line in Fig. 1, which shows the intensity and polarization profiles computed with two different values for the elastic collision cross-section. The SrI line is formed in the solar photosphere, in regions where the density is large, and where the rate of depolarizing collisions, denoted by $D^{(2)}$, may take values on the order of or larger than the radiative deexcitation rate. Depolarizing collisions have the same effect as the Hanle effect due to a weak, mixed polarity magnetic field. The diagnostics of these fields by means of the Hanle effect thus requires a very good knowledge of the depolarizing collision rate. This is the reason why precise quantum mechanical calculations of this quantity have been performed recently, using an accurate form for the inter-atomic potential (Faurobert-Scholl et al., 1995). The situation is very different in the case of the Ca I line, because the line core is formed in the low chromosphere, in regions where the density is low. The line core polarization is thus not affected by elastic collisions.

Let us briefly recall the results on the elastic collision cross-sections. For electron temperatures between 4000 and 10,000 K the broadening due to elastic collisions, denoted by Γ_C, and the depolarizing collision rates may be represented by the following analytical expressions: For the Sr I line

$$\Gamma_C = \gamma_w N_H (T/5000)^{0.16}, \text{ with } \gamma_w = 2.73\,10^{-8}, \quad D^{(2)} \simeq 0.5\Gamma_C. \tag{5}$$

For the Ca I line

$$\Gamma_C = \gamma_w N_H (T/5000)^{0.20}, \text{ with } \gamma_w = 2.90\,10^{-8}, \quad D^{(2)} \simeq 0.6\Gamma_C. \tag{6}$$

N_H denotes the number density of neutral hydrogen. We notice that these results are quite different from the Van der Waals approximation, the numerical coefficients γ_w being twice as large and the temperature dependence lower.

2.2.4. *Velocity fields*

Microturbulent and macroturbulent motions also affect the intensity and the polarization profiles. The microturbulent velocity is usually taken from the solar model. The effect of the macroturbulent velocity field has been represented by a smearing of both the intensity and the polarization profiles. We have determined the macroturbulence by fitting the observed center-to-limb variations of the line width and central intensity. The line width of Sr I is dominated by macroturbulence, so our procedure puts strong constraints on this quantity. However, the width of the polarization peak in the line core is generally smaller than the width of the intensity profile, and this observed property is not accounted for in the present calculations. A more refined treatment of turbulence may be necessary to recover this property. We will discuss this point further below.

The sensitivity of both the intensity and polarization to a change in the microturbulent velocity is illustrated in Fig. 2, which shows the profiles without and with smearing due to macroturbulence, at $\mu = 0.09$, for two different models of microturbulence. One is taken from the VAL3C solar model, while the other corresponds to a depth-independent microturbulence of 1 km/s (MIC1 model). The two models differ mainly in the upper photosphere, where the VAL3C microturbulence decreases to values on the order of 0.5 km/s.

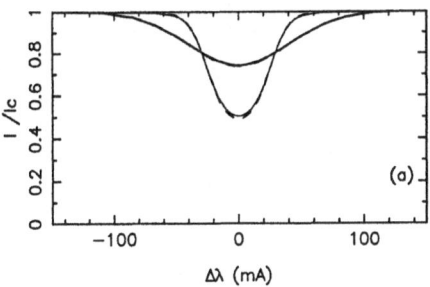

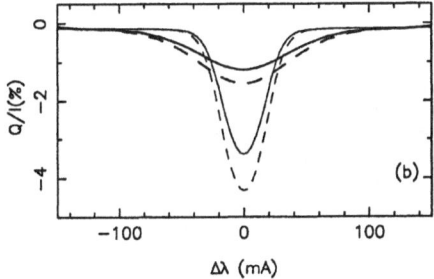

Fig. 1.   Intensity and polarization profiles of the Sr I 4607 Å line at $\mu = 0.09$, for two different values of γ_w (Eq. (5)). Full lines $\gamma_w = 2.8\,10^{-8}$, dashed lines: $\gamma_w = 1.4\,10^{-8}$. Thin lines: profiles obtained without macroturbulence, thick lines: profiles smeared by macroturbulent motions.

The smeared intensity profile is not very sensitive to the microturbulent model, in contrast to the line core polarization. We find that the resonance polarization in the Sr I line obtained after smearing is somewhat larger for the MIC1 model. The comparison with the observed rates then leads to somewhat stronger turbulent magnetic fields than with the VAL3C model (cf. Fig. 3).

For the Ca I line the procedure we have just described could not be applied, because the line width is not controlled by macroturbulence. The macroturbulence was determined from the intensity at line center only. As the line center intensity observed by Stenflo et al. (1980) increases strongly towards the solar limb, we have been led to consider a very anisotropic macroturbulence with a vertical component of 1.5 km/s and a horizontal component of 3.5 km/s. This allows us to recover the line center intensity observed up to $\mu = 0.2$. We note that the value of the vertical component is in agreement with a previous estimate by Lites (1974). The horizontal component had, as far as we know, not been determined before. This strong apparent anisotropy might also reflect a height increase of the macroturbu-

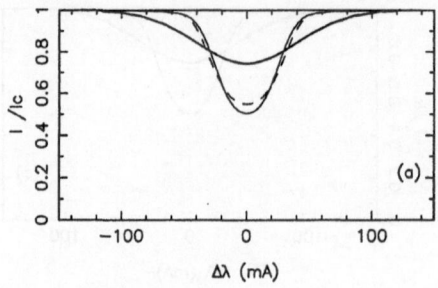

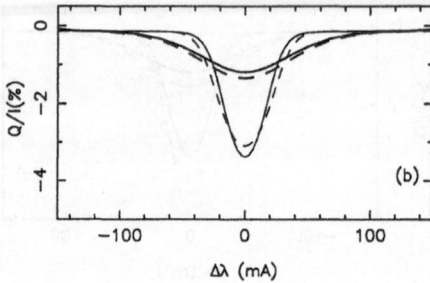

Fig. 2.   Intensity and polarization profiles at $\mu = 0.09$ of the Sr I 4607 Å line. Full lines: VAL3C microturbulent velocity. Dashed lines: Depth-independent microturbulence of 1 km/s. Thin lines: No macroturbulence. Thick lines: With macroturbulent smearing.

lence, because the intensity observed near the solar limb is formed higher in the chromosphere.

Let us stress that the effect of an anisotropic or depth-dependent macro-turbulent velocity fields on resonance polarization could be quite different from a simple smearing of the polarization peak. Theoretical investigations of this issue should be done in the near future, because this has strong consequences for the diagnostic methods based on the Hanle effect. Some work has recently been done by Briquez and Sahal-Bréchot, in the case of the Ca I line (private communication). The polarization formation is treated with the last scattering approximation using the observed center-to-limb variations of the line intensity. A moving atom scatters the incident line radiation field, which is identified with the observed line intensity, and the polarization of the re-emitted radiation is calculated for the resonant frequency, assuming that the atom absorbs and reemits radiation at this frequency only. Then an average over the anisotropic maxwellian velocity distribution of the atoms is performed. This procedure assumes that the polarization is formed higher in the atmosphere than the intensity profile, and does not take into account any frequency coupling due to scattering. In the absence of macroturbulent velocities, the resonance polarization at line center is overestimated by a factor of two as compared with a full polarized transfer calculation. It may be used to get a first qualitative estimate of the effect of an anisotropic macroturbulent velocity. The results are the following. When the anisotropy is along the vertical direction, the line core polarization is increased with respect to the case where there is no macroturbulent motions. In contrast, if the anisotropy is horizontal the line polarization decreases. The reason is that the limb-darkening of the line radiation field "seen" in the rest frame of a moving atom is larger (respectively smaller) if its velocity is perpendicular (respectively parallel) to the solar surface. In the first case, the Doppler shift increases the incident intensity seen at $\mu = 1$ for the resonant frequency of the atom, in the second case it increases the incident intensity seen at $\mu = 0$. Multiple scattering has to be treated in detail to get quantitative estimates of these effects.

2.2.5. *Inhomogeneities in the solar atmosphere*

The effect of modifications of the solar model on the polarization profiles has not been systematically studied. We believe that the "average" atmospheric structure is now quite well known in quiet regions of the sun. We have used the model C of Fontenla et al. (1993), which is an improved version of the VAL3C model (Vernazza et al., 1981). This average semi-empirical model was established from intensity observations in continua and in some spectral lines. However, as the presence of inhomogeneities in the solar atmosphere may affect the intensity profiles and the polarization profiles very differently, the average solar model may not be well adapted to represent polarization data. One way of checking this point would be to compute the spatial average of intensity and polarization profiles obtained from 2D atmospheric models representing some inhomogeneities in the solar atmosphere. For the Sr I line, a two-component model of the photosphere, modeling the solar granulation, could be used. This idea will be investigated in the future. It requires 2D radiative transfer calculations with polarization. Numerical codes designed for such 2D calculations are now under development.

2.3. HANLE EFFECT

The Hanle effect due to a weak, mixed polarity magnetic field is computed under the assumption that the correlation length of the field is smaller than a typical photon mean free path. In this case its effect may be calculated in the micro-turbulent limit. We then only need to replace in the transfer equation the Hanle phase matrix by its average over the distribution of magnetic field strengths and directions.

An isotropic distribution of field vectors is used to model the turbulent magnetic field in the solar photosphere. This is in agreement with previous estimates by Stenflo (1989). For the chromospheric canopy we consider horizontal fields with random azimuth. In the first case the average Hanle phase matrix is

$$\langle \mathbf{P_H}(\mu, \mu', B) \rangle = \mathbf{P_{is}} + \frac{3}{4} \left[1 - 0.4 \left(S_I^2 + S_{II}^2 \right) \right] W_2 \mathbf{P_0^{(2)}}(\mu, \mu'). \qquad (7)$$

The magnetic field strength enters only in the expressions for S_I and S_{II}:

$$S_I = \frac{\gamma_H}{\sqrt{1 + \gamma_H^2}}, \qquad S_{II} = \frac{2\gamma_H}{\sqrt{1 + 4\gamma_H^2}}, \qquad (8)$$

with

$$\gamma_H = 0.88 g_J \frac{B}{\Gamma_R + D^{(2)}}. \qquad (9)$$

Here B denotes the field strength, measured in G, while g_J is the Landé factor of the upper level. The coefficients Γ_R and $D^{(2)}$ are given in units of $10^7 \ \mathrm{s}^{-1}$.

We recall that W_2 is a constant that depends on the quantum numbers J and J' of the lower and upper levels of the transition. For a normal triplet $W_2 = 1$. The matrix $\mathbf{P_{is}}$ is the isotropic phase matrix (first position unity, remaining positions zero), while

$$\mathbf{P_0^{(2)}}(\mu, \mu') = \frac{1}{2} \begin{pmatrix} \frac{1}{3}(1 - 3\mu^2)(1 - 3\mu'^2) & (1 - 3\mu^2)(1 - \mu'^2) \\ (1 - \mu^2)(1 - 3\mu'^2) & 3(1 - \mu^2)(1 - \mu'^2) \end{pmatrix} \qquad (10)$$

When the magnetic field strength is stochastic and described by a distribution function $f(B)$, the average values of S_I^2 and S_{II}^2 depend not only on the mean value, but also on the higher moments of the field strength distribution. For the sake of simplicity, we assume that the distribution function is a Dirac delta function. The value of the field strength that we will determine may be considered as the "effective" magnetic field strength.

In the case of a horizontal magnetic field with random azimuths,

$$\langle \mathbf{P_H}(\mu, \mu', B) \rangle = \mathbf{P_{is}} + \frac{3}{4} (1 - 0.75 \, S_{II}^2) \, W_2 \mathbf{P_0^{(2)}}(\mu, \mu'). \qquad (11)$$

The determination of the weak magnetic fields is made by comparison of the calculated depolarization at line center, denoted by Q/Q_0, with the same quantity derived from the observed polarization, Q_{obs}/Q_0. This is illustrated in Figs. 3a, b in the case of the Sr I line. The results have been obtained with a microturbulent velocity of $1 \ \mathrm{km/s}$, as well as with the VAL3C microturbulence. We notice that for this spectral line the Hanle depolarization Q/Q_0 is almost insensitive to the adopted microturbulence, and only weakly sensitive to the smearing due to macroturbulence. In constrast, the ratio Q_{obs}/Q_0 is very sensitive to both quantities because of the sensitivity

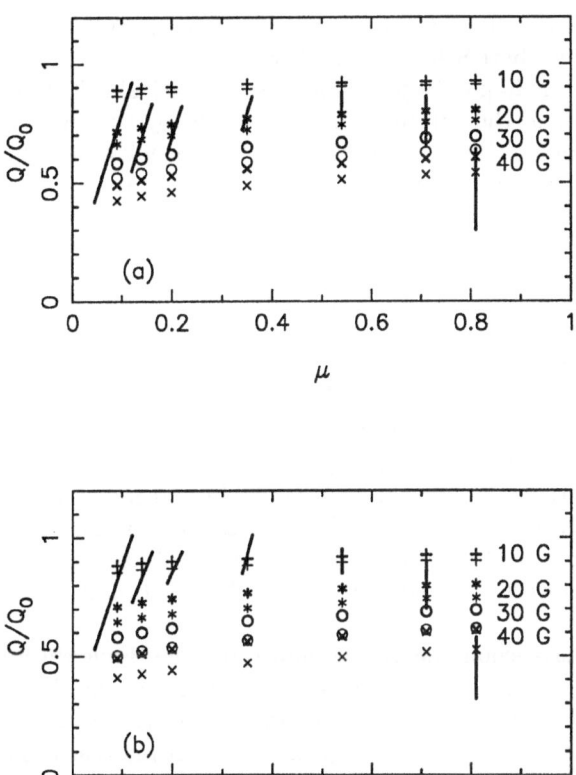

Fig. 3.   Hanle depolarization in the Sr I line core, due to depth-independent turbulent magnetic fields. Thin symbols show results obtained without macroturbulence, while for the thick symbols macroturbulent smearing has been applied. The thick bars give Q_{obs}/Q_0. Fig. 3a : microturbulent velocity of 1 km/s. Fig. 3b: VAL3C microturbulent velocity.

of the non-magnetic resonance polarization Q_0. As a consequence the determination of the strength of the turbulent magnetic field also depends on both quantities.

3.   First results

3.1.   TURBULENT MAGNETIC FIELD IN THE SOLAR PHOTOSPHERE

The polarization observed in the Sr I line core is significantly smaller than the resonance polarization calculated in the absence of a magnetic field. It is consistent with the presence of a turbulent magnetic field with an effective strength between 10 and 20 G in the region between 200 and 400 km above $\tau_{5000} = 1$ (Faurobert-Scholl et al., 1995).

This result is in good agreement with theoretical calculations of the passive transport of the turbulent magnetic field by turbulent motions in the solar photosphere (Petrovay, 1994, Faurobert-Scholl et al., 1995). According to these calculations the turbulent magnetic field decreases smoothly from values on the order of 30–50 G at $\tau_{5000} = 1$ to less than 10 G at 500 km above $\tau_{5000} = 1$.

3.2. CHROMOSPHERIC MAGNETIC CANOPY

The observations of the linear polarization in the Ca I line core may be used to investigate weak magnetic fields in the low chromosphere where the line core is formed. The comparison of the observed polarization rates with the calculated resonance polarization gives a surprising result. Observations performed near the solar limb ($\mu \leq 0.3$) are in agreement with non-magnetic calculations, whereas closer to disk center the observed polarization is significantly larger. Let us note that this comparison has been made for calculated profiles which were smeared to represent the effect of macroturbulent velocity fields.

In Faurobert-Scholl (1994) it is shown that the Hanle effect due to a magnetic canopy lying below the height where the line core is formed actually causes an increase of the Ca I line core polarization observed at $\mu \geq 0.3$. This surprising effect is a consequence of multiple scattering in the solar chromosphere. The line core polarization goes through a positive maximum between $z = 700$ and 900 km and becomes negative (parallel to the solar limb) at higher altitudes, where it does not change anymore. The Hanle effect of a weak magnetic field located in the region of the positive maximum locally decreases the positive contribution, so that the emergent negative polarization gets enhanced as compared with the non-magnetic situation. Let us note that a positive polarization maximum in the region where the line optical depth is between 1 and 10 is also present for lines formed in isothermal atmospheres. This is a property of multiple Rayleigh scattering.

4. Conclusions

Diagnostic methods based on the Hanle effect are at a beginning stage. They require, as a first step, reliable calculations of the resonance polarization in the absence of a magnetic field. As resonance polarization appears only in lines formed under non-LTE conditions, we need very good knowledge of the physical processes which play a role in the line formation. Furthermore resonance polarization is very sensitive to the anisotropy of the line radiation field, which is a higher order quantity as compared with the intensity itself. Some processes affect more drastically the polarization than the intensity. This is the case for example for elastic collisions, partial frequency redistribution, velocity fields, and anisotropies in the solar atmosphere. Theoretical studies on the effects of velocity fields and anisotropies in the solar atmosphere should be done in the near future.

Let us note that until now only the polarization observed at line center has been used for Hanle effect diagnostics. The reason is that the observations have been performed with relatively low spectral resolution. The analysis of new observations should aim at the interpretation of the whole polarization profile. This would allow us to control the consistency of the results. Whenever possible several lines with different sensitivities to the Hanle effect should be used together. Several strong resonance lines, such as the Sr II line at 4078 Å and the Ba II line at 4554 Å, which are formed in the chromosphere, show significant resonance polarization. Photospheric lines sensitive to the Hanle effect are quite rare because collisions destroy the resonance polarization, except for lines with very strong oscillator strengths, such as the Sr I line.

References

Faurobert, M.: 1987, *Astron. Astrophys.* **178**, 269

Faurobert-Scholl, M.: 1991, *Astron. Astrophys.* **246**, 469

Faurobert-Scholl, M.: 1992, *Astron. Astrophys.* **258**, 511

Faurobert-Scholl, M.: 1993, *Astron. Astrophys.* **268**, 765

Faurobert-Scholl, M.: 1994, *Astron. Astrophys.* **285**, 655

Faurobert-Scholl, M., Feautrier, N., Machefert, F., Petrovay, K., Spielfiedel, A.: 1995, *Astron. Astrophys.* **298**, 289

Fontenla, J.M., Avrett, E.H., Loeser, R.: 1993, *Astrophys. J.* **406**, 319

Lites, B.W.: 1974, *Astron. Astrophys.* **30**, 297

Mihalas, D.: 1978, *Stellar Atmosphere*, Freeman, San Francisco

Petrovay, K.: 1994, in C.J. Schrijver & R.J. Rutten (eds), *Solar Surface Magnetism*, NATO ASI Series C 433, Kluwer, Dordrecht, p.415

Briquez, C., Sahal-Bréchot, S.: 1995, private communication

Stenflo, J.O.: 1980, *Astron. Astrophys.* **84**, 68

Stenflo, J.O.: 1982, *Solar Phys.* **80**, 209

Stenflo, J.O.: 1989, *Astron. Astrophys. Rev.* **1**, 3

Stenflo, J.O., Baur, T.G., Elmore, D.F.: 1980, *Astron. Astrophys.* **84**, 68

Stenflo, J.O., Twerenbold, D., Harvey, J.W.: 1983a, *Astron. Astrophys. Suppl. Ser.* **52**, 161

Stenflo, J.O., Twerenbold, D., Harvey, J.W., Brault, J.W.: 1983b, *Astron. Astrophys. Suppl. Ser.* **54**, 505

Vernazza, J.E., Avrett, E.H., Loeser, R.: 1981, *Astrophys. J. Suppl.* **45**, 635

THE FIRST AND SECOND ORDER MOMENTS OF THE POLARIZATION PROFILES OF HYDROGEN LINES

R. CASINI

Dipartimento di Astronomia e Scienza dello Spazio, Università di Firenze, Largo E. Fermi 5, I-50125 Firenze, Italy

and

E. LANDI DEGL'INNOCENTI*

Instituto de Astrofísica de Canarias, E-38200 La Laguna, Tenerife, España

Abstract. The main properties of the first- and second-order moments of polarized hydrogen lines, forming in the presence of stationary electric and magnetic fields, are reviewed. The analytical results presented here apply directly to the case of optically-thin emission lines in the LTE regime. Some applications of such results to electric- and magnetic-field diagnostics in (solar) plasmas are then briefly considered.

Key words: Spectropolarimetry – Zeeman Effect – Stark Effect – Hydrogen Lines

1. Introduction

The detection and measurement of (stationary) electric and magnetic fields is a main issue in the understanding of the equilibrium condition and dynamical evolution of solar plasmas.

Historically, however, the magnetic field always played a dominant role in solar physics investigation. This is essentially due to the fact that, on the sun, relatively high-intensity magnetic fields exist (e.g., in sunspots), which may determine easily detectable circular-polarization signature in magnetic-sensitive lines, because of the linear Zeeman effect.

On the contrary, though the existence of macroscopic electric fields in the solar atmosphere was long ago suggested (Wien, 1916), and is presently provided by most of MHD modelling theories of solar structures (like flares), very few attempts of measuring them—through the linear polarization induced by the Stark effect on electric-sensitive lines—have been made so far (e.g., Dravins, 1973; Foukal *et al.*, 1988; Foukal and Behr, 1995). Such more involute stage of the electric-field diagnostics may be ascribed to different reasons.

First of all, the attempts of measuring electric fields in many cases could only determine upper limits to the electric-field intensities possibly occurring in different structures of the solar atmosphere. Those limits show that the electric fields on the sun might be relatively weak (e.g., Foukal and Hinata,

* On leave from the *Dipartimento di Astronomia e Scienza dello Spazio, Università di Firenze, Largo E. Fermi 5, I-50125 Firenze, Italy*

Solar Physics **164**: 91–96, 1996.

© 1996 *Kluwer Academic Publishers.*

1991; Foukal and Behr, 1995). Low electric-field intensities ask for a refinement of the electrograph techniques, and require the choice of particularly electric-sensitive lines.

Beside this, the Stark effect does not contribute to first order of perturbation (i.e., there is no a *linear* Stark effect) for most of the observable lines in the solar atmosphere, with the only (important) exception of lines which are originated in transitions of hydrogen or hydrogen-like atoms. On the other hand, the *quadratic* Stark effect usually becomes important (i.e., observable) only at much higher electric-field intensities than the ones which are of concern in solar plasma investigation.

Thus, the possibility of an electric-field diagnostics based on the Stark effect ultimately depends on the choice of highly electric-sensitive hydrogen (or hydrogen-like) lines, and on their actual observability in the solar atmosphere (Casini and Foukal, 1995).

Anyway, joint electric- and magnetic-field diagnostics is needed to test the field topologies provided by MHD models of solar plasma structures (Foukal and Hinata, 1991; Foukal and Behr, 1995). The results presented here for the hydrogen lines can be used to conceive a possible observing procedure finalized to that purpose, at least for the most simple field topologies (Casini, 1995b).

2. First and Second Order Moments of Hydrogen lines

The theory of radiative transfer for polarized radiation was formalized within the framework of (non relativistic) quantum electrodynamics by Landi Degl'Innocenti (1983). Its application to the problem of formation of hydrogen lines in the presence of simultaneous, stationary electric and magnetic fields was considered only a decade later by the authors (Casini and Landi Degl'Innocenti, 1993). In this last work, one of the main concerns was that of writing a numerical code for the complete calculation of the Stokes profiles of hydrogen lines.

Until now such code has only been applied to the case of *optically-thin emission* lines, forming in a *LTE atmosphere*. Under such simplifying assumptions the atomic density matrix of any level ℓ, $\rho_{\lambda\lambda'}^{(\ell)}$, is simply proportional to the unit matrix of rank $2\ell^2$ (taking into account the electronic spin), and does not need to be determined by solving the statistical equilibrium equations for the atomic system (see the contribution of Landi Degl'Innocenti in these proceedings). Moreover, since in the optically-thin case the observed intensity of a line is simply proportional to the optical depth of the layer in which the line is formed (plus a possible intensity contribution from background sources), nor is there a need for solving the radiative-transfer problem for polarized radiation. Then, the numerical code provides directly the (normalized) Stokes profiles of such a line.

In spite of these rude simplifications, the solution can still be a rather time consuming one, if very high transitions of the hydrogen atom are involved. For instance, the complete calculation of the 15–9 transition would require the calculation of the four Stokes parameters for the 72 900 ($= 2^2 \times 15^2 \times 9^2$) fine-structure components of the line.

For purposes which do not explicitly require the knowledge of the frequency dependence of the Stokes profiles, the information provided by integral properties of those profiles, like the first- and second-order frequency moments, may be sufficient. In fact, those moments already determine the center of gravity of the Stokes profiles and their characteristic width— estimated by the standard deviation. Both these quantities are affected by the external fields, so that their estimate can in principle be applied to the problem of the diagnostics of electric and magnetic fields in astrophysical plasmas. In addition, by determining the analytical form of the moments, the contribution of the external fields to the polarization of hydrogen lines can be quantified—at least to lowest orders—in a simple and straightforward way.

In the absence of the broadening mechanisms typically present in a thermal plasma (i.e., for infinitely narrow line components), the qth-order frequency moments—with respect to a reference frequency $\overline{\omega}$—of the Stokes profiles of an optically-thin emission line in the LTE regime can be written in the form

$$\langle \omega^q(i) \rangle = \sum_\alpha \tilde{s}_\alpha(i) \left(\omega_\alpha - \overline{\omega} \right)^q , \qquad i = 0, 1, 2, 3 , \tag{1}$$

where the index i enumerates the four Stokes parameters I, Q, U, V. In the above equation, the $\tilde{s}_\alpha(i)$ are the relative strengths—for the four Stokes parameters—of the line component at the frequency ω_α, the summation being extended to all the line components.

If we choose as the reference frequency $\overline{\omega}$ the center of gravity of the intensity profile ($i = 0$) in the absence of external fields, one finds (Casini and Landi Degl'Innocenti, 1994a,b)

$$\langle \omega(i) \rangle = -\delta_{i3} \frac{\mu_0}{\hbar} B \cos \vartheta_B , \qquad i = 0, 1, 2, 3 , \tag{2}$$

and

$$\langle \omega^2(0) \rangle = \langle \omega^2(0) \rangle_{\text{ff}} + \frac{e_0^2 a_0^2 E^2}{\hbar^2} \left[A_0(n, m) + \frac{1}{2} \left(1 - 3 \cos^2 \vartheta_E \right) A_2(n, m) \right]$$
$$+ \frac{1}{2} \frac{\mu_0^2 B^2}{\hbar^2} \left(1 + \cos^2 \vartheta_B \right) , \tag{3}$$

$$\langle \omega^2(1) \rangle = \frac{3}{2} \frac{e_0^2 a_0^2 E^2}{\hbar^2} \sin^2 \vartheta_E \cos 2(\varphi_E - \gamma) A_2(n, m)$$

$$- \frac{1}{2} \frac{\mu_0^2 B^2}{\hbar^2} \sin^2 \vartheta_B \cos 2(\varphi_B - \gamma) , \qquad (4)$$

$$\langle \omega^2(2) \rangle = \langle \omega^2(1) \rangle \left\{ \begin{array}{l} \cos 2(\varphi_E - \gamma) \to \sin 2(\varphi_E - \gamma) \\ \cos 2(\varphi_B - \gamma) \to \sin 2(\varphi_B - \gamma) \end{array} \right\} , \qquad (5)$$

$$\langle \omega^2(3) \rangle = 0 . \qquad (6)$$

(Actually a contribution to $\langle \omega(i) \rangle$, for $i = 0, 1, 2$, is present if one accounts for the quadratic Zeeman effect. Such contribution, negligible for magnetic-field intensities typical of the solar atmosphere, is evaluated in Casini and Landi Degl'Innocenti, 1994a.)

In the above equations, the field polar angles, ϑ, are measured from the line-of-sight. The field azimuthal angles, φ, and the position angle of the reference direction for Q-polarization, γ, are measured from an arbitrarily chosen direction on the plane of the sky (normal to the line-of-sight). The dimensionless coefficients $A_0(n, m)$ and $A_2(n, m)$, and the second-order moment of the intensity profile in the field-free case, $\langle \omega^2(0) \rangle_{ff}$, are tabulated for all the hydrogen transitions up to the level $\ell = 50$ (Casini, 1995a).

3. Discussion and Conclusions

From inspection of Equations (2)–(6), one can draw some immediate conclusions.

To first order, only the circular-polarization profile is affected by a (longitudinal) magnetic field. The fact that the electric field does not bring any contribution to first order tells us that the magnetic-field measurements by longitudinal magnetographs are still reliable in the presence of electric fields.

To second order, both the electric and the magnetic field contribute to the intensity and linear-polarization profiles. In particular, we see that only the transverse components of the fields contribute to linear polarization. Instead, the circular-polarization profile is unaffected. Another important property is that the dependence of the second-order moments on the fields is *purely* quadratic, i.e., there are no *mixed* terms of the form $E B$.

However, the most important fact is that, for the hydrogen lines, only the electric-field contribution to line polarization is dependent on the line considered—through the coefficients $A_0(n, m)$ and $A_2(n, m)$. Instead, the magnetic field always brings the *same* contribution to the Stokes (frequency) profiles of hydrogen lines. In particular, the introduction of the effective Landé factor, g_{eff}, does not make sense for hydrogen lines, since it has been shown (Casini and Landi Degl'Innocenti, 1994a) that $g_{\text{eff}} = 1$ for any hydrogen line. This difference in the behavior of the electric field and of the magnetic field in determining the polarization signature of hydrogen lines is particularly important since on it relies the possibility of a joint electric-and magnetic-field diagnostics (Casini, 1995b).

From another standpoint, the fact that the electric-field contribution to line polarization is line-dependent gives a meaning to the concept of electric-sensitivity (or Stark-sensitivity) for the hydrogen lines. In particular, one can try to define a Stark-sensitivity parameter for hydrogen lines which could help in the choice of transitions particularly suited to electric-field investigation. A first rigorous attempt in that sense, based on the results for the second-order moments, has been made by Casini (1995b).

A main advantage of the integral approach to the problem of electric-field diagnostics, as compared to the complete calculation of the Stokes profiles, is the very fast computation of the moments from Equations (2)–(6) (though some relevant coefficients, namely $A_2(n, m)$ and $\langle \omega^2(0) \rangle_{\text{ff}}$, must still be calculated numerically; see Casini, 1995a). Indeed, this fact enables one to consider very high transitions of the hydrogen atom, which are particularly suited to electric-field investigation due to their high Stark-sensitivity (Casini, 1995b).

A final remark should be made on the fact that Equations (2)–(6) hold only in the absence of broadening mechanisms. The generalization of such equations to the case of arbitrarily broadened profiles has however been dealt with in Casini and Landi Degl'Innocenti (1994b).

In a recent work (Casini and Landi Degl'Innocenti, 1995), the given expressions for the first- and second-order moments have also been applied to the generalization of the weak-field solution of the LTE radiative-transfer problem for polarized hydrogen lines in the case in which *both* the electric and the magnetic field are present. It is shown that the lowest non-vanishing orders (in a Taylor expansion of the line profile with respect to frequency) of the circular- and linear-polarization profiles are proportional respectively to the first and second derivatives (with respect to frequency) of the lowest-order intensity profile. More remarkably, such proportionality relations take the same form as those already known in the purely magnetic case, which have been widely used in the calibration of solar magnetographs so far. Eventually, the only difference would lie in the structure of the proportionality factors, which must now account for the general presence of electric as well as magnetic fields. Indeed, these factors turn out to be precisely the first-and second-order moments given by Eqs. (2)–(6).

These results are particularly important in conceiving reliable calibration procedures for vector magnetographs. In particular, it follows from the above—as it has already been stated in the opening of this section—that the circular-polarization profile is not affected (up to second order) by the presence of electric fields, thus longitudinal magnetographs can be safely calibrated through the well-known relation between V and the first derivative of I even in the presence of electric fields. On the contrary, one should consider the contribution of the electric field in the relation between Q (U) and the second derivative of I, since this would affect the reliability of the calibration procedure for transverse magnetographs.

References

Casini, R.: 1995a, *Astron. Astrophys. Suppl.*, in press
Casini, R.: 1995b, *Astron. Astrophys.*, submitted
Casini, R. and Foukal, P.: 1995, *Solar Phys.*, in press
Casini, R. and Landi Degl'Innocenti, E.: 1993, *Astron. Astrophys.* **276**, 289
Casini, R. and Landi Degl'Innocenti, E.: 1994a, *Astron. Astrophys.* **291**, 668
Casini R. and Landi Degl'Innocenti, E.: 1994b, *Astron. Astrophys.*, in press
Casini R. and Landi Degl'Innocenti, E.: 1995, *Astron. Astrophys.*, submitted
Dravins, D.: 1973, *Astrophys. Letters* **13**, 243
Foukal, P. and Behr, B.: 1995, *Solar Phys.* **156**, 293
Foukal, P. and Hinata, S.: 1991, *Solar Phys.* **132**, 307
Foukal, P., Little, R., and Gilliam, L.: 1988, *Solar Phys.* **114**, 65
Landi Degl'Innocenti, E.: 1983, *Solar Phys.* **85**, 3
Wien, W.: 1916, *Ann. Phys.* **49**, 842

MAGNETIC POLARIMETRIC REFRACTION IN THE SOLAR CORONA

Y. N. GNEDIN

Central Astronomical Observatory at Pulkovo,
196140 Saint-Petersburg, Russia

and

E.D. LOPEZ

Central Astronomical Observatory at Pulkovo,
196140 Saint-Petersburg, Russia;
Escuela Politecnica Nacional, 22650 Quito, Ecuador

Abstract. We present expressions which describe the angular displacement of radio sources due to refraction in a magnetized plasma. The main objective of the present paper is to take into account the combined effect of gradients of the electron density and the magnetic field. We use the geometrical optics approximation for the determination of the angular broadening of the radiation. The expressions obtained are applied to the case of the solar corona.

Key words: Refraction – Plasma – Solar corona

1. Introduction

The inhomogeneous character of space plasmas radically changes the main characteristics of radiative transfer. For example, the dispersion in the interstellar medium causes a variation of the pulsar radiation including the timing of the pulses. The solar corona produces an angular broadening of radio sources.

Dispersion can play an important role in changing the spatial, temporal and polarization properties of the radiation emitted by a distant radio source, as it travels through the inhomogeneous plasma. The spatial displacements and temporal fluctuations of such radiation are due to various inhomogeneities. Electron density fluctuations and variations of the external magnetic field give rise to highly anisotropic regions in the plasma. As the radio waves travel a long distance through this kind of system, a certain degree of angular broadening can be expected. For example, in the case of the solar corona, the angular broadening is mainly due to the presence of inhomogeneities in the electron number density.

If a ray traverses a homogeneous layer with a uniform refractive index, it will go undistorted. If there is a gradient in the medium perpendicular to the path, then the radio source seems to be displaced in the direction of the gradient. The presence of gradients in the electron density produces an angular broadening of the rays propagating in an inhomogeneous plasma (de Pater and Ip, 1984).

Solar Physics **164**: 97–102, 1996.
© 1996 *Kluwer Academic Publishers.*

In the presence of a magnetic field the plasma becomes anisotropic and its electromagnetic properties are characterized by a complex tensor for the dielectric permittivity. This can be significant even in weak magnetic fields. The magnetic field produces an additional contribution to the refraction of the radiation, the so-called "magnetic refraction".

Here we have studied the propagation of radio waves in a plasma containing both a gradient in the electron density and in the magnetic field. We derive an expression for the angular displacement of a radio source in an inhomogeneous, magnetized plasma. Taking into account dispersion, the radiation is displaced from its normal direction of propagation. This effect is observed for instance: a) in the case of stellar radiation passing through the inner solar corona (Soboleva and Timofeeva, 1983), b) in occultations of radio sources by comets (de Pater and Ip, 1984), and c) in the occultation of stars by planets of the solar system (Baum and Code, 1953).

2. Magnetic Refraction

The angular deviation of the radiation from a radio source, which is traversing a distance L through some rarefied plasma (refractive index ≈ 1), is given according to Wright and Nelson (1979) by:

$$\Delta\alpha \approx L\nabla_\perp n \tag{1}$$

where $\Delta\alpha$ is the angular displacement of the beam, n the refractive index, and L the characteristic length of the dispersion area.

A magnetized plasma with inhomogeneities can be described by the theory of geometric optics only if the system has small gradients of its physical parameters. Then it is possible to define limited regions where the classical theory for homogeneous media is applicable. One can obtain an expression for the refractive index of the waves in such media, as done by Ginsburg (1967). The refractive index, neglecting absorption, is described by the following expression:

$$n_{1,2}^2 = 1 - 2w_o^2(w^2 - w_o^2)\Big/\Big[2(w^2 - w_o^2)w^2 - w_h^2 w^2 \sin^2\alpha$$
$$\pm \sqrt{w^4 w_h^4 \sin^4\alpha + 4w_h^2 w^2(w^2 - w_o^2)^2 \cos^2\alpha}\,\Big] \tag{2}$$

where $w_o = \sqrt{4\pi e^2 N/m}$ is the plasma frequency, $w_h = eH/mc$ is the gyrofrequency, α the angle between the propagation vector $\mathbf{k}$ and the magnetic field $\mathbf{H}$, w the radiation frequency, e the electron charge, m the electron mass, N the electron density, H the magnetic field intensity, and c the velocity of light.

It is possible to deduce from this equation some particular cases for the refractive index: for the purely transversal and longitudinal propagations (relative to the direction of the magnetic field), and for the ordinary and extraordinary waves.

With the condition $n_{1,2} \sim 1$ (rarefied plasma) and in the case of quasi-longitudinal propagation, the angular displacement of extraordinary and ordinary waves is given by the relations:

$$\Delta\alpha_1 = L\left[\frac{w_o\,\nabla_\perp w_o}{w(w - w_l)} + \frac{w_o^2\,\nabla_\perp w_l}{2w(w - w_l)^2}\right] \tag{3}$$

$$\Delta\alpha_2 = L\left[\frac{w_o\,\nabla_\perp w_o}{w(w + w_l)} - \frac{w_o^2\,\nabla_\perp w_l}{2w(w + w_l)^2}\right] \tag{4}$$

where $w_l = w_h\cos\alpha$.

Let us introduce the average angular displacement

$$\Delta\alpha = \frac{1}{2}|\Delta\alpha_1 + \Delta\alpha_2| \quad, \tag{5}$$

and the difference of angular displacements between waves with different polarization states

$$\delta\alpha = |\Delta\alpha_1 - \Delta\alpha_2| \quad. \tag{6}$$

For quasilongitudinal propagation these parameters take the form:

$$\Delta\alpha = L\left|\frac{w_o\nabla_\perp w_o}{(w^2 - w_l^2)} + \frac{w_l w_o^2\nabla_\perp w_l}{(w - w_l)^2(w + w_l)^2}\right| \tag{7}$$

$$\delta\alpha = L\left|\frac{2w_o w_l\nabla_\perp w_o}{w(w^2 - w_l^2)} + \frac{(w^2 + w_l^2)w_o^2\nabla_\perp w_l}{w(w - w_l)^2(w + w_l)^2}\right| \quad. \tag{8}$$

Let us consider the high-frequency dispersion of the electromagnetic waves. With the condition $w_l^2 \ll w^2$, Eqs. (7) and (8) become

$$\Delta\alpha = \frac{2\pi e^2 L}{mw^2}\left|\nabla_\perp N + \frac{2e^2\cos^2\alpha}{m^2c^2w^2}NH\nabla_\perp H\right| \tag{9}$$

$$\delta\alpha = \frac{4\pi e^3 L}{cm^2w^3}\left|\cos\alpha(H\nabla_\perp N + N\nabla_\perp H)\right| \quad. \tag{10}$$

When the path of the waves is perpendicular to the direction of the magnetic field (quasitransversal propagation), with the additional condition $w_t^2 \ll (w^2 - w_o^2)$, where $w_t = w_h\sin\alpha$, the plasma dispersion parameters acquire the following form:

$$\Delta\alpha = \frac{2\pi e^2 L}{mw^2}\left|\nabla_\perp N + \frac{e^2\sin^2\alpha N H \nabla_\perp H}{m^2 c^2 w^2} + \frac{e^2\sin^2\alpha}{2m^2 c^2 w^2}H^2\nabla_\perp N\right| \qquad (11)$$

$$\delta\alpha = \frac{4\pi e^4\sin^2\alpha L}{m^3 c^2 w^4}\left|N H \nabla_\perp H + \frac{1}{2}H^2\nabla_\perp N\right| \qquad . \qquad (12)$$

If the refraction occurs in a weak magnetic field, for example in the solar corona, the contribution to the refraction that arises from the terms containing the factor $H^2\nabla_\perp N$ is very small and can be neglected. It is important to note that our equations give the absolute value of the angular displacements.

3. The Case of the Solar Corona

In this section the previous relations for the average angular displacement $\Delta\alpha$ and the relative angular displacement $\delta\alpha$ are computed for the solar corona. In these calculations we have considered the radial variation of the solar parameters. Approximate values of the electron density can be found in a numerical formula, due to Hollweg, quoted in Stelzried et al. (1970). Expressing the density in cm^{-3},

$$N(r) = 10^8\left[1.55(\frac{R_\odot}{r})^6 + 0.01(\frac{R_\odot}{r})^2\right] \qquad , \qquad (13)$$

while, for the magnetic field, we have used the formula (Parker, 1965)

$$H(r) = H_s\left(\frac{R_\odot^2}{r^2}\right) \qquad (14)$$

where H_s is the surface magnetic field of the Sun, $R_\odot$ the solar radius and r the radial distance from the Sun.

The electron density and magnetic field gradients can be calculated by differentiation of Eqs. (13) and (14), respectively. For H_s a value of 1.5 G was used. We have carried out the integration of $\Delta\alpha$ and $\delta\alpha$ along the path of the radio waves, which are passing through the solar corona to the observer. The propagating distance L considered was about $40R_\odot$.

Results of our calculations of angular displacements, for two values of the wavelength ($\lambda = 20$ cm and $\lambda = 13$ cm), at three different distances from the center of the solar disk ($d = 4R_\odot, 6R_\odot, 10R_\odot$), are presented in Table I and Table II.

The theory of geometric optics can be applied only in the case when abrupt gradients are absent in the medium. In this case it is possible to evaluate the angular broadening for a plasma containing gradients of both

TABLE I
WAVELENGTH $\lambda = 20cm$

	$d(R_\odot)$	$\Delta\alpha$(sec.)	$\delta\alpha$(sec.)
Quasilongitudinal	4	0.1	2×10^{-5}
	6	3×10^{-2}	2×10^{-6}
	10	9×10^{-3}	3×10^{-7}
Quasitransversal	4	0.1	5×10^{-7}
	6	3×10^{-2}	3×10^{-8}
	10	9×10^{-3}	1×10^{-9}

TABLE II
WAVELENGTH $\lambda = 13cm$

	$d(R_\odot)$	$\Delta\alpha$(sec.)	$\delta\alpha$(sec.)
Quasilongitudinal	4	5×10^{-2}	5×10^{-6}
	6	1×10^{-2}	6×10^{-7}
	10	4×10^{-3}	7×10^{-8}
Quasitransversal	4	5×10^{-2}	8×10^{-8}
	6	1×10^{-2}	5×10^{-9}
	10	4×10^{-3}	2×10^{-10}

the electron density and the magnetic field. If there are abrupt gradients of
the physical parameters, reflection of the electromagnetic waves may occur
in these regions. In general, for an arbitrary angle between the direction
of the wave propagation and the external magnetic field, the electromagnetic waves have elliptical polarization. In the special case of longitudinal
propagation, the polarization becomes circular (the ordinary wave has right-
handed circular polarization, the extraordinary has left-handed). In the case
of transversal propagation the polarization is linear for the ordinary wave
and elliptical for the extraordinary. In the present paper, the orthogonal
waves are considered to be mutually independent, i.e. the interaction between the two types of waves is neglected, and the transfer of polarization
from one mode to the other is not considered.

The values presented show that the angular displacements decrease with
increasing distance from the center of the sun, as expected because the

interaction with the solar plasma is greater where the density of the corona is larger.

Our calculations give small but quite measurable values for the angular displacements. Interferometric methods of the VLBI type might be used for the measurement of these small angular deviations. This new method allows the determination of gradients of the magnetic field by measuring only the angular deviation of radiation that propagates through a magnetized astrophysical plasma.

Acknowledgements

This work was supported by grants INTAS-93-2478 and Federal Program "Astronomy-95".

References

Baum, W.A. and Code, A.D.: 1953, *Astron. Journ.* **58**, 108.
de Pater, I. and Ip, W.-H.: 1984,, *Astrophys. Journ.* **283**, 895.
Ginsburg, B.L.: 1967, *Electromagnetic Waves in Plasmas*, Science, Moscow.
Parker, E.N.:1965, *Interplanetary Dynamical Processes*, Mir, Moscow.
Soboleva, N.S. and Timofeeva, G.M.: 1983, *Soviet Astron. Lett.* **9**, 409.
Stelzried, C.T., Levy, G.S., Sato, T., Rusch, W.V.T., Ohlson, J.E., Schatten, K.H., and
    Wilcox, J.M.:1970, *Solar Physics* **14**, 440.
Wright, C.S. and Nelson, G.J.: 1979, *Icarus* **38**, 123.

NUMERICAL METHODS IN
POLARIZED RADIATIVE TRANSFER

D.E. REES and G. GEERS

Signal and Imaging Technology Program
CSIRO Division of Radiophysics, PO Box 76, Epping NSW 2121, Australia

Abstract. This paper looks at three aspects of numerical methods for solving polarized radiative transfer problems associated with spectral line formation in the presence of a magnetic field. First we prove "Murphy's law for Stokes evolution operators" which is the basis of the efficient algorithm used in the SPSR software package to compute the Stokes line depression contribution functions. Then we use a *two–stream* model to explain the efficacy of the *field–free* method in which the non–LTE line source function in a uniform magnetic field is approximated by the source function neglecting the magnetic field. Finally we introduce a totally new and computationally efficient approach to solving non–LTE problems based on a method of sparsely representing integral operators using wavelets. As an illustration, the wavelet method is used to solve the source function integral equation for a two–level atomic model in a finite atmosphere with coherent scattering, ignoring polarization.

Key words: Polarization – Radiative Transfer – Magnetic fields – Wavelets

1. Introduction

Before launching into the collective "we", the first author begs from his co–author, Glenn Geers, some space to reminisce a little. I began research in Solar Polarization three decades ago. At that time the major developments in the theory of spectral line formation in a magnetic field were to be found in scientific literature from the then Soviet Union. I remember waiting eagerly, Russian–English dictionary at the ready, for the next issue of *Izvestiya Krymsk. Astrofiz. Obs.*, and especially for the latest ground–breaking contribution of D.N. Rachkovsky. Numerical solution of Rachkovsky's (1963) formulation of this theory for the case of non–coherent scattering led to the introduction of the *field–free* method (Rees, 1969) in which the non–local thermodynamic equilibrium (non–LTE) atomic level populations computed in the absence of a magnetic field are used as input to the polarized radiative transfer equations for the Stokes parameters I, Q, U, V, including the Zeeman effect. A number of papers have confirmed the validity of the *field–free* approximation for the case of a uniform magnetic field (see Table I).

In the late 1960s and early 1970s another researcher in the Eastern Bloc, J. Staude, also taught me much about spectral line formation in a magnetic field, both in his published papers and in correspondence. Staude's (1969) reformulation of the vector differential equation of transfer as an integral

Solar Physics **164**: 103–116, 1996.
© 1996 *Kluwer Academic Publishers.*

TABLE I

Validation of Field–Free Method for N–Level Atom in Uniform Magnetic Field

Reference	N	Numerical Method
Domke (1970)	2	H functions
Domke and Staude (1973)	2	H functions
Auer, Heasley and House (1976)	5	Complete linearization
Stenholm and Stenflo (1978)	2	Core saturation
Takeda (1991)	2	Accelerated lambda iteration

equation was the key to the development of the recursive *Diagonal Element Lambda Operator* (DELO) method (Rees *et al.*, 1989). DELO is currently the fastest technique for formal numerical integration of the vector transfer equation.

The SPSR (Stokes Profile Synthesis Routine) software package (Murphy and Rees, 1990) takes user–supplied line opacities and source functions derived from LTE or *field–free* non–LTE atomic level populations and uses DELO to solve for the Stokes vector for an arbitrary Zeeman multiplet, allowing depth variations in all physical parameters, including the magnetic field. SPSR and codes similar to it are widely used by the Solar Polarization community (see Table II). However, recent breakthroughs in the self–consistent numerical solution of the coupled non–LTE equations of polarized radiative transfer and statistical equilibrium (Trujillo Bueno and Landi Degl'Innocenti, 1996; Bruls and Trujillo Bueno, 1996) have demonstrated the inadequacy of the *field–free* approximation for lines formed in the presence of a magnetic field gradient.

The challenge in preparing this lecture was to find something new to say about numerical methods in polarized radiative transfer, given that I left this field of research over four years ago. Three ideas came to mind.

The first idea is not really new, as it appeared in Murphy's (1990) PhD thesis but, to my knowledge, the details, given in Section 2, have not been published in the open literature. I'm sure Graham will forgive me if I call it "Murphy's Law for Evolution Operators". In this context the law has a positive connotation as it permits efficient computation of the line depression contribution functions of the Stokes parameters.

The second idea is again not new. I recall discussing it with Bruce Lites and Andy Skumanich in 1985 when I worked with them on HAO's Advanced Stokes Polarimeter Project. In Section 3 a *two–stream* model of non–LTE

TABLE II

Examples of Usage of SPSR and Similar Software

Reference	Spectra
Rees et al. (1989)	Ca II H & K
Lites et al. (1987, 1988)	Mg I b & 4571Å
Briand and Solanki (1995)	Mg I b & 4571Å
Bruls et al. (1995)	Mg I $12\mu m$
Hewagama et al. (1993)	Mg I $12\mu m$
Ruiz Cobo and Del Toro Iniesta (1992)	Fe I 6302.5Å
Solanki et al. (1992 a,b)	Fe I $1.56\mu m$
Solanki et al. (1994)	Fe I $1.56\mu m$
Solanki and Montavon (1993)	Fe I 6302.5Å

line formation is used to illustrate why the *field–free* approximation works so well in a uniform magnetic field.

The third idea is new and, I hope, will inspire colleagues to take a new perspective on numerical solution of non–LTE radiative transfer problems. It is the outcome of my interaction with Glenn Geers who is applying wavelets to the numerical solution of electromagnetic problems for antenna design. Originally developed for siesmic signal processing, wavelets are causing a revolution in signal and image processing, and numerical analysis. Wavelet bases for function decomposition fall within the province of functional (generalized Fourier) analysis. Daubechies' (1988) theory of the "multiplier 2" compactly supported wavelet transform, which forged the link between digital signal processing and functional analysis, is regarded by Resnikoff (1992) as "one of the most important mathematical contributions of the second half of the 20th century".[*]

Powerful new algorithms have been developed recently for efficiently solving Fredholm integral equations of the second kind using wavelets (Alpert et al., 1993). Wavelets were applied for the first time in radiative transfer by Gortler et al. (1993) to solve the radiosity equation, a Fredholm integral equation used in computer graphics to provide a realistic rendering of the illumination in a scene. In traditional solutions, e.g. the Galerkin technique,

[*] An excellent way to keep abreast of the rapidly expanding field of wavelets is to access the World Wide Web Homepage http://www.mathsoft.com/wavelets.html. Information is available there on how to subscribe to the *Wavelet Digest* which is maintained by Wim Sweldens.

the integral operator is represented as a *dense* matrix (all elements non–zero) of large dimension. Applying a wavelet transform leads to a *sparse* representation for the integral operator, a property which is central to the design of a fast solution algorithm. Alpert *et al.* give an example where inverting the sparse version of an 8192×8192 matrix operator, preserving three–digit accuracy, took five minutes on a Sparcstation compared with an estimated 95 days to invert the dense version by the Gauss–Jordan method.

We illustrate the wavelet method in Section 4 by solving a simple coherent scattering problem, neglecting Zeeman splitting. The wavelet approach to solving integral equations has resonances with accelerated lambda iteration methods which enable efficient numerical solution of multilevel non–LTE radiative transfer problems in multidimensional model atmospheres (Auer *et al.*, 1994). In Section 5 we conclude with some speculations on the potential of wavelet methods to tackle such complex non–LTE problems.

2. Murphy's Law

We consider the standard model for non–LTE line formation in a Zeeman split spectral line in which Zeeman state coherences and atomic level polarization are neglected, and scattered line photons are completely redistributed in frequency and angle (see Rees *et al.* 1989 for full details). The *total* populations of the upper and lower level of the line transition then suffice to compute the line centre opacity (for zero line damping and corrected for stimulated emission) κ_0 and the frequency and angle independent line source function S; Zeeman splitting and magneto–optical effects enter via a 4×4 line absorption matrix $\boldsymbol{\Phi}$ which is a generalization of the line absorption profile ϕ in the absence of a magnetic field. The continuum polarization, which is negligible compared with Zeeman–induced polarization, is characterized by the continuum opacity κ_c and source function $B = B_\nu(T_e)$, the Planck function at the local electron temperature T_e.

The radiative transfer equation for the Stokes vector $\mathbf{I} = (I, Q, U, V)^\dagger$, ($\dagger$ = transpose) is

$$\frac{d\mathbf{I}}{d\tau} = \mathbf{K}\mathbf{I} - \mathbf{j}, \qquad (1)$$

where τ is line–of–sight continuum optical depth, $\mathbf{K}$ is the total opacity matrix,

$$\mathbf{K} = 1 + \eta_0 \boldsymbol{\Phi}, \qquad (2)$$

and $\mathbf{j}$ is the total emission vector,

$$\mathbf{j} = (B1 + \eta_0 S \boldsymbol{\Phi})\mathbf{I}_0, \qquad (3)$$

with $\eta_0 = \kappa_0/\kappa_c$, $\mathbf{1} = 4 \times 4$ identity matrix, and $\mathbf{I}_0 = (1, 0, 0, 0)^\dagger$.

Equation (1) can be integrated using the evolution operator $\mathbf{O}(\tau, t)$ which satisfies (see Landi Degl'Innocenti, 1987)

$$\frac{d\mathbf{O}(\tau, t)}{d\tau} = \mathbf{K}(\tau)\mathbf{O}(\tau, t). \tag{4}$$

In particular, the emergent Stokes vector at the surface $\tau = 0$ of a semi–infinite atmosphere is

$$\mathbf{I}(0) = \int_0^\infty \mathbf{C}(\tau)d\tau, \tag{5}$$

where $\mathbf{C}(\tau)$ is the Stokes vector contribution function,

$$\mathbf{C}(\tau) = \mathbf{O}(0, \tau)\mathbf{j}(\tau). \tag{6}$$

The line depression for the Stokes vector is defined as

$$\mathbf{R} = \mathbf{I}_0 - \mathbf{I}/I_c, \tag{7}$$

where I_c is the continuum intensity which satisfies the scalar transfer equation,

$$\frac{dI_c}{d\tau} = I_c - B. \tag{8}$$

The transfer equation for $\mathbf{R}$ is

$$\frac{d\mathbf{R}}{d\tau} = \mathbf{K_R}\mathbf{R} - \mathbf{j_R}, \tag{9}$$

where

$$\mathbf{K_R} = (B/I_c)\mathbf{1} + \eta_0\mathbf{\Phi}, \tag{10}$$

and

$$\mathbf{j_R} = (1 - S/I_c)\eta_0\mathbf{\Phi}\mathbf{I}_0. \tag{11}$$

Analogously to $\mathbf{O}(\tau, t)$ we can define an evolution operator $\mathbf{O_R}(\tau, t)$ such that

$$\frac{d\mathbf{O_R}(\tau, t)}{d\tau} = \mathbf{K}(\tau)\mathbf{O_R}(\tau, t), \tag{12}$$

and so the emergent line depression from a semi–infinite atmosphere is

$$\mathbf{R}(0) = \int_0^\infty \mathbf{C_R}(\tau)d\tau, \tag{13}$$

where $\mathbf{C_R}(\tau)$ is the depression contribution function,

$$\mathbf{C_R}(\tau) = \mathbf{O_R}(0, \tau)\mathbf{j_R}(\tau). \tag{14}$$

Rees *et al.* (1989) determined $\mathbf{C_R}(\tau)$ by applying the DELO method to Equation (9). Murphy (1990) later found a more efficient method which is incorporated in the SPSR code.

Using Equations (2), (8), (10) and (12) we can show that

$$\frac{d}{d\tau}\left[I_c(\tau)\mathbf{O_R}(\tau,t)\right] = \mathbf{K}(\tau)\left[I_c(\tau)\mathbf{O_R}(\tau,t)\right]. \tag{15}$$

Comparing this with Equation (4) and applying the boundary conditions $\mathbf{O}(\tau,\tau) = \mathbf{O_R}(\tau,\tau) = 1$, we obtain Murphy's law for the Stokes evolution operators,

$$I_c(\tau)\mathbf{O_R}(\tau,t) = I_c(t)\mathbf{O}(\tau,t). \tag{16}$$

It follows from Equations (11), (14) and (16) that

$$\mathbf{C_R}(\tau) = \mathbf{O}(0,\tau)(I_c(\tau) - S(\tau))\eta_0\Phi/I_c(0), \tag{17}$$

and so both contribution functions are obtained essentially for the price of solving Equation (1) only or, equivalently, determining $\mathbf{O}(0,\tau)$, which is a by–product of the DELO method. Note that $I_c(\tau)$ may be obtained by solving Equation (1) with $\eta_0 = 0$.

3. An Exact Case of the Field–Free Approximation

To gain insight into the reason for the efficacy of the *field–free* approximation in a uniform magnetic field, we consider a highly simplified *two–stream* model of non–LTE spectral line radiative transfer for a standard collisionally dominated two–level atom in a semi–infinite atmosphere with no external radiation incident on the surface. Photons travel in one dimension only, *up* or *down*. In the absence of a magnetic field, the *field–free* line source function is

$$S = (1 - \epsilon)\bar{J} + \epsilon B, \tag{18}$$

where ϵ is the usual photon destruction probability, and $\bar{J}$ is the mean integrated intensity,

$$\bar{J} = \int_{-\infty}^{\infty} dv\phi(v)(I^u(v) + I^d(v)), \tag{19}$$

ϕ is the line absorption profile, v is the wavelength measured from line centre in Doppler width units, and $I^{u,d}(v)$ denote the intensities in the *up* and *down* directions respectively.

Now suppose that a uniform magnetic field of strength H is imposed parallel to the axis of photon flow. The atomic transition is assumed to be a normal triplet with Zeeman splitting v_H in Doppler widths, giving rise to *blue*– and *red*–shifted absorption profiles,

$$\phi_b(v) = \phi(v + v_H) \text{ and } \phi_r(v) = \phi(v - v_H). \tag{20}$$

Two spectral lines are formed with orthogonal circular polarizations, shifted to the *blue* and *red* by v_H, with *up* and *down* flowing intensities equal to half the intensities in the absence of the magnetic field, i.e.

$$I_b^{u,d}(v) = \tfrac{1}{2}I^{u,d}(v + v_H) \text{ and } I_r^{u,d}(v) = \tfrac{1}{2}I^{u,d}(v - v_H). \tag{21}$$

The mean integrated intensity for the *total* radiation field in this case is

$$\bar{J}_H = \int_{-\infty}^{\infty} dv \left[\phi_b(v)(I_b^u(v) + I_b^d(v)) + \phi_r(v)(I_r^u(v) + I_r^d(v)) \right], \tag{22}$$

and the line source function is

$$S_H = (1 - \epsilon)\bar{J}_H + \epsilon B. \tag{23}$$

Transforming dummy variables in the integrals in Equation (22) reveals that $\bar{J}_H = \bar{J}$ and hence $S_H = S$.

Thus, for this two–stream model, the *field–free* approximation is *exact*.

4. Wavelets in Mainstream Numerical Radiative Transfer

The multilevel non–LTE line formation problem, including polarization, can be formulated as a set of coupled integral equations where the unknowns are the atomic density matrix elements or, in the restricted case discussed in Section 2, the line source functions for the various atomic transitions. In general, the problem is nonlinear because the integral operators are themselves functions of the unknowns. This numerical problem has been solved by Trujillo Bueno and Landi Degl'Innocenti (1996) using accelerated lambda iteration, a method based on the ideas of Chris Cannon (1973a, b). According to his widow, Anne, Chris had quite a struggle getting his revolutionary method accepted into the mainstream of astrophysical radiative transfer. It is gratifying to note that Cannon–type techniques are now *the main stream*. The purpose of this section is to hint at the possibility that wavelets add extra power to this main stream.

For simplicity we ignore polarization and focus again on the standard plane–parallel atmosphere, two–level atom, non–LTE problem. The line source function satisfies an integral equation,

$$S(\tau) = (1 - \epsilon(\tau))\Lambda(\tau, t)S(t) + \epsilon(\tau)B(\tau), \tag{24}$$

where $\Lambda(\tau, t)$ denotes an integral operator

$$\Lambda(\tau, t) = \int_0^\infty K(\tau, t) dt \cdots, \tag{25}$$

with kernel $K(\tau, t)$, τ being the optical depth at line centre, for example. More generally, in a three–dimensional inhomogeneous medium, e.g. a model flux tube in which physical variables vary both vertically and horizontally,

$$S(\mathbf{r}) = (1 - \epsilon(\mathbf{r}))\Lambda(\mathbf{r}, \mathbf{s})S(\mathbf{s}) + \epsilon(\mathbf{r})B(\mathbf{r}), \tag{26}$$

where $\mathbf{r}$ and $\mathbf{s}$ denote position vectors in three dimensions and

$$\Lambda(\mathbf{r}, \mathbf{s}) = \int_{\mathbf{s}^3} K(\mathbf{r}, \mathbf{s}) d^3\mathbf{s} \cdots. \tag{27}$$

Steiner (1991) combined accelerated lambda iteration with a *multigrid* method to solve for the non–LTE source function in a finite plane–parallel atmosphere with coherent scattering and zero magnetic field. Wavelet decomposition of the integral operator has *built–in multiresolution* properties which obviate the need to formulate the multigrid equations explicitly. These properties were exploited by Gortler *et al.* (1993) in their solution of the radiosity equation.

To illustrate the wavelet approach we consider a very simple problem, similar to that solved by Steiner. In a plane parallel atmosphere of total optical thickness T, the source function integral equation is

$$S(\tau) = (1 - \epsilon) \int_0^T K(|t - \tau|)S(t)dt + \epsilon B. \tag{28}$$

For coherent scattering, the kernel $K(|t - \tau|)$ is the exponential integral function $E_1(|t - \tau|)$ (Chandrasekhar, 1950) which has a mild logarithmic singularity at $t = \tau$. As a further simplification we replace $E_1(|t - \tau|)$ by an exponential function,

$$K(|t - \tau|) = \frac{1}{2}e^{-|t-\tau|}, \tag{29}$$

and suppose that ϵ and B are constants, so that Equation (28) has the analytic solution (Avrett and Hummer, 1965):

$$S(\tau) = B\left[1 - \frac{e^{-\sqrt{\epsilon}\tau} + e^{-\sqrt{\epsilon}(T-\tau)}}{1/(1 - \sqrt{\epsilon}) + e^{-\sqrt{\epsilon}T}/(1 + \sqrt{\epsilon})}\right]. \tag{30}$$

Following Alpert *et al.* (1993) we solve Equation (28) for $S(\tau)$ numerically on an equispaced optical depth grid $\tau_i = (i - 1)T/(N - 1)$, $i = 1, \cdots, N$, replacing the integral by a simple quadrature. The discrete approximation to Equation (28) is

$$\mathbf{S} = (1 - \epsilon)\Lambda\mathbf{S} + \epsilon\mathbf{B}, \tag{31}$$

where $S = (S(\tau_1), \cdots S(\tau_N))^\dagger$, $B = B(1, 1, \cdots, 1)^\dagger$, and $\Lambda_{ij} = \alpha_j K(|\tau_j - \tau_i|)$, with quadrature weights α_j based on a modified Simpson's rule.

Clearly we can write the solution of Equation (31) as

$$S = [1 - (1 - \epsilon)\Lambda]^{-1} (\epsilon B). \tag{32}$$

The point of the discussion below is to address the issue of solving for S for large N where this direct solution is infeasible.

The matrix operator Λ is dense. Techniques such as accelerated lambda iteration seek sparse approximate representations of Λ. Wavelets provide an alternative approach to achieving sparsity.

We now introduce a discrete wavelet transform. The choice of wavelet is not unique. Here we use Daubechies' $D4$ wavelet filter coefficients $c_0, \cdots, c_3$, as described in Press et $al.$ (1992) to construct an $N \times N$ transformation matrix W with $N = 2^n$ (n integer):

$$W = \begin{pmatrix} c_0 & c_1 & c_2 & c_3 & & & & & \\ c_3 & -c_2 & c_1 & -c_0 & & & & & \\ & & c_0 & c_1 & c_2 & c_3 & & & \\ & & c_3 & -c_2 & c_1 & -c_0 & & & \\ \vdots & \vdots & & & & & \ddots & & \\ & & & & & & c_0 & c_1 & c_2 & c_3 \\ & & & & & & c_3 & -c_2 & c_1 & -c_0 \\ c_2 & c_3 & & & & & & & c_0 & c_1 \\ c_1 & -c_0 & & & & & & & c_3 & -c_2 \end{pmatrix}, \tag{33}$$

which is orthogonal, i.e. $WW^\dagger = 1$. An example of the D4 wavelet function is given in Figure 1. It is a fractal curve. A complete set of orthonormal wavelet basis functions on the real line can be constructed by scalings and translations of this function.

Applying W to Equation (31) we obtain

$$\tilde{S} = (1 - \epsilon)\tilde{\Lambda}\tilde{S} + \epsilon\tilde{B}, \tag{34}$$

where $\tilde{S} = WS$, $\tilde{B} = WB$ and

$$\tilde{\Lambda} = W\Lambda W^\dagger. \tag{35}$$

In Figure 2 (left) $\tilde{\Lambda}$ is visualized as a greyscale image for the case $T = 10$ and $N = 2^7 = 128$ (we have also done this for $N = 2^{12} = 4096$ with comparable results). At this stage the representation $\tilde{\Lambda}$ is still dense. However, many elements are very small and by setting those elements below an assigned threshold δ to zero, i.e.,

$$\tilde{\Lambda}_{ij} \to 0 \text{ if } |\tilde{\Lambda}_{ij}| < \delta, \tag{36}$$

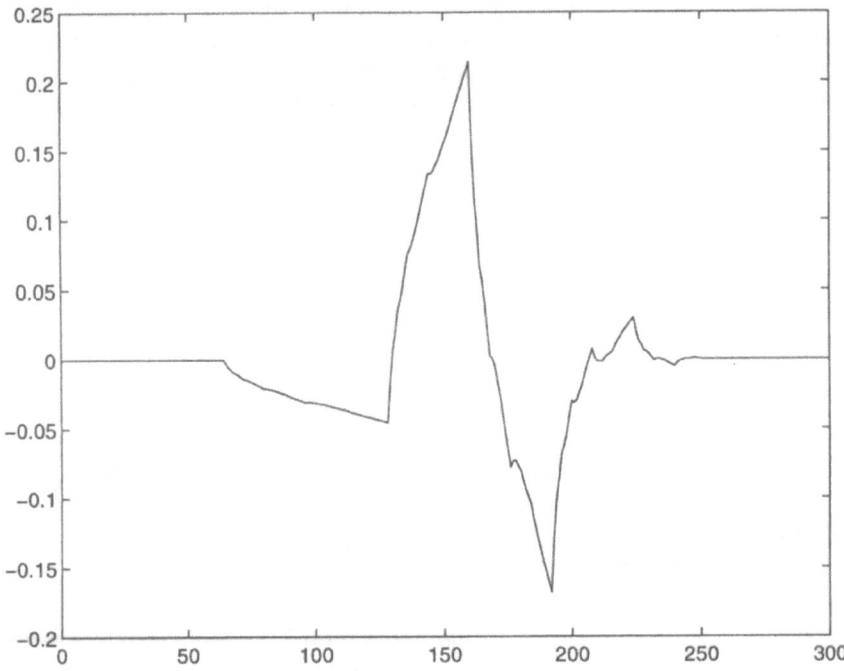

Fig. 1. An example of Daubechies' D4 wavelet family

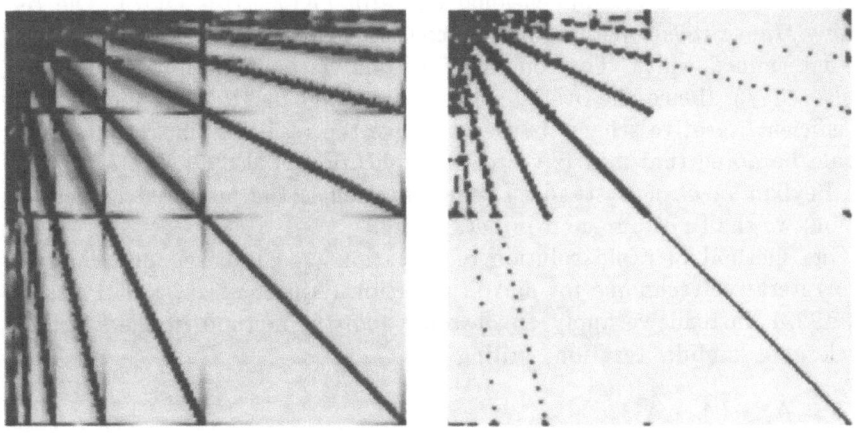

Fig. 2. Dense (left) and sparse (right) versions of $\tilde{\Lambda}$.

we obtain a sparse representation. This is shown in Figure 2 (right) for $\delta = 10^{-4}$.

An alternative and computationally much more attractive approach is to expand the kernel function and the source function in the wavelet basis at the start and use a numerical quadrature rule to compute *only* those elements of $\tilde{\Lambda}$ that are above the prescribed threshold (in absolute value). Such algorithms are available if the kernel, $K(x, y)$, is in the class of operators known as Caldéron-Zygmund, in which case they satisfy

$$|K(x, y)| \leq \frac{1}{|x - y|} \tag{37}$$

$$|\partial_x^M K(x, y)| + |\partial_y^M K(x, y)| \leq \frac{C_M}{|x - y|^{1+M}} \tag{38}$$

for some integer $M \geq 1$ and some positive constant C_M. The kernels occurring in the radiative transfer problem are of this type.

In wavelet jargon the matrix $\tilde{\Lambda}$ is said to be in the standard form and is a representation of the matrix Λ in the chosen wavelet basis. The direct computation of the standard form of an operator is complicated by mixing resolutions and, in general, this form is dropped in favour of the so-called non-standard form (Beylkin *et al.*, 1991). This is an over-representation of the operator and is not the representation of the operator in *any* basis. In the non–standard form, only basis functions from the same resolution level are required and this leads to much simpler numerical computation.

Beylkin *et al.* show that for a Caldéron-Zygmund operator the number of non-zero matrix elements above a prescribed threshold in the non-standard representation of a one–dimensional operator $(K(x, y))$ is $O(N)$. The resulting transformed operator matrix consists of distinct non-zero bands of pre-determined width. The computation time via a simple quadrature rule is also $O(N)$. Hence the overall solution time will be $O(N)$ assuming that an efficient iterative scheme is used to solve the resulting linear system (it should be noted that matrix-vector multiplication is also $O(N)$). Although the Beylkin *et al.* non–standard representation is the preferred numerical option, we shall not pursue it in detail here.

One method of rapid solution of Equation (34) is based on Schultz's $O(N)$ iterative technique for matrix inversion (Alpert *et al.*, 1993; Press *et al.*, 1992). Instead we apply an *operator splitting* method in the spirit of accelerated lambda iteration, writing

$$\tilde{\Lambda} = \tilde{\Lambda}^* + (\tilde{\Lambda} - \tilde{\Lambda}^*). \tag{39}$$

Many choices for $\tilde{\Lambda}^*$ are possible. Here we use the tridiagonal matrix composed of the elements on and adjacent to the diagonal of $\tilde{\Lambda}$. We then solve iteratively as follows (here n denotes the iterate number):

$$\tilde{S}^{(n+1)} = \left[1 - (1 - \epsilon)\tilde{\Lambda}^*\right]^{-1} \left[(1 - \epsilon)(\tilde{\Lambda} - \tilde{\Lambda}^*)\tilde{S}^{(n)} + \epsilon\tilde{B}\right]. \tag{40}$$

The inversion of the tridiagonal matrix is done by the Thomas algorithm. The corresponding iterates of the original source function are

$$\mathbf{S}^{(n)} = \mathbf{W}^{\dagger}\tilde{\mathbf{S}}^{(n)}. \tag{41}$$

The convergence of this method is illustrated in Figure 3 with initial estimate $\mathbf{S}^{(0)} = \mathbf{0}$ for $\epsilon = 10^{-2}$ and $B = 1$.

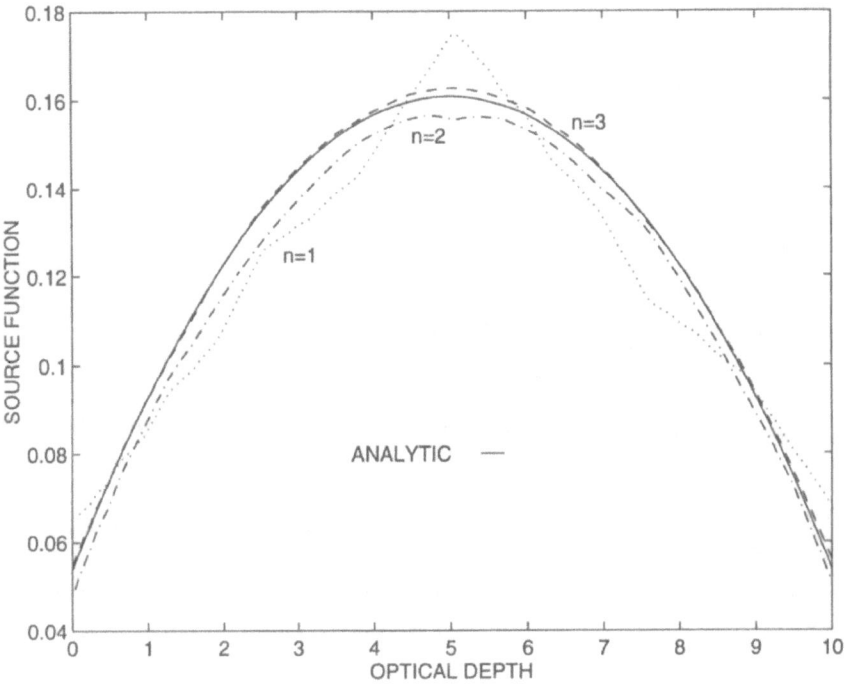

Fig. 3.   Iterates of the line source function.

5. Conclusion

This paper began with a nostalgic return to the Crimean roots of the *field–free* method. This method has survived since 1969 as a way of avoiding the numerical realities of self–consistent solution of the non–LTE equations of polarized radiative transfer and statistical equilibrium. Why the method works at all, at least in the uniform magnetic field case, was revealed in Section 3. However, magnetic field gradients are the rule rather than the exception in the solar atmosphere, and the *field–free* method should gracefully step aside in favour of more powerful methods reported at this Workshop

(Trujillo Bueno and Landi Degl'Innocenti, 1996; Bruls and Trujillo Bueno, 1996).

The DELO method still stands as the preferred method of formally integrating the vector transfer equation, and those who still believe that contribution functions are useful in Stokes spectral diagnostics should take heed of Murphy's law as discussed in Section 2, and use the SPSR code (Murphy and Rees, 1990) on which DELO is based.

Researchers with a serious knowledge of the pitfalls of non–LTE computations may be less than convinced by our wavelet solution to the toy problem in Section 4. We unashamedly admit that the example given is a "hit–and–run" exercise, intended to alert colleagues to the potential of wavelets. A number of questions need to be addressed, such as

- how does one handle depth grids that are not equispaced?
  The "lifting scheme" (Sweldens, 1995) is designed to tackle this problem.
- how can wavelets handle the fact that in "real" non–LTE problems the integral equations are not written explicitly (i.e. there are no explicit kernels), such as in the accelerated lambda iteration method?
  The answer probably lies in the fact that differential equations also may be solved using wavelet decompositions (Bacry *et al.*, 1994). Moreover, these decompositions are naturally *multigrid*, a fact that could be used to advantage in multidimensional model problems such as line formation in magnetic flux tubes.

We conclude by affirming our strong belief that wavelets are indeed worthy of serious consideration in non–LTE, polarized radiative transfer. This belief is founded on our experience in an entirely different context, electromagnetic modelling, where wavelets are poised to make possible the numerical design and construction of new classes of antennae not amenable to standard numerical methods.

The "algebraic Annapurna" (Rees, 1987) has been conquered by the methods reported elsewhere at this Workshop. Perhaps wavelets may permit future expeditions up this peak "without oxygen".

Acknowledgements

David Rees thanks the organizers of the Workshop for the opportunity to share in the celebration of the "coming of age" of Solar Polarization in St Petersburg. He also gratefully acknowledges the financial support of the CSIRO Division of Radiophysics for travel to the Workshop. Finally both authors give credit to the other members of the "Wavelet Conspiracy", a regular Friday–afternoon "skunkworks" Radiophysics seminar activity where the plot to marry radiative transfer and wavelets was hatched.

References

Alpert, B., Beylkin, G., Coifman, R., and Rokhlin, V.: 1993, *SIAM J. Sci. Statistical Comp.* **14**, 159

Auer, L.H., Heasley, J.N., and House, L.L.: 1976, *Astrophys. J.* **216**, 531

Avrett, E., and Hummer, D.G.: 1965, *Monthly Notices Roy. Astron. Soc.* **130**, 295

Beylkin, G., Coifman, R., and, Rokhlin, V.: 1991, *Commun. Pure Appl. Math.* **XLIV**, 909

Bacry, E., Mallat, S., and Papanicolaou, G.: 1994, *RAIRO Math. Modelling Num. Analysis* **26**, 7

Briand, C., and Solanki, S.K.: 1995, *Astron. Astrophys.*, in press

Bruls, J.H.M.J, Solanki, S.K., Rutten, R.J., and Carlsson, M.: 1995, *Astron. Astrophys.* **293**, 225

Bruls, J.H.M.J, and Trujillo Bueno, J.: 1996, these proceedings

Cannon, C.J.: 1973a, *J. Quant. Spectrosc. Radiative Transfer* **13**, 627

Cannon, C.J.: 1973b, *Astrophys.J.* **185**, 621

Chandrasekhar, S.: 1950, *Radiative Transfer*, Oxford University Press, UK

Daubechies, I.: 1988, *Commun. Pure Appl. Math.* **XLI**, 909

Domke, H.: 1970, *Astrofizika* **5**, 525

Domke, H., and Staude, J.: 1973, *Solar Phys.* **31**, 279

Gortler, S.J., Schroder, P., Cohen, M.F., and Hanrahan, P.: 1993, *Computer Graphics, Annual Conference Series*, August 1993, Siggraph, p. 222.

Hewagama, T., Deming, D., Jennings, D.E., Osherovich, V., Wiedemann, G., Zipoy, D., Mickey, D.L., and Garcia, H.: 1993, *Astrophys. J. Suppl.* **86**, 313

Landi Degl'Innocenti, E.: 1987, *Numerical Radiative Transfer*, Ed. W. Kalkofen, Cambridge University Press, Cambridge UK, p. 265

Lites, B.W., Skumanich, A., Rees, D.E., Murphy, G.A., and Carlsson, M.: 1987, *Astrophys. J.* **318**, 930

Lites, B.W., Skumanich, A., Rees, D.E., and Murphy, G.A.: 1988, *Astrophys. J.* **330**, 493

Murphy, G.A.: 1990, *The Synthesis and Inversion of Stokes Spectral Profiles*, NCAR Cooperative Thesis No. 124, HAO, Boulder, Colorado

Murphy, G.A., and Rees, D.E.: 1990, *Operation of the Stokes Profile Synthesis Routine*, NCAR Technical Note NCAR/TN-348+IA, HAO, Boulder, Colorado

Press, W.H., Teukolsky, S.A., Vetterling, W.T., and Flannery, B.P.: 1992, *Numerical Recipes in FORTRAN*, 2nd Edition, Cambridge University Press, UK, p. 584

Rachkovsky, D.N.: 1963, *Izv. Krymsk. Astrofiz. Obs.* **30**, 267

Rees, D.E.: 1969, *Solar Phys.* **10**, 268.

Rees, D.E.: 1987, *Numerical Radiative Transfer*, Ed. W. Kalkofen, Cambridge University Press, Cambridge UK, p. 213

Rees, D.E., Murphy, G.A., and Durrant, C.J.: 1989, *Astrophys. J.* **339**, 1093

Resnikoff, H.L.: 1992, *Optical Eng.* **31**, 1229

Ruiz Cobo, B., and Del Toro Iniesta, J.C.:1992, *Astrophys.J.* **398**, 375

Solanki, S.K., Rüedi, I., and Livingston, W.: 1992a, *Astron. Astrophys.* **263**, 312

Solanki, S.K., Rüedi, I., and Livingston, W.: 1992b,*Astron. Astrophys.* **263**, 339

Solanki, S.K., and Montavon, C.A.P.: 1993, *Astron. Astrophys.* **275**, 283

Solanki, S.K., Montavon, C.A.P., and Livingston, W.: 1994,*Astron. Astrophys.* **283**, 221

Staude, J.: 1969, *Solar Phys.* **8**, 264

Steiner, O.: 1991, *Astron. Astrophys.* **242**, 290

Stenholm, L.J., and Stenflo, J.O.: 1978, *Astron. Astrophys.* **67**, 33

Sweldens, W.: 1995,"The Lifting Scheme: A Custom-Design Construction of Biorthogonal Wavelets.", available via ftp as ftp.math.edu/pub/imi_94/imi94_7.ps.

Takeda, Y.: 1991, *Publ. Astron. Soc. Japan* **43**, 719

Trujillo Bueno, J., and Landi Degl'Innocenti, E.: 1996, these proceedings

NON-LTE POLARIZED RADIATIVE TRANSFER
IN INTERMEDIATE MAGNETIC FIELDS:
NUMERICAL PROBLEMS AND RESULTS

V. BOMMIER

Laboratoire 'Atomes et Molécules en Astrophysique', CNRS URA 812 – DAMAp,
Observatoire de Paris, Section de Meudon, F-92195 Meudon Cedex, France

and

E. LANDI DEGL'INNOCENTI*

Instituto de Astrofísica de Canarias, E-38200 La Laguna, Tenerife, España

Abstract. This paper presents some numerical results relative to a solution, based on the density matrix formalism, of the non-LTE, polarized radiative transfer problem for a two-level atom. The results concern the atomic upper level population and alignment, and the emergent radiation Stokes profiles, for a plane-parallel, static, isothermal atmosphere embedded in a magnetic field of intermediate strength, such that the Zeeman splitting has to be taken into account in the line profile. Zeeman coherences are neglected, whereas magneto-optical effects are taken into account, resulting in a full 4×4 absorption matrix. Induced emission is neglected and complete frequency redistribution, in the rest and laboratory frames, is assumed. Pure Doppler absorption profile (gaussian shape) has also been assumed. The presentation of the results is preceded by a brief discussion of their accuracy and of the numerical difficulties that were met in the solution of the problem.

Key words: Density Matrix – Stokes Profiles – Polarized Radiative Transfer – Intermediate Magnetic Fields

1. Introduction

This paper is aimed at presenting some of the results obtained with a numerical code that was developed by the authors in order to solve, within the density-matrix formalism, the basic problem of non-LTE polarized radiative transfer in magnetized stellar atmospheres. The problem, which is of an extreme complexity in its more general formulation, was solved in a rather schematic situation described by the following set of assumptions:

a) we consider a spectral line that is formed by a two-level atom composed by a lower level, having angular momentum quantum number J and Landé factor g_J, and an upper level, with corresponding quantities J' and $g_{J'}$;

b) the ground level is supposed to be unpolarized and the effect of stimulated emission is neglected;

---

* On leave from the *Dipartimento di Astronomia e Scienza dello Spazio, Università di Firenze, Largo E. Fermi 5, I-50125 Firenze, Italia*

Solar Physics **164**: 117–133, 1996.

c) the spectral line is formed in a semi-infinite, plane-parallel, static atmosphere in the presence of a constant magnetic field;

d) the atoms are affected by thermal motions with a maxwellian distribution of velocities characterized by a Doppler width Δv_D, supposed constant with optical depth;

e) correlations between velocities and density-matrix elements are neglected;

f) the magnetic field is sufficiently strong, so that the Zeeman splitting between any of the magnetic sublevels of the upper level is much larger than the natural broadening of the sublevels themselves. This implies that coherences (the off-diagonal elements of the density matrix) are zero in the reference system having its z-axis directed along the magnetic field. Apart from this restriction, the Zeeman splitting is arbitrary and can vary from a fraction of to many Doppler broadenings (intermediate field regime). In this paper, the magnetic field intensity is parametrized through the quantity

$$\gamma = \frac{v_L}{\Delta v_D} \quad , \tag{1}$$

where v_L is the Larmor frequency. The direction of the magnetic field is specified by the angle ψ_B defined in Figure 1. Formally, one can also consider the case of zero magnetic field ($\gamma = 0$), although this has to be regarded as the limiting case where the field is so weak to introduce negligible splitting (with respect to Δv_D) but, at the same time, sufficiently strong to destroy coherences. The results for $\gamma = 0$ obviously depend on ψ_B. The real case of a plane-parallel atmosphere devoid of magnetic field is obtained, in this formalism, by setting $\gamma = 0$ and $\psi_B = 0$.

g) the atoms are affected by collisions with a collection of perturbers having a maxwellian distribution of velocities, characterized by the temperature T. The effect of collisions is twofold: excitation (and de-excitation) of the upper level and depolarization of the upper level;

h) the emission coefficient in the four Stokes parameters is supposed to be uncorrelated with the spectrum of the exciting radiation field (hypothesis of the complete redistribution in frequency).

The density matrix formalism which is at the basis of the present work is presented in Bommier (1996) and in Landi Degl'Innocenti (1996) where some of the assumptions itemized above are also discussed. By means of this formalism one finds, for the case of a plane-parallel atmosphere, a set of integral equations which couple the upper-level density-matrix elements (expressed in the representation of the statistical tensors) at any given depth in the atmosphere to the same quantities at different depths, the coupling being described by suitable kernels. The relevant equations are given in Landi Degl'Innocenti, Bommier, and Sahal-Bréchot (1991a,b).

The numerical code that we have developed solves this set of integral equations. The solution is achieved by introducing a grid of depth-points

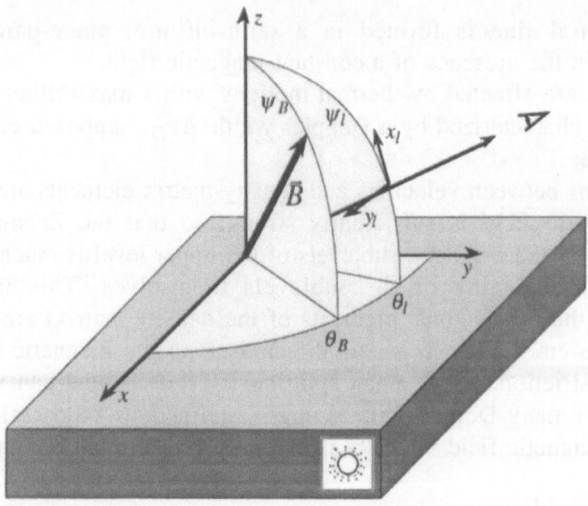

Fig. 1. Geometry specifying the direction of the magnetic field and that of the line-of-sight. x_l is the reference direction for defining the positive Q Stokes parameter.

and by considering as unknowns the values of the statistical tensors at any given grid-point. The run with depth of each statistical tensor between two adjacent grid-points is supposed to be described by a linear interpolation (or by more sophisticated interpolations involving 2-nd order or cubic spline polynomials). By so doing, the integrals containing the kernels can be evaluated analytically, and the set of integral equations reduces to a system of linear equations. The system is finally solved by standard techniques, and the value of the statistical tensors at each grid-point is thus obtained.

In Section 2 we report on the accuracy attained in our numerical code and on some of the difficulties that have been met in the solution. In Section 3 we give some results for the run with depth of the statistical tensors. Finally, in Section 4 we present some results for the emerging Stokes parameters. All these results refer to the particular case of an isothermal and static atmosphere with pure gaussian profiles.

2. Accuracy and numerical difficulties

It is difficult to give general statements on the accuracy of the results of the numerical code because such results, being mostly original, cannot be obviously compared with pre-existing ones. Moreover, they depend on a large number of parameters and the parameter-space is quite large. However, it is possible to perform a rather stringent test by taking into account an analytical property concerning the value of the statistical tensors at the

TABLE I

Convergence of the numerical solution

points per decade	interpolation			
	linear	2-nd order (a)	2-nd order (b)	cubic spline
3	0.37330130	1.0012101	1.0476226	1.0167244
5	0.52948550	0.98251048	1.0107510	1.0046362
10	0.73004583	0.98680385	1.0015857	1.0006369
15	0.81997291	0.99053579	1.0006390	1.0003726
20	0.86889163	0.99271745	1.0004068	1.0002673
25	0.89895703	0.99411577	1.0003243	1.0002766
30	0.91901064	0.99508337	1.0002883	1.0002496

Quantity σ (whose analytical limit is 1, see text), for various interpolations of the atomic density matrix, as a function of the number of grid-points. These results refer to the case $\varepsilon' = 10^{-4}$. They are nearly independent of γ and ψ_B.

surface of an isothermal atmosphere. This property, in its simplest form, can be expressed as

$$\sigma = 1 \qquad (2)$$

(Landi Degl'Innocenti and Bommier, 1994), where

$$\sigma = \sqrt{\sum_K [S_K(0)]^2} \Big/ \sqrt{\varepsilon'} \qquad , \qquad (3)$$

$S_K(0)$ being the statistical source function (proportional to the statistical tensor) evaluated at the top of the atmosphere, and where ε' is the standard parameter of the non-LTE theory.

Table I shows that our numerical code satisfies the test of Eq. (2) with a high degree of accuracy (the relative error on σ being smaller than 3×10^{-4}), although, to obtain such an accuracy, quadratic or cubic spline interpolations are necessary.

Table I also shows the convergence of the solution, as a function of the number of grid-points (logarithmically equispaced in the line optical depth τ) for various interpolations: it can be seen that the linear interpolation converges very slowly; this is consistent with the result of Frisch and Froeschlé (1977), who show that the interpolation must ensure the continuity of the derivative of the unknown function. To get faster convergence we have then tried a 2-nd order interpolation. This can be done at two different levels of sophistication. The coupling coefficients which connect the density matrix at point τ_i with the density matrix at each of the other grid-points can be obtained either by assuming a parabolic behavior of the density matrix only within the interval $[\tau_{i-1}, \tau_{i+1}]$ – as suggested by Frisch and

Froeschlé (1977) – or by assuming a parabolic behavior in all the intervals $...[\tau_{i-3}, \tau_{i-1}], [\tau_{i-1}, \tau_{i+1}], [\tau_{i+1}, \tau_{i+3}]...$. These two possibilities are referred to as 2-nd order (a) and 2-nd order (b) in the Table. Finally, we have also considered a cubic spline interpolation, which ensures the continuity of the derivative at any point. It can be seen from the Table that the 2-nd order (a) interpolation is convenient in the case of a small number of grid-points, but its convergence is slow when the number of points is increased, even if a factor of 10 in accuracy (for 10 points) is got with respect to the linear interpolation. With the 2-nd order (b) interpolation, another factor of 10 in accuracy is obtained, and, with the cubic spline interpolation, the accuracy is further increased by a factor of 2 only. Unfortunately, on usual computers running in double precision, the cubic spline interpolation does not work for large magnetic field strengths; quadruple precision is needed for $\gamma > 1$ values.

To reach such a high degree of numerical accuracy it was also necessary to solve, in a rather careful way, some numerical difficulties that are typical of this particular problem. Before discussing them, let us mention that, in our computations, the full 4×4 absorption matrix $\mathbf{H}$ for the Stokes parameters vector, which can be written as[1]

$$\mathbf{H} = \frac{1}{\eta^{(0)}} \begin{pmatrix} \eta_I & \eta_Q & \eta_U & \eta_V \\ \eta_Q & \eta_I & \rho_V & -\rho_U \\ \eta_U & -\rho_V & \eta_I & \rho_Q \\ \eta_V & \rho_U & -\rho_Q & \eta_I \end{pmatrix} \quad , \tag{4}$$

has been scaled to a relative absorption matrix $\mathbf{H}_{rel}$, whose elements are ≤ 1

$$\mathbf{H}_{rel} = \frac{1}{\Xi} \mathbf{H} \quad , \tag{5}$$

where the normalization factor is given by

$$\Xi = \frac{1}{\eta^{(0)}} \sqrt{\eta_I^2 + \eta_Q^2 + \eta_U^2 + \eta_V^2 + \rho_Q^2 + \rho_U^2 + \rho_V^2} \quad . \tag{6}$$

A first series of numerical difficulties arises from the fact that the determinant of the relative absorption matrix $\mathbf{H}_{rel}$, which is of the order of 1 at line center, becomes very small with respect to 1 in the line wings (see Fig. 2); as a consequence of this behavior, the matrix product $\left(\mathbf{H}_{rel}\right)^{-1}\mathbf{H}_{rel}$ cannot be accurately computed in the line wings, whatever the method of computation of $\left(\mathbf{H}_{rel}\right)^{-1}$ is; as the result of this operation is however well known, the solution to such a problem obviously consists in avoiding the

---

[1] $\quad \eta^{(0)} = \frac{h\nu_0}{4\pi} B N_J \quad ,$

is the line absorption coefficient; ν_0 is the line center frequency, B is the Einstein absorption coefficient and N_J is the lower level population (in the present computations, $N_J = 1$ is assumed).

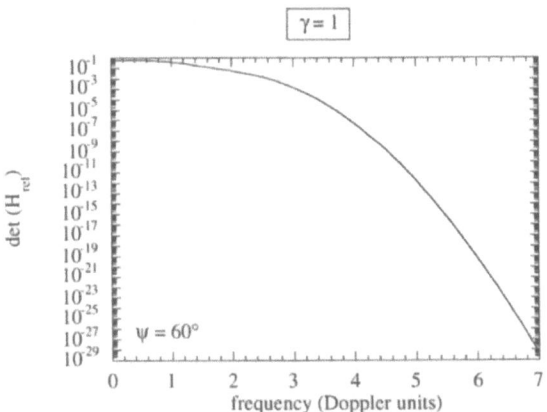

Fig. 2. Determinant of the relative absorption matrix $H_{rel}(\nu,\Omega)$, as a function of frequency, for $\gamma = 1$ and for a direction Ω forming an angle of 60° with the magnetic field.

numerical computation of this matrix product everywhere it appears: $(H_{rel})^{-1}$ appears when one integrates the evolution operator $O = \exp(-\tau H)$ over the optical depth, whereas H_{rel} appears in the computation of O itself (see the analytical expression of O in Landi Degl'Innocenti and Landi Degl'Innocenti, 1985). Moreover, the first column (= first row) of H_{rel} also appears in the final summation of the computation (Landi Degl'Innocenti, Bommier, and Sahal-Bréchot, 1991b, Eq. (25)). This summation contains then some elements of the matrix product $(H_{rel})^{-1} H_{rel}$ that have to be conveniently evaluated.

The introduction of a continuous absorption coefficient is insufficient to eliminate such problems: even though, in this case, H_{rel} tends towards the identity matrix in the far wings, some problems still remain, because the determinant may become very small at intermediate frequencies between line center, where $\det(H_{rel}) \approx 1$, and the far wings, where $\det(H_{rel}) = 1$.

In Figure 3, we have plotted the quantity $\Xi \det(H_{rel})$, which is proportional to the 'mean absorption probability'[2], as a function of the angle ψ between the incident ray and the magnetic field direction, for

---

[2] It can be shown that the quantity $\tau \Xi \det(H_{rel})$ plays here the same role as $\tau_\nu = \tau \phi_\nu$ in the unpolarized and unmagnetized case (ϕ_ν is the standard absorption profile). We can then consider the quantity $\Xi \det(H_{rel})$ as a kind of mean (in the sense of 'polarization averaged') absorption probability. Note that, in the limit $\gamma \to 0$ one gets $\Xi \det(H_{rel}) = \phi_\nu$.

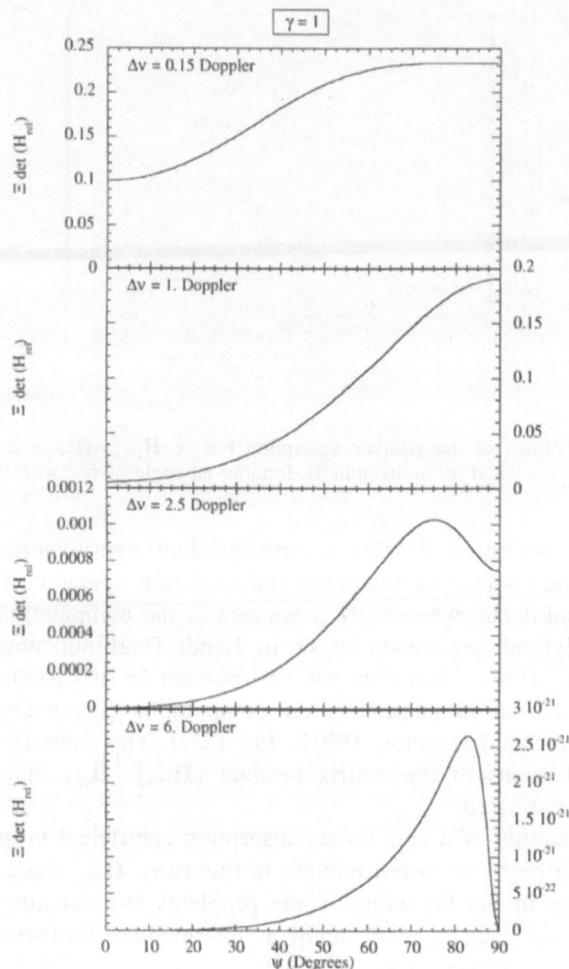

Fig. 3. 'Mean absorption probability' (see text) as a function of the angle ψ between the incident ray direction Ω and the magnetic field direction, for various frequencies, and for a relative field strength $\gamma = 1$. The absorption probability is larger for incident rays that are highly inclined with respect to the magnetic field.

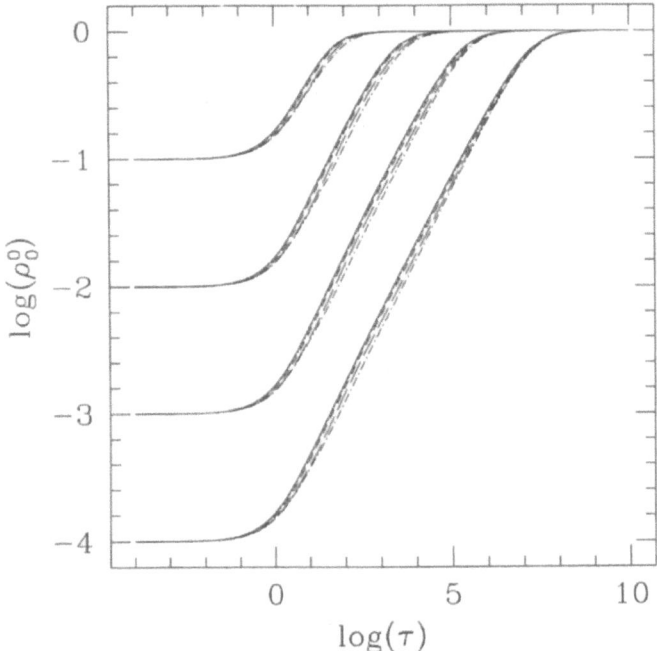

Fig. 4. Upper level population (normalized to 1 for $\tau \to \infty$) as a function of optical depth, for various relative field strengths γ (full line $\gamma = 0$ – small dashes $\gamma = 0.5$ – medium dashes $\gamma = 1$ – large dashes $\gamma = 2$ – dot-dashes $\gamma = 3$). The different curves refer to various values of the parameter ε'. From top to bottom: $\varepsilon' = 10^{-2}, 10^{-4}, 10^{-6}, 10^{-8}$, respectively. The line is a normal Zeeman triplet line with $J = 0$, $J' = 1$, $g_{J'} = 1$.

various frequencies; this Figure shows that, for $\gamma = 1$, the mean absorption probability is larger for incident rays that are highly inclined with respect to the magnetic field. This feature appears when the Zeeman components begin to separate, namely when $\gamma \geq 1$.

A second series of numerical difficulties arises when one uses second (and more) order interpolations: in the wings, the computation of the contribution of each optical depth interval leads to the evaluation of some differences that become too small to be accurately computed; this problem can be solved by using Taylor expansions of the contributions in the wings.

3. Results for the statistical tensors as a function of depth

The results of the solution of the integral equations are given in Figures 4-9, showing the atomic density matrix elements (upper level population and

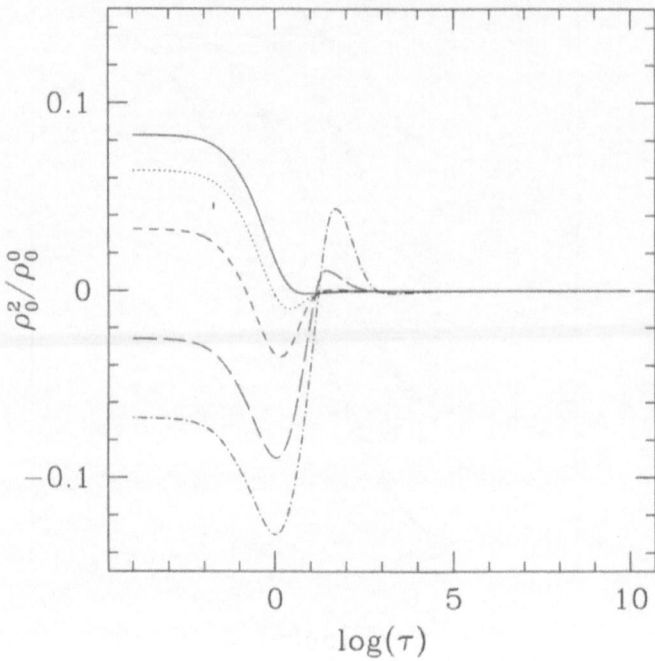

Fig. 5. Upper level relative alignment as a function of optical depth, for various relative field strengths γ (full line $\gamma = 0$ – small dashes $\gamma = 0.5$ – medium dashes $\gamma = 1$ – large dashes $\gamma = 2$ – dot-dashes $\gamma = 3$). The curves are obtained for the set of parameters $\psi_B = 0°$, $\varepsilon' = 10^{-4}$, $\delta^{(2)} = 0$ (no depolarizing collisions). The line is a normal Zeeman triplet line with $J = 0$, $J' = 1$, $g_{J'} = 1$.

alignment – see Bommier, 1996, for a definition of these quantities) for various values of the parameters

$$\varepsilon = \frac{C_{J'J}}{A} \quad \Rightarrow \quad \varepsilon' = \frac{\varepsilon}{1+\varepsilon} = \frac{C_{J'J}}{A+C_{J'J}}$$

$$\delta^{(2)} = \frac{D^{(2)}}{A} \tag{7}$$

where A is the Einstein coefficient for spontaneous emission, $C_{J'J}$ is the inelastic collisions deexcitation rate, $D^{(2)}$ is the alignment destruction rate due to elastic collisions (depolarizing collisions), and for various magnetic field strengths and directions.

The result for the upper level population is given in Figure 4, which refers to the case of a vertical magnetic field $\psi_B = 0°$ with no depolarizing collisions. This Figure is not modified in a visible way in any other case. Either changing the magnetic field direction, or introducing depolarizing collisions, or even neglecting a-priori atomic alignment, the differences that

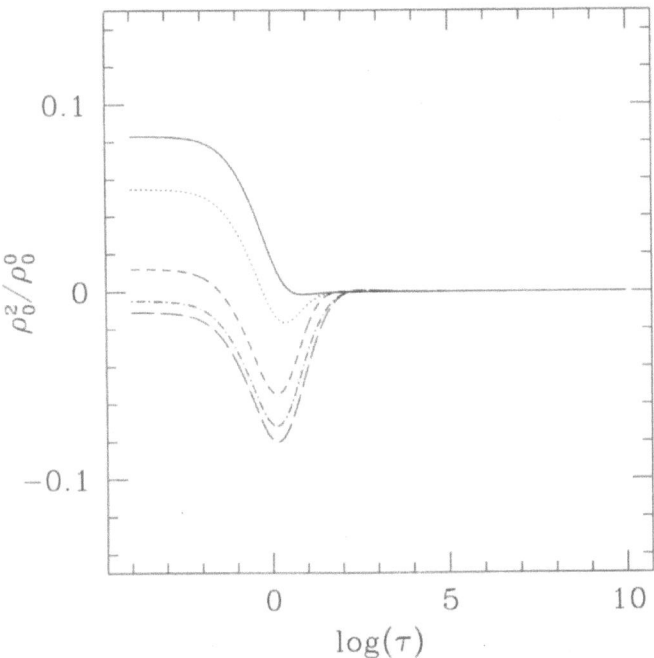

Fig. 6. Same as Fig. 5, but neglecting magneto-optical effects.

are found in ρ_0^0 result in being too small to be perceived in the Figure. The behavior of the various curves as a function of γ can be qualitatively understood by considering that, when the Zeeman components begin to separate ($\gamma \geq 1$), the absorption probability presents wide angular variations. This implies a stronger coupling (with respect to the non-magnetic case) between points that are far apart in optical depth, and a weaker coupling between close points. The net result is, for a fixed value of τ, a decrease of the upper level population. Results very similar to those shown in Fig. 4 have been obtained by Trujillo Bueno and Landi Degl'Innocenti (1996) by means of a completely different numerical approach. In this last paper, however, atomic alignment is disregarded.

Figures 5-9 give the results for the upper level atomic alignment, for various cases. The atomic alignment is the signature, at the atomic level, of the anisotropy of the medium. It has to be stressed that, for $\psi_B = 0°$, a positive alignment corresponds to incident radiation more intense (and/or more absorbed) along the vertical, and that a negative alignment corresponds to incident radiation more intense (and/or more absorbed) in the horizontal plane.

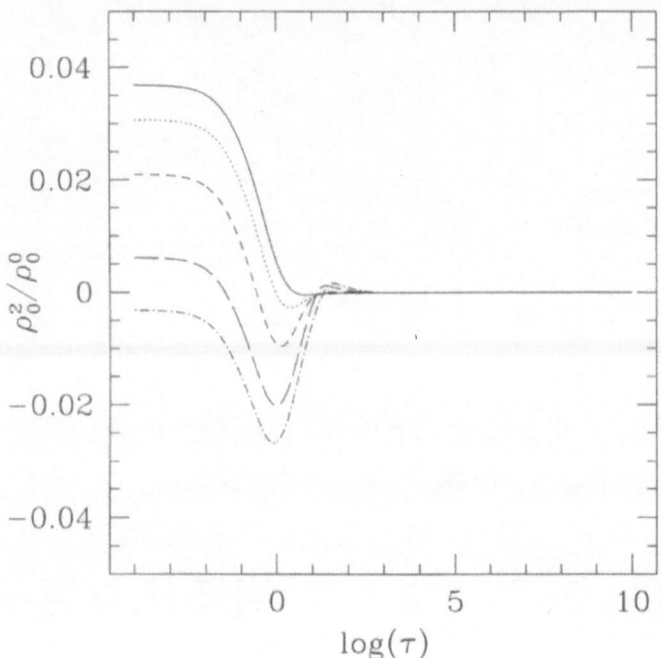

Fig. 7. Same as Fig. 5, but assuming now the presence of depolarizing collisions ($\delta^{(2)} = 1$).

Figure 5 refers to the case of a vertical magnetic field, for a normal Zeeman triplet, without any depolarizing collisions effects. This Figure shows that the modification of the anisotropy of the medium due to the presence of the magnetic field is important: the alignment, which is always positive in zero magnetic field, as expected, can become negative in the presence of the field due to the fact that, when the Zeeman components begin to separate, the absorption probability is larger for an incident ray perpendicular to the magnetic field, than for a parallel one (see Fig. 3). Note also, for $\gamma = 3$, the bump of positive alignment that occurs in the region $10^1 \leq \tau \leq 10^3$. This Figure shows the complexity of the medium anisotropy in the presence of a magnetic field in the intermediate regime.

It is important to mention that, for zero magnetic field, the ratio ρ_0^2/ρ_0^0 in Fig. 5 attains, in the limit $\tau \to 0$, the value $\rho_0^2/\rho_0^0 = 8.294\%$. This implies, for a tangential line-of-sight ($\psi_l = 90°$ in Fig. 1 – limb polarization), a value $Q/I = -9.063\%$. This result has been obtained assuming $\varepsilon' = 10^{-4}$. In the limit $\varepsilon' \to 0$ (no collisions, conservative scattering), the surface values and limb polarization given by our code (in the absence of a magnetic field) are $\rho_0^2/\rho_0^0 = 8.631\%$ and $Q/I = -9.443\%$, in full agreement with the result by

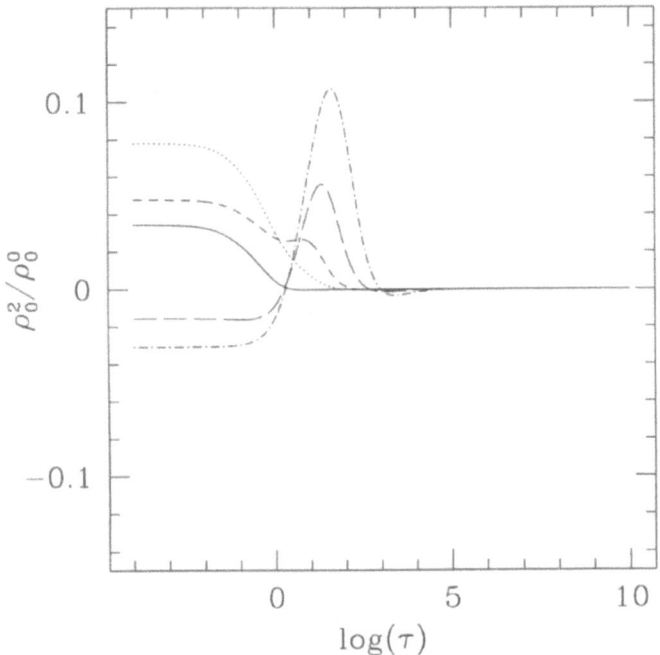

Fig. 8. Same as Fig. 5, but now for a Zeeman multiplet line, with: lower level $J = 2$, upper level $J' = 3$, Landé factors $g_J = 0.5$, $g_{J'} = 1.5$.

Ivanov (1990, 1991)[3]. Also, for the case of Rayleigh scattering, which is obtained in our formalism by substituting the line profile with a Dirac's delta, we fully recover the classical exact result of Chandrasekhar (1960, p. 248), namely $Q/I = -11.713\%$ (corresponding to $\rho_0^2/\rho_0^0 = 10.628\%$).

Figure 6 is the same as Figure 5, but neglecting magneto-optical effects (the $\rho_{Q,U,V}$ elements of the absorption matrix $\mathbf{H}$ are now set to 0); the comparison between the two Figures shows the important role of those effects on the upper level alignment; in particular, the bump of alignment in the region $10^1 \leq \tau \leq 10^3$, for $\gamma = 3$, has now disappeared.

Figure 7 is the same as Figure 5, but introducing now some depolarizing collisions: this results in an obvious decrease of the atomic alignment, and also in an important modification of the curves. Note, for instance, the depression of the positive bumps and the sign change of the surface value of ρ_0^2 in a wide interval of γ values.

---

[3] The value $Q/I = -9.443\%$ was reported by Ivanov in his oral presentation at the workshop, and also as a private communication to one of the authors (VB) in 1993. In the papers quoted, the same value is given with only two significant digits as $Q/I = -9.5\%$.

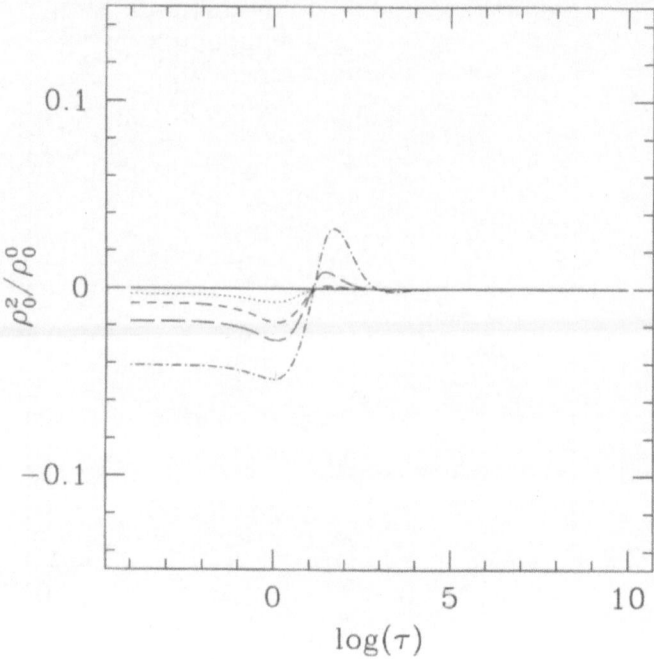

Fig. 9. Same as Fig. 5, but now for a magnetic field inclined at the Van Wleck angle ($3\cos^2 \psi_B = 1$).

Figure 8 refers to the case of a Zeeman multiplet different from a Normal triplet; the comparison with Figure 5 shows that the Zeeman structure of the line plays an important role for the atomic alignment, which results in being very different in the two cases.

Finally, Figure 9 shows the relative alignment in the case of a non-vertical magnetic field inclined at the Van Wleck angle ($3\cos^2 \psi_B = 1$). Note that ρ_0^2 is defined in the reference system of the magnetic field, so that the anisotropy of the radiation field plays now a quite different role. The Figure shows that the alignment is rigorously zero in the absence of magnetic field[4] (a property that can be proven analytically) and that this property is approximately satisfied up to $\gamma \approx 1$. This result confirms (in this field regime) an approximation that was introduced by Landi Degl'Innocenti and Bommier (1993) in a paper where a spectroscopic method was proposed for solving the 180° azimuth ambiguity in magnetic field measurements. As the

---

[4] neglecting, nevertheless, Zeeman coherences (see the discussion in Section 1, point *f*).

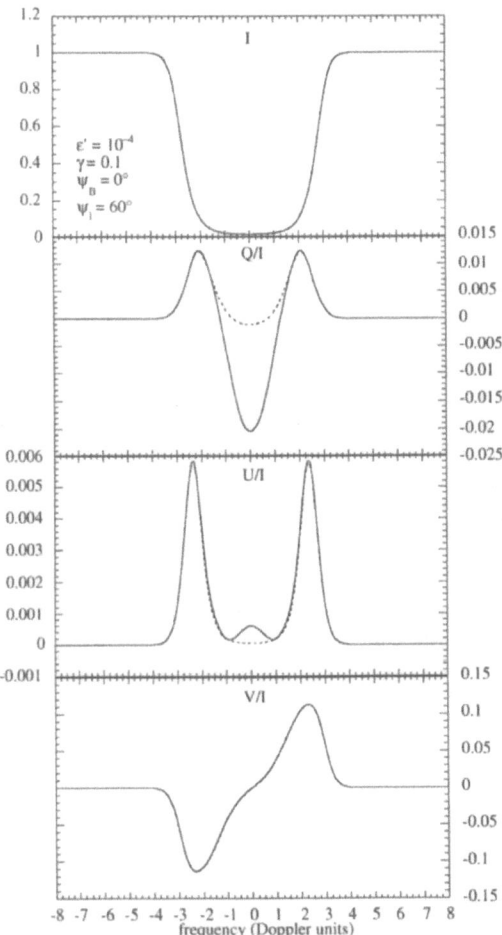

Fig. 10. Emerging Stokes parameters computed for a normal Zeeman triplet with the following set of parameters: ; $\gamma = 0.1$, $\psi_B = 0°$, $\psi_l = 60°$, $\varepsilon' = 10^{-4}$, $\delta^{(2)} = 0$. Full line: atomic alignment is taken into account; dotted line: atomic alignment is neglected (see text).

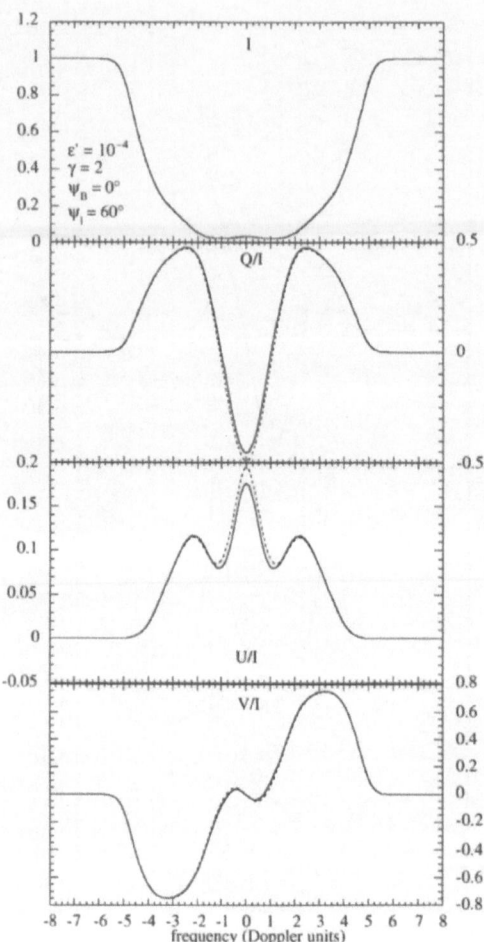

Fig. 11. Same as Fig. 10, but for another magnetic field strength $\gamma = 2$.

linear polarization of resonance lines is sensitive to atomic polarization only in the regime $\gamma < 1$ (see next section), the approximate expressions derived in the paper just quoted can be conveniently applied in all cases of interest.

4. Results for the Stokes profiles

Figures 10 and 11 show the Stokes parameters emerging from the isothermal atmosphere for two values of the magnetic field ($\gamma = 0.1$ and $\gamma = 2$) corresponding to two extreme cases. In the first case the Zeeman splitting is practically negligible, while in the second case the Zeeman components are well separated. Both Figures refer to a normal Zeeman triplet with a vertical magnetic field. The line of sight is inclined at 60° from the vertical and the reference direction for positive-Q is contained in the plane defined by the vertical and the line-of-sight (see Fig. 1).

Both Figures show a comparison between the nominal case (full lines) and the case where atomic polarization is neglected (broken lines). The profiles corresponding to the first case are evaluated through the solution of the statistical tensors obtained for the given magnetic field strength and in the absence of depolarizing collisions ($\delta^{(2)} = 0$). The profiles corresponding to the second case are evaluated through a different solution of the statistical tensors obtained for the same value of γ but in the limit $\delta^{(2)} \to \infty$ (obviously, in this case, only the statistical tensor ρ_0^0 is $\neq 0$).

Figure 10 shows that, for $\gamma = 0.1$, the effect of atomic alignment on the signature of the linear polarization profiles is quite important. In particular, the shape of the Q-profile is strongly affected, atomic alignment being responsible for the appearance of a negative central lobe. In contrast, for $\gamma = 2$, the profiles are only weakly affected (see Fig. 11). Indeed, a comparison of the two Figures shows that the contribution of atomic alignment to linear polarization is of the same order of magnitude in both cases; however, for $\gamma = 2$, the polarization due to the Zeeman effect is much larger, so that the effect of alignment becomes "hidden" by the Zeeman effect. A possible way of extracting a clearer signature of the presence of atomic polarization is the one of considering *integrated profiles* as suggested in Landi Degl'Innocenti and Bommier (1993).

Finally, we want to stress that the results presented in this paper refer to an isothermal atmosphere with a constant magnetic field. The effect of stratified atmospheres and/or variable magnetic fields has still to be investigated.

References

Bommier, V.: 1996 (this conference)
Chandrasekhar, S.: 1960, *Radiative Transfer*, Dover, New York
Frisch, H., and Froeschlé, Ch.: 1977, *M.N.R.A.S.* **181**, 281
Ivanov, V.V.: 1990, *Astron. Zh.* **67**, 1233; *Sov. Astron.* **34**, 621
Ivanov, V.V.: 1991, in L. Crivellari et al. (eds), *Stellar Atmospheres: Beyond Classical Models*, Kluwer Academic Publishers, Dordrecht, pp. 81–104
Landi Degl'Innocenti, E.: 1996 (this conference)
Landi Degl'Innocenti, E., and Landi Degl'Innocenti, M.: 1985, *Solar Phys.* **97**, 239
Landi Degl'Innocenti, E., Bommier, V., and Sahal-Bréchot, S.: 1991a, *Astron. Astrophys.* **244**, 391
Landi Degl'Innocenti, E., Bommier, V., and Sahal-Bréchot, S.: 1991b, *Astron. Astrophys.* **244**, 401
Landi Degl'Innocenti, E., and Bommier, V.: 1993, *Ap. J.* **411**, L49
Landi Degl'Innocenti, E., and Bommier, V.: 1994, *Astron. Astrophys.* **284**, 865
Trujillo Bueno, J., and Landi Degl'Innocenti, E.: 1996 (this conference)

THE POLARIZATION-FREE APPROXIMATION

J. TRUJILLO BUENO and E. LANDI DEGL'INNOCENTI*

Instituto de Astrofísica de Canarias, 38200 La Laguna, Tenerife, Spain

Abstract. The development of effective iterative methods capable of accurately solving NLTE Stokes transfer problems is of considerable importance for the investigation of solar and stellar magnetic fields. After briefly indicating the *iterative* approach which is being presently pursued for the exact solution of such problems, the particular regime where polarization signals can only be due to the Zeeman effect is considered in some detail. By means of NLTE Stokes transfer calculations for a two-level atomic model it is first shown that the currently-used *field-free* approximation (Rees, 1969) cannot be safely applied in the presence of magnetic field gradients. Such gradients lead to changes in the shape and width of the line profiles and they can produce non-negligible effects on the atomic level populations and line source functions. A new approximate method is then proposed, which does not require the actual solution of the Stokes vector transfer equation and is practically as fast as the *field-free* one. This *polarization-free* approximation provides a fairly good account of the effects of *homogeneous* and *inhomogeneous* magnetic fields on the statistical equilibrium and is very easy to implement in any existing non-magnetic, multi-level transfer code.

Key words: NLTE Stokes transfer – Spectropolarimetry – Solar and Stellar magnetic fields

1. Introduction

Within the framework of the generation and transfer of *polarized* radiation the non-local thermodynamic equilibrium (NLTE) problem may be formulated as follows. Given the temperature, the density, the macroscopic velocity and the magnetic field vector as functions of the spatial coordinates, the problem is to determine the excitation and ionization state of chemical species of given abundance that is *consistent* with the *polarization properties* of the radiation field they produce. As in the standard unpolarized case (e.g. Mihalas, 1978) this requires a *self-consistent* solution of two coupled sets of equations: the statistical equilibrium (SE) and the radiative transfer (RT) equations. The SE equations describe the excitation and ionization state, while the RT equations govern the behavior of the radiation field. However, in the *polarized* case these two sets of equations are substantially more complicated due to the following reason.

For a given atomic model, the description of its excitation state is extremely more complex in polarized transfer because one has to take into account that each atomic level of total angular momentum J has associated

* On leave from the *Dipartimento di Astronomia e Scienza dello Spazio, Università di Firenze, Largo E. Fermi 5, I-50125 Firenze, Italia*

Solar Physics **164**: 135–153, 1996.
© 1996 *Kluwer Academic Publishers.*

with it $(2J+1)$ sublevels. The *populations* of these sublevels sensitively depend on the polarization and anisotropy state of the radiation field at each point within the medium and, if the Zeeman splitting is not much larger than the natural width of the atomic levels, there appear quantum interferences or *coherences* among the sublevels themselves. These *coherences* must also be properly quantified in order to fully specify the excitation state.

In the polarized case, instead of the standard scalar RT equation for the specific intensity $I(\nu, \Omega)$ one has to solve, in general, the following *vectorial* transfer equation for the Stokes vector $I(\nu, \Omega) = (I, Q, U, V)^\dagger$ (with $\dagger$=transpose)

$$\frac{d}{ds}I = j - KI , \tag{1}$$

where s measures the path-length along the ray with direction Ω, K is the 4×4 absorption matrix and j is the emission vector. Moreover, instead of the conventional SE equations for the atomic level populations one has, in general, the master equation for the atomic density matrix whose diagonal elements measure the *population* of each atomic sublevel and non-diagonal elements quantify the degree of *coherence* among the sublevels themselves (Landi Degl'Innocenti, 1983).

As in the standard unpolarized case, the combination of the SE and RT equations leads in polarized transfer to a coupled system of equations of the form

$$Ax = b , \tag{2}$$

where A is an operator which depends on the collisional and radiative rates, b is a known vector and x is the unknown. Depending on the complexity of the problem considered, the unknown vector x may be formed for instance, by the density matrix elements, or just by the populations of the Zeeman sublevels, or only by the line source function at each spatial grid-point.

It is possible to select particular *two-level* atomic models (all of them sharing the assumption that there is no atomic polarization in the lower level) for which the NLTE problem is *linear*, i.e. such that the operator A *does not* depend on the unknown x. Because of this linearity one may apply direct matrix inversion solution methods (e.g. Faurobert-Scholl, 1991; Bommier, Landi Degl'Innocenti and Sahal-Bréchot, 1991; Bommier and Landi Degl'Innocenti, 1996). However, *direct* solution methods are of little practical interest because the computing time needed to invert a matrix scales as the cube of the number of unknowns and for large-scale problems (e.g. for 2D and 3D geometries) the solution itself soon becomes polluted with rounding errors. Therefore, in order to be able to improve the diagnostic techniques which are being presently applied within the framework of the two-level atomic model for the investigation of solar magnetic fields, it is important to develop very fast and accurate *iterative* methods of solution. An example that helps to clarify why this is so important is the following: the computing time required to solve directly a system of linear algebraic equations with 10^3 unknowns is an order-of-magnitude larger than the time needed by a basic iterative method like Jacobi's.

In general, however, the NLTE problem is *non-linear*, i.e. the operator **A** does depend on **x**. The reason is the following. The radiative rates are integrals of particular combinations of the Stokes parameters. Because the absorption matrix (**K**) and the emission vector (**j**) which appear in the Stokes transfer Equation (1) depend on **x**, the Stokes parameters also depend on **x**; hence **x** depends non-linearly on itself via the transfer equation. Because of this non-linearity, iterative algorithms are necessary to solve the system of Equations (2). Note that, for the same reasons mentioned above, iterative methods that require the construction and inversion of large matrices are useless. It is therefore imperative to develop effective iterative methods capable of solving NLTE polarization problems with complex atomic models and geometries.

The simplest procedure one might think of to solve the coupled system of Equations (2) is Picard's or Λ-iteration: Using the unknown **x** from the current iterative step, evaluate the elements of the absorption matrix **K** and those of the emission vector **j**, solve the Stokes vector transfer equation and compute the Stokes parameters in all transitions, then with the radiative rates obtained, solve the SE equations *at each point* of the spatial grid *independently* and obtain a new estimate of the unknown vector **x**. Unfortunately, this simple method is not effective. The flaw with this iterative procedure is that, as mentioned above, the radiative rates depend on the unknown **x** and, thus, the equations which determine **x** depend implicitly on **x**. In other words, with the Λ-iteration method the only way for the SE equations to produce changes in **x** is via a subsequent solution of the RT equation. As a result, under typical NLTE conditions in optically thick media, the method converges extremely slowly, yielding very small changes in **x** between successive iterations. In fact, unless one is lucky to initialize the calculation using an estimate for **x** which is already very close to the "exact" solution, stopping a Λ-iteration calculation just because those changes are found to be very small, is no guarantee at all that the exact solution has been found.

It is certainly regrettable that this Λ-iteration scheme has such an extremely slow convergence rate under typical NLTE conditions. With this simple iterative scheme the resulting SE equations at each iterative step turn out to be automatically linearized because the radiative rates are simply being taken equal to those corresponding to the previous iteration. Accordingly, at the current iterative step such equations can be solved directly for obtaining **x** *at each spatial grid-point* independently, i.e. one point after the other. Therefore, since at each iterative step of this Λ-iteration procedure the resulting SE equations are not spatially coupled, there is no need of inverting large matrices and the computing time per iteration is minimal. Fortunately, very effective alternative iterative schemes can indeed be developed where everything goes as simply as in the Λ-iteration procedure, but for which the convergence rate is *extremely* high. The reader is referred to Trujillo Bueno and Fabiani Bendicho (1995), where such novel iterative methods are developed for the standard unpolarized RT case, and to the paper by Trujillo Bueno (1995), where the generalization of such very efficient methods to the NLTE transfer of polarized radiation is presented in detail.

In this paper we will concentrate on introducing an approximation which allows one to solve NLTE Stokes transfer problems with a computational cost similar to that required by the standard unpolarized transfer case. This approximation, that we call the *polarization-free* approximation, is suitable for the regime where NLTE polarization signals can be produced *only* through the Zeeman effect. In this regime, the Zeeman splitting is assumed to be sufficiently strong (magnetic fields B larger than $\sim$100 G) so that the magnetic sublevels can be treated as independent because there are no *coherences* among them, and the effect of *depolarizing* collisions is assumed to be so strong that the *populations* of the magnetic sublevels pertaining to the same level are equal. Therefore, this is the regime where one has NLTE Zeeman-line transfer *without* atomic polarization. For our presentation below it is convenient to note at this stage that, in this regime, the emission vector **j** and the absorption matrix **K** of Equation (1) can be written as

$$\mathbf{K} = \kappa_c \mathbf{1} + \kappa_l \mathbf{\Phi} , \tag{3}$$

$$\mathbf{j} = \kappa_c S_c \mathbf{U} + \kappa_l S_l \mathbf{\Phi} \mathbf{U} , \tag{4}$$

where $\mathbf{U} = (1,0,0,0)^{\dagger}$, $\mathbf{\Phi}$ is the 4×4 line absorption matrix, S_l the line source function of the considered transition, and **1** the unit matrix; see Rees, Murphy and Durrant (1989) for the definition of the remaining symbols used.

In the last few years some diagnostic investigations of solar magnetic fields have been carried out assuming the validity of the above-mentioned regime, but using the so-called *field-free* approximation (Rees, 1969). This approximation consists in using populations calculated with a standard non-magnetic NLTE code and then simply apply a formal solution of Equation (1) to obtain the emergent Stokes parameters accounting for the Zeeman effect. However, the *field-free* approximation has been tested *only* for the case of *height-independent* magnetic fields (Rees, 1969; Domke and Staude,

1973; Auer, Heasley and House, 1977; Stenholm and Stenflo, 1978), while realistic solar magnetic fields do have important vertical gradients. These spatial variations of the Zeeman splitting lead to changes in the shape and width of the line profile. From standard *unpolarized* transfer physics, it has been known for a long time that such profile changes alter the way photons get redistributed into the line wings and their probability of escape (Mihalas, 1978). Therefore, since the *field-free* approximation is *insensitive* to the magnetic field, there are *a priori* well-founded reasons to question its validity. In this paper we will show that the *field-free* approximation cannot be used safely in the presence of magnetic field gradients, but we will develop an equally fast approximation (the *polarization-free* approximation), which seems to be sufficiently accurate for most practical purposes.

The outline of this paper is as follows. In Section 2 we briefly describe the equations which correspond to the above-mentioned strong-field regime without atomic polarization, showing some results of two-level atom numerical calculations for the exact case where the polarization of the radiation field is fully taken into account. A physical interpretation of such results will lead us in Section 3 to introduce the *polarization-free* (POF) approximation, and to compare it with the exact results. In Section 4 we present some analytic calculations for a Milne-Eddington atmosphere which allow us to gain physical insight for understanding why our POF approximation turns out to be so reliable. Finally, Section 5 provides some concluding remarks.

2. NLTE Zeeman-line transfer without atomic polarization

The basic theory for the NLTE transfer of polarized radiation in the presence of a magnetic field has beed developed by Domke and Staude (1973), Dolginov and Pavlov (1973), Šidlichovský (1974), House and Steinitz (1975), and Landi Degl'Innocenti, Landolfi and Landi Degl'Innocenti (1976). If one takes the SE equations as formulated in this last publication and assumes that the populations of the magnetic Zeeman sublevels pertaining to the same level are equal (i.e. no atomic polarization) one then obtains a system of Equations (2) where all is exactly as in the standard unpolarized case, but where instead of having the quantity

$$\bar{J}_{\text{unpol}} = \frac{1}{4\pi} \int d\nu \int d\Omega \phi_\nu I(\nu, \mathbf{\Omega}) , \tag{5}$$

(with ϕ_ν the Voigt absorption profile) one has

$$\bar{J}_{\text{pol}} = \frac{1}{4\pi} \int d\nu \int d\Omega [\phi_I I + \phi_Q Q + \phi_U U + \phi_V V] , \tag{6}$$

where ϕ_X (with X=I, Q, U or V) are the profiles of the first row (or column) of the 4×4 absorption matrix $\mathbf{\Phi}$, which depend on the Zeeman splitting, on the frequency, and on the direction of the magnetic field vector with respect to the direction Ω of each ray. For instance, for the particular case of a two-level atomic model the line source function is given by (e.g. Domke and Staude, 1973; Stenflo, 1994)

$$S = (1 - \epsilon)\bar{J}_{\mathrm{pol}} + \epsilon B_\nu , \qquad (7)$$

which is exactly the same expression of the standard unpolarized case (with ϵ the collisional de-excitation probability and B_ν the Planck function), but with $\bar{J}_{\mathrm{pol}}$ instead of $\bar{J}_{\mathrm{unpol}}$.

The exact numerical solution of this type of NLTE Stokes transfer problems (both for *two-level* and *multi-level* atomic models) can indeed be obtained very efficiently by applying new iterative RT methods (see Trujillo Bueno, 1995), which have been developed recently for the unpolarized case (Trujillo Bueno and Fabiani Bendicho, 1995). For a successful application of these iterative techniques to the *polarized* case including magneto-optical effects we had to develop an effective formal solution method of the Stokes vector transfer equation (see Trujillo Bueno and Landi Degl'Innocenti, 1995), which is capable of calculating (*a*) the diagonal elements of the true Λ-operator (which in the polarized case becomes sensitive to the magnetic field), and (*b*) the quantity $\bar{J}_{\mathrm{pol}}$ (cf. Equation (6)) at each spatial grid-point. All the numerical results presented in this paper have been obtained using such novel iterative schemes, but the details concerning the methods and calculations themselves will be presented in a forthcoming publication (Trujillo Bueno, 1995).

2.1. THE HOMOGENEOUS MAGNETIC FIELD CASE

Figure 1 shows some examples of the effect of *homogeneous* magnetic fields on the line source function of a two-level atomic model for the case of an isothermal atmosphere. The spectral line chosen is a Zeeman triplet without continuum opacity. The line source function S_l is shown for three values of the NLTE parameter ϵ (i.e. for 10^{-4}, 10^{-6}, and 10^{-8}). For each of these ϵ-values, the figure shows the line source function behavior for increasing values of the Zeeman splitting. The solid curves correspond to a Zeeman splitting $v_B = 0$, which is the *field-free* solution. The figure shows that, as the magnetic field increases, the S_l-curves slightly displace towards deeper layers till a saturation effect takes place. This occurs once the σ and π components are completely separated so that the escape probability of photons cannot be modified by further enhancements of the magnetic field. The conclusion is that, for the *homogeneous* magnetic field case, the effect of the Zeeman splitting on the excitation equilibrium of an *isothermal* atmosphere is rather small, in accordance with Rees (1969) and Domke and Staude (1973).

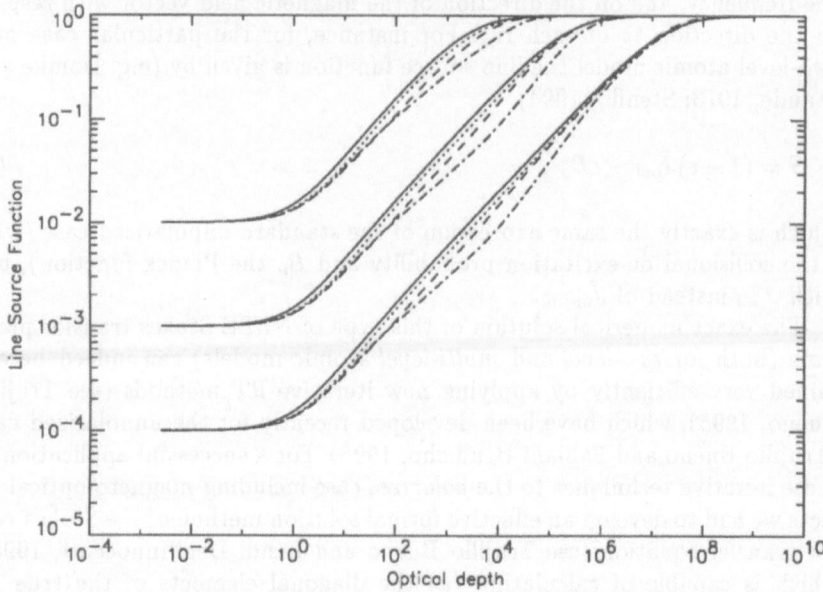

Fig. 1.   The run of the line source function with line-integrated optical depth for a line without continuum opacity, and for three values of the NLTE parameter ϵ. The absorption profile used is Gaussian. Each group of curves corresponds to ϵ equal to 10^{-4}, 10^{-6}, and 10^{-8}, from top to bottom, respectively. The isothermal atmosphere considered is permeated by a *constant* magnetic field which is oriented perpendicularly to the surface. For each ϵ-value the line source function S_l is shown for increasing values of the Zeeman splitting v_B. The solid lines are for $v_B = 0$, i.e. they give the *field-free* solutions. The remaining curves within each ϵ-group are for $v_B=1.5$, 3, and 6 in units of the Doppler width.

2.2. THE CASE OF A MAGNETIC FIELD CHANGING WITH HEIGHT

Consider, for example, the scenario suggested for sunspots with a magnetic canopy which merges at a given continuum optical depth τ_c. Scenarios like this provide the justification for the consideration of the following two possible cases for the variation of the magnetic field with height: (a) magnetic field *decreasing* outwards (for the sunspot umbra center) and (b) magnetic field *increasing* outwards (for the sunspot outer part).

Figure 3 shows the exact solution for the line source-function in an isothermal atmosphere, but where the Zeeman splitting (measured in units of the Doppler width) is varying with height in the interval [0,3] according to $v_B = \alpha + \beta \exp(-\gamma \tau_c)$ (see the three particular variations in Figure 2). The solid line in Figure 3 is the *field-free* solution. The dashed-dotted line lying below the *field-free* solution is the exact solution corresponding to the

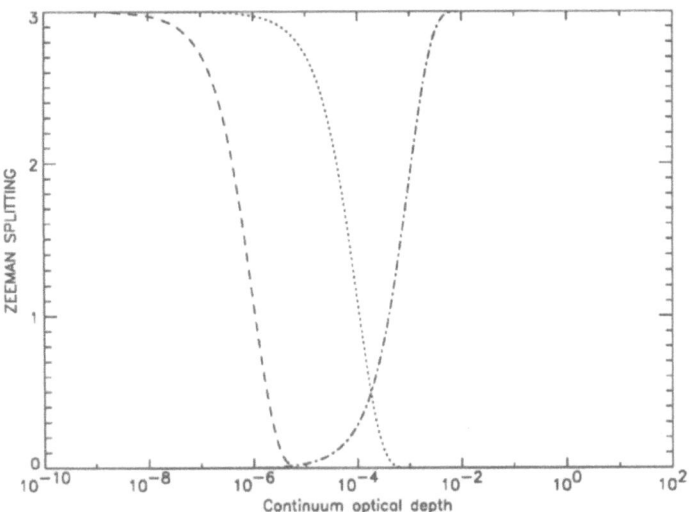

Fig. 2.   The three height variations of the Zeeman splitting corresponding to the results shown in Figures 3 and 4. The values (α, β, γ) of the formula for v_B given in the text are as follows: For the dashed-dotted line $(3,-3,10^3)$, for the dotted-line $(0,3,10^4)$, and for the dashed-line $(0,3,10^6)$.

case of a *magnetic field decreasing outwards* as depicted by the corresponding dashed-dotted line of Figure 2. In this case, note that everywhere in the atmosphere the outward decrease in the Zeeman splitting pushes the line source function well below the *field-free* values. The dashed line and the dotted line of Figure 3, situated above the *field-free* solid line, are the exact solutions for the case of a *magnetic field increasing outwards* as indicated by the corresponding lines of Figure 2. Note that, in these cases, the outward increase in the Zeeman splitting pushes the line source function well above the *field-free* values.

The above effects of a height-dependent magnetic field on the excitation equilibrium are very similar to those produced in standard *unpolarized* transfer by Doppler-width variations of the absorption profile (e.g. Athay, 1972). In the present magnetic context the physical interpretation is the following: in the first case the outward *decrease* of the Zeeman broadening of the absorption profile greatly *enhances* the escape of photons leading to a *decrease* of the line source function with respect to the *field-free* values; in the second case, where the magnetic field *increases* outwards, the opposite takes place and photon escape from the lower layers is greatly *inhibited*, leading to an increase of the line source function with respect to the *field-free* solution.

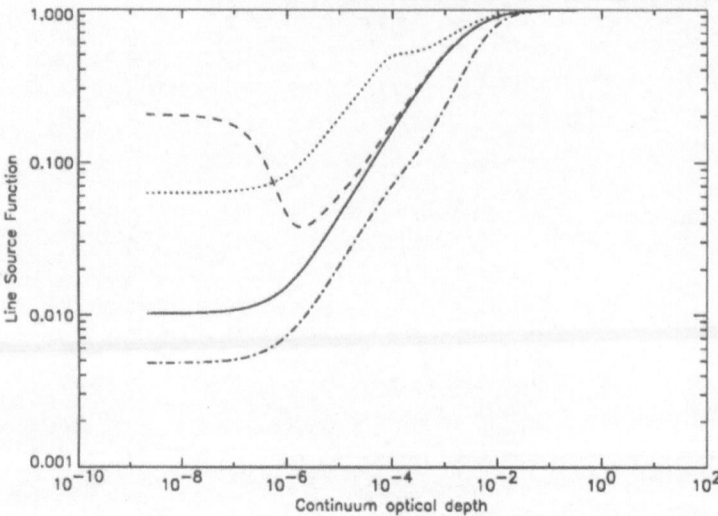

Fig. 3. The line source functions for a Zeeman triplet line of strength $r = \kappa_l/\kappa_c = 10^6$ and $\epsilon = 10^{-4}$. The absorption profile used is Gaussian. Results are shown for the three height variations of the Zeeman splitting indicated in Figure 2. The solid line shows the *field-free* result. The atmosphere is isothermal as in Figure 1 with the magnetic field vector perpendicular to the surface.

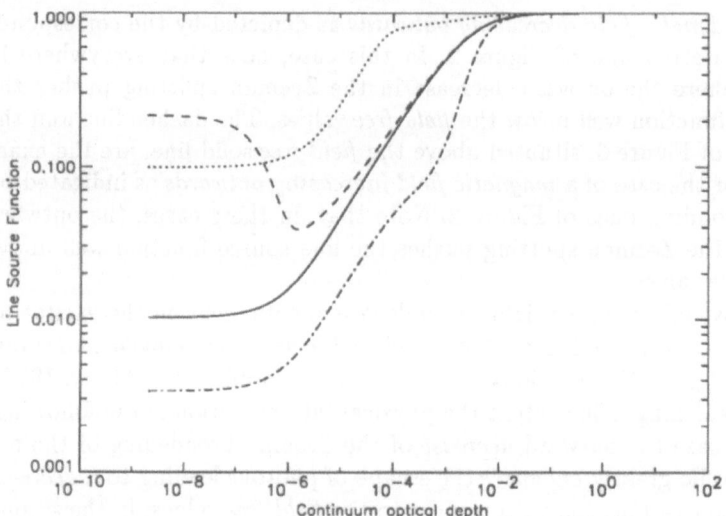

Fig. 4. As in Figure 3, but using the POF approximation

3. The polarization-free approximation

Although, as clarified below, for very large Zeeman splittings the contribution to $\bar{J}_{pol}$ (cf. Equation (6)) due to Q, U and V is not completely negligible, the main physical mechanism responsible for the above results (cf. Figure 3) is the Zeeman broadening of the Stokes-I profile. This suggests the introduction of the *polarization-free* (POF) approximation. It consists in solving the coupled system of Equations (2) using, instead of the exact quantity $\bar{J}_{pol}$ defined in Equation (6), the followingm one

$$\bar{J}_{POF} = \frac{1}{4\pi} \int d\nu \int d\Omega \phi_I I(\nu, \Omega), \tag{8}$$

where the specific intensity $I(\nu, \Omega)$ is to be calculated by solving only the following *scalar* transfer equation

$$\frac{d}{ds}I = (\kappa_c S_c + \kappa_l \phi_I S_l) - (\kappa_c + \kappa_l \phi_I)I . \tag{9}$$

Note that this *scalar* RT equation for the specific intensity $I(\nu, \Omega)$ has been obtained from the I-component of the Stokes transfer Equation (1), simply by omitting the terms containing the coupling with Stokes Q, U and V. Therefore, with our POF approximation everything is exactly as in the standard *unpolarized* case, but instead of the Voigt absorption profile ϕ_ν (which is *insensitive* to the magnetic field) one has

$$\phi_I = \frac{1}{2}\phi_p \sin^2\gamma + \frac{1}{4}(\phi_r + \phi_b)(1 + \cos^2\gamma) , \tag{10}$$

which is precisely the diagonal element of the line absorption matrix $\boldsymbol{\Phi}$ (cf. Equations (1) and (3)), with γ the angle which the ray under consideration makes with the direction of the magnetic field. Note that this profile ϕ_I is *sensitive* to the magnetic field. In fact, calculations done with our POF approximation for the *homogeneous* magnetic field case are in qualitative agreement with the exact results of Figure 1, i.e. they also show an inward displacement of the line source function curves as the Zeeman splitting is increased.

The crucial check of any approximation to the NLTE Stokes transfer problem under consideration must be done however, for the case of a *height-dependent* magnetic field. Figure 4 shows the POF results for the same *extreme* Zeeman-splitting variations depicted in Figure 2. The line source functions are in qualitative agreement with the exact results (cf. Figure 3). Quantitatively the POF approximation leads to a underestimation of the line source function for the case of a magnetic field decreasing outwards and to an overestimation for the opposite case. Note, however, that for the case in

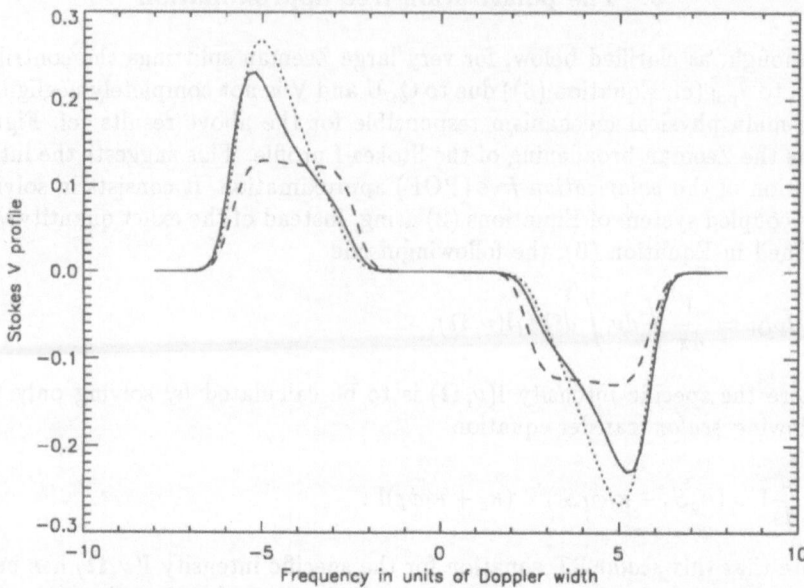

Fig. 5. The emergent Stokes-V profile corresponding to the case of a magnetic field *decreasing* outwards as indicated by the dashed-dotted line of Figure 2. The solid line gives the exact result, the dotted line the POF result, and the dashed-line the *field-free* one. The line of sight makes an angle of 55 degrees with respect to the normal direction to the surface.

which the magnetic field increases outwards as indicated by the dashed-line of Figure 2 the agreement between our POF and exact results is perfect.

Figures 5 and 6 show the emergent Stokes V and Q profiles corresponding to the exact, *field-free*, and POF line source function variations shown in Figures 3 and 4. The dashed lines give the *field-free* results, the dotted lines the POF ones, and the solid lines the exact results. The V-profiles of Figure 5 are for the case *"magnetic field decreasing outwards"*, while the Q-profiles of Figure 6 refer to the case *"magnetic field increasing outwards"* as indicated by the dashed-line variation of Figure 2. As it can be seen the POF approximation generally leads to much better agreement with the exact results, giving a good account of the effects of inhomogeneous magnetic fields on the excitation equilibrium.

4. Some analytical insight

As explained above, the differences between the exact, the *field-free* (FF) and the POF solutions lie in the quantity $\bar{J}$, which is used in the calculation

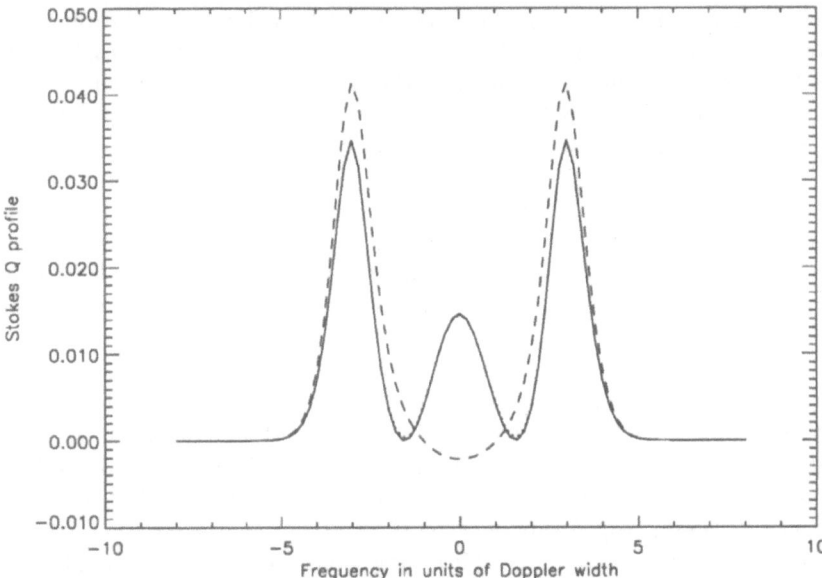

Fig. 6.   The emergent Stokes-Q profile corresponding to the case of a magnetic field *increasing* outwards as indicated by the dashed line of Figure 2. The positive Q-direction is in the plane defined by the magnetic field vector and the line of sight. Note that the POF result (dotted line) is practically indistinguishable from the exact result (solid line). See the caption of Figure 5 for additional information.

of the radiative rates at each iterative step. For a given initialization of the unknown $\mathbf{x}$ the ensuing quantities $\bar{J}_{\mathrm{pol}}$, $\bar{J}_{\mathrm{POF}}$ and $\bar{J}_{\mathrm{FF}}$ will not be exactly equal. Each *iterative* calculation (either with the exact, with the POF or with the FF method) will then lead to different new current estimates of $\mathbf{x}$. Therefore, once convergence is reached, the differences in the three final solutions for $\mathbf{x}$ (obtained with the exact, with the POF and with the FF methods) may well turn out to be larger than the differences between their respective $\bar{J}$-quantities corresponding to the initialization. In this section, however, our approach is to try to gain some analytical insight by just comparing the exact, the POF and the FF $\bar{J}$-quantities calculated for a given initialization.

For the exact case $\bar{J} = \bar{J}_{\mathrm{pol}}$ (cf. Equation (6)), which can be written in terms of its following four contributions

$$\bar{J}_{\mathrm{pol}} = \bar{J}_{\mathrm{I}} + \bar{J}_{\mathrm{Q}} + \bar{J}_{\mathrm{U}} + \bar{J}_{\mathrm{V}} \ . \tag{11}$$

Note that, in particular, the specific intensity (I) that has to be used for the calculation of the $\bar{J}_{\mathrm{I}}$-contribution is exactly the Stokes-I parameter which reflects the coupling with Q, U and V. This is because the I-parameter which is relevant for Equation (11) is the one obtained via the solution of the *vectorial* transfer equation (cf. Equation (1)). With the POF approximation $\bar{J} = \bar{J}_{\text{POF}}$ (cf. Equation (8)), but here the relevant specific intensity is that calculated from the *scalar* transfer equation (cf. Equation (9)), which *does not* contain the coupling with the linear and circular polarization profiles.

When the Zeeman splitting is very large, the Q, U and V contributions to the exact $\bar{J}_{\text{pol}}$ are not negligible. Since the POF results turn out to be in fairly good agreement with the exact ones even for such strong-field cases, it would be interesting to gain some analytical insight with the aim of clarifying to some extent why our POF approximation can lead to such a good agreement. Before presenting some analytical calculations we first call the attention to the numerical results of Figure 7, which correspond to an atmosphere where the temperature has a chromospheric rise and the Zeeman splitting is $v_B = 6$ Doppler widths. This figure shows the total $\bar{J}_{\text{pol}}$ (cf. Equation (11)) and its various I, Q and V contributions; (we note that, in this example, $\bar{J}_U = 0$ because we chose for each Ω-direction the preferred reference direction, which implies that $\phi_U = 0$). As it can seen in Figure 7 $\bar{J}_Q$ and $\bar{J}_V$ are not negligible, which shows that the *exact* $\bar{J}_I$-contribution (with the Stokes-I obtained by solving Equation (1); see the dotted line) is not a good approximation to the total $\bar{J}_{\text{pol}}$, although it certainly gives the main contribution. However, as indicated by the dashed-dotted line, the POF quantity $\bar{J}_{\text{POF}}$ actually provides an excellent approximation to the exact $\bar{J}_{\text{pol}}$.

In order to gain some insight as to why $\bar{J}_{\text{POF}}$ can be so close to the exact $\bar{J}_{\text{pol}}$ we have calculated *analytically* the exact, the *field-free* and the POF quantities $\bar{J}$ for a spectral line *without* a background continuum in a magnetized Milne-Eddington (ME) atmosphere *without* magneto-optical effects and for a constant magnetic field vector oriented perpendicularly to the surface. The line source function is assumed to vary linearly with the optical depth, i.e. $S_l = S_0 + S_1 \tau$ (with $d\tau = -\kappa_l dz$). Since for a ME atmosphere the evolution operator reduces to the exponential of the line absorption matrix Φ (cf. Landi Degl'Innocenti and Landi Degl'Innocenti, 1985), the formal solution expressions for the *outgoing* and *incoming* Stokes vector at a given optical depth τ in the atmosphere can be written as

$$\mathbf{I}^{\text{out}}(\tau, \mu, \nu) = \int_\tau^\infty \exp[-(\frac{t-\tau}{\mu})\Phi](S_0 + S_1 t)\Phi\mathbf{U}\frac{dt}{\mu}, \tag{12}$$

for *outgoing* rays with $\mu = \cos\theta > 0$ (with θ the angle the ray makes with the normal to the surface) and

$$\mathbf{I}^{\text{in}}(\tau, \mu, \nu) = \int_0^\tau \exp[-(\frac{\tau-t}{|\mu|})\Phi](S_0 + S_1 t)\Phi\mathbf{U}\frac{dt}{|\mu|}, \tag{13}$$

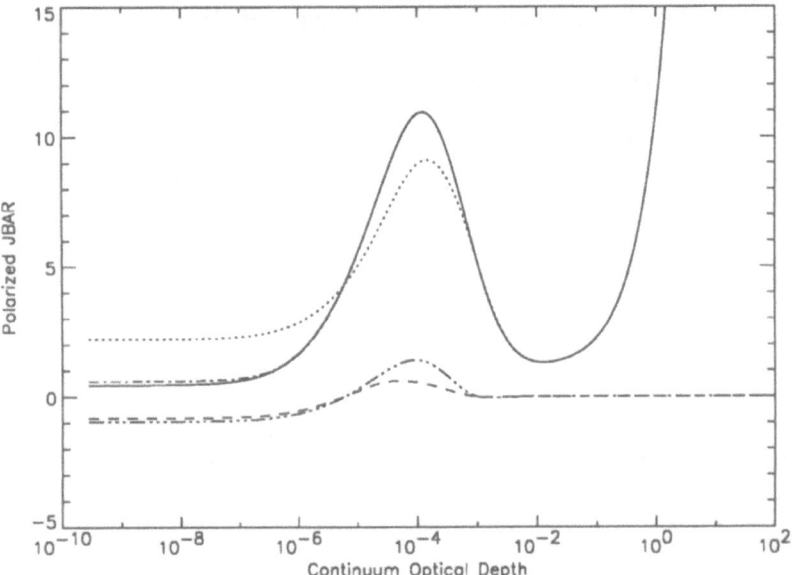

Fig. 7. This figure shows, for a strong Zeeman-triplet line with $\epsilon = 10^{-4}$ and a constant splitting $v_B = 6$, the exact quantity $\bar{J}_{\text{pol}}$ (solid line) and the one obtained with the POF approximation (which is the dashed-dotted line almost fully coincident with the solid line). It also gives the exact contributions $\bar{J}_I$ (dotted-line), $\bar{J}_Q$ (dashed-line) and $\bar{J}_V$ (the dashed-double dotted line). The input atmospheric model has a chromospheric temperature rise. The magnetic field vector is oriented along the normal direction to the surface. The line source function S_l from which these J-quantities have been calculated via a formal solution of the relevant RT equation (i.e. by solving the *vectorial* Equation (1) for the exact J and the *scalar* Equation (7) for the POF J) is that of the fully converged solution corresponding to the *field-free* approximation.

for *incoming* rays with $\mu < 0$.

These two expressions can be integrated straightforwardly to yield

$$\mathbf{I}^{\text{out}}(\tau, \mu, \nu) = (S_0 + S_1\tau)\mathbf{U} + \mu S_1 \mathbf{\Phi}^{-1}\mathbf{U}, \tag{14}$$

and

$$\mathbf{I}^{\text{in}}(\tau, \mu, \nu) = (S_0 + S_1\tau)\mathbf{U} - |\mu|S_1\mathbf{\Phi}^{-1}\mathbf{U} - S_0\exp[-(\frac{\tau}{|\mu|})\mathbf{\Phi}]\mathbf{U} + \tag{15}$$
$$|\mu|S_1\mathbf{\Phi}^{-1}\exp[-(\frac{\tau}{|\mu|})\mathbf{\Phi}]\mathbf{U}.$$

The inverse of the line absorption matrix Φ can also be calculated analytically. The results for the *non-zero* elements $[\Phi^{-1}]_{ij}$ are

$$[\Phi^{-1}]_{11} = \frac{\phi_I^2}{\Delta} \qquad [\Phi^{-1}]_{12} = -\frac{\phi_Q \phi_I}{\Delta} \qquad [\Phi^{-1}]_{14} = -\frac{\phi_V \phi_I}{\Delta} \qquad (16)$$

$$[\Phi^{-1}]_{21} = [\Phi^{-1}]_{12} \qquad [\Phi^{-1}]_{22} = \frac{\phi_I^2 - \phi_V^2}{\Delta} \qquad [\Phi^{-1}]_{24} = \frac{\phi_Q \phi_V}{\Delta} \qquad (17)$$

$$[\Phi^{-1}]_{33} = \frac{\phi_I^2 - \phi_Q^2 - \phi_V^2}{\Delta} \qquad (18)$$

$$[\Phi^{-1}]_{41} = [\Phi^{-1}]_{14} \qquad [\Phi^{-1}]_{42} = [\Phi^{-1}]_{24} \qquad [\Phi^{-1}]_{44} = \frac{\phi_I^2 - \phi_Q^2}{\Delta} \qquad (19)$$

In the calculation of these elements, besides neglecting the magneto-optical terms, we have used $\phi_U = 0$, simply because we selected the preferred reference system. The quantity Δ is given by

$$\Delta = \phi_I(\phi_I^2 - \phi_Q^2 - \phi_V^2) . \qquad (20)$$

Taking the analytical expression for the evolution operator given by Landi Degl'Innocenti and Landi Degl'Innocenti (1985), but assuming that there are no magneto-optical effects we find (with $\omega = \sqrt{\phi_Q^2 + \phi_V^2}$)

$$\exp[-(\frac{\tau}{|\mu|})\Phi] = e^{-\frac{\phi_I}{|\mu|}\tau}[\frac{1}{\omega^2}\mathbf{A} + \cosh(\frac{\omega}{|\mu|}\tau)\frac{1}{\omega^2}\mathbf{B} - \sinh(\frac{\omega}{|\mu|}\tau)\frac{1}{\omega}\mathbf{C}] , \qquad (21)$$

where the matrices $\mathbf{A}$, $\mathbf{B}$ and $\mathbf{C}$ are given by

$$\mathbf{A} = \begin{pmatrix} 0 & 0 & 0 & 0 \\ 0 & \phi_V^2 & 0 & -\phi_Q\phi_V \\ 0 & 0 & \omega^2 & 0 \\ 0 & -\phi_Q\phi_V & 0 & \phi_Q^2 \end{pmatrix} ,$$

$$\mathbf{B} = \begin{pmatrix} \omega^2 & 0 & 0 & 0 \\ 0 & \phi_Q^2 & 0 & \phi_Q\phi_V \\ 0 & 0 & 0 & 0 \\ 0 & \phi_Q\phi_V & 0 & \phi_V^2 \end{pmatrix} , \qquad \mathbf{C} = \begin{pmatrix} 0 & \phi_Q & 0 & \phi_V \\ \phi_Q & 0 & 0 & 0 \\ 0 & 0 & 0 & 0 \\ \phi_V & 0 & 0 & 0 \end{pmatrix} .$$

For the purpose of obtaining the analytical expressions for the *incoming* Stokes parameters of Equation (15) it is only necessary to take into account the first column of each of these three matrices. To this end, matrix $\mathbf{A}$ does not play any role because its first column has all its elements equal to zero, while the first column of matrix $\mathbf{B}$ is $(\omega^2, 0, 0, 0)^\dagger$, and that of matrix $\mathbf{C}$ is $(0, \phi_Q, 0, \phi_V)^\dagger$. After a somewhat tedious but straightforward calculation it is possible to obtain the following results.

First, for a given *outgoing* ray-direction and frequency, the integrand of Equation (6) (i.e. $\mathcal{J}_{\text{pol}} = \phi_I I + \phi_Q Q + \phi_U U + \phi_V V$) is given by

$$\mathcal{J}_{\text{pol}}^{\text{out}}(\tau) = \phi_I S_0 + \phi_I S_1 \tau + \mu S_1. \tag{22}$$

In order to obtain the POF and the *field-free* expressions of $\mathcal{J}$ (i.e. the integrands of Equations (8) and (5), respectively) it is only necessary to make $\phi_Q = \phi_U = \phi_V = 0$ in the exact solution expressions, but still retaining ϕ_I for the POF case, while putting $\phi_I = \phi_\nu$ (i.e. ϕ_I equal to the standard Voigt profile) for the *field-free* (FF) one. Accordingly, one finds

$$\mathcal{J}_{\text{POF}}^{\text{out}}(\tau) = \mathcal{J}_{\text{pol}}^{\text{out}}(\tau), \tag{23}$$

and

$$\mathcal{J}_{\text{FF}}^{\text{out}}(\tau) = \phi_\nu S_0 + \phi_\nu S_1 \tau + \mu S_1. \tag{24}$$

This shows that, for each frequency and *outgoing* ray-direction, our POF approximation already provides the exact contribution. However, since the quantity of interest is the average of $\mathcal{J}$ over all frequencies and directions, we see that the *field-free* approximation also provides the exact *outgoing* contribution to $\bar{J}$ because the integrals over frequency of both ϕ_I and ϕ_ν are normalized to unity.

The above results show that for a ME atmosphere possible differences between the FF and the POF approximations with respect to the exact $\bar{J}$ are to be attributed to differences in their respective *incoming* contributions. In order to better compare their corresponding expressions it is convenient to write $\omega = \sqrt{\phi_Q{}^2 + \phi_V{}^2} = a\phi_I$, where a is a function which, for a given frequency and ray direction, varies between zero, when there is no magnetic field, and unity, for very large Zeeman splittings. For the *incoming* contributions to $\mathcal{J}$ we obtain

$$\begin{aligned} \mathcal{J}_{\text{pol}}^{\text{in}}(\tau) = {} & \phi_I S_0 + \phi_I S_1 \tau - |\mu| S_1 \\ & + \frac{|\mu| S_1}{2} [e^{-(1-a)\frac{\phi_I}{|\mu|}\tau} + e^{-(1+a)\frac{\phi_I}{|\mu|}\tau}] \\ & - \phi_I \frac{S_0}{2} [(1-a)e^{-(1-a)\frac{\phi_I}{|\mu|}\tau} + (1+a)e^{-(1+a)\frac{\phi_I}{|\mu|}\tau}], \end{aligned} \tag{25}$$

$$\mathcal{J}_{\text{POF}}^{\text{in}}(\tau) = \phi_I S_0 + \phi_I S_1 \tau - |\mu| S_1 + |\mu| S_1 e^{-\frac{\phi_I}{|\mu|}\tau} - \phi_I S_0 e^{-\frac{\phi_I}{|\mu|}\tau}, \tag{26}$$

and

$$\mathcal{J}_{FF}^{in}(\tau) = \phi_\nu S_0 + \phi_\nu S_1 \tau - |\mu| S_1 + |\mu| S_1 e^{-\frac{\phi_\nu}{|\mu|}\tau} - \phi_\nu S_0 e^{-\frac{\phi_\nu}{|\mu|}\tau}. \tag{27}$$

In order to obtain a compact expression for the angle and frequency averaged quantities $\bar{J}$ one can use a one-point angular quadrature with $\mu = \pm 1/\sqrt{3}$ (i.e. to select Eddington's approximation; see Mihalas, 1978) and write (with x the frequency from line center measured in units of the Doppler width)

$$\bar{J} = \frac{1}{2} \int [\mathcal{J}^{out} + \mathcal{J}^{in}] dx . \tag{28}$$

The resulting expressions for $\bar{J}_{pol}$, $\bar{J}_{POF}$, and $\bar{J}_{FF}$ are the following

$$\bar{J}_{pol}(\tau) = S_0 + S_1\tau + \frac{S_1}{2\sqrt{3}} \int \frac{[e^{-(1-a)\phi_I\sqrt{3}\tau} + e^{-(1+a)\phi_I\sqrt{3}\tau}]}{2} dx \tag{29}$$
$$- \frac{S_0}{2} \int \frac{[(1-a)\phi_I e^{-(1-a)\phi_I\sqrt{3}\tau} + (1+a)\phi_I e^{-(1+a)\phi_I\sqrt{3}\tau}]}{2} dx ,$$

$$\bar{J}_{POF}(\tau) = S_0 + S_1\tau + \frac{S_1}{2\sqrt{3}} \int e^{-\phi_I\sqrt{3}\tau} dx - \frac{S_0}{2} \int \phi_I e^{-\phi_I\sqrt{3}\tau} dx , \tag{30}$$

and

$$\bar{J}_{FF}(\tau) = S_0 + S_1\tau + \frac{S_1}{2\sqrt{3}} \int e^{-\phi_\nu\sqrt{3}\tau} dx - \frac{S_0}{2} \int \phi_\nu e^{-\phi_\nu\sqrt{3}\tau} dx . \tag{31}$$

In the expressions for $\bar{J}_{POF}$ and $\bar{J}_{FF}$ we identify the probability that a photon emitted at τ escapes from the medium in a single direct flight. First, $e^{-\phi_x\tau/\mu}$ (with the index x = I or x = ν, respectively) gives the probability that a photon of a given frequency escapes along a ray at angle $\cos^{-1}\mu$ from the given point to the boundary of the medium. Second, the average of this probability over frequency, with weight ϕ_x, gives the escape probability for photons along the ray with the emitted frequency distribution. Finally, in the Eddington approximation the average over the directions of one hemisphere is done by just taking $\mu = 1/\sqrt{3}$ as angular quadrature point. Therefore, the integrals of the last terms in Equation (30) (with $\phi_x = \phi_I$) and of Equation (31) (with $\phi_x = \phi_\nu$) give the escape probability for photons emitted in all directions with the given frequency distribution.

In order to interpret physically the exact expression for $\bar{J}$ (cf. Equation (29)) it is first convenient to remember Stepanov's contribution. Stepanov (1958 a, b) diagonalized the absorption matrix in the absence of magneto-optical effects and transformed the transfer equation to a new polarization basis in which the radiation field is characterized by two mutually orthogonal, elliptically polarized beams: the first with an absorption profile $\phi_+ = \phi_I + \omega = (1 + a)\phi_I$ and the second with an absorption profile $\phi_- = \phi_I - \omega = (1 - a)\phi_I$. Therefore, in the exact magnetized case the escape probability which appears in Equation (29) can be interpreted as just the average of the escape probabilities corresponding to each of these two mutually orthogonal, non-interacting elliptically polarized beams.

The above expressions for $\bar{J}$ show that, for a ME atmosphere, both the POF and the FF approximations give the exact result for $\bar{J}$ in atmospheric regions where the optical depth τ is very small or very large. However, such expressions also help to better understand why our POF approximation is, in general, more suitable than the FF one, namely because the POF expression for $\bar{J}$ (Equation (30)) contains the profile ϕ_I which, as can be seen in the exact expression for $\bar{J}$ (Equation (29)), plays a key role in establishing the probability of escape and the way photons get redistributed in the line wings in the presence of magnetic fields.

5. Concluding remarks

By means of NLTE Stokes transfer calculations for a two-level atomic model we have shown that the *field-free* approximation (Rees, 1969) cannot be used safely in the presence of magnetic field gradients. As an alternative, we have developed the *polarization-free* approximation, which is capable of giving a fairly good account of the effect of inhomogeneous magnetic fields on the excitation equilibrium *without* having to solve any Stokes vector transfer equation.

In order to carry out RT calculations with our POF approximation it is *only* neccessary to use, instead of the standard Voigt absorption profile, the profile ϕ_I which appears as the diagonal element of the line-absorption 4×4 matrix Φ. This ϕ_I-profile, which is sensitive to the magnetic field vector, can be easily incorporated in any existing *non-magnetic* scalar *multi-level* RT code, either *one-dimensional* (e.g. Carlsson, 1986; see Bruls and Trujillo Bueno, 1996) or *multi-dimensional* (see Auer, Fabiani Bendicho and Trujillo Bueno, 1994). In this respect, our POF approximation should be particularly useful for investigating the effects of spatially inhomogeneous magnetic fields (with both *vertical* and *horizontal* variations) on the statistical equilibrium of 2D and 3D MHD models of the solar (stellar) atmosphere.

Finally, we think that our *polarization-free* approximation can also be considered as the best *initialization* estimate for the iterative solution of the full NLTE *multi-level* Stokes transfer problem, which can now be done efficiently by means of the application of recently-developed, novel iterative RT schemes (see Trujillo Bueno and Fabiani Bendicho, 1995; Trujillo Bueno, 1995).

Acknowledgements

This work was initiated while the first author of this paper was visiting the Dipartimento di Astronomia e Scienza dello Spazio (Università di Firenze). We are grateful to Maurizio Landi Degl'Innocenti and Marco Landolfi for many fruitful and helpful discussions. Thanks are also due to Jo Bruls for more recent stimulating discussions at the IAC. Partial support by the Spanish DGICYT (project PB 91-0530) and by the EC is gratefully acknowledged. The second author also acknowledges the support of the DGICYT for a sabatical stay at the IAC.

References

Athay, R.: 1972, *Radiation Transport in Spectral Lines*, Dordrecht: Reidel

Auer, L.H., Heasley, J.N., and House, L.L.: 1977, *Astrophys. J.* **216**, 531

Auer, L.H., Fabiani Bendicho, P., and Trujillo Bueno, J.: 1994, *Astron. Astrophys.* **292**, 599

Bruls, J.H.M.J. and Trujillo Bueno, J.: 1996, *Solar Phys.*, this issue

Bommier, V., Landi Degl'Innocenti, E., and Sahal-Bréchot, S.: 1991, *Astron. Astrophys.* **244**, 383

Bommier, V. and Landi Degl'Innocenti, E.: 1996, *Solar Phys.*, this issue

Carlsson, M.: 1986, *A Computer Program for Solving Multi-level Non-LTE Radiative Transfer Problems in Moving or Static Atmospheres*, Report No. 33, Uppsala Astronomical Observatory.

Dolginov, A.Z. and Pavlov, G.G.: 1973, *Astron. Zhurnal* **50**, 762

Domke, H. and Staude, J.: 1973, *Solar Phys.* **31**, 279

Faurobert-Scholl, M.: 1991, *Astron. Astrophys.* **246**, 469

House, L.L. and Steinitz, R.: 1975, *Astrophys. J.* **195**, 235

Landi Degl'Innocenti, E.: 1983, *Solar Phys.* **85**, 3

Landi Degl'Innocenti, E. and Landi Degl'Innocenti, M.: 1985, *Solar Phys.* **97**, 239

Landi Degl'Innocenti, M., Landolfi, M., and Landi Degl'Innocenti, E.: 1976, *Il Nuovo Cimento* **35**, 1

Mihalas, D.: 1978, *Stellar Atmospheres*, Freeman and Company

Rees, D.E.: 1969, *Solar Phys.* **10**, 268

Rees, D.E., Murphy, G.A., and Durrant, C.J.: 1989, *Astrophys. J.* **339**, 1093

Šidlichovský, M.: 1974, *Bull. Astron. Inst. Czech.* **25**, 198

Stenholm, L.J. and Stenflo, J.O.: 1978, *Astron. Astrophys.* **67**, 33

Stenflo, J.O.: 1994, *Solar Magnetic Fields: Polarized Radiation Diagnostics*, Kluwer Academic Publishers

Stepanov, V.E.: 1958a, *Izv. Krimsk. Astrofiz. Obs.* **18**, 136

Stepanov, V.E.: 1958b, *Izv. Krimsk. Astrofiz. Obs.* **19**, 20

Trujillo Bueno, J.: 1995, in preparation

Trujillo Bueno, J. and Fabiani Bendicho, P.: 1995, *Astrophys. J.*, in press

Trujillo Bueno, J. and Landi Degl'Innocenti, E.: 1995, in preparation

THE POLARIZATION-FREE APPROXIMATION APPLIED TO MULTI-LEVEL NON-LTE RADIATIVE TRANSFER

J.H.M.J. BRULS and J. TRUJILLO BUENO

Instituto de Astrofísica de Canarias, La Laguna, Tenerife, Spain

Abstract. The polarization-free (POF) approximation (Trujillo Bueno and Landi Degl'Innocenti, 1996) is capable of accounting for the approximate influence of the magnetic field on the statistical equilibrium, without actually solving the full Stokes vector radiative transfer equation. The method introduces the Zeeman splitting or broadening of the line absorption profile ϕ_I in the scalar radiative transfer equation, but the coupling between Stokes I and the other Stokes parameters is neglected. The expected influence of the magnetic field is largest for strongly-split strong lines and the effect is greatly enhanced by gradients in the magnetic field strength. Formally the interaction with the other Stokes parameters may not be neglected for strongly-split strong lines, but it turns out that the error in Stokes I obtained through the POF approximation to a large extent cancels the neglect of interaction with the other Stokes parameters, so that the resulting line source functions and line opacities are more accurate than those obtained with the field-free approach. Although its merits have so far only been tested for a two-level atom, we apply the POF approximation to multi-level non-LTE radiative transfer problems on the premise that there is no essential difference between these two cases. Final verification of its validity in multi-level cases still awaits the completion of a non-LTE Stokes vector transfer code.

For two realistic multi-level cases (CaII and MgI in the solar atmosphere) it is demonstrated that the POF method leads to small changes, with respect to the field-free method, in the line source functions and emergent Stokes vector profiles (much smaller than for a two-level atom). Real atoms are dominated by strong ultraviolet lines (only weakly split) and continua, and most lines with large magnetic splitting (in the red and the infrared) are at higher excitation energies, i.e. they are relatively weak and unable to produce significant changes in the statistical equilibrium. We find that it is generally unpredictable by how much the POF results will differ from the field-free results, so that it is nearly always necessary to confirm predictions by actual computations.

The POF approximation provides more reliable results than the field-free approximation without significantly complicating the radiative transfer problem, i.e. without solving any extra equations and without excessive computational resource requirements, so that it is to be preferred over the field-free approximation.

Key words: Line formation – Magnetic fields – Polarization – Radiative transfer – Sun: atmosphere – Sun: magnetic fields

1. Introduction

In the framework of multi-level non-LTE radiative transfer in a magnetized atmosphere the field-free approximation (Rees, 1969) is often invoked to avoid having to deal with the possible influence of the magnetic field on the statistical equilibrium of a multi-level atom. Contrary to general belief, this approximation is not well-founded; to our knowledge its accuracy has only been assessed for one particular case, namely the CaII H & K resonance lines

Solar Physics **164**: 155–168, 1996.

© 1996 *Kluwer Academic Publishers.*

(Auer *et al.*, 1977). And even in that case only homogeneous magnetic fields were considered.

Computations for a two-level atom (Trujillo Bueno and Landi Degl'Innocenti, 1996) have shown that the magnetic field can have a non-negligible influence on the statistical equilibrium, especially in the presence of magnetic field gradients. The magnetic field most sensitively affects the statistical equilibrium through a strongly-split strong line.

Determining the exact impact of a magnetic field requires consistent solution of the multi-level non-LTE polarized radiative transfer and statistical equilibrium equations, which is a difficult problem. Although this type of problems can be solved via the application of recently-developed very efficient iterative schemes (Trujillo Bueno and Fabiani Bendicho, 1995), it remains a difficult problem.

As an intermediate step towards this consistent solution Trujillo Bueno and Landi Degl'Innocenti (1996) proposed the polarization-free (POF) approximation, which takes into account in an approximate manner the influence of the magnetic field on the statistical equilibrium without solving the full Stokes vector transfer equation. This is done by incorporating into the conventional scalar radiative transfer equation the magnetic splitting of the line absorption profile ϕ_I, while still neglecting the interaction of Stokes I with the other Stokes parameters (Q, U, V).

Relying on its proven accuracy for a two-level atom, this paper discusses the implementation of the POF approximation in Carlsson's (1986) radiative transfer code MULTI and subsequent application to realistic multi-level radiative transfer problems. It is set up as follows: Section 2 outlines the principles of the POF method, Section 3 discusses the essentials of its implementation in MULTI, Section 4 shows a few examples of applications and Section 5 summarizes the results and discusses the usefulness of the method.

2. Principles of the POF approximation

The formulation of the polarization-free approximation starts from the I-component of the Stokes vector radiative transfer equation (see Rees *et al.*, 1989, for the notation):

$$\frac{dI}{dz} = (\kappa_c S_c + \kappa_0 S_L \phi_I) - (\kappa_c + \kappa_0 \phi_I)I - \kappa_0(\phi_Q Q + \phi_U U + \phi_V V). \quad (1)$$

For many cases, in particular for weakly-split lines and for not too strong lines in general, Stokes Q, U and V are one or two orders of magnitude smaller than I and the absorption coefficients ϕ_X $(X = \{Q, U, V\})$ are at most of the same order as ϕ_I, so that we may safely neglect the last term in this equation. Neglecting that term, but still preserving the magnetic field dependence of ϕ_I leads to the polarization-free equation of transfer (Trujillo Bueno and Landi Degl'Innocenti, 1996). That equation only involves the I-component of the Stokes vector, it takes into account the magnetic splitting pattern of the line absorption profile ϕ_I, and it reduces to the standard non-magnetic radiative transfer equation in case of zero fields. It reads:

$$\frac{dI_{\mathrm{PF}}}{dz} = (\kappa_c S_c + \kappa_0 S_L \phi_I) - (\kappa_c + \kappa_0 \phi_I) I_{\mathrm{PF}}, \tag{2}$$

with the total line absorption profile

$$\phi_I = \frac{1}{2} \phi_p \sin^2 \gamma + \frac{1}{4}(\phi_b + \phi_r)(1 + \cos^2 \gamma). \tag{3}$$

In the latter formula, ϕ_p, ϕ_b and ϕ_r signify the generalized absorption profiles for the different polarization states and γ is the angle between the magnetic field vector and the ray under consideration.

For strongly-split strong lines, however, the POF approximation seems equally inappropriate as the field-free approach, since then the interaction of Stokes I with the other Stokes parameters should be strong and the last term of Equation (1) becomes significant. Fortunately, as shown by comparison with the exact solution for a two-level atom (Trujillo Bueno and Landi Degl'Innocenti, 1996), inclusion of the magnetic splitting of the line absorption profile (in the scalar radiative transfer equation) introduces an error in I that largely cancels the missing interaction with Stokes Q, U and V, so that the final result for the line source function and line opacity is significantly better than with the field-free approach.

3. Implementation in a multi-level radiative transfer code

We have implemented the modified expression for the line absorption profile of the POF approximation in version 2.1 of Carlsson's (1986) radiative transfer code MULTI. Relatively few modifications to the existing code were necessary and only minor amounts of additional code needed to be written:

- The input routine ATOM now needs to read the level Landé factors g and orbital quantum numbers S, L and J (with an option to compute g assuming LS-coupling).
- A new routine (RDFIELD) was introduced to read the magnetic field configuration and interpolate it to the appropriate depth grid.
- The line absorption profile, variable PHI, instead of being a single Voigt profile now consists of the sum of several Voigt functions each with their own weight and wavelength shift. This sum is computed by means of a slightly modified version of the generalized Voigt function routines of the

Stokes Profile and Synthesis Routine (SPSR) of Murphy and Rees (1990). This change affects routines PROFIL and TAUNYQ.

- In MULTI the frequency quadrature points are close and equidistant in frequency in the first few Doppler widths from the line core (input parameter Q0) and beyond that they are equidistant in the logarithm of the frequency. This assures that the integrals in the rate equations (Scharmer and Carlsson, 1985, Equation (3.20)) are computed fairly accurately (but see Stift and Moser (1993) on the use of adaptive frequency grids). Due to magnetic splitting the line profile widths increase and one has to modify the frequency grid accordingly to obtain accurate representation of all absorption components. Without resorting to adaptive frequency grids one generally requires an increase of Q0 that depends on the splitting characteristics and wavelength of the line and on the maximum expected value of the magnetic field strength. An increase of the number of frequency points (NQ) may then be needed in order to avoid too large spacing of the grid points.

- The CPU-time per iteration is dominated by the time required to update the radiation field and the rate matrix. This time scales linearly with the total number of angle-frequency points in all transitions, so that occasional large increases of NQ (for long-wavelength lines) do not excessively burden the computations. A magnetic field makes the line absorption profiles direction-dependent (just as a macroscopic velocity does), but it also requires the sum of a number of Voigt profiles to be computed. The initialization of the profiles therefore becomes considerably more time-consuming, but it remains only a small fraction of the total CPU-time. The direction-dependence of the line profiles and the associated additional read operations do not markedly increase the CPU-time per iteration.

The current implementation of the POF approximation in MULTI is valid for vertical magnetic fields, for which case the line absorption profile only depends on the inclination of the line of sight, i.e. the μ value of the angle-quadrature points. For inclined fields the line absorption profile also depends on the azimuth angle of the line of sight. A future paper will deal with the details of how to treat inclined fields. In any case, it can be shown (e.g. Trujillo Bueno, 1995; in preparation) that the field strength plays a much more important role than the inclination, so that the use of vertical fields is not a significant limitation here.

Computations for a simplified model atmosphere and a two-level atom served to test the implementation of the method. The results can be summarized as follows (see Trujillo Bueno and Landi Degl'Innocenti (1996) for more details). A depth-independent magnetic field has only very limited influence on the line source function S_L: it starts to drop below the Planck function B_ν slightly deeper in the atmosphere than without magnetic field.

Once the field is strong enough to completely split the line into its Zeeman components, further increase of B has no effect on S_L: the components of the line behave as if they were completely independent lines of comparable strength and the net effect is simply a small inward shift of their formation; the line profiles are only slightly affected. Gradients in B, in particular the ones that cause significant changes of the line splitting within the line formation region (e.g. unsplit at one end and completely split at the other end), produce significantly larger line source function and line profile changes: in this case there is a competition between increased line photon losses due to decreased line opacity (line split into several components) and trapping of continuum photons due to the depth-dependence of the wavelength shifts of the individual line components. Provided the gradients occur in the line formation region, strong outward decrease of B causes S_L to drop below its field-free value and strong outward increase produces an enhancement of S_L. Mixed or even completely opposite behavior may occur if the regions of line formation and large field strength gradients only partially coincide.

4. Applications

Below we briefly show application of the POF approximation to two realistic multi-level atoms. One should bear in mind that the findings for a two-level atom need not always hold for multi-level cases: interlocking line systems may even produce opposite behavior for some lines. In such cases a strong gradient in the field strength may cause a source function decrease in one line and an increase in another. Additionally, ultraviolet lines, which nearly always dominate the overall excitation equilibrium, are relatively insensitive to magnetic splitting, so that the influence of the magnetic field on the level populations is expected to be much smaller than for a two-level atom.

Although not explicitly shown here, the strength of the field is much more important than its inclination (Domke and Staude, 1973; Trujillo Bueno, 1995, in preparation); in practice this means that one may perform a non-LTE solution for a vertical field of suitable strength and afterwards use its results to perform a formal Stokes vector solution for any line of sight and any desired orientation of the magnetic field vector, even for configurations with a gradient in the magnetic field inclination.

4.1. APPLICATION TO CaII LINES

We start with this application, because the validity of the field-free approximation has only been demonstrated for the CaII atom in combination with a depth-independent magnetic field (Auer et al., 1977). Significant sensitivity of the statistical equilibrium to the magnetic field is not to be expected for

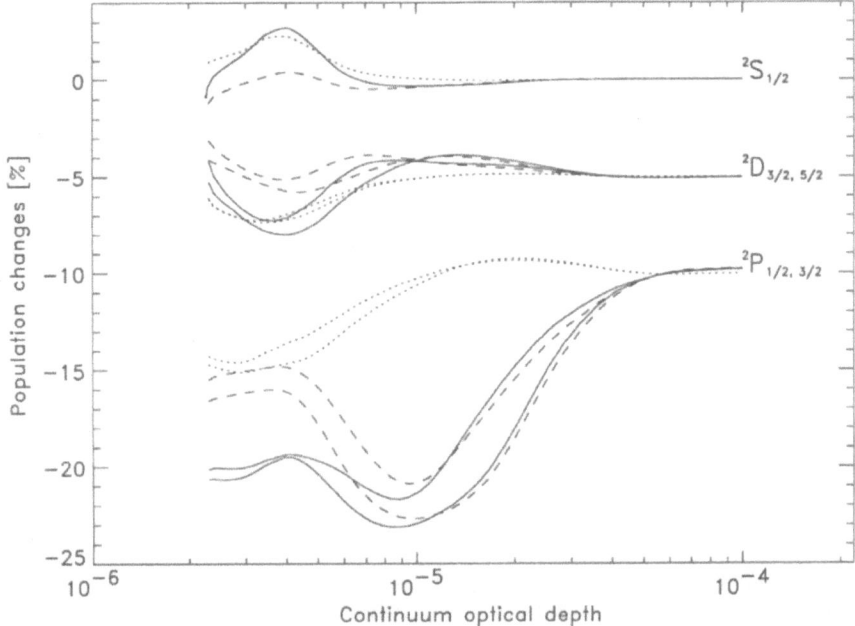

Fig. 1.   Level population changes, in a quiet Sun model atmosphere, due to a uniform magnetic field of 3000 G, relative to the field-free populations. The curves for the ^{2}D (^{2}P) levels have been shifted down by 5 (10) percent to avoid confusion. Three different cases are compared: all five lines subject to magnetic splitting (solid lines), magnetic splitting only for the H & K lines (dotted), and magnetic splitting only for the IR triplet lines (dashed)

this case, because these lines are only very weakly split. This weak split— very wide lines in the ultraviolet—implies that the field-free approach should be quite accurate.

We employ the standard 5-level Ca$^+$ atom, with the H & K resonance doublet and the IR triplet lines, together with the Ca^{2+} ground state, which suffices to accurately describe the non-LTE statistical equilibrium. We assume complete frequency redistribution of photons in order to demonstrate the role of the magnetic field, but partial frequency redistribution of photons needs to be taken into account for a more correct description of the resonance lines. Following Auer *et al.*, (1977), we first experimented with rather strong depth-independent fields in a quiet Sun model (Maltby *et al.*, 1986), without accounting for magnetic pressure effects.

Figure 1 shows the population changes of all five Ca$^+$ levels, induced by a magnetic field of 3000 G, relative to their field-free populations. Figure 1 not only displays the population changes for the case that all five lines are subject to magnetic splitting (case A, solid lines), but also the ones that

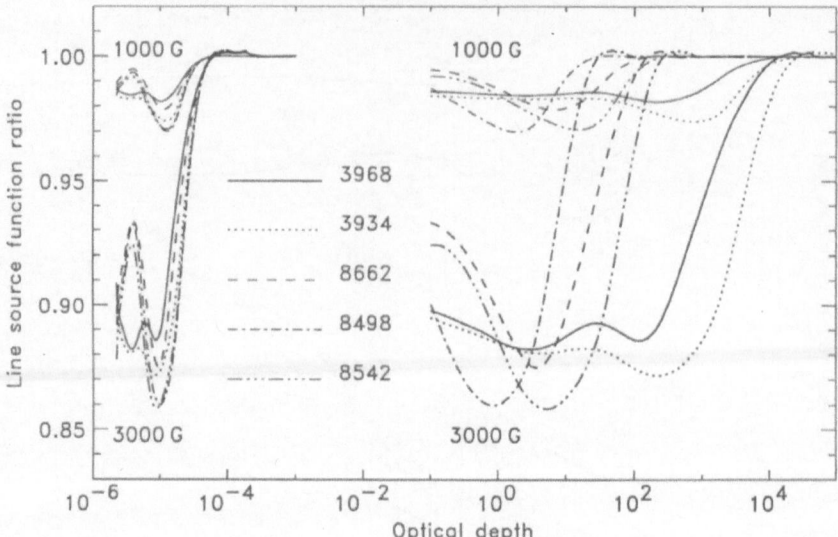

Fig. 2. POF line source functions in a quiet Sun model atmosphere, for uniform magnetic fields of 1000 and 3000 G, relative to the field-free line source functions. Left part: as functions of the standard continuum optical depth at 5000 Å; right part: as functions of the respective (unsplit) line center optical depths

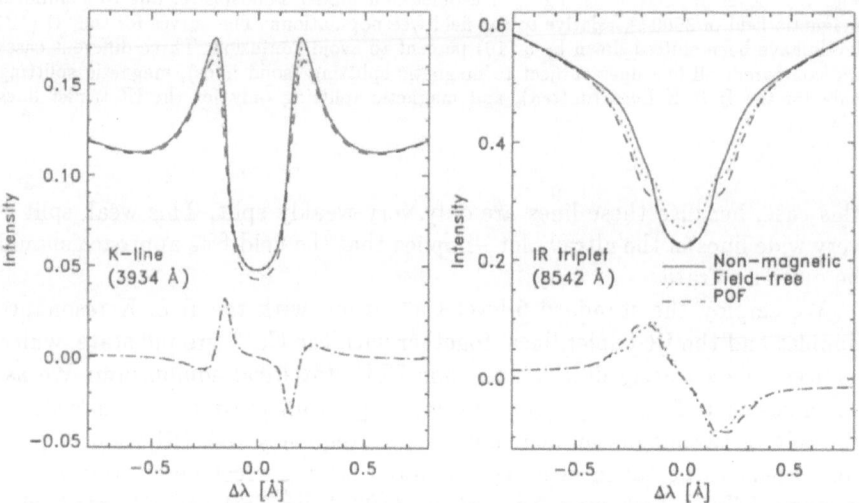

Fig. 3. Stokes I and V profiles of the K-line and the strongest member of the IR triplet for a quiet Sun model atmosphere. Only the center parts of the lines are shown and the intensities are normalized to the local continuum values. The solid lines represent the profiles for $B = 0$; the dotted and dashed lines represent the Stokes vector formal solutions obtained respectively from field-free and POF non-LTE statistical equilibrium computations for a LOS inclination of 45° (from vertical) and a 3000 G field with 45° inclination with respect to the LOS

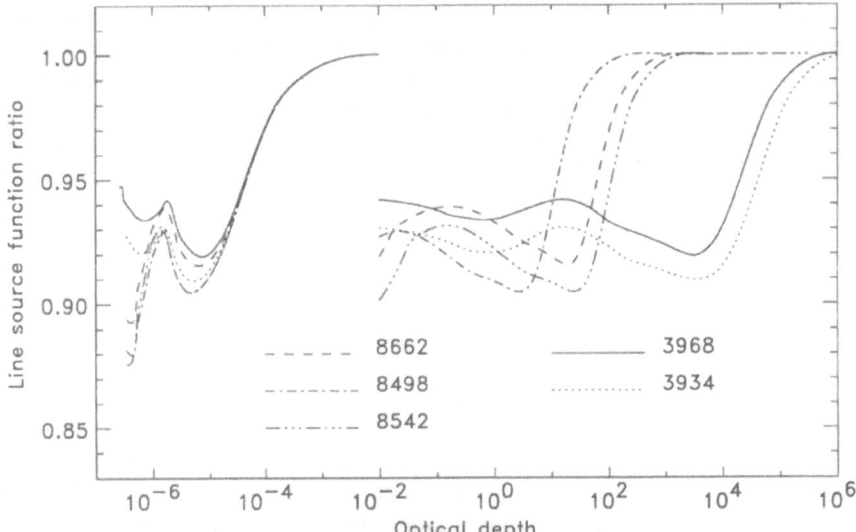

Fig. 4.   POF line source functions, relative to their field-free values, for an umbra model
atmosphere, with a magnetic field of 3000 G at $\tau_c = 1$ and an outward decrease of 3 G/km.
Left part: as functions of the standard continuum optical depth; right part: as functions
of the respective (unsplit) line center optical depths

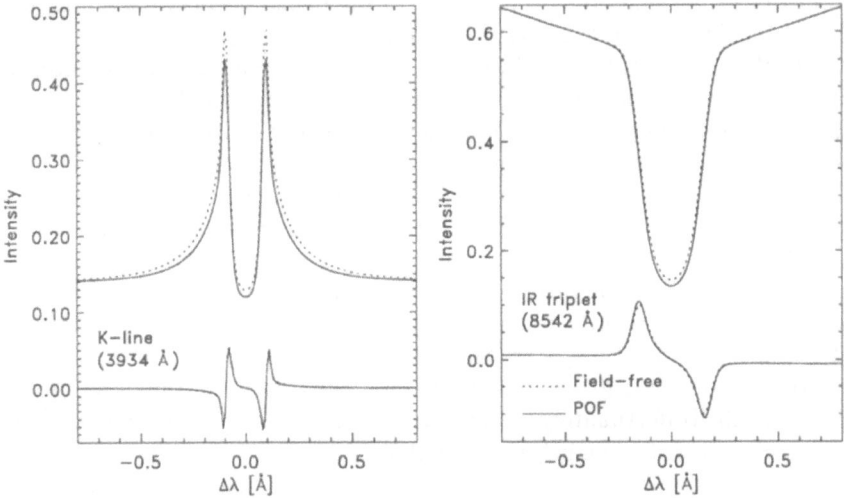

Fig. 5.   Stokes I and V profiles of the K-line and the strongest member of the IR triplet
for an umbra model atmosphere, with a magnetic field of 3000 G at $\tau_c = 1$ and an outward
decrease of 3 G/km. Only the center parts of the lines are shown and the intensities are
normalized to the local continuum values. The dotted and solid lines represent the Stokes
vector formal solutions obtained respectively from field-free and POF non-LTE statistical
equilibrium computations for a LOS inclination of 45° (from vertical) and a field with 45°
inclination with respect to the LOS

result if only the resonance lines are subject to magnetic splitting (case B, dotted lines) or if only the IR triplet lines are split (case C, dashed lines). This shows that the population changes in case A are equal to the sum of the changes in cases B and C, and that the largest contribution comes from the IR triplet lines; only where those become optically thin (in the extreme upper atmosphere) does the contribution of the resonance lines become comparable to that of the triplet.

The line opacities, determined by the populations of the lower level of each line, are much less affected than the line source functions (Figure 2), which are determined by the ratios of the upper and lower level populations. By far the largest changes occur in the populations of the $4p^2P$ levels, the upper levels of all five lines, so that all line source functions undergo virtually the same changes. For comparison, Figure 2 also shows the significantly smaller line source function changes due to a field of 1000 G.

Since the line source functions only change in the upper atmosphere, largest line profile changes are to be expected for oblique lines of sight (LOS). Figure 3 therefore shows a comparison between field-free and POF profiles for the K-line and the strongest IR triplet line (8542 Å) obtained through a formal Stokes vector solution for a 45° inclination of the line of sight and a 3000 G field with 45° inclination with respect to the LOS. These profiles are also compared to the line profile that results in the absence of any field, from which they differ only very little. For the K-line the maximum difference between field-free and POF results occurs near the K_2 peaks and it is at most about 1% of the continuum intensity or about 6% percent of the intensity at those wavelengths. The degrees of linear and circular polarization (Stokes Q, U and V) differ only by a few tenths of a percent. For the triplet lines the profile differences are enhanced due to their longer wavelengths and near the line core they amount to about 3% of the continuum intensity or 12% of the actual core intensity. The differences in Stokes V are somewhat smaller, but most important is the shift in the position of the peaks.

Two-level atom computations show that gradients in the magnetic field can produce significantly larger changes in the populations. A realistic application would be a superpenumbral canopy, but unless one assumes its location extremely high in the atmosphere, the strong CaII lines are not affected at all. Unfortunately, a high location implies a smaller field strength jump, so that again the impact is limited.

Another obvious case with a field-strength gradient is provided by an umbra; there one may expect a magnetic field gradient $\partial B/\partial z$ of several Gauss per kilometer to exist over a significant height interval. We use the Maltby et al. (1986) umbra model M and assume a field gradient of 3 G/km. Figure 4 shows the line source function changes for this case, as function of continuum optical depth and of non-magnetic line-center optical depth. The differences are very similar in character to the ones for a uniform magnetic field, but the maximum deviations from the field-free values are only about 9% as compared to 12% for the above case with uniform field of 3000 G. The line profile differences (Figure 5) are nevertheless larger than in the previous case and they are now more pronounced for the K-line than for the IR triplet, up to 8% difference in Stokes I at the K_2 peaks; the linear and circular polarization do not change significantly.

We conclude that the field-free approach is sufficiently accurate for the CaII H & K lines in case the magnetic field is homogeneous. In the presence of magnetic field gradients, especially for oblique lines of sight, a better approximation, such as the polarization-free approach, is desirable. The IR triplet lines call for a more accurate method in the presence of homogeneous magnetic fields, but are already described accurately by the field-free approach for the umbra field model used here.

4.2. APPLICATION TO MgI 12 μM LINES

The interest in this application is based on the presence of a number of lines in the near and far infrared, some of them rather strong, that as a whole contribute significantly in setting the statistical equilibrium of neutral magnesium. Due to the nature of their emission, the 12 μm lines are very sensitive to even small changes in the populations that may be induced by magnetic fields. The 12 μ lines themselves are weak and will not markedly influence the statistical equilibrium, even though they are completely split at field strengths of only a few hundred Gauss. We use the full 66-level model atom of Carlsson et al. (1992), with a few typing error corrections as noted by Bruls et al. (1995). Given that the 12 μm lines disappear in umbrae, penumbrae are a prime choice here. We performed non-LTE radiative transfer computations using the line-blanketed radiative equilibrium model atmosphere T5000 (Kurucz, 1991) with effective temperature $T_{eff} = 5000$ K, and a magnetic field of 1500 G at the 12 μm line formation height ($\tau_c \approx 10^{-3}$) with a gradient of 3 G/km, which is close to the maximum observed penumbral field strength gradient.

Differences between the field-free and POF results can best be expressed in terms of line source function ratios. Figure 6 shows these ratios for the 12.32 μm line and for the lines identified by Carlsson et al. (1992, Figure 12) as the ones to which the 12.32 μm emission is most sensitive. In this case line source function changes (top panel) are more instructive than population changes (bottom panel), because the role of stimulated emission increases with wavelength: small population changes may induce large line source function changes. Population changes of all neutral Mg levels, except for a few low-excitation ones which are irrelevant, are very similar, but the 12.32 μm line source function is clearly more enhanced than the ones of the driving lines, which all have shorter wavelengths.

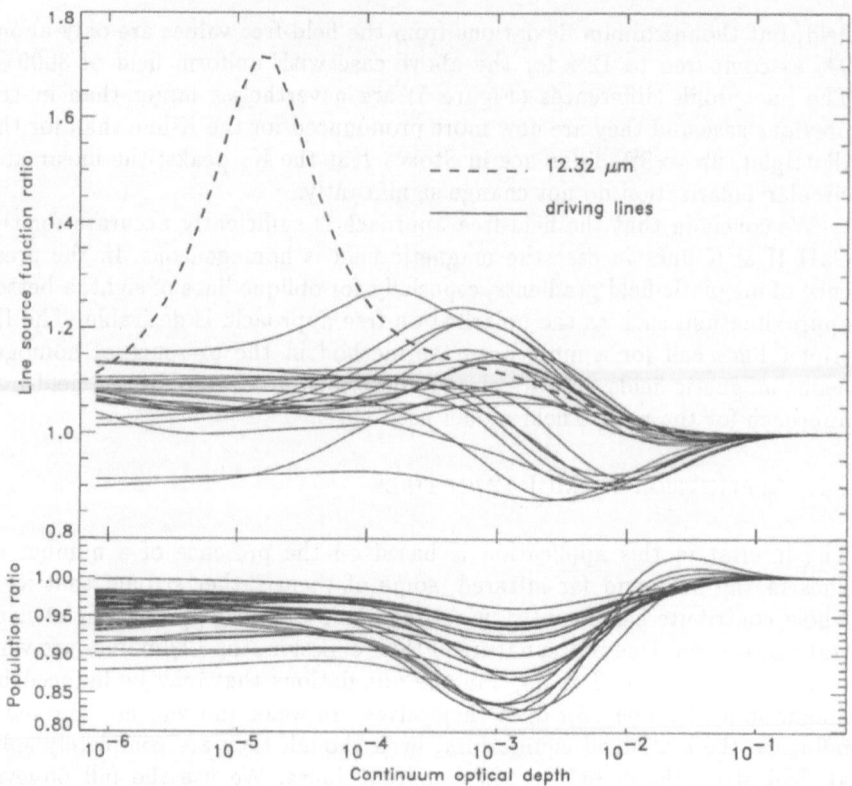

Fig. 6. Line source functions, relative to their field-free values, for the T5000 model atmosphere and a magnetic field of 1500 G at $\tau_c = 10^{-3}$ with a gradient of 3 G/km. The dashed line refers to the MgI 12.32 μm line source function and the solid ones to the source functions of the lines to which the amount of emission in the 12.32 μm line is most sensitive (Carlsson *et al.*, 1992). The lower part of the figure shows the populations, relative to their field-free values, for all but the lowest 8 levels of Mg which behave slightly differently

Figure 7 displays the profile of the 12.32 μm and 9412 Å lines for a 80° inclined line of sight, to exploit the increase of the source function differences with height. The 9412 Å line (6f–3d) is one of the few observable ones of the 'driving lines'. The behavior displayed by both lines is typical of the whole set of lines: small changes in all Stokes parameters, but no remarkable changes of the profile shapes. For the 12.32 μm lines the effect on Q and V seems to be comparable to a multiplication by a factor close to unity, whereas for the 9412 Å line the differences are more subtle, including a slight wavelength shift of the Stokes V peak positions. Given the significant changes of the 12.32 μm line emission, diagnostic applications that rely on the absolute

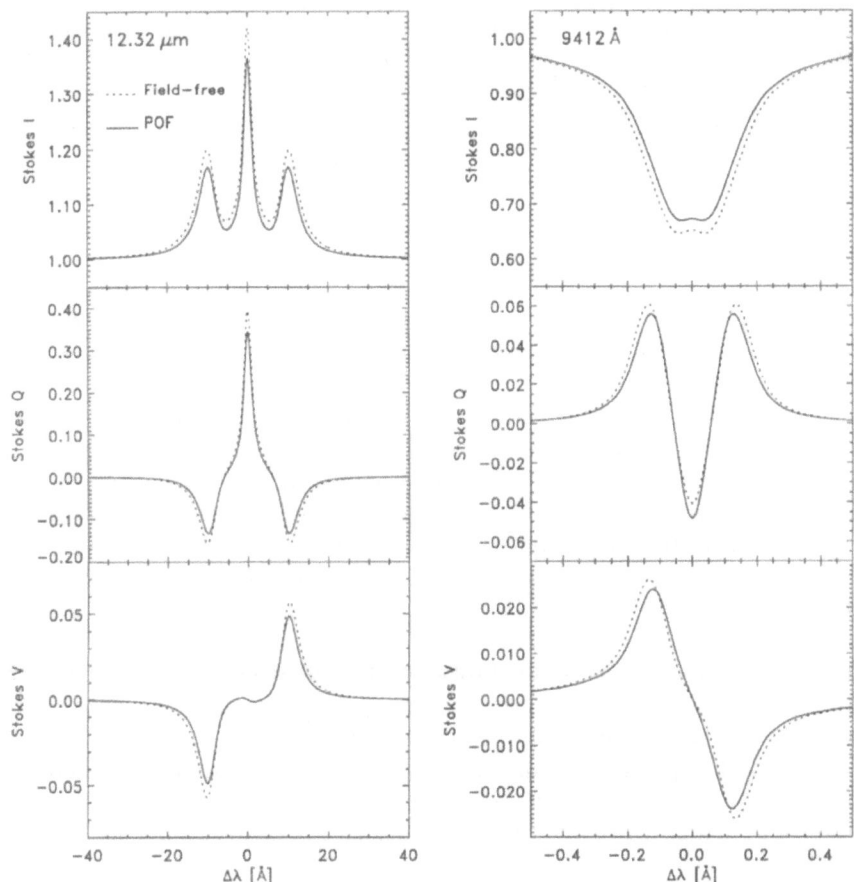

Fig. 7. Stokes I, Q and V profiles for the 12.32 μm line (left column) and for the 6f–3d 9412 Å line, one of the most important 'driving lines' and one of the few that are actually observable (Wallace *et al.*, 1993), for the T5000 model atmosphere and a magnetic field of 1500 G at $\tau_c = 10^{-3}$ with a gradient of 3 G/km. The dotted and solid lines represent the Stokes vector formal solutions obtained respectively from field-free and POF non-LTE statistical equilibrium computations for a LOS inclination of 80° (from vertical) and a field with 80° inclination with respect to the LOS

amount of emission in this line should be analyzed taking into account the magnetic field.

For a canopy type field configuration (in this case $B = 800$ G above $\tau_c = 10^{-3}$ and $B = 0$ below that height) populations change by no more than 5% due to the magnetic field. The 12.32 μm line source function increases by up to 5%, and even for a 45° inclined LOS the Stokes Q and V peak values, which are small anyway, increase by 5% only. The size of the emission peaks in Stokes I, however, increases by about 20%, which is easily measurable on observed profiles.

5. Discussion

We have implemented the polarization-free approximation, which describes in an approximate way the influence of a magnetic field on the radiative transfer, in Carlsson's (1986) multi-level non-LTE radiative transfer code. This required relatively few modifications to the existing code and the performance is only slightly degraded: the initialization of the line absorption profiles is more involved and has to be performed for each grid point of the angle-frequency quadrature instead of only once per frequency point. In addition, a slightly higher total number of frequency points is generally needed in the presence of a magnetic field, with significant increases only required for lines in the far infrared. The CPU-time per iteration scales linearly with the number of frequency points and the convergence rate is not affected.

Furthermore, since this approximation can also easily be implemented in 2-D or 3-D multi-level radiative transfer codes, such as the one recently developed by Auer *et al.* (1994), it provides a fast and sufficiently reliable means of computing non-LTE Stokes profiles as long as a fully consistent 3-D multi-level non-LTE Stokes vector transfer code is not available, and even beyond that time it may prove to be the preferred method.

The most important reason why for realistic model atoms the differences between the field-free and the polarization-free approximation are so small, is given by the presence of strong (interlocking) ultraviolet lines that are crucial in setting the statistical equilibrium. Those lines are rather insensitive to magnetic fields and they also strongly suppress population changes induced by more sensitive (infra)red lines. This property is inherent to multi-level atoms and has nothing to do with the polarization-free approximation: consistent solution of the Stokes vector radiative transfer and statistical equilibrium equations, which will ultimately be necessary to evaluate the accuracy of the polarization-free approximation, will not change this situation.

The applications to realistic multi-level radiative transfer problems show that the differences between POF and field-free line profiles, though much smaller than for the two-level atom case, are still important enough to warrant a critical look and that the decision whether or not to use POF instead of field-free depends on each particular problem considered.

We finally conclude that the present work presents a valuable tool that for many problems may turn out to be preferable above consistent Stokes vector solutions (due to its simplicity and accuracy), but that it also supports the validity of the field-free approximation, for the CaII and MgI problems we have studied here, given a reasonable error margin.

Acknowledgements

Partial support of the Spanish DGICYT (project PB 91-0530) is gratefully acknowledged.

References

Auer, L. H., Fabiani Bendicho, P. and Trujillo Bueno, J.: 1994, *Astron. Astrophys.* **292**, 599

Auer, L. H., Heasley, J. N. and House, L. L.: 1977, *Astrophys. J.* **216**, 531

Bruls, J. H. M. J., Solanki, S. K., Rutten, R. J. and Carlsson, M.: 1994, *Astron. Astrophys.* **293**, 225

Carlsson, M.: 1986, A Computer Program for Solving Multi-Level Non-LTE Radiative Transfer Problems in Moving or Static Atmospheres, Report No. 33, Uppsala Astronomical Observatory

Carlsson, M., Rutten, R. J. and Shchukina, N. G.: 1992, *Astron. Astrophys.* **253**, 567

Domke, H. and Staude, J.: 1973, *Solar Phys.* **31**, 279

Kurucz, R. L.: 1991, in L. Crivellari, I. Hubeny and D. G. Hummer (eds.), Stellar Atmospheres: Beyond Classical Models, NATO ASI Series C-341, Kluwer, Dordrecht, p. 441

Maltby, P., Avrett, E. H., Carlsson, M., Kjeldseth-Moe, O., Kurucz, R. L. and Loeser, R.: 1986, *Astrophys. J.* **306**, 284

Murphy, G. A. and Rees, D. E.: 1990, Operation of the Stokes Profile Synthesis Routine, NCAR Technical Note NCAR/TN−348+IA, High Altitude Observatory, Boulder

Rees, D. E.: 1969, *Solar Phys.* **10**, 268

Rees, D. E., Murphy, G. A. and Durrant, C. J.: 1989, *Astrophys. J.* **339**, 1093

Scharmer, G. B. and Carlsson, M.: 1985, *J. Comput. Phys.* **59**, 56

Stift, M. J. and Moser, G.: 1993, *Astron. Astrophys.* **268**, 617

Trujillo Bueno, J. and Fabiani Bendicho, P.; 1995, *Astrophys. J.* in press

Trujillo Bueno, J. and Landi Degl'Innocenti, E.: 1996, *Solar Phys.* this issue

Wallace, L., Hinkle, K. and Livingston, W.: 1993, An Atlas of the Photospheric Spectrum from 8900 to 13600 cm^{-1} (7350 to 11230 Å), NSO Technical Report # 93-001, National Solar Observatory, NOAO, Tucson

STOKES PROFILES INVERSION TECHNIQUES

J.C. DEL TORO INIESTA and B. RUIZ COBO

Instituto de Astrofísica de Canarias, E-38200 La Laguna, Tenerife, Spain

e-mail: jti@iac.es & brc@iac.es

Abstract. Inversion techniques of the radiative transfer equation for polarized light are presented as one of the best current procedures to infer the vector magnetic field, as well as other quantities governing the physical state of the atmospheric layers that photons are coming from. Several characteristics of the various available inversion procedures are pointed out. They are mostly based on the diagnostic contents of the spectral lines as well as on the main hypotheses assumed in these procedures. In particular, the role of gradients in the atmospheric quantities is emphasized as of paramount importance in any diagnostic analysis and, hence, in any interpretation of inversion results.

Key words: Spectropolarimetry – Magnetic field diagnostics – Radiative Transfer

1. Introduction

Nobody can measure physical quantities of the solar atmosphere. Although this statement may sound too strong, it is a common characteristic of the whole of Astrophysics. By language abuse we indistinguishably employ the word *measurement* for both the process of really measuring the physical properties of the radiation coming from the solar surface and that of *inferring* relevant solar physical parameters. We cannot measure solar temperatures, magnetic fields, or velocities because we do not have thermometers, magnetometers, or speedometers that can directly probe the solar atmosphere. The only *ruler* we have is the radiative transfer equation (RTE). Therefore, the physical meaning of the results depends on the use we make of this *ruler* and on the available information. By the use of the RTE we mean the set of assumptions, simplified models or solutions which help to interpret, in some cases, the observations in terms of physical parameters of the solar atmosphere. Obviously, the available information is either the four Stokes parameters measured as functions of wavelength or any function of them (the Stokes parameters). For example, everybody knows that the wavelength separation between the Stokes V-peaks of a completely Zeeman split spectral line is a good estimate of the magnetic field strength *if and only if* this is constant through the whole atmosphere. What happens, then, when the hypotheses fail and/or no additional observational constraints are considered? (see a related topic in an interesting discussion between full Stokes profiles analysis and magnetograph-type measurements in Lites, Martínez Pillet, & Skumanich 1994; see also Ruiz Cobo & del Toro Iniesta 1994 and

Sánchez Almeida, Ruiz Cobo, & del Toro Iniesta 1995). In this regard, we should recognize that what we usually call *measurements* would be better described as *estimates* and the associated *errors* as *uncertainties*.

Usually, by opposition to the so-called synthesis approach, we mean by inversion the process of finding the physical quantities which, under a given set of hypotheses, are able to generate Stokes profiles that reproduce the observations. In this broad sense, in which inversion and inference are equated, any *measurement* technique can be considered an inversion procedure. In particular, the sequential synthesis of Stokes profiles with varying model atmospheres until a best-fit to the observed profiles is reached could be called a *manual* inversion. The community, however, has agreed in calling inversion techniques (hereafter referred to as ITs) those procedures that *automatically* compute as an output a model atmosphere with which synthetic Stokes profiles match to or fit, up to a given degree of accuracy, the observed ones. We would like to add another major characteristic of ITs in their definition, namely, their aim of *self-consistently* determining as many solar physical quantities as possible, within the framework of an explicitly established set of hypotheses. It is precisely in self-consistency that the reliability of the results of these techniques is firmly rooted.

In order to deal with several atmospheric quantities at the same time, ITs are obliged to pursue an optimum description of the involved (nonlinear) dependence of the Stokes profiles on each of such quantities governing the physical state of the atmosphere.

2. Perturbation Analysis of the RTE

Determination of the nonlinear dependences of the observed Stokes spectrum on the various atmospheric quantities is the cornerstone of any interpretation of solar observations. Linearization of the RTE is, perhaps, the most straightforward way to attack this problem. In the field of polarized radiative transfer, such linearization was carried out for the first time by Landi Degl'Innocenti & Landi Degl'Innocenti (1977), following former analyses for non-polarized light (the original idea can be found in the "weight functions" introduced by Mein 1971). As a consequence of such a linearization analysis, it can easily be proven that the modifications, $\delta\mathbf{I}$, of the observed Stokes spectrum, $\mathbf{I} \equiv (I, Q, U, V)$, can be expressed in terms of the perturbations, δx_i, to the m physical quantities, $\{x_i\}_{i=1,\dots,m}$:

$$\delta\mathbf{I} = \sum_{i=1}^{m} \int_0^\infty \mathbf{R}_i(\tau)\, \delta x_i(\tau)\, d\tau \tag{1}$$

(Ruiz Cobo & del Toro Iniesta 1994), where the response functions (hereafter referred to as RFs), $\mathbf{R}_i$, as functions of the optical depth, τ, have the explicit definition (Sánchez Almeida 1992)

$$\mathbf{R}_i(\tau) = \mathbf{O}(0,\tau)\left[\mathbf{K}(\tau)\frac{\partial \mathbf{S}}{\partial x_i} - \frac{\partial \mathbf{K}}{\partial x_i}(\mathbf{I}(\tau) - \mathbf{S}(\tau))\right]. \tag{2}$$

In Equation (2), $\mathbf{K}$ is the 4×4 absorption matrix, $\mathbf{S}$ is the source function vector, and $\mathbf{O}$ is the 4×4 evolution operator (Landi Degl'Innocenti & Landi Degl'Innocenti 1985). RFs provide the sensitivity of the observed Stokes profiles to each of the atmospheric parameters relevant to line formation, up to a first-order approximation, as shown by Equation (1). Noteworthy of this equation is the fact that a given $\delta\mathbf{I}$ can conceivably be produced by many different $\delta x_i(\tau)$ combinations. Therefore, any failure to consider the effects of some of the x_i's can induce significant errors in the remainder. This is the reason why the simultaneous account for the different atmospheric parameters is so important (see §2.2). Special strategies (observational and/or numerical) have been developed to this aim (§4).

2.1. EVALUATION OF THE STOKES PROFILES SENSITIVITIES

In a Milne-Eddington (ME) atmosphere, the calculation of RFs poses no difficulties since both the evolution operator and the derivatives in Equation (2) are analytic (recall that the solution of the RTE is analytic as well). This is not, unfortunately, the case in more general situations (Landi Degl'Innocenti 1987). The absence of analytic expressions for $\mathbf{O}$ and $\mathbf{K}$ hinders the RFs calculation enough for them not to have been numerically evaluated until recently (Ruiz Cobo & del Toro Iniesta 1992, 94). These authors use the $\mathbf{O}$-matrix as provided through the DELO (Rees, Murphy, & Durrant 1989) integration of the RTE. Figure 1 shows, as an example, the RFs of Stokes V (also shown in the figure) to temperature at wavelengths ± 12 pm apart from line center (solid and dashed lines, respectively), as calculated in a model atmosphere described in Equation (10) of the paper by Ruiz Cobo & del Toro Iniesta (1992). The interpretation of these RFs is easy: an enhancement of temperature in deep layers produce larger absolute V-values at $\Delta\lambda = \pm 12$ pm; the opposite occurs if such temperature enhancement is applied to high atmospheric layers. We will speak about the differences between the two RFs of Figure 1 in §5.2, where the importance of atmospheric gradients is discussed.

A simple approach for evaluating the $\mathbf{R}_i$'s is provided by Equation (1). If we just deal with a given quantity, x_k, and $\delta x_k(\tau) = \delta(\tau - \tau_0)$ (Dirac's delta, whose physical meaning would be a perturbation in a small interval around τ_0), then $\delta\mathbf{I} = \mathbf{R}_k(\tau_0)$. This calculation should thus be repeated for all the physical quantities and for each of the depth grid points. One can

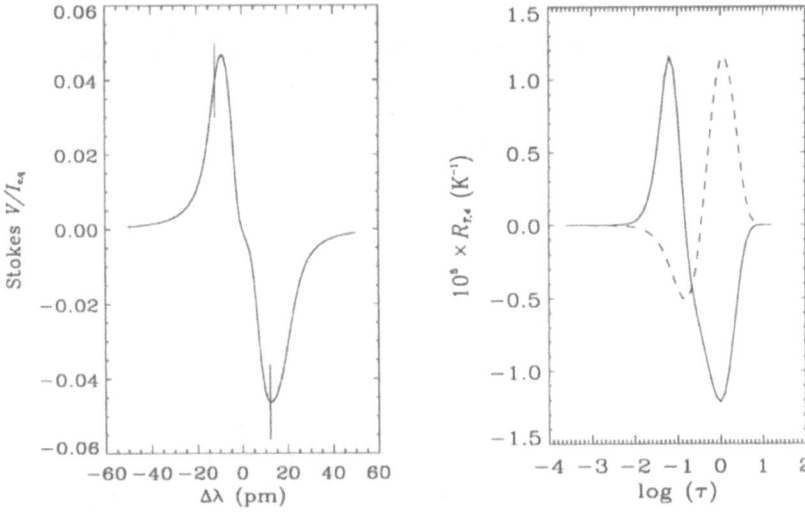

Fig. 1. Left panel: Stokes V-profile of the Fe I line at 6302.5 Å in units of the quiet Sun continuum intensity, $I_{c,q}$. Right panel: RFs of V to temperature variations at the wavelengths marked with vertical bars in the left panel ($\Delta\lambda = +12$ pm, solid line; $\Delta\lambda = -12$ pm, dashed line).

easily conceive that such procedure (sometimes used, e.g., Wittmann 1974, and for non-polarized light analyses, e.g., Grigoryeva, Turova, & Teplitskaya 1991) is enormously tedious and precludes most authors from employing it. Rather, in order to have an idea about how the Stokes profiles *feel* the variations of temperature or another atmospheric parameter, some researchers use a prescribed grid of models (e.g., Keller et al. 1990; Solanki, Rüedi, & Livingston 1992), or interpolations between two "standard" models (Solanki, Montavon, & Livingston 1994, Bernasconi et al. 1995), among which that parameter is found. This approach certainly simplifies the diagnostics, provided that some more physics has been assumed for constraining the results. Special care, however, is needed when prescribing the grid of models so that it covers a broad enough range of parameter variations.

2.2. ACCOUNTING FOR PHYSICAL PARAMETERS

The number of atmospheric parameters whose determination a given IT is aimed at depends on its specific hypotheses (see §3). Of course, observations impart constraints to such number as well. On the one hand, the number of observables (e.g., wavelength samples) should be significantly larger than the free parameters of the fit. On the other hand, a given physical quantity cannot be inferred if its modifications do not induce observable variations on the spectrum.

Once the precise set of sought unknowns is established, the simultaneous account for the sensitivities of the Stokes spectrum to the physical quantities is crucial, according to Equation (1). An illustrative example of the risks one can run by sequentially searching and locking each of the unknowns is presented in Figure 2. In it, the Stokes V-profile of the Fe I line at 6302.5 Å has been synthesized (open circles) with the E-model atmosphere by Maltby et al. (1986) with a magnetic field strength $B = 1200$ G and a macroturbulence velocity $v_{\mathrm{mac}} = 1.0$ km s^{-1}(open circles); this profile has been fitted (solid line) by sequentially calculating T, B, and v_{mac}. The resulting model differs from the former (at all optical depths) by $\delta T = +300$ K, $\delta B = -215$ G, and $\delta v_{\mathrm{mac}} = +0.69$ km s^{-1}. No apparent differences can be discerned between the two profiles although the two models are by no means the same. Some doubts may be raised about the ability of a single V-profile to simultaneously constrain T, B, and v_{mac}. However, the results obtained with this very profile after using an inversion technique like that by Ruiz Cobo & del Toro Iniesta (1992, see §3) differ from the original model (irrespective of the initial guess) some 20 times less than the results from the sequential fit. This indicates that this single V-profile contains, in fact, the required information.

3. Hypotheses on the RTE and/or the Model Atmosphere

Typically, the inversion of the RTE is a nonlinear procedure for minimizing the squared differences between observed and synthetic Stokes profiles. The seminal paper by Auer, Heasley, & House (1977) laid down the basis for such minimization. The Marquardt algorithm (e.g. Press et al. 1990) they used is still in use by all the current ITs. Their method was based on Unno's (1956) solution of the RTE. It thus assumed that a Milne-Eddington model is able to describe the basic thermodynamics, as well as that the magnetic field vector and the velocity along the line-of-sight (LoS) are constant with depth.

The technique was generalized by Landolfi, Landi Degl'Innocenti, & Arena (1984) to account for magneto-optics [Rachkovsky's (1962) solution of the RTE] and damping effects. A similar extension was later carried out by Skumanich & Lites (1987). These authors, besides dealing with magneto-optics and damping, avoided the normalization of Stokes profiles and include a free parameter for the influence of scattered light, and/or the unresolved character of the observations, to be fully accounted for. This technique (hereafter referred to as Unno-fitting) is currently being used as the routine analysis procedure for the data obtained with the Advanced Stokes Polarimeter

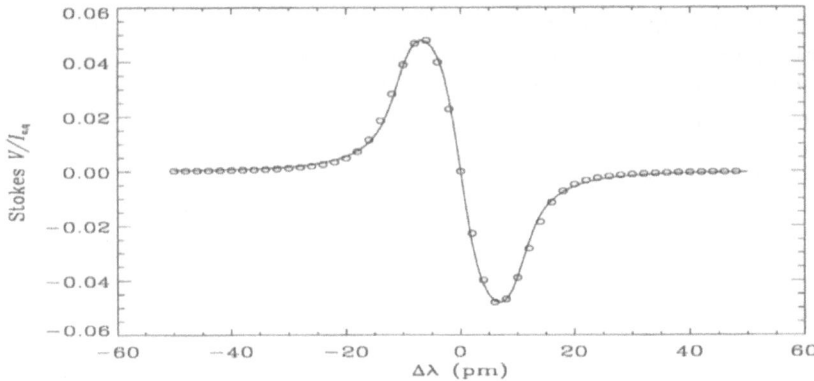

Fig. 2.   Stokes V-profiles of the Fe I line at 6302.5 Å synthesized with the E-model atmosphere by Maltby et al. (1986) with $B = 1200$ G and $v_{mac} = 1.0$ km s^{-1}(open circles); and with a model differing from it by $\delta T = +300$ K, $\delta B = -215$ G, and $\delta v_{mac} = +0.69$ km s^{-1}(solid line).

(ASP) of the High Altitude Observatory (see Skumanich, Lites, & Martínez Pillet 1995, for a summary of the main results obtained). The technique was later extended to account for some NLTE effects by Lites et al. (1988).

The calculation of the RF of **I** to velocity fluctuations in a ME atmosphere led Landi Degl'Innocenti & Landolfi (1983) to draw important hints about recovering velocity gradients from observations. The qualitative discussion by Landolfi, Landi Degl'Innocenti, & Arena (1984) on the possible influence of magnetic and velocity gradients crystallized in the proposal by Landolfi (1987) of a new IT in which information about gradients is extracted from the residuals between observed and Unno-fitting profiles. In this method, estimates of the magnetic vector gradients and of the LoS velocity gradient are obtained separately.

Increasing efforts to develop ITs that include a more comprehensive integration of the RTE than the ME solution have been appearing in the literature during the last five years. Keller et al. (1990) built up an inversion procedure intended to reproduce selected observables calculated from the Stokes V-profile and not to reproduce the full profile itself. In the technique proposed by Solanki, Rüedi, & Livingston (1992), the temperature stratification is inferred among 5 fixed models from the literature. Instead of fitting some observables, they use full Stokes I- and V-profiles. Besides the temperature, they try to infer single-valued magnetic field strengths and inclinations, and a stray-light (filling factor) parameter. No variation of the LoS velocity with depth is considered. A slight modification of this procedure has been used by Solanki, Montavon, & Livingston (1994) and by Bernasconi et al. (1995), in which, instead of using pre-fixed temperature stratifications, the $\log \tau$ dependence of T is allowed to be obtained by interpolation between two closely parallel models.

Ruiz Cobo & del Toro Iniesta (1992) proposed a new IT (which we will hereafter call SIR, an acronym of Stokes Inversion based on Response functions). This technique allows us to obtain the stratification through the atmosphere of all the physical quantities considered relevant to the line formation. Explicitly, the output model atmosphere contains the run with depth of T, P_e, the electronic pressure, v, the LoS velocity, B, γ, ϕ, and v_{mic}, plus v_{mac} which is assumed to be constant with depth. Of course, the reliability of the results is limited to those layers from which physical information is available; in other words, from those layers in which RFs are significantly different from zero. As a matter of fact, RFs permit to estimate the uncertainties of the results.

Figures 3 and 4 illustrate the quality of the fits and the output model obtained after the application of SIR to a single point of a full ASP-map of a sunspot with the pair of Fe I lines at 6301.5 and 6302.5 Å. This point is located in the inner penumbra of the spot. The Unno-fitting results for B, γ, and ϕ were used as initialization. The error bars are located at the *nodes* of the inversion (particular depth grid points), whose number vary for each of the atmospheric quantities. The formal error of microturbulence, which is assumed constant with depth, is less than 1% of the value derived and, thus, not distinguishable in the plot. Notice the large error bars at the outer boundaries of the photosphere for all the parameters other than temperature. They correspond to the insensitivity of Stokes profiles to these quantities at these layers. Information from the whole atmosphere is kept, however, even when only values at the nodes are sought (see the concept of equivalent RFs introduced in Appendix A of Ruiz Cobo & del Toro Iniesta 1992). This is the reason, for instance, for the fact that, although the insensitivity of the Stokes profiles to the temperature of the very bottom layers, the formal error of T at the deepest node ($\log \tau = 1.2$) is not larger than 100 K. If such uncertainty were larger, the gradient of T at the following node (at $\log \tau = -0.25$), where sensitivity is significant, would be altered. See §5.1 for details.

4. Disentangling the Effects of Different Physical Quantities

The combined and involved effects of the several physical quantities certainly hinder the success of ITs, mainly aimed at inferring estimates of them all. For example, in some instances, an increase of temperature and a decrease of magnetic field strength may have the same observable consequences. Equation (1) can be considered as an analytic demonstration of such a problem

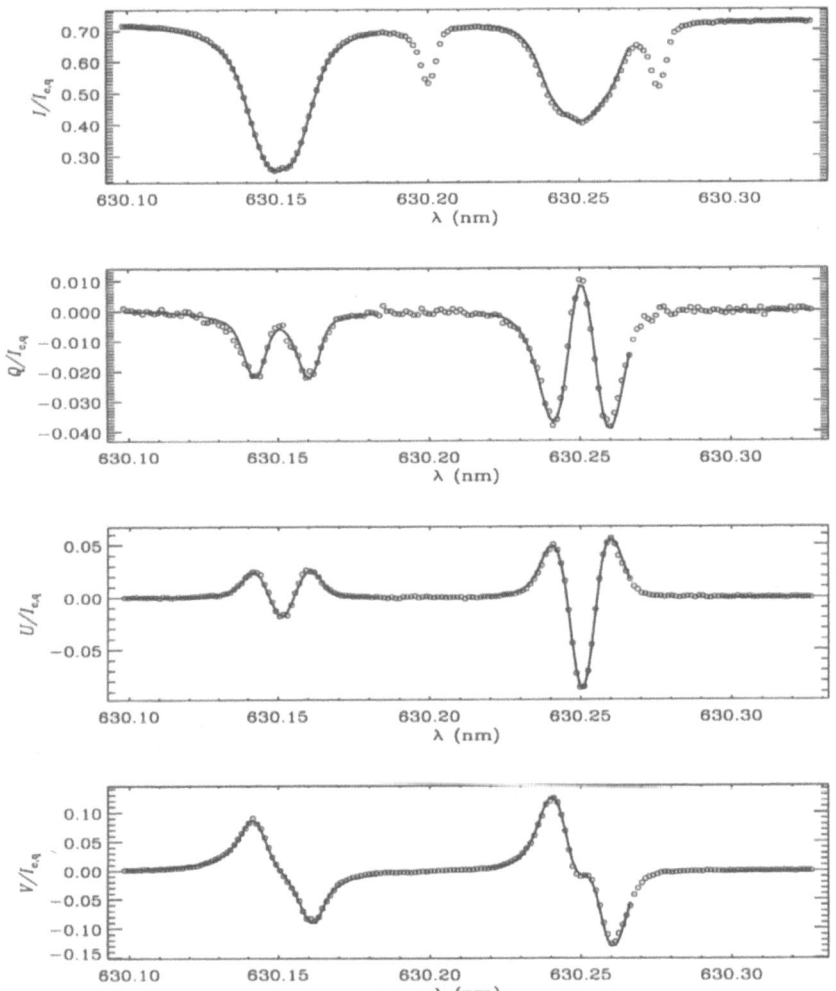

Fig. 3. Stokes profiles, normalized to the quiet Sun continuum intensity, of the Fe I lines at 6301.5 and 6302.5 Å as observed with the ASP (open circles) and as a result of the SIR technique (solid line).

if the summation over the m physical parameters is not accounted for. The real challenge, thus, of any IT is to disentangle these effects.

As mentioned in the preceding section, all the current ITs are based on the Marquardt algorithm. The algorithm lies in the solution of a linear system of equations and, hence, in the inversion of a matrix, the covariance matrix, whose elements are given by products of first-order derivatives of

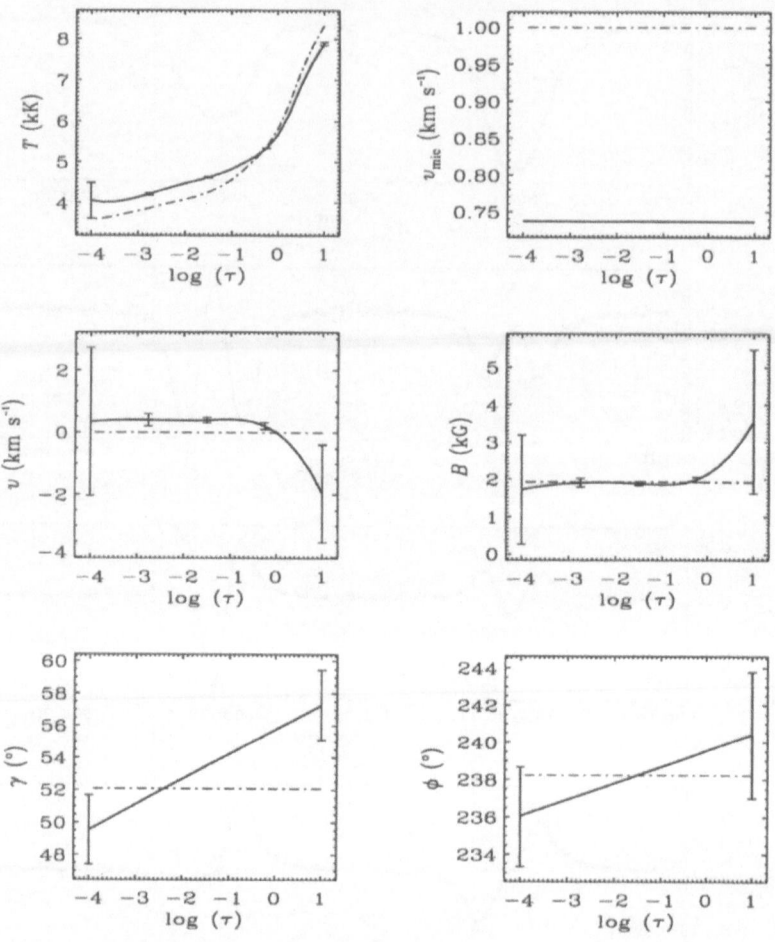

Fig. 4. Output model atmosphere from SIR, after inversion of the Stokes profiles of Figure 3 (solid lines). The initialization is plotted in dashed-dotted lines (Unno-fitting values for the magnetic parameters).

χ^2 (products of RFs in SIR). On the one hand, this matrix is often ill-conditioned (i.e. quasi-singular). Because of the problem described in the paragraph above, some rows can be close to proportional. On the other hand, those rows containing null sensitivities (for instance, the RF of I to B in layers below, say, $\log \tau = 0$) are null. This problem has been overcome with one of two possible strategies, namely, the use of a well selected set of spectral lines (as in Unno-fitting or in the techniques by Keller et al. 1990 and by Solanki, Rüedi, & Livingston 1992) or of a modified version of the Singular Value Decomposition technique for inverting singular matrices (as described in Appendix B of Ruiz Cobo & del Toro Iniesta 1992). The observational approach roots in enhancing the diagnostics without increasing the number of free parameters of the fit. This is reached, for instance, within the Unno-fitting technique by simultaneously using pairs of lines like Fe I at 6301.5 and 6302.5 Å or Mg I *b* at 5172.7 and 5183.6 Å. Each pair belongs to a given multiplet so that no more free parameters are needed for the ME model insofar as the ratio of oscillator strengths of the two lines are properly known, and the Doppler widths and damping parameters are the same for the two lines (Lites et al. 1988). The more sophisticated, numerical approach used by SIR ensures versatility of the IT since no specific observations are required to properly disentangle the influence of the different physical quantities. The number of free parameters is only limited by the diagnostic contents of the data. (See further details in the original paper –Ruiz Cobo & del Toro Iniesta 1992).

5. The Importance of Gradients in Stokes Profiles Diagnostics

Atmospheric gradients are present everywhere in the Sun. The first, intuitive strategy for obtaining information about them is the use of several spectral lines, under the assumption that they are formed in distinct layers. However, spectral lines do not form at specific narrow layers. Instead, the sensitivities (RFs) of a given spectral line to the various atmospheric quantities are rich, varied, and span a fairly broad range of depths. Since the RFs depend on the atmosphere the line has been generated through, the diagnostic content of a given spectroscopic parameter varies when it is measured in different solar structures; for instance, in the inner or the outer parts of the penumbra (del Toro Iniesta, Tarbell, & Ruiz Cobo 1994). The concept of *height of formation* of a spectral line is meaningless (Ruiz Cobo & del Toro Iniesta 1994; Sánchez Almeida, Ruiz Cobo, & del Toro Iniesta 1995). Far from being a drawback in the interpretation of Stokes profiles, the presence of gradients provides a priceless help for probing broader ranges of optical depths in the atmosphere (see, e.g., the top panel of Figure 12 by del Toro Iniesta, Tarbell, & Ruiz Cobo 1994). Advantages in diagnostics come out after properly taking care of gradients (del Toro Iniesta, & Ruiz Cobo 1995). The aim of this section is to further our understanding of why and how we can learn after including atmospheric gradients into spectropolarimetric analyses. In particular, we try to emphasize that a single spectral line *does* contain much more information than is usually expected.

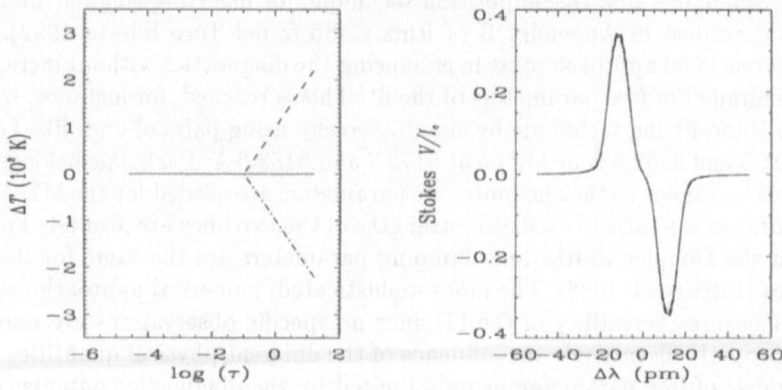

Fig. 5.   Left panel: Differences in the temperature stratification with respect to that of the HSRA model (Gingerich et al. 1971) of the three models with which the Stokes profiles are synthesized. The remaining atmospheric quantities are the same for the three. $B = 2500$ G. Right panel: Stokes V-profile of the Fe I line at 6302.5 Å in units of the **local** continuum intensity. Line types correspond to left panel.

5.1. Gradients of Temperature

Temperature is by far the most important quantity in line formation. The RFs of the Stokes profiles to temperature are larger than to any other quantity (Ruiz Cobo & del Toro Iniesta 1994). As a matter of fact, the very existence of lines in the spectrum is due to temperature gradients, as far as spectral lines are formed in LTE. Unfortunately, this precious information is sometimes lost or obliterated.

Either because of observational requirements (as in FTS or in scanning-spectrographs cases), or because of inherited tradition, it is customary to deal with Stokes profiles normalized to the *local* continuum intensity. If possible, normalization should be avoided, mainly if we want to infer temperatures. This assertion is fairly well illustrated by the example of Figure 5. We have synthesized the Fe I line at 6302.5 Å with the HSRA model (Gingerich et al. 1971) –solid line, and other two whose temperature stratifications differ from the former as shown in the left panel of the figure. The remaining atmospheric parameters are the same. In particular, $B = 2500$ G throughout the photosphere. The differences in temperature practically are not reflected into the Stokes V-profiles, once normalized to the *local continuum intensity* (right panel of the figure). Prior to normalization, such V-profile differences are approximately 4 times larger. Of course, this is a specific, spatially resolved case. Other Fe I lines, with lower excitation potential, could show different Stokes profiles even after normalization. This is the reason why most authors claim for the need of using a number of spectral lines, simultaneously observed, for obtaining reliable estimates of physical quantities. Such multiline analyses with normalized Stokes profiles, however, just allow to better constrain the upper photospheric layers. The information about temperature gradients around $\tau = 1$, where the continuum is mainly formed, is severely reduced. Extremely unrealistic models can result from normalized line profiles as that for penumbrae by Ding & Fang (1989); see a discussion in del Toro Iniesta, Tarbell, & Ruiz Cobo (1994). The improvements found by Skumanich & Lites (1987) in the stability of convergence of their IT as compared with that of Auer, Heasley, & House (1977) is a good example in favor of our arguments. They avoid normalization and have the gradient of the ME source function as a free parameter, by using an average of the nearby quiet Sun continuum as a standard candle.

5.2. Gradients of the Remaining Physical Quantities

Magnetic field gradients, as well as those of any atmospheric quantity other than temperature, also leave their tracks on the observed Stokes spectrum so that we can expect to be able to infer them. In this section, we want to point out two basic hints concerning the inference of such gradients: the importance of a simultaneous $T(\tau)$ inference, and the invaluable diagnostic provided by LoS velocity gradients.

Diagnostics about parameters like B, γ, ϕ, and v are only available from layers above, say, $\tau = 1$. This can easily be understood after a glance at Equation (2). Only the second term on the right hand side applies to the RFs of Stokes profiles to variations of such quantities. Since $|\mathbf{I} - \mathbf{S}| \longrightarrow 0$ as $\tau \longrightarrow \infty$, these RFs drop to zero much faster than RFs to temperature variations. Also conspicuous from Equation (2) is the paramount role of $T(\tau)$ in the diagnostics of the remaining parameters, since $\mathbf{O}(0, \tau)$, $\mathbf{K}(\tau)$, $\mathbf{I}(\tau)$, and $\mathbf{S}(\tau)$ obviously depend on T. This also applies to single-valued estimates of these quantities since the heights at which the measurements can be ascribed strongly depend on the precise model atmosphere (Sánchez Almeida, Ruiz Cobo, & del Toro Iniesta 1995).

The useful signatures that a gradient of the LoS velocity leave in the Stokes profiles are not widely exploited yet. This is an unfortunate situation because such gradients double the number of observables from single spectral lines to be matched. Since the symmetries of the absorption matrix are broken down, each point through the profile contains information about different atmospheric properties. A clear demonstration of this effect can be found in Figure 1, where the two RFs shown are different, although they correspond to symmetric wavelengths with respect to line center. Had the velocity gradient been absent from the model atmosphere, they would have been identical but of opposite sign. As a matter of fact, when no velocity gradients are present one could, in principle, neglect half of the profiles.

Magnetic gradients, acting alone, are unable to alter the symmetry properties of Stokes profiles. Nevertheless, they produce a differential broadening of the Stokes profiles which can be detected. Besides, their combined action with velocity gradients significantly modify the asymmetries. Therefore, the importance of simultaneous determinations of all the physical quantities is again to be stressed both for single- or multi-line analyses.

6. Concluding Remarks

Should a given physical parameter not alter the observed spectrum it would remain undetectable from it. Hence, the inference of atmospheric quantities obviously depends on how their perturbations are able to induce observable modifications of the Stokes profiles. Inversion techniques can be classified into three groups according to the three approaches devised to account for the relationships between atmospheric and observable parameters. For the Unno-fitting technique, such relationships are analytic at the expense of ME assumptions. For the present structure of the codes proposed by Keller et al. (1990) and by Solanki, Rüedi, & Livingston (1992), they are obtained after the analysis of a prescribed grid of models. We have shown that RFs explicitly provide the sensitivity of the spectrum to perturbations of the atmosphere, within the LTE approach and up to first-order approximation. RFs are at the basis of the SIR technique.

One of the most important features supplied by SIR to Stokes diagnostics (preceeded by the technique by Landolfi 1987) is the possibility to deal with atmospheric gradients. Their importance has been emphasized in the present paper. The concept of height of formation of a given spectral line has been shown to be meaningless (Ruiz Cobo & del Toro Iniesta 1994; Sánchez Almeida, Ruiz Cobo, & del Toro Iniesta 1995). Therefore, the gradients inferred by using single-valued estimates from several spectral lines can result in large uncertainties. This problem worsens if the Stokes profiles are normalized to the local continuum intensity. Far from being a drawback in the interpretation of Stokes profiles, the presence of gradients provides a priceless help for probing broader ranges of optical depths in the atmosphere. For example, the loss of the symmetry properties of Stokes profiles, due to a LoS velocity gradient, doubles the number of observables to be interpreted. A single spectral line contains a richer information than is usually expected.

Nobody can measure physical quantities of the solar atmosphere, but we believe to have shown that inversion techniques can be considered as a good step forward within the set of inferential techniques.

Acknowledgements

M. Collados, L.R. Bellot Rubio, and C. Westendorp Plaza carefully read the manuscript and suggested modifications. We also thank E. Landi Degl'Innocenti for his accurate suggestions. The analysis of ASP data shown in §3 is part of the thesis work of C. Westendorp Plaza within a joint IAC/HAO project. This work has been funded by the Spanish DGICYT (project PB 91-0530) and CICYT (project ESP95-1023-E).

References

Auer, L.H., Heasley, J.N., & House, L.L.: 1977, *Solar Phys.* **55**, 47.
Bernasconi, P.N., Keller, C.U., Povel, H.P., & Stenflo, J.O.: 1995, *Astron. Astrophys.* in press.
Ding, M.D. & Fang, C.: 1989, *Astron. Astrophys.* **225**, 204.
Gingerich, O., Noyes, R.W., Kalkofen, W., & Cuny, Y.: 1971, *Solar Phys.* **18**, 347.
Grigoryeva, S.A., Turova, I.P, & Teplitskaya, R.B.: 1991, *Solar Phys.* **135**, 1.
Keller, C.U., Solanki, S.K., Steiner, O., & Stenflo, J.O.: 1990, *Astron. Astrophys.* **238**, 583.
Landi Degl'Innocenti, E. 1987, in W. Kalkofen (ed.), Numerical Radiative transfer, Cambridge University Press, Cambridge, p. 265
Landi Degl'Innocenti, E. & Landi Degl'Innocenti, M.: 1977, *Astron. Astrophys.* **56**, 11.
Landi Degl'Innocenti, E. & Landi Degl'Innocenti, M.: 1985, *Solar Phys.* **97**, 239.
Landi Degl'Innocenti, E. & Landolfi, M.: 1983, *Solar Phys.* **87**, 221.
Landolfi, M.: 1987, *Solar Phys.* **109**, 287.
Landolfi, M., Landi Degl'Innocenti, E., & Arena, P.: 1984, *Solar Phys.* **93**, 269.
Lites, B.W., Martínez Pillet, V., & Skumanich, A.: 1994, *Solar Phys.* **155**, 1.
Lites, B.W., Skumanich, A., Rees, D.E., & Murphy, G.A.: 1988, *Astrophys. J.* **330**, 493.
Maltby, P., Avrett, E.H., Carlsson, M., Kjeldseth-Moe, O., Kurucz, R.L., & Loeser, R.: 1986, *Astron. Astrophys.*, **306**, 284.
Mein, P.: 1971, *Solar Phys.* **20**, 3.
Press, W.H., Flannery, B.P., Teukolsky, S.A., & Vetterling, W.T.: 1990, *Numerical Recipes*, Cambridge University Press.
Rachkovsky, D.N.: 1962, *Izv. Krymsk. Astrofiz. Obs.* **28**, 259.
Rees, D.E., Murphy, G.A., & Durrant, C.J.: 1989, *Astrophys. J.* **339**, 1093.
Ruiz Cobo, B. & del Toro Iniesta, J.C.: 1992, *Astrophys. J.* **398**, 375.
Ruiz Cobo, B. & del Toro Iniesta, J.C.: 1994, *Astron. Astrophys.* **283**, 129.
Sánchez Almeida, J.: 1992, *Sol. Phys.* **137**, 1.
Sánchez Almeida, J., Ruiz Cobo, B., & del Toro Iniesta, J.C.: 1995, *Astron. Astrophys.*, submitted.
Skumanich, A. & Lites, B.W.: 1987, *Astrophys. J.* **322**, 473.
Skumanich, A., Lites, B.W., & Martínez Pillet, V.: 1995, in N. Mein & S. Sahal (eds.), *La polarimétrie, outil pour l'étude de l'activité magnétique solaire et stellaire*, p. 115. Observatoire de Paris.
Solanki, S.K., Rüedi, I., & Livingston, W.: 1992, *Astron. Astrophys.* **263**, 339.
Solanki, S.K., Montavon, C.A.P., & Livingston, W.: 1994, *Astron. Astrophys.* **283**, 221.
del Toro Iniesta, J.C., Tarbell, T.D., & Ruiz Cobo, B.: 1994, *Astrophys. J.* **436**, 400.
del Toro Iniesta, J.C. & Ruiz Cobo, B.: 1995, in N. Mein & S. Sahal-Bréchot (eds.), *La polarimétrie, outil pour l'étude de l'activité magnétique solaire et stellaire*, p. 127. Observatoire de Paris.
Unno, W.: 1956, *Pub. Astron. Soc. Japan* **8**, 108.
Wittmann, A.: 1974, *Solar Phys.* **35**, 11.

CONTRIBUTION FUNCTIONS FOR POLARIZED
RADIATIVE TRANSFER

J. STAUDE

Astrophys. Institut Potsdam, Sonnenobservatorium Einsteinturm, D-14473 Potsdam,
Germany

Abstract. The concepts of contribution functions (CF) and of mean depths of line formation of unpolarized light as well as of Stokes profiles will be critically discussed. After having outlined the historical development arguments are given in favour of the use of directly observable quantities such as the emergent line intensity or the polarized components seen through polarization optics only. The arguments are provided by a probability interpretation of the CF; the ambiguities of line depression CF as well as some physically strange features in Stokes profiles are avoided if the rules based on this interpretation are observed. Some problems of the interpretation of measurements in chromospheric lines will be discussed as well.

Key words: Contribution functions – Polarization – Solar spectrum

1. Introduction

Remote sensing of the physical state of the solar (or a stellar) atmosphere is provided by spectroscopic diagnostics, that is, the information searched for is encoded in the observable spectrum of the electromagnetic radiation, in spectral line profiles in particular, and must be extracted from there. The demand for full information, including that on anisotropic structures such as magnetic fields, requires to analyze the complete state of polarization of the radiation field described by the four Stokes parameters (I, Q, U, V).

In order to approach this 'translation problem' of information, for several decades astrophysicists have been looking for methods to determine the contribution of different atmospheric layers to the observed radiation. Before 1969 all of these efforts were focused on unpolarized light, but most of the methods have subsequently been generalized and applied to polarized radiation in lines. Without claiming completeness, a few steps of this development will be outlined here:

More than six decades ago Unsöld (1932) introduced the concept of 'weighting functions' for the special case of weak absorption lines; it has been generalized to arbitrarily strong absorption lines by Pecker (1951) and de Jager (1952) and applied to the emergent intensity as well as to the relative line intensity with respect to the neighbouring continuum, that is to the line depression.

The weighting functions are closely related to the contribution functions (CF), the latter including additionally the effect of line absorption and thus

Solar Physics **164**: 183–190, 1996.
© 1996 *Kluwer Academic Publishers.*

comprising the whole integrand of the solution of the radiative transfer problem for the emergent intensity (see Equation (1)): the CF describe the contribution of different optical depths τ_λ in the solar atmosphere to the intensity $I_\lambda(\mu)$ emergent from the solar surface at a given wavelength λ in the line into a direction $\mu = \cos\theta$ (where θ is the angle between the vertical direction and the line of sight). The CF form the basis for defining a mean depth of line formation $\bar{\tau}_\lambda$ (a more detailed definition of CF and of $\bar{\tau}_\lambda$ is given in the subsequent section).

While the definition of a CF for the emergent line intensity is unique, this is not the case for the CF of the line depression which is a linear combination of intensities (difference between continuum and line intensity): integrations by parts result in different integrands which provide different forms of CF but the same emergent line depression after integration. Already three decades ago that ambiguity has provoked lively discussions about the 'correct' or most suitable form of a CF for the line depression (e.g., Elste, 1955, 1963; Gussmann, 1967; Ruhm and Elste, 1969); Van't Veer (1966) even doubted any physical meaning of $\bar{\tau}_\lambda$. The discussion has been continued until recently (Magain, 1986; Achmad et al., 1991).

In the meantime another approach has been provided: a response function (RF) describes the response of the observed intensity to modifications of a special physical quantity described by the linear variations with respect to that quantity. This method has been used by Mein (1971) already before the name RF was introduced in papers by Beckers and Milkey (1975) and by Caccin et al. (1977) generalizing the concept of RF.

A first approach to a $\bar{\tau}_\lambda$ for polarized line radiation is due to Rachkovsky (1969). He concluded that 'the line polarization is formed in higher layers than those where the intensity originates'. This has been doubted by Staude (1971 b), because $\bar{\tau}_\lambda$ for the total intensity I on the one hand, and for the Stokes parameters (Q, U, V) on the other hand have been determined by different methods. He introduced a probability concept for the contribution of different layers to polarized radiation (Staude, 1972); that concept will be outlined in the following section in order to give a foundation for the interpretation and discussion in the present paper.

In many subsequent papers the various concepts of CF and RF for unpolarized light have been generalized and applied to polarized line radiation: Grigorjev and Katz (1975) did this for the CF, Wittmann (1974) and Landi Degl'Innocenti and Landi Degl'Innocenti (1977) for the RF, and so on. Grossmann-Doerth et al. (1988) adapted the line depression CF of Magain (1986), and recently Solanki and Bruls (1994) have adapted the CF of Achmad et al. (1991) to Stokes profiles.

2. The Probability Interpretation of the Contribution Functions

The specific intensity of radiation $I_\lambda(\mu)$ is given by

$$I_\lambda(\mu) = \int_0^\infty \exp(-\tau_\lambda/\mu) S_\lambda(\tau_\lambda) d\tau_\lambda/\mu, \tag{1}$$

where $\tau_\lambda = \tau_l + \tau_c$ is the total optical depth, τ_l that for the line and τ_c that for the neighbouring continuum. The quantity $S_\lambda(\tau_\lambda)$ is the source function which equals the Planck function $B_\lambda(T(\tau_\lambda))$ in the case of LTE, where T is the temperature. Following Staude (1972) the integrand of Equation (1) normalized to $I_\lambda(\mu)$ will be interpreted as the probability $p^*(\tau_\lambda)$ that photons produced at a depth τ_λ reach the surface and are emitted from there:

$$p^*(\tau_\lambda) = \exp(-\tau_\lambda/\mu) S_\lambda(\tau_\lambda)(\mu I_\lambda(\mu))^{-1}. \tag{2}$$

For practical applications the scale of integration is usually transformed from τ_λ to τ_c, resulting in

$$p(\tau_c) = p^*(\tau_\lambda)(1 + d\tau_\lambda/d\tau_c), \tag{3}$$

or to its logarithmic value $x = \log(\tau_c)$, that is to $p(x) = (\ln 10)\tau_c p(\tau_c)$; in a first approximation x is proportional to the geometric depth in the atmosphere. The probability interpretation of Equation (3) allows us to define the mean depth or the expectation value of photon formation $\bar{\tau}_c$ by the first moment of τ_c with respect to $p(\tau_c)$

$$\bar{\tau}_c = \int_0^\infty \tau_c p(\tau_c) d\tau_c, \tag{4}$$

while the variance (the second moment) σ^2 will give some information on the extent of the layers contributing to $I_\lambda(\mu)$:

$$\sigma^2 = \int_0^\infty (\tau_c - \bar{\tau}_c)^2 p(\tau_c) d\tau_c. \tag{5}$$

Higher order moments could be used to describe the contribution in more detail, for instance, the asymmetry could be characterized by the third moment.

For polarized line radiation the emergent Stokes vector $\mathbf{I}_\lambda = (I, Q, U, V)^T$ (the superscript T describes the transpose of a matrix) can be written in a form similar to Equation (1) if atomic coherences in the absorption and stimulated emission are neglected:

$$\mathbf{I}_\lambda(\mu) = \int_0^\infty \exp(-\mathbf{K}(\tau_c)/\mu) \cdot \mathbf{S}_\lambda(\tau_c) d\tau_c/\mu \equiv \int_0^\infty \mathbf{C}(\tau_c) d\tau_c. \tag{6}$$

The total absorption matrix is given by

$$\mathbf{K}(\tau_c) = \int_0^{\tau_c} (\boldsymbol{\eta}(\tau_c') + \mathbf{E}) d\tau_c', \tag{7}$$

where $\mathbf{E}$ is the unity matrix. The matrix η consists of symmetric elements determined by line absorption effects as well as of antisymmetric elements due to both anomalous dispersion in the line (magneto-optical effects) and variations of the azimuth of the magnetic field vector $\mathbf{B}$ with depth. A detailed definition of the elements of η is given, e. g., by Staude (1982) and Stenflo (1994). For LTE in both the continuum and the line the source vector $\mathbf{S}_\lambda(\tau_c)$ is given by $(\eta + \mathbf{E})$ multiplied by $\mathbf{S}_\lambda(\tau_c) = B_\lambda(T(\tau_c))\mathbf{1}$, where $\mathbf{1}^T = (1, 0, 0, 0)$.

For simplicity we consider now the special case that η is independent of τ_c. This is possible, e.g., if $\mathbf{B}$ as well as the Doppler width and the damping parameter of the line are independent of τ_c, but a linear gradient of the azimuth of $\mathbf{B}$ may be allowed for. Then $\mathbf{K}(\tau_c) = \tau_c(\eta + \mathbf{E}) \equiv \tau_c \mathbf{A}$, and an analytic solution of Equation (6) can be obtained for any analytically given $B_\lambda(\tau_c)$. For example, a solution for

$$B_\lambda(\tau_c) = B_0(1 + \beta_0\tau_c) + A_0 \exp(-\alpha\tau_c), \tag{8}$$

has been presented in our earlier paper (Staude, 1971 a); the constants B_0, β_0, A_0, and α can be fitted to a semi-empirical model atmosphere.

The probability interpretation of a scalar product of Stokes vectors will now be used to extend the definitions given in the Equations (3–5) to the case of polarized light: we describe a special polarization equipment by a normalized Stokes vector $\mathbf{I}_N$ and the intensity I_i observed through such optics by the projection of the incident Stokes vector $\mathbf{I}_\lambda$ onto $\mathbf{I}_N$:

$$I_i = \mathbf{I}_N^T \cdot \mathbf{I}_\lambda. \tag{9}$$

For instance, the total intensity $I_i = I$ is obtained by $\mathbf{I}_N^T = (1, 0, 0, 0)$ (no polarization optics), a circular analyzer $\mathbf{I}_N^T = (1, 0, 0, \pm 1)$ produces the signal $I_i = I \pm V$, and so on. The probability that photons passing through the polarization filter $\mathbf{I}_N$ and being observed as I_i originate at the depths τ_c to $\tau_c + d\tau_c$ is given by

$$p(\tau_c) = \mathbf{I}_N^T \cdot \mathbf{C}(\tau_c)/I_i, \tag{10}$$

the related expectation value is analogous to Equation (4)

$$\bar{\tau}_i = (I_i)^{-1}\mathbf{I}_N^T \cdot \int_0^\infty \tau_c \mathbf{C}(\tau_c)d\tau_c, \tag{11}$$

and the variance is given by

$$\sigma_i^2 = (I_i)^{-1}\mathbf{I}_N^T \cdot \int_0^\infty (\tau_c - \bar{\tau}_i)^2 \mathbf{C}(\tau_c)d\tau_c. \tag{12}$$

For a constant η and the source function of Equation (8) we obtain the following analytic solutions, using Equations (6), (9), (11), and (12):

$$I_i(\mu) = \mathbf{I}_N^T \cdot [B_0\mathbf{E} + \mu(\beta_0 B_0 \mathbf{A}^{-1} + A_0 \mathbf{A}_1^{-1})] \cdot \mathbf{1} \tag{13}$$

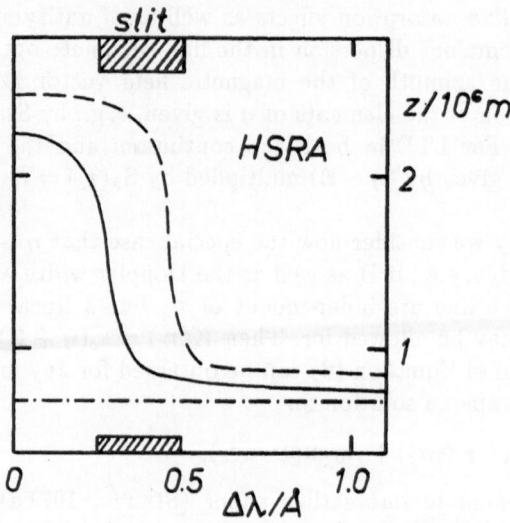

Fig. 1. Total optical depths $\tau_\lambda = 1$ (dashed curve) and 0.1 (full curve) versus height z across the Hα line profile. The dash-dot line shows $\tau_c = 1$ for comparison. The photometer slit positions of the Crimean magnetograph are marked by hatching.

(this equation corresponds to the solution of Staude, 1971 a),

$$\bar{\tau}_i = \mu (I_i)^{-1} \mathbf{I}_N^T \cdot [B_0 \mathbf{A}^{-1} + A_0 \mathbf{A}_1^{-1} + 2\mu\beta_0 B_0 \mathbf{A}^{-2}] \cdot \mathbf{1} , \tag{14}$$

and

$$\sigma_i^2 = 2\mu^2 (I_i)^{-1} \mathbf{I}_N^T \cdot [B_0 \mathbf{A}^{-2} + A_0 \mathbf{A}_1^{-2} + 3\mu\beta_0 B_0 \mathbf{A}^{-3}] \cdot \mathbf{1} - \bar{\tau}_i^2 , \tag{15}$$

where $\mathbf{A}_1 = \mathbf{A} + \alpha\mu\mathbf{E}$. Numerical examples for a sunspot model and the line Fe I 6302.5 Å are given by Staude (1972).

3. Some Problems of Chromospheric Lines

In a simple 'zero-field approximation' non-LTE effects in the chromosphere can easily be incorporated into the LTE formalism as described in Section 2 (Staude, 1982) by introducing the factors b_i describing deviations from the LTE population of the atomic energy levels i. The b_i factors are obtained from a multi-level non-LTE calculation neglecting both magnetic splitting and deviations from complete frequency redistribution. The coefficients of line absorption and emission (or the source function) are corrected for the non-LTE effects by applying the b_i factors of the lower and upper levels of the considered transition as usually done.

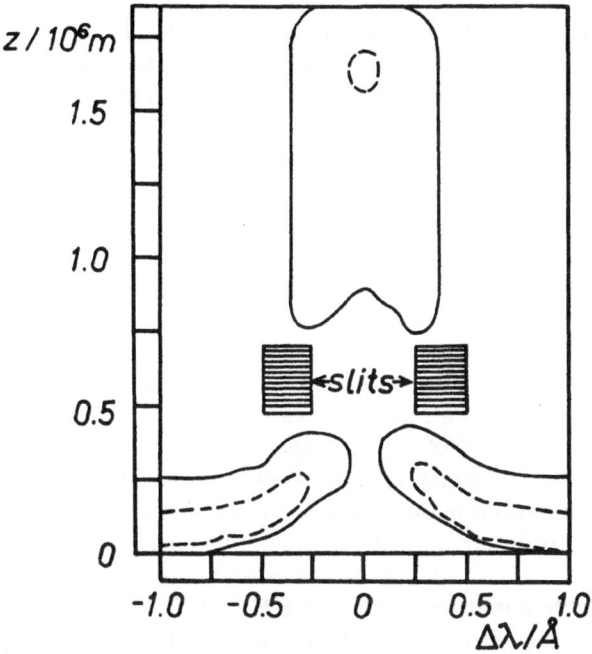

Fig. 2.   The curves encircle the regions where each layer of $\Delta z = 25$ km contributes more than 1 % (full curve) or 5 % (dashes curve) to the $H\alpha$ line formation. The zero point is now defined by $\tau_c(z = 0) = 1$.

Strong chromospheric lines show a complicated behaviour with respect to the contribution of different layers to the line formation, making the interpretation of polarization measurements in such lines rather difficult. For instance, several times the $H\alpha$ line has been used for measurements by means of Babcock-type magnetographs (the positions of the photometer slits of the Crimean magnetograph have been indicated in the following figures). Figure 1 shows the optical depths $\tau = 1$ and 0.1 versus the geometrical height z in the atmospheric model HSRA (Gingerich *et al.*, 1971) for different positions in the line ($\Delta\lambda$ is the distance from the line center); the data for the $H\alpha$ line are taken from Zelenka (1974). In Figure 2 the regions are encircled where each layer of a thickness of $\Delta z = 25$ km contributes more than a certain percentage to the line formation. We see, that the polarimeter integrates over contributions from both the photosphere and a large part of the chromosphere, while the contribution from the region around the minimum of $T(z)$ is negligible.

4. Discussion

The ambiguities in the definitions of line depression CF and $\bar{\tau}_c$ for unpolarized as well as for polarized radiation do not mean that such concepts are of no use and without any physical significance. Each of the different forms can be interpreted in a special way as it has been demonstrated already in some earlier papers on the subject (see, e.g., Gussmann, 1967; Ruhm and Elste, 1969).

On the contrary, the definitions of CF and $\bar{\tau}_c$ are unambigous for the emergent intensity. The probability interpretation of the related CF provides the basis for a physically reasonable and clear concept of the expectation value or mean depth of photon formation, and of the extent of the layers contributing to the observed radiation as well. For strong chromospheric lines the interpretation becomes more complicated due to the presence of two maxima in the CF: we have to take into account contributions from both chromospheric and photospheric layers.

The problems of ambiguity in the line depression CF are caused by the use of linear combinations of directly observable quantities such as the intensities. Analogous to these problems we meet difficulties with the probability interpretation unless the analysis of polarized components is restricted to directly measurable quantities such as $I \pm V$, $I \pm Q$, and $I \pm U$. The traditional Stokes parameters Q, U, and V as well as other forms of generalized Stokes parameters however, are linear combinations of the directly observable quantities; though all of them have the dimensions of intensity, they exhibit peculiar features such as changes of the sign which looks strange for a quantity such as intensity. Of course, such features are also reflected in the related CF. Restricting to the directly observable parameters as defined in Section 2 (we cannot observe 'polarization without intensity'), the probability interpretation can be applied in a straightforward way, and the results are unambigous.

However, due to the dominant role of temperature, the CF and $\bar{\tau}$ reflect mainly the temperature stratification and often to a minor extent the influence of other physical parameters. The diagnostics of such physical quantities are achieved by the use of RF rather than the CF. Such RF have been successfully incorporated into an effective inversion code for Stokes profiles (Ruiz Cobo and del Toro Iniesta, 1992, 1994).

Acknowledgements

The original concept of the probability interpretation of contribution functions for Stokes profiles has been developed many years ago after stimulating discussions with Helmut Domke which is gratefully acknowledged. Critical comments by K. N. Nagendra and Sami Solanki helped to improve the presentation of the present paper.

References

Achmad, L., de Jager, C., and Nieuwenhuijzen, H.: 1991, *Astron. Astrophys.* **250**, 445

Beckers, J.M. and Milkey, R.W.: 1975, *Solar Phys.* **43**, 289

Caccin, R., Gómez, M.T., Marmolino, C., and Severino, G.: 1977, *Astron. Astrophys.* **54**, 227

de Jager, C.: 1952, *The hydrogen spectrum of the Sun.* Thesis. Rijksuniversiteit Utrecht

Elste, G.: 1955, *Z. Astrophys.* **37**, 184

Elste, G.: 1963, *J.Q.S.R.T.* **3**, 157

Gingerich, O., Noyes, R.W., Kalkofen, W., and Cuny, Y.: 1971, *Solar Phys.* **18**, 347

Grigorjev, V.M. and Katz, J.M.: 1975, *Solar Phys.* **42**, 21

Grossmann-Doerth, U., Larson, B., and Solanki, S.K.: 1988, *Astron. Astrophys.* **204**, 266

Gussmann, E.A.: 1967, *Z. Astrophys.* **65**, 456

Landi Degl'Innocenti, E. and Landi Degl'Innocenti, M.: 1977, *Astron. Astrophys.* **56**, 111

Magain, P.: 1986, *Astron. Astrophys.* **163**, 135

Mein, P.: 1971, *Solar Phys.* **20**, 3

Pecker, J.C.: 1951, *Ann. d'Astrophys.* **14**, 115

Rachkovsky, D.N.: 1969, *Izv. Krymsk. Astrofiz. Observ.* **40**, 127

Ruhm, H. and Elste, G.: 1969, *Astron. Astrophys.* **3**, 277, 279

Ruiz Cobo, B. and del Toro Iniesta, J.C.: 1992, *Astrophys. J.* **398**, 375

Ruiz Cobo, B. and del Toro Iniesta, J.C.: 1994, *Astron. Astrophys.* **283**, 129

Solanki, S.K. and Bruls, J.H.M.J.: 1994, *Astron. Astrophys.* **286**, 269

Staude, J.: 1971a, *Solar Phys.* **18**, 22

Staude, J.: 1971b, *Solar Phys.* **18**, 24

Staude, J.: 1972, *Solar Phys.* **24**, 255

Staude, J.: 1982, *HHI-STP-Report* **14**, Berlin, 24

Stenflo, J.O.: 1994, *Solar Magnetic Fields. Polarized Radiation Diagnosics.* Kluwer Acad. Publ., Dordrecht

Unsöld, A.: 1932, *Z. Astrophys.* **4**, 339

Van't Veer, F.: 1966, *Ann. d'Astrophys.* **29**, 223

Wittmann, A.: 1974, *Solar Phys.* **35**, 11

Zelenka, A.: 1974, *Astron. Mitt. Zürich* **327**,

NET CIRCULAR POLARIZATION IN MAGNETIC SPECTRAL LINES PRODUCED BY VELOCITY GRADIENTS: SOME ANALYTICAL RESULTS

M. LANDOLFI

Osservatorio di Arcetri, Largo E. Fermi 5, I-50125 Firenze, Italia

and

E. LANDI DEGL'INNOCENTI[*]

Instituto de Astrofísica de Canarias, E-38200 La Laguna, Tenerife, España

Abstract. The net circular polarization in a spectral line due to the combined effect of magnetic fields and velocity gradients is analyzed for a few schematic situations. In some particular cases, its dependence on the magnetic field, velocity field and line parameters can be expressed analytically.

Key words: Magnetic fields – Polarization

1. Introduction

Let us consider an atomic spectral line formed in a static stellar atmosphere permeated by a magnetic field. Provided the line is unblended and atomic orientation can be neglected, the circular polarization Stokes V parameter is antisymmetrical about the central wavelength λ_0 of the line (see e.g. Landi Degl'Innocenti and Landi Degl'Innocenti, 1981). It follows that the net (i.e. wavelength-integrated) circular polarization (NCP) is zero. However, the broadband observations by Illing, Landman, and Mickey (1974), as well as the spectropolarimetric observations by Stenflo *et al.* (1984) clearly show that this is not the case for sunspots and solar active regions.

In this paper the NCP produced by velocity gradients is investigated. The NCP has a very complicated dependence both on the parameters specifying the model atmosphere (including the velocity gradient and the magnetic field) and on the spectral line itself. Here we will not attempt a systematic investigation of this dependence, but will rather restrict attention to a few schematic situations where some analytical results can be derived.

The interpretation of observational data generally requires more sophisticated models than those considered in this paper. However, we believe that an approach based on simple models is the most appropriate to give a deeper insight into the involved topic of radiative transfer in the presence of magnetic and velocity fields. This is why in the following we consider,

---

[*] On leave from Dipartimento di Astronomia e Scienza dello Spazio, Università di Firenze, Largo E. Fermi 5, I-50125 Firenze, Italia

Solar Physics **164**: 191–202, 1996.
© 1996 *Kluwer Academic Publishers.*

for instance, a model where the NCP is generated by a variable velocity field and a constant magnetic field, although such a model is believed to be inadequate to reproduce the observations (see e.g. Skumanich and Lites, 1987).

2. The Transfer Equation for Polarized Radiation

The transfer equation for polarized radiation in a magnetic spectral line is well established (see e.g. Landi Degl'Innocenti and Landi Degl'Innocenti, 1972, 1975). We rewrite it here in order to fix the notations that will be needed later.

Consider an isolated spectral line formed in an atmosphere where a magnetic field and a macroscopic velocity field are present. The line – characterized by the rest wavelength λ_0 – originates from the (electric-dipole) transition between two energy levels having angular momentum quantum numbers and Landé factors (J_ℓ, g_ℓ) and (J_u, g_u), respectively. At each optical depth, the two levels are split according to the local value of the magnetic field. We assume that the Zeeman effect regime holds and that no atomic polarization is present: i.e. the magnetic sublevels of each J-level are evenly populated and the off-diagonal elements of the atomic density matrix are zero. Under these assumptions, the transfer equation for the Stokes vector $\mathbf{I} = (I, Q, U, V)^\dagger$ (defined according to Shurcliff, 1962) characterizing a radiation beam travelling along the direction Ω is

$$\frac{d\mathbf{I}}{d\tau} = \mathbf{C}\,\mathbf{I} - \mathbf{j}\,, \tag{1}$$

where τ is the continuum optical depth at the line wavelength measured along Ω, and where the propagation matrix $\mathbf{C}$ and the emission vector $\mathbf{j}$ are given by

$$\mathbf{C} = \begin{pmatrix} 1 + k_I & k_Q & k_U & k_V \\ k_Q & 1 + k_I & f_V & -f_U \\ k_U & -f_V & 1 + k_I & f_Q \\ k_V & f_U & -f_Q & 1 + k_I \end{pmatrix}, \qquad \mathbf{j} = \begin{pmatrix} S_c + k_I S_L \\ k_Q S_L \\ k_U S_L \\ k_V S_L \end{pmatrix}.$$

Here S_c and S_L are the continuum and line source functions, while the k and f coefficients, related to absorption and anomalous dispersion, respectively, can be written in the form

$$k_I = \kappa_L \tfrac{1}{2} \left[\eta_p \sin^2 \vartheta + \tfrac{1}{2}(\eta_b + \eta_r)(1 + \cos^2 \vartheta) \right]$$

$$k_Q = \kappa_L \tfrac{1}{2} \left[\eta_p - \tfrac{1}{2}(\eta_b + \eta_r) \right] \sin^2 \vartheta \cos 2\varphi$$

$$k_U = \kappa_L \tfrac{1}{2} \left[\eta_p - \tfrac{1}{2}(\eta_b + \eta_r) \right] \sin^2 \vartheta \sin 2\varphi$$

$$k_V = \kappa_L \tfrac{1}{2} \left[\eta_r - \eta_b \right] \cos \vartheta \tag{2}$$

$$f_Q = \kappa_L \tfrac{1}{2} \left[\rho_p - \tfrac{1}{2}(\rho_b + \rho_r) \right] \sin^2 \vartheta \cos 2\varphi$$

$$f_U = \kappa_L \tfrac{1}{2} \left[\rho_p - \tfrac{1}{2}(\rho_b + \rho_r) \right] \sin^2 \vartheta \sin 2\varphi$$

$$f_V = \kappa_L \tfrac{1}{2} \left[\rho_r - \rho_b \right] \cos \vartheta$$

with

$$\eta_b = \eta_{-1}, \quad \eta_p = \eta_0, \quad \eta_r = \eta_{+1}, \quad \rho_b = \rho_{-1}, \quad \rho_p = \rho_0, \quad \rho_r = \rho_{+1},$$

and (using 3-j symbols)

$$\eta_q = \sum_{M_\ell M_u} 3 \begin{pmatrix} J_u & J_\ell & 1 \\ -M_u & M_\ell & -q \end{pmatrix}^2 \frac{1}{\sqrt{\pi}} H(v - v_A + v_B(g_u M_u - g_\ell M_\ell), a),$$

$$\rho_q = \sum_{M_\ell M_u} 3 \begin{pmatrix} J_u & J_\ell & 1 \\ -M_u & M_\ell & -q \end{pmatrix}^2 \frac{1}{\sqrt{\pi}} L(v - v_A + v_B(g_u M_u - g_\ell M_\ell), a), \tag{3}$$

where $q = M_\ell - M_u = 0, \pm 1$. In these expressions, κ_L is the ratio between line and continuum absorption coefficient, ϑ and φ the inclination and azimuth angles specifying the magnetic field direction (cf. Landi Degl'Innocenti and Landi Degl'Innocenti, 1972, Figure 1), H the Voigt function and L the associated dispersion profile

$$H(v, a) = \frac{a}{\pi} \int_{-\infty}^{\infty} \frac{e^{-y^2}}{(v - y)^2 + a^2} \, dy, \quad L(v, a) = \frac{1}{\pi} \int_{-\infty}^{\infty} \frac{e^{-y^2}(v - y)}{(v - y)^2 + a^2} \, dy,$$

a being the damping constant. Note that $L(v, a) = 2F(v, a)$, where F is generally referred to as Faraday-Voigt function. Finally, v, v_A and v_B are the wavelength distance from line center, the Doppler shift due to the line-of-sight velocity w_A of the ambient medium, and the Zeeman splitting, all expressed in Doppler width units

$$v = \frac{\lambda - \lambda_0}{\Delta \lambda_D}, \qquad v_A = \frac{\Delta \lambda_A}{\Delta \lambda_D} = \frac{\lambda_0}{\Delta \lambda_D} \frac{w_A}{c},$$

$$v_B = \frac{\Delta \lambda_B}{\Delta \lambda_D} = \frac{\lambda_0^2 e_0}{4\pi m c^2 \Delta \lambda_D} |\mathbf{B}|, \tag{4}$$

with e_0 the absolute value of the electron charge, m the electron mass, and c the velocity of light. In accordance with the astrophysical sign convention, a positive w_A value means a red-shifted line.

3. The Net Circular Polarization in two simple cases

In order to characterize the NCP in a spectral line, we define the dimension-less parameter v to be the ratio between the wavelength-integrated emerging V Stokes parameter and the fraction of continuum subtracted by the line,

$$v = \int V(\tau = 0, \lambda)\, d\lambda \; \Big/ \; \int [I_c(\tau = 0) - I(\tau = 0, \lambda)]\, d\lambda \,,$$

where I_c is the continuum intensity at the line wavelength and the integrals extend over the line profile. As apparent from the definition, the v parameter is unaffected by a global shift of the V and I profiles. This means that, as far as NCP is concerned, what is important is not the line-of-sight velocity itself, but the line-of-sight velocity *gradient*.

3.1. THE MILNE-EDDINGTON ATMOSPHERE WITH A SMALL VELOCITY GRADIENT

Under the Milne-Eddington approximation (plane-parallel atmosphere with $\mathbf{C} = \text{const.}$ and $S_c = S_L = B_P = B_0(1 + \beta t)$), Equation (1) reduces to

$$\mu \frac{d\mathbf{I}}{dt} = \mathbf{C}\,(\mathbf{I} - B_P\mathbf{U})\,, \tag{5}$$

where μ is the cosine of the heliocentric angle, t the continuum optical depth measured along the vertical, and $\mathbf{U} = (1, 0, 0, 0)^{\dagger}$. All the parameters on which the matrix $\mathbf{C}$ depends are independent of optical depth; in particular, the reduced Doppler shift v_A defined in Equations (4) is constant (or zero if $w_A = 0$).

Let us now suppose that a small velocity gradient $\delta w_A(t)$ is present, where small means

$$\frac{\lambda_0}{\Delta\lambda_D} \frac{\delta w_A(t)}{c} \ll 1$$

for any t. We can expand the propagation matrix $\mathbf{C}$ to first order in the gradient

$$\mathbf{C} \to \mathbf{C} + \left(\frac{\partial\mathbf{C}}{\partial w_A}\right)\delta w_A(t) = \mathbf{C} - \frac{\lambda_0}{c}\left(\frac{\partial\mathbf{C}}{\partial\lambda}\right)\delta w_A(t)\,,$$

and look for a perturbative solution of the form $\mathbf{I} + \delta\mathbf{I}$. Substituting the above expression for $\mathbf{C}$ into Equation (5) and equating the zero and first order terms, we obtain a zero order equation (which is identical to Equation (5)) corresponding to the static (or constant-velocity) atmosphere, and a first order equation

$$\mu \frac{d}{dt}\delta\mathbf{I} = \mathbf{C}\,\delta\mathbf{I} - \frac{\lambda_0}{c}\frac{\partial\mathbf{C}}{\partial\lambda}(\mathbf{I} - B_P\mathbf{U})\,\delta w_A(t)\,, \tag{6}$$

which has the same form as the usual transfer equation, with an emission vector depending on the solution of the zero order equation. The latter is given by the well-known Unno-Rachkovsky formula

$$\mathbf{I}(t, \mu) = B_0 \left[(1 + \beta t)\mathbf{1} + \beta \mu \, \mathbf{C}^{-1} \right] \mathbf{U} \,, \tag{7}$$

where $\mathbf{1}$ is the unit matrix and $\mathbf{C}^{-1}$ the inverse of the matrix $\mathbf{C}$. The solution to Equation (6) can be written, using the evolution operator defined in Landi Degl'Innocenti and Landi Degl'Innocenti (1985), in the form

$$\delta \mathbf{I}(t, \mu) = \int_t^\infty e^{-\frac{1}{\mu}(t'-t)} \, \mathbf{C} \, \frac{\lambda_0}{c} \frac{\partial \mathbf{C}}{\partial \lambda} \, (\mathbf{I} - B_P \mathbf{U}) \, \delta w_A(t') \, \frac{dt'}{\mu} \,, \tag{8}$$

the exponential factor being defined by its Taylor expansion.

If we now make the further assumption that the velocity gradient is *linear* in the optical depth ($\delta w_A(t) = w_A^{(1)} t$), the integral in Equation (8) can be performed analytically. Since for any constant matrix $\mathbf{M}$, we can write

$$\int_a^b e^{-x\mathbf{M}} x \, dx = \left\{ [a\mathbf{1} + \mathbf{M}^{-1}] e^{-a\mathbf{M}} - [b\mathbf{1} + \mathbf{M}^{-1}] e^{-b\mathbf{M}} \right\} \mathbf{M}^{-1} \,,$$

we obtain for the first order correction to the emerging Stokes parameters

$$\delta \mathbf{I}(0, \mu) = -B_0 \beta \mu^2 \lambda_0 \frac{w_A^{(1)}}{c} \mathbf{C}^{-1} \frac{\partial \mathbf{C}^{-1}}{\partial \lambda} \mathbf{U} \,. \tag{9}$$

It should be noticed that the integrand in Equation (8) is, apart from the factor $\delta w_A(t')$, the response function for the Stokes vector $\mathbf{I}$ to a velocity perturbation, evaluated for a Milne-Eddington atmosphere. The concept of response function for the Stokes parameters has been introduced by Landi Degl'Innocenti and Landi Degl'Innocenti (1977). Expressions for $\delta \mathbf{I}(0, \mu)$, basically equivalent to Equation (9), have been obtained by Landolfi (1987) and by Sánchez Almeida (1992). The derivation presented here provides however a much simpler expression for this quantity.

Direct calculation of the matrix product in the right-hand side of Equation (9) leads to the following expression for the NCP parameter

$$v = \mu \frac{\lambda_0}{\Delta \lambda_D} \frac{w_A^{(1)}}{c} \frac{\int \Delta^{-2} \mathcal{A} \, dv}{\int \left\{ 1 - \Delta^{-1}(1 + k_I)[(1 + k_I)^2 + f_Q^2 + f_V^2] \right\} dv} \,, \tag{10}$$

where v is the reduced wavelength defined in Equation (4) and

$$\Delta = (1 + k_I)^4 + (1 + k_I)^2(f_{\tilde{Q}}^2 + f_V^2 - k_{\tilde{Q}}^2 - k_V^2) - (k_{\tilde{Q}}f_{\tilde{Q}} + k_V f_V)^2$$

$$\begin{aligned}
\mathcal{A} = &\left\{ (1 + k_I)^3(k_{\tilde{Q}}k_V + f_{\tilde{Q}}f_V) \right.\\
&\left. + (1 + k_I)[-k_{\tilde{Q}}k_V(f_{\tilde{Q}}^2 - f_V^2) + f_{\tilde{Q}}f_V(k_{\tilde{Q}}^2 - k_V^2)] \right\}\frac{\partial k_{\tilde{Q}}}{\partial v}\\
&- (1 + k_I)^3(k_{\tilde{Q}}^2 + f_{\tilde{Q}}^2)\frac{\partial k_V}{\partial v}\\
&+ \left\{ (1 + k_I)^3(k_V f_{\tilde{Q}} - k_{\tilde{Q}}f_V) \right.\\
&\left. + (1 + k_I)[k_{\tilde{Q}}f_V(k_V^2 + f_{\tilde{Q}}^2) + k_V f_{\tilde{Q}}(k_{\tilde{Q}}^2 + f_V^2)] \right\}\frac{\partial f_{\tilde{Q}}}{\partial v}\\
&- (1 + k_I)(k_{\tilde{Q}}f_{\tilde{Q}} + k_V f_V)(k_{\tilde{Q}}^2 + f_{\tilde{Q}}^2)\frac{\partial f_V}{\partial v} ,
\end{aligned}$$

with

$$k_{\tilde{Q}} = k_Q(\varphi = 0), \qquad f_{\tilde{Q}} = f_Q(\varphi = 0) .$$

It can be seen that the v parameter is independent of the azimuth angle φ.

A symmetry relation concerning the dependence on the angle ϑ follows from Equations (10) and (2),

$$v(\pi - \vartheta) = -v(\vartheta) .$$

Hence $v(\pi/2) = 0$. Moreover, $v(0) = v(\pi) = 0$. Thus the NCP parameter is zero if the magnetic field is perpendicular or parallel to the propagation direction. The former result is obvious, since Equations (2) give $k_V = f_V = 0$ for $\vartheta = \pi/2$, so that the quantity $\mathcal{A}$ in Equation (10) vanishes. In other words, for $\vartheta = \pi/2$ the profile $V(\lambda)$ is identically zero. The latter result has been obtained by various authors (see e.g. Skumanich and Lites, 1987). Under the assumption of small velocity gradient, Sánchez Almeida, Collados, and del Toro Iniesta (1989) have proved that it always holds in LTE.

For weak magnetic field ($v_B \ll 1$), the η_q, ρ_q profiles defined in Equations (3) can be expanded in power series of v_B (see Landi Degl'Innocenti and Landi Degl'Innocenti, 1973). To the lowest order we obtain from Equation (10)

$$v = -\frac{1}{16}\mu\frac{\lambda_0}{\Delta\lambda_D}\frac{w_A^{(1)}}{c}v_B^5\,\bar{g}\,\bar{G}^2\sin^4\vartheta\cos\vartheta\,\frac{\mathcal{I}_5(\kappa_L, a)}{\mathcal{I}_1(\kappa_L, a)} , \tag{11}$$

where $\bar{g}$ is the effective Landé factor of the line and

$$\begin{aligned}
\bar{G} = \bar{g}^2 - \frac{1}{80}(g_\ell - g_u)^2 &\left\{ 16\left[J_\ell(J_\ell + 1) + J_u(J_u + 1)\right] \right.\\
&\left. - 7\left[J_\ell(J_\ell + 1) - J_u(J_u + 1)\right]^2 - 4 \right\},
\end{aligned}$$

$$\mathcal{I}_1(\kappa_L, a) = \int\frac{\kappa_L\eta}{1 + \kappa_L\eta}\,dv ,$$

$$\mathcal{I}_5(\kappa_L, a) = \int\frac{\kappa_L^5[(\eta'\eta'' + \rho'\rho'')\eta''' - (\eta''^2 + \rho''^2)\eta'' + (\eta''\rho'' - \eta''\rho')\rho''']}{(1 + \kappa_L\eta)^5}\,dv,$$

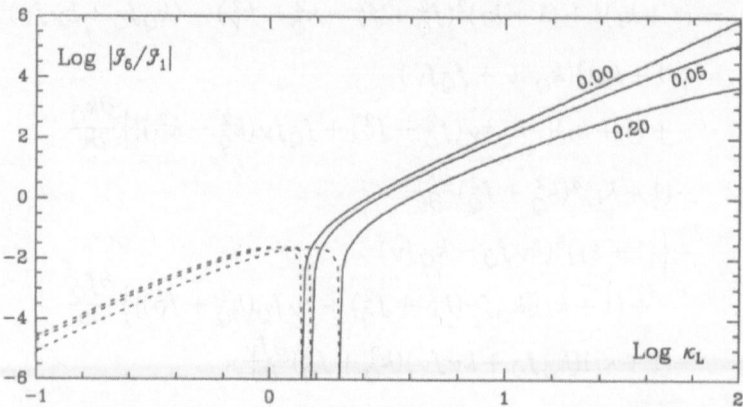

Fig. 1.   The absolute value of the ratio $\mathcal{I}_5(\kappa_L, a)/\mathcal{I}_1(\kappa_L, a)$ is plotted against κ_L on a Log-Log scale. The three curves are labeled by the value of a. The ratio $\mathcal{I}_5(\kappa_L, a)/\mathcal{I}_1(\kappa_L, a)$ is positive on the right of the figure (full lines) and negative on the left(dashed lines).

with

$$\eta = H(v, a)/\sqrt{\pi}\,, \qquad \rho = L(v, a)/\sqrt{\pi}\,,$$

where the primes in the integrand of $\mathcal{I}_5(\kappa_L, a)$ denote the derivatives with respect to the reduced wavelength v.

According to Equation (11), the NCP is a strongly increasing function of ϑ which attains the maximum value for $\cos\vartheta = 1/\sqrt{5}$ (i.e. $\vartheta \simeq 63°$). As to the dependence on the Zeeman pattern, it can be shown that for assigned $\bar{g}$, the $\bar{G}$ factor – and, therefore, the NCP – is maximum for Zeeman triplets. Finally, the dependence of the NCP on κ_L and a is contained in the ratio $\mathcal{I}_5/\mathcal{I}_1$. The integral $\mathcal{I}_1$ is a positive quantity (proportional to the equivalent width of the line), while the sign of $\mathcal{I}_5$ cannot be established without a numerical evaluation of the integral. This shows that $\mathcal{I}_5$ is positive (for all a values) provided κ_L is larger than about 1.5. It follows that, apart from very weak lines, the NCP originating from an atmosphere with $w_A^{(1)} > 0$ obeys, in the weak field regime, the sign rule

$$\text{sign}(v) = \text{sign}(-\bar{g}\cos\vartheta)\,.$$

It is important to notice that just the *opposite* sign rule would be obtained if the ρ terms in the expression for $\mathcal{I}_5(\kappa_L, a)$ were neglected. This is a striking example of the significance of magneto-optical effects.

The order of magnitude of the NCP predicted by Equation (11) can be estimated with the help of Figure 1. It can be seen that v scales approximately as κ_L^3.

Another result, concerning the case of very strong magnetic field, can be derived from Equation (10): provided the H and L profiles corresponding to the transition between *each* couple of magnetic sublevels $(J_\ell M_\ell, J_u M_u)$ do not overlap, v is zero. However, this is true only for really strong fields, because $L(v,a)$ is a weakly decreasing function of v $(L(v,a) \sim v^{-1}$, while $H(v,a) \sim v^{-2})$.

3.2. A DISCONTINUOUS MODEL WITH VELOCITY AND MAGNETIC FIELD GRADIENTS

The next case we consider is a discontinuous model formed by a thin slab laying on top of a semi-infinite atmosphere. We consider vertical propagation $(\mu = 1)$ and assume that a (constant) line-of-sight velocity $w_A^{(s)}$ is present in the slab, while the underlying atmosphere is static; the slab and the atmosphere are permeated by different magnetic fields, $\mathbf{B}^{(s)} \equiv (B_s, \vartheta_s, \varphi_s)$ and $\mathbf{B} \equiv (B, \vartheta, \varphi)$ respectively, also assumed as constant with optical depth. This is essentially the model considered by Illing, Landman, and Mickey (1975).

The transfer of radiation across the slab is described by the equation

$$\frac{d\mathbf{I}}{dt} = \mathbf{C}^{(s)}\,\mathbf{I} - \mathbf{j}^{(s)}\,,$$

whose formal solution is

$$\mathbf{I}(0) = \int_0^{t_s} \mathbf{O}(0,t)\,\mathbf{j}^{(s)}\,dt + \mathbf{O}(0,t_s)\,\mathbf{I}^{(b)}\,, \qquad (12)$$

where t_s is the optical thickness of the slab, $\mathbf{O}$ the evolution operator, and $\mathbf{I}^{(b)}$ the boundary Stokes vector, characterizing the radiation emerging from the underlying atmosphere. Since $t_s \ll 1$, we can expand the evolution operator in power series up to first order,

$$\mathbf{O}(0,t) = \mathrm{e}^{-t\,\mathbf{C}^{(s)}} \simeq 1 - t\,\mathbf{C}^{(s)} \qquad (0 \le t \le t_s)\,.$$

To the lowest order in t_s we get for the NCP parameter, the expression

$$v = -t_s \frac{\int \left[k_V^{(s)} I^{(b)} + f_U^{(s)} Q^{(b)} - f_Q^{(s)} U^{(b)} + k_I^{(s)} V^{(b)}\right] d\lambda}{\int \left[I_c^{(b)} - I^{(b)}\right] d\lambda}\,, \qquad (13)$$

where $I^{(b)}$, $Q^{(b)}$, $U^{(b)}$, $V^{(b)}$ are centered at the rest wavelength λ_0 while $k_I^{(s)}$, $k_V^{(s)}$, $f_Q^{(s)}$, $f_U^{(s)}$ are centered at $\lambda_0(1 + w_A^{(s)}/c) = \lambda_0 + \Delta\lambda_A^{(s)}$.

In order to further simplify the problem, we assume that the boundary Stokes parameters are given by the Seares formulae

$$I^{(b)} = B_0[(1 + \beta) - \beta k_I] \qquad U^{(b)} = -B_0\beta k_U$$
$$Q^{(b)} = -B_0\beta k_Q \qquad\qquad V^{(b)} = -B_0\beta k_V . \tag{14}$$

These expressions – which are indeed rough approximations – can be derived from Equation (7) under the limit of weak spectral line ($\kappa_L \ll 1$). From Equations (13), (14) and (2) we have

$$
\begin{aligned}
\mathbf{v} = \frac{\kappa_L^{(s)} t_s}{4\,\Delta\lambda_D} \Big\{ &\tfrac{1}{2}\Big[\cos\vartheta_s(1 + \cos^2\vartheta) + \cos\vartheta(1 + \cos^2\vartheta_s)\Big] \\
&\times \int (\eta_r^{(s)}\eta_r - \eta_b^{(s)}\eta_b)\,d\lambda \\
&+ \cos\vartheta_s \sin^2\vartheta \int (\eta_r^{(s)} - \eta_b^{(s)})\eta_p\,d\lambda \\
&+ \cos\vartheta \sin^2\vartheta_s \int (\eta_r - \eta_b)\eta_p^{(s)}\,d\lambda \\
&+ \tfrac{1}{2}\Big[\cos\vartheta_s(1 + \cos^2\vartheta) - \cos\vartheta(1 + \cos^2\vartheta_s)\Big] \\
&\times \int (\eta_r^{(s)}\eta_b - \eta_b^{(s)}\eta_r)\,d\lambda \\
&- \sin^2\vartheta_s \sin^2\vartheta \sin 2(\varphi - \varphi_s) \\
&\times \int \Big[\rho_p^{(s)} - \tfrac{1}{2}(\rho_b^{(s)} + \rho_r^{(s)})\Big]\Big[\eta_p - \tfrac{1}{2}(\eta_b + \eta_r)\Big]d\lambda \Big\},
\end{aligned}
\tag{15}
$$

where all the quantities with the index (s) refer to the slab and the other quantities to the underlying atmosphere. We will now consider the three special cases which are obtained from this equation by setting 2 of the 3 magnetic field parameters to the same value. The case ($\vartheta_s = \vartheta, \varphi_s = \varphi$) will be denoted as ΔB-effect (the fourth and fifth term in Equation (15) vanish); similarly, we will consider the $\Delta\vartheta$-effect ($B_s = B, \varphi_s = \varphi$; terms 1 and 5 are zero) and the $\Delta\varphi$-effect ($B_s = B, \vartheta_s = \vartheta$; the first four terms are zero). For simplicity, we restrict attention to Zeeman triplets and assume the Doppler width and the damping constant in the slab, and in the atmosphere to be the same.

Equation (15) contains convolutions of H functions and of H and L functions. These can be evaluated analytically, yielding

$$
\begin{aligned}
\int_{-\infty}^{\infty} H(v - v_0, a)H(v - v_0', a)\,dv &= \sqrt{\tfrac{\pi}{2}}\, H\left(\tfrac{v_0 - v_0'}{\sqrt{2}}, \sqrt{2}a\right) \\
\int_{-\infty}^{\infty} H(v - v_0, a)L(v - v_0', a)\,dv &= \sqrt{\tfrac{\pi}{2}}\, L\left(\tfrac{v_0 - v_0'}{\sqrt{2}}, \sqrt{2}a\right) .
\end{aligned}
\tag{16}
$$

The following expressions for the three effects mentioned above are obtained:

ΔB-effect

$$\mathbf{v} = \frac{\kappa_{\mathrm{L}}^{(\mathrm{s})} t_{\mathrm{s}}}{4\sqrt{2\pi}} \cos\vartheta \left\{ (1 + \cos^2\vartheta)\Big[H(v_{\mathrm{rr}}, \hat{a}) - H(v_{\mathrm{bb}}, \hat{a})\Big] \right.$$
$$\left. + \sin^2\vartheta \Big[H(v_{\mathrm{rp}}, \hat{a}) + H(v_{\mathrm{pr}}, \hat{a}) - H(v_{\mathrm{bp}}, \hat{a}) - H(v_{\mathrm{pb}}, \hat{a})\Big] \right\}, \tag{17}$$

$\Delta\vartheta$-effect

$$\mathbf{v} = \frac{\kappa_{\mathrm{L}}^{(\mathrm{s})} t_{\mathrm{s}}}{4\sqrt{2\pi}} \left\{ \Big[\cos\vartheta_{\mathrm{s}} \sin^2\vartheta - \cos\vartheta \sin^2\vartheta_{\mathrm{s}}\Big]\Big[H(v_{\mathrm{rp}}, \hat{a}) - H(v_{\mathrm{bp}}, \hat{a})\Big] \right.$$
$$\left. + \tfrac{1}{2}\Big[\cos\vartheta_{\mathrm{s}}(1+\cos^2\vartheta) - \cos\vartheta(1+\cos^2\vartheta_{\mathrm{s}})\Big]\Big[H(v_{\mathrm{rb}}, \hat{a}) - H(v_{\mathrm{br}}, \hat{a})\Big] \right\}, \tag{18}$$

$\Delta\varphi$-effect

$$\mathbf{v} = \frac{\kappa_{\mathrm{L}}^{(\mathrm{s})} t_{\mathrm{s}}}{4\sqrt{2\pi}} \sin^4\vartheta \sin 2(\varphi - \varphi_{\mathrm{s}})$$
$$\times \Big[\tfrac{3}{2}L(v_{\mathrm{pp}}, \hat{a}) - L(v_{\mathrm{rp}}, \hat{a}) - L(v_{\mathrm{bp}}, \hat{a}) + \tfrac{1}{4}L(v_{\mathrm{rb}}, \hat{a}) + \tfrac{1}{4}L(v_{\mathrm{br}}, \hat{a})\Big], \tag{19}$$

where

$$v_{\mathrm{rr}} = \frac{1}{\sqrt{2}\,\Delta\lambda_{\mathrm{D}}}\Big[\Delta\lambda_{A}^{(\mathrm{s})} + \bar{g}(\Delta\lambda_{B}^{(\mathrm{s})} - \Delta\lambda_{B})\Big]$$

$$v_{\mathrm{bb}} = \frac{1}{\sqrt{2}\,\Delta\lambda_{\mathrm{D}}}\Big[\Delta\lambda_{A}^{(\mathrm{s})} - \bar{g}(\Delta\lambda_{B}^{(\mathrm{s})} - \Delta\lambda_{B})\Big]$$

$$v_{\mathrm{pp}} = \frac{1}{\sqrt{2}\,\Delta\lambda_{\mathrm{D}}}\,\Delta\lambda_{A}^{(\mathrm{s})}$$

$$v_{\mathrm{rp}} = \frac{1}{\sqrt{2}\,\Delta\lambda_{\mathrm{D}}}\Big[\Delta\lambda_{A}^{(\mathrm{s})} + \bar{g}\Delta\lambda_{B}^{(\mathrm{s})}\Big]$$

$$v_{\mathrm{bp}} = \frac{1}{\sqrt{2}\,\Delta\lambda_{\mathrm{D}}}\Big[\Delta\lambda_{A}^{(\mathrm{s})} - \bar{g}\Delta\lambda_{B}^{(\mathrm{s})}\Big]$$

$$v_{\mathrm{pr}} = \frac{1}{\sqrt{2}\,\Delta\lambda_{\mathrm{D}}}\Big[\Delta\lambda_{A}^{(\mathrm{s})} - \bar{g}\Delta\lambda_{B}\Big]$$

$$v_{\mathrm{pb}} = \frac{1}{\sqrt{2}\,\Delta\lambda_{\mathrm{D}}}\Big[\Delta\lambda_{A}^{(\mathrm{s})} + \bar{g}\Delta\lambda_{B}\Big]$$

$$v_{\mathrm{rb}} = \frac{1}{\sqrt{2}\,\Delta\lambda_{\mathrm{D}}}\Big[\Delta\lambda_{A}^{(\mathrm{s})} + \bar{g}(\Delta\lambda_{B}^{(\mathrm{s})} + \Delta\lambda_{B})\Big]$$

$$v_{\mathrm{br}} = \frac{1}{\sqrt{2}\,\Delta\lambda_{\mathrm{D}}}\Big[\Delta\lambda_{A}^{(\mathrm{s})} - \bar{g}(\Delta\lambda_{B}^{(\mathrm{s})} + \Delta\lambda_{B})\Big]$$

$$\hat{a} = \sqrt{2}\,a.$$

As far as the ΔB-effect is concerned, the sign of the first line of Equation (17) is easily evaluated. Since $H(v, a)$ decreases monotonically with increasing $|v|$, we have $H(v_{\mathrm{bb}}, \hat{a}) > H(v_{\mathrm{rr}}, \hat{a})$ when $\Delta\lambda_{A}^{(\mathrm{s})}$ and $\bar{g}(\Delta\lambda_{B}^{(\mathrm{s})} - \Delta\lambda_{B})$ have the same sign. For slightly inclined magnetic fields (i.e., $\vartheta \simeq 0$ or $\vartheta \simeq \pi$) we thus obtain the sign rule

$$\mathrm{sign}(\mathbf{v}) = \mathrm{sign}\big[\bar{g}\,\Delta\lambda_{A}^{(\mathrm{s})}\cos\vartheta\,(\Delta\lambda_{B} - \Delta\lambda_{B}^{(\mathrm{s})})\big] \qquad (\Delta B - effect). \tag{20}$$

By similar reasoning, and observing that the two factors in Equation (18) depending on the inclination angles have the same sign as $(\vartheta - \vartheta_{\mathrm{s}})$, we get

$$\text{sign}(v) = \text{sign}\left[-\bar{g}\,\Delta\lambda_A^{(s)}\,(\vartheta - \vartheta_s)\right] \qquad (\Delta\vartheta - effect). \qquad (21)$$

Comparison of Equations (20) and (21) shows that an increase in the optical depth of the transverse component of the magnetic field affects the NCP parameter in the same way as a decrease of the magnetic field strength, and vice versa. A relation similar to Equation (20) has been derived by Pahlke and Solanki (1986) and Solanki and Pahlke (1988). Sánchez Almeida, Collados, and del Toro Iniesta (1989) have demonstrated that it always holds in LTE, provided the magnetic field gradient and the velocity gradient are small. A relation similar to Equation (21) is given in Solanki and Montavon (1993).

As for the $\Delta\varphi$-effect, Equation (19) shows that v has a complicated dependence on the Doppler shift and Zeeman splitting. Numerical evaluation of the expression in square brackets leads to the following result: provided the Doppler shift is sufficiently small ($\Delta\lambda_A^{(s)} < 1.5\,\Delta\lambda_D$), we can write

$$\text{sign}(v) = \text{sign}\left[\Delta\lambda_A^{(s)}\,(\varphi - \varphi_s)\right] \qquad (\Delta\varphi - effect). \qquad (22)$$

It should be pointed out that Equation (22), unlike Equations (20) and (21), is independent of the Landé factor (the transformation $\bar{g} \to -\bar{g}$ merely changes v_{rp} into v_{bp} and v_{rb} into v_{br}). Therefore, the use of two spectral lines with a positive and a negative Landé factor respectively, allows in principle to distinguish the $\Delta\varphi$-effect from the ΔB and $\Delta\vartheta$-effects.

Finally, Equations (20), (21), and (22), as well as Equation (10), predict that a sign switch of the velocity gradient produces a sign switch of the NCP. Sánchez Almeida and Lites (1992) have proved that this is always the case under LTE conditions.

Acknowledgements

We are grateful to J. Sánchez Almeida who gave us several suggestions for improving the manuscript.

References

Illing, R.M.E., Landman, D.A., and Mickey, D.L.: 1974, *Astron. Astrophys.* **35**, 327.
Illing, R.M.E., Landman, D.A., and Mickey, D.L.: 1975, *Astron. Astrophys.* **41**, 183.
Landi Degl'Innocenti, E. and Landi Degl'Innocenti, M.: 1972, *Solar Phys.* **27**, 319.
Landi Degl'Innocenti, E. and Landi Degl'Innocenti, M.: 1973, *Solar Phys.* **31**, 299.
Landi Degl'Innocenti, E. and Landi Degl'Innocenti, M.: 1975, *Il Nuovo Cimento* **27B**, 134.
Landi Degl'Innocenti, E. and Landi Degl'Innocenti, M.: 1977, *Astron. Astrophys.* **56**, 111.
Landi Degl'Innocenti, E. and Landi Degl'Innocenti, M.: 1981, *Il Nuovo Cimento* **62B**, 1.
Landi Degl'Innocenti, E. and Landi Degl'Innocenti, M.: 1985, *Solar Phys.* **97**, 239.

Landolfi, M.: 1987, *Solar Phys.* **109**, 287.
Pahlke, K.D. and Solanki, S.K.: 1986, *Mitt. Astron. Gesellschaft* **65**, 162.
Sánchez Almeida, J.: 1992, *Solar Phys.* **137**, 1.
Sánchez Almeida, J., Collados, M., and del Toro Iniesta, J.C.: 1989, *Astron. Astrophys.* **222**, 311.
Sánchez Almeida, J. and Lites, B.W.: 1992, *Astrophys. J.* **398**, 359.
Shurcliff, W.A.: 1962, *Polarized Light*, Harvard University Press, Cambridge.
Skumanich, A. and Lites, B.W.: 1987, *Astrophys. J.* **322**, 483.
Solanki, S.K. and Montavon, C.A.P.: 1993, *Astron. Astrophys.* **275**, 283.
Solanki, S.K. and Pahlke, K.D.: 1988, *Astron. Astrophys.* **201**, 143.
Stenflo, J.O., Harvey, J.W., Brault, J.W., and Solanki, S.K.: 1984, *Astron. Astrophys.* **131**, 333.

MICRO-STRUCTURED MAGNETIC ATMOSPHERES

J. SÁNCHEZ ALMEIDA and E. LANDI DEGL'INNOCENTI*

Instituto de Astrofísica de Canarias, E-38200 La Laguna, Tenerife, Spain

Abstract. Asymmetrical Stokes profiles are produced if the photospheric magnetic and velocity fields are structured on scales smaller than the mean-free-path of the photons. Here we put forward a compact analytical expression for the radiative transfer equation in this case. Explicitly, micro-variations of the magnetic field strength and the velocity are considered. The existence of micro-structures might have serious implications on the techniques currently used to measure solar magnetic fields. For example, we show the failure of the relationship employed to calibrate magnetographs.

Key words: Line asymmetries – Polarized radiation transfer – Structure of the photospheric magnetic fields

1. Introduction

With the possible exception of those formed in the umbrae of sunspots, the polarization of the typical photospheric spectral lines has no symmetry about the line center. This conspicuous and ubiquitous feature is commonly denoted as the asymmetry of the Stokes profiles. Asymmetries are observed in the penumbrae of sunspots (e.g., Sánchez Almeida and Lites, 1992), in plage and network regions (e.g., Stenflo *et al.*, 1984), and even in the quiet sun (Lites *et al.*, 1995). It is often difficult to account for such observations within the framework of the standard radiative transfer equation for polarized radiation (Unno-Rachkovsky equation); the simplest gradients of velocity and magnetic field which reproduce observations tend to be unphysical (e.g., Sánchez Almeida, Collados and del Toro Iniesta, 1988). One needs to resort to more sophisticated modeling in order to explain the observations (e.g., Bünte, Solanki and Steiner, 1993). Of course the latter modeling may be correct, however, one is always tempted to seek for simpler solutions. In particular, it may well be that the present use of the Unno-Rachkovsky equation is missing a basic property of the photosphere. In Sánchez Almeida *et al.* (1995; Paper I), we tried to answer this question by considering the structuring of magnetic fields over scales much smaller than the mean-free-path of visible photons (say, 100 km). Atmospheres having such a property were denoted by MISMAs, i.e., MIcro-Structured Magnetic Atmospheres. It turned out that this *ad-hoc* ingredient produces line asymmetries of the kind observed in penumbrae and plages. According to the scenarios conjectured in Paper I, the penumbrae of sunspots may be made of magnetic

---

* On leave from the *Dipartimento di Astronomia e Scienza dello Spazio, Università di Firenze, Largo E. Fermi 5, I-50125 Firenze, Italy*

Solar Physics **164**: 203–210, 1996.
© 1996 *Kluwer Academic Publishers.*

fields having different motions and different inclinations with respect to the vertical direction. Inclinations and motions are assumed to be tightly correlated, so that the more horizontal the field lines, the larger the flow along them. The scenario just pushes to very small scales the common wisdom that the Evershed flow occurs in the dark horizontal penumbral filaments (see, e.g., Thomas and Weiss, 1992; except for the size and the piling up of structures, this model is also similar to the *uncombed fields* proposed by Solanki and Montavon, 1993). As far as network and facular regions are concerned, observed asymmetries are reproduced by almost vertical micro-magnetic structures embedded into a non-magnetic medium with downflows. Again, this scenario is rather similar to the standard picture (e.g., Solanki, 1989), the main difference being the scale of the magnetic concentrations, and the horizontal component of the background motions. We supposed a correlation between the deviation of the magnetic field from the vertical direction and the deviation of the external velocity from a downflow. Such a correlation might be due to local granular motions which produce a net horizontal drag force and a subsequent inclination of the magnetic fields (like in Schüssler, 1985).

The synthesis of the emerging Stokes profiles can be carried out once the details of the MISMA are known (Section 2). Consequently, a guided-by-intuition trial-and-error may allow to reproduce the observed asymmetries. This approach is of little practical application, though. For routine interpretation of measurements in the framework of the MISMA hypothesis, one needs to write down the transfer equation in terms of a small set of parameters describing the micro-structure of the irregular atmosphere. One would then like to include such a parametrization into an inversion procedure which automatically modifies the model atmosphere so that a *best* solution is found (see the review in this proceedings by del Toro Iniesta and Ruiz Cobo). This paper presents a first step towards this goal, i.e., a practical parametrization of the radiative transfer equation depending on a limited set of variables which describe the MISMA. Specifically, we will be concerned only with micro-scale variations of the magnetic field strength and the velocity, allowing also for correlations between them (Section 3).

Current techniques to determine average properties of the solar magnetic fields neglect the possibility of micro-structure. However, the existence of micro-structure modifies the relationship between the measured parameters (polarization) and the magnetic properties to be probed. Consequently, one would expect that conventional techniques need an update if MISMAs turn out to exist. This problem can be illustrated using solutions of the radiative transfer equations which incorporate micro-structure. Using the parametrization developed in Section 3, we show in Section 4 how the equation used to calibrate magnetographs is no longer valid for a MISMA.

2. Radiative transfer through a micro-structured magnetic atmosphere

The traditional radiative transfer equation for polarized light can be expressed as (Landi Degl'Innocenti, 1992; hereafter LD)

$$\frac{d\mathbf{I}}{ds} = -\mathbf{K}(\mathbf{I} - \mathbf{S}). \tag{1}$$

$\mathbf{K}$, $\mathbf{S}$, $\mathbf{I}$, and s stand, respectively, for the absorption matrix, the vector source function, the Stokes vector $(I, Q, U, V)^t$, and the coordinate along the line-of-sight. As usually, I, Q, U and V represent the intensity, the two independent types of linear polarization, and the circular polarization. The Stokes vector emerging from an atmosphere having irregularities over scales much smaller than the mean-free-path of the photons (i.e., a MISMA), is found after solving

$$\frac{d\mathbf{I}}{ds} = -<\mathbf{K}>(\mathbf{I} - \mathbf{S}'),$$
$$\mathbf{S}' = <\mathbf{K}>^{-1}<\mathbf{K}\mathbf{S}>, \tag{2}$$

where the symbol $<\ >$ represents an average over the micro-structures existing at each depth. For details on the derivation of Equation (2), see Paper I. To illustrate the MISMA concept, we sketch in Figure 1 an atmosphere whose outcoming polarization can be computed by integration of Equation (2).

Equations (1) and (2) are formally identical. However, there is a subtle distinction which makes them different for producing line asymmetries. The coefficients of $\mathbf{K}$ are either even or odd functions of the wavelength about the line center. Consequently, the only way for Equation (1) to produce asymmetrical profiles is by introducing a gradient of the line-of-sight component of the bulk velocity (associated or not with a gradient of the magnetic field). Conversely, correlations between velocities and magnetic fields in the micro-structures destroy the symmetries of $<\mathbf{K}>$, and so, Equation (2) automatically renders asymmetrical solutions. Of course, the asymmetries generated by MISMAs are ultimately caused by variations along the line-of-sight. However, the gradients occur over sub-mean-free-path scales, and so, MISMAs may well present uniform average kinetic and magnetic properties.

3. Average absorption matrix for varying magnetic field strength and velocity

The absorption matrix $\mathbf{K}$ can be explicitly expressed in terms of the magnetic field vector, temperature, pressure and velocity (see LD). Inversion

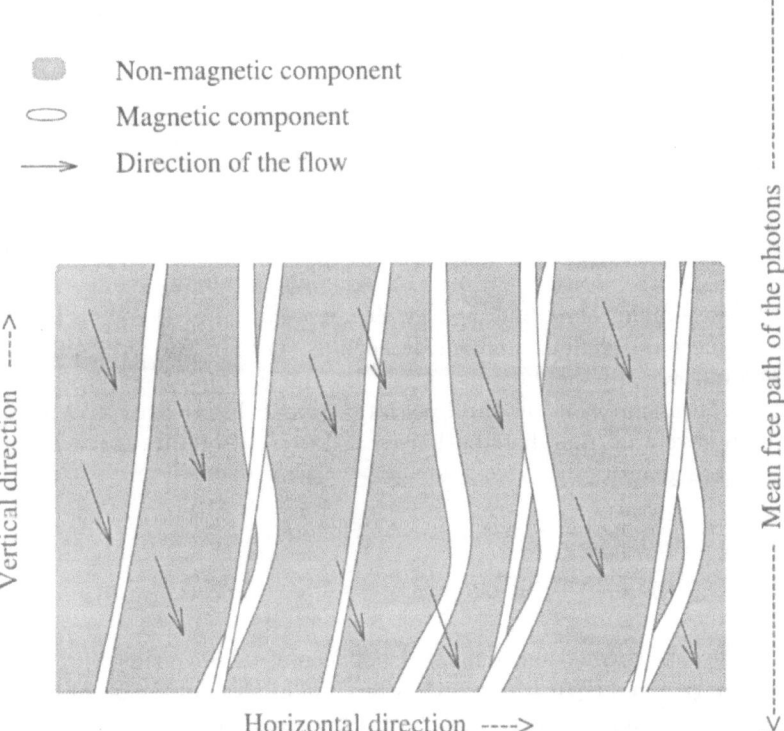

Fig. 1.   Example of a MISMA, i.e., an atmosphere whose properties vary over scales smaller than the mean-free-path of the photons. It represents an unmagnetized downflowing atmosphere pierced by static magnetic fingers.

procedures based on Equation (1) allow us to decipher the variation of theses parameters along the atmosphere (see, e.g., Ruiz Cobo and del Toro Iniesta, 1992). If MISMAs indeed play a role in the solar photosphere, then Equation (2) should replace Equation (1) when retrieving physical information from the Stokes profiles. However, there is no available expression for the average absorption matrix that is suitable for inversion. Here we write down a closed analytical form for $< K >$, assuming that just magnetic field strengths and velocities show micro-variations. Temperatures, pressures etc. are supposed to vary over larger scales. Note that, under LTE conditions, the previous assumption simplifies the computation of S' in Equation (2). The weighted average source function S' is just that given by the local temperature of the atmosphere. Velocity and magnetic field strength micro-variations will be described by a bivariate normal distribution. This hypothesis implies that, when independently considered, both the local magnetic field strength and the velocity present a range of values according to a normal distribution. When jointly considered, there may be correlations between the individual values of the strength and the velocity (see, e.g., Martin, 1971, for the properties of this type of probability density functions).

The magnetic field strength and the velocity affect the absorption matrix K by shifting the various Zeeman components of the line according to,

$$\Delta\lambda = \lambda_U + p\lambda_B, \tag{3}$$

λ_U being the Doppler shift and λ_B the Zeeman shift. For the time being, just normal Zeeman triplets are considered; p is zero for the π component, $+1$ for the red σ component, and -1 for the blue σ component (see LD). Because of the assumption on the bivariate normal distribution, $\Delta\lambda$ itself turns out to have a normal distribution (see Martin, 1971). Its mean μ and variance σ^2 are given by

$$\mu = \mu_U + p\mu_B, \tag{4}$$
$$\sigma^2 = \sigma_U^2 + p^2\sigma_B^2 + 2p\sigma_U\sigma_B C_{UB}. \tag{5}$$

C_{UB} stands for the correlation coefficient between λ_U and λ_B ($|C_{UB}| = 1$ for perfect correlation and $C_{UB} = 0$ for no correlation at all). On the other hand, μ_U, σ_U^2, μ_B and σ_B^2 represent, respectively, the mean and the variance of the Doppler and Zeeman shifts. Because $\Delta\lambda$ follows a normal distribution, the effect of this particular MISMA on the transfer equation will be to change the Doppler widths of the Voigt and Faraday-Voigt profiles associated with each component (see LD). The matrix $< K >$ becomes formally like K but, according to Equation (5), each Zeeman component has its own width $\sigma(p)$. Different broadenings for different components lead to asymmetrical Stokes profiles; see the Stokes V profile in Figure 2(a).

The extension of Equations (4) and (5) to anomalous Zeeman patterns is straightforward. These equations keep their form except that each component has a value of p which, in general, differs from 0 or ±1. Now p depends on the particular Zeeman splitting of the component (see LD).

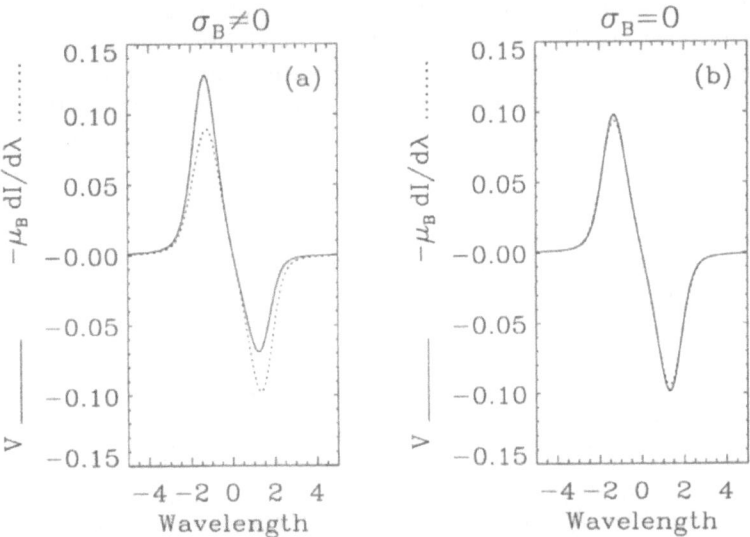

Fig. 2. Failure of the equation $V \simeq -\mu_B(dI/d\lambda)$ for a MISMA. Such a relationship holds for an atmosphere having a single magnetic field strength ($\sigma_B = 0$; case b). However, it fails in case (a), where the magnetic field has a distribution of values correlated with the velocities (see text for details). Wavelengths are given in units of the Doppler width σ_v.

4. Need for upgrading current methods to determine magnetic fields

We pointed out in the Introduction that the existence of MISMAs may have serious implications on the diagnostic techniques presently used to infer properties of the solar photosphere. Here we present a brief study which supports our claim. It shows that standard magnetographic techniques have problems to determine the average field strength of a MISMA. Specifically, the equation used to calibrate the longitudinal magnetographs,

$$V \simeq -\lambda_B \cos\theta \frac{dI}{d\lambda}, \tag{6}$$

is shown to be no longer valid when correlations between velocity and magnetic field are present, i.e.,

$$V \not\simeq -\mu_B \cos\theta \frac{dI}{d\lambda}. \tag{7}$$

Equation (6) transforms the circular polarization observed in the wing of a line into the longitudinal magnetic field $\lambda_B \cos\theta$ (θ represents the angle between the magnetic field vector and the line-of-sight). The so-called calibration constant $dI/d\lambda$ is obtained from the observed intensity profile. There are well known physical reasons which limit the applicability of Equation (6) to solar observations. The case of a magnetograph is used here as a schematic example, though. It illustrates how the existence of micro-structure may induce large errors on the measurements. In other words, magnetographic measurements are considered just as a prototype of magnetic field determination.

Figure 2 overplots the left-hand-side and right-hand-side of Equation (7), as obtained by solving Equation (2) under the approximations described in Section 3. For the sake of simplicity we adopted a Milne-Eddington atmosphere with longitudinal magnetic fields (i.e., $\cos\theta = 1$). In case (a) the parameters which describe the micro-structure are; $\mu_U = 0$, $\mu_B = 0.2$ and $\sigma_B = 0.05$, all these quantities being expressed in units of the Doppler width σ_U. In addition, we consider a perfect correlation between magnetic fields and velocities $C_{UB} = 1$. Other parameters of the Milne-Eddington atmosphere (line absorption coefficient, damping constant, etc., see LD) have been chosen to represent the Fe I 6302.5 Å line in the quiet sun. The line and the atmosphere are the same in case (b), except for σ_B which is now set to zero. Equation (6) is correct to a high degree of approximation when the magnetic field strength is unique, i.e., case (b). On the contrary, a modest range of possible field strengths produces significant differences between the two sides of Equation (7) ($\sigma_B/\mu_B = 25\%$ in case a). A magnetograph working on the red flank of the profile in Figure 2(a) would yield a +35% erroneous value for the average magnetic field μ_B.

5. Conclusions

Sánchez Almeida et al. (1995) argued that the structuring of photospheric magnetic fields over scales smaller than the mean-free-path of the photons may explain the observed Stokes line asymmetries. Should these micro-structures exist, the techniques used to infer properties of the photosphere from observations need an upgrading. For example, Section 4 shows how the equation used to calibrate longitudinal magnetographs does not hold for MISMAs (i.e., MIcro-Structured Magnetic Atmospheres).

To incorporate this type of atmospheres into the diagnostic techniques, the transfer equation for a MISMA, Equation (2), has to be written down in terms of a small set of parameters. Here we have introduced an analytical expression for the equation which accounts for micro-variations of the magnetic field strength and the velocity. The micro-structure is, in this particular case, described by the average velocity μ_U, the average magnetic field strength μ_B, the velocity broadening σ_U, the magnetic broadening σ_B, and the correlation coefficient between magnetic fields and velocities C_{UB}. The advantage of the present way of describing a MISMA is that it is straightforward to compute the emerging polarization employing exiting synthesis codes based on the traditional Unno-Rachkovsky equation (1). One simply has to incorporate different Doppler widths for the different Zeeman components, as given by Equation (5).

Acknowledgements

Thanks are due to J. Trujillo Bueno and V. Martínez Pillet for exciting discussions during the development of the work. ELD acknowledges the support of the Spanish DGYCIT for a sabbatical stay at the IAC. The work has been partly funded by the DGYCIT under project PB 91-0530.

References

Bünte, M., Solanki,S. K., and Steiner, O.: 1993, *Astron. Astrophys.* **268**, 736
del Toro Iniesta, J. C. and Ruiz Cobo, B.: 1996, *Solar Phys.*, this proceedings
Landi Degl'Innocenti, E.: 1992, in F. Sánchez, M. Collados, and M. Vázquez (eds.), *Solar Observations: Techniques and Interpretation*, Cambridge Univ. Press, 73 (LD)
Lites, B. W., Leka, K. D., Skumanich, A., Martínez Pillet, V., and Shimizu, T.: 1995, *Astrophys. J.*, submitted
Martin, B. R.: 1971, *Statistics for Physicists*, Academic Press, London
Ruiz Cobo, B. and del Toro Iniesta, J. C.: 1992, *Astrophys. J.* **398**,375
Sánchez Almeida, J., Collados, M., and del Toro Iniesta, J. C.: 1988, *Astron. Astrophys.* **201**, L37
Sánchez Almeida, J., Landi Degl'Innocenti, E., Martínez Pillet, V., and Lites, B. W.: 1995, *Astrophys. J.*, submitted (Paper I)
Sánchez Almeida, J. and Lites, B. W.: 1992, *Astrophys. J.* **398**, 359
Schüssler, M.: 1985, in W. Deinzer, M. Knölker, and H. H. Voigt (eds.), *Small-Scale Magnetic Flux Concentrations in the Solar Atmosphere*, Vandenhoeck and Ruprecht, 103
Solanki, S. K.: 1989, *Astron. Astrophys.* **224**, 225
Solanki, S. K. and Montavon, C. A. P.: 1993, *Astron. Astrophys.* **275**, 283
Stenflo, J. O., Harvey, J. W., Brault, J. W., and Solanki, S. K.: 1984, *Astron. Astrophys.* **131**, 333
Thomas, J. H. and Weiss, N. O.: 1992, in J. H. Thomas and N. O. Weiss (eds.), *Sunspots: Theory and Observations*, Kluwer, 3

A COMPLEX DIAGNOSTIC OF SOLAR PROMINENCES

P. HEINZEL

Observatoire de Meudon, DASOP-URA326, F-92195 Meudon Cedex, France
Astronomical Institute, Academy of Sciences of the Czech Republic
CZ-25165 Ondřejov, Czech Republic

V. BOMMIER

Observatoire de Meudon, DAMAP-URA812, F-92195 Meudon Cedex, France

and

J.-C. VIAL

Institut d'Astrophysique Spatiale, Université Paris XI, Bat. 121, F-91405 Orsay, France

Abstract. We use the polarimetric and intensity measurements of Hα and HeI D₃ lines in solar prominences to derive the *true* geometrical thickness for several quiescent prominences. The electron densities, derived from the collisional depolarization in Hα by Bommier *et al.* (1994), are used to evaluate the thickness from the emission measure. The emission measure was obtained from the theoretical correlation with the Hα integrated intensity, according to Gouttebroze, Heinzel, and Vial (1993). Theoretical electron densities obtained by latter authors are also compared with those of Bommier *et al.* (1994) and we find a very good agreement between them. The prominence geometrical thickness exhibits a relatively large range of values from about 100 km up to a few 10^4 km. The plasma densities vary by almost two orders of magnitude in the observed structures, but the total column mass in the direction perpendicular to the prominence sheet seems to be fairly constant for the set of prominences studied.

Key words: Prominence diagnostic – Polarization – Emission measure

1. Introduction

The diagnostic of solar prominences has focussed on the physical conditions inside (see e.g. Priest, 1989; Ruždjak and Tandberg-Hanssen, 1990 and more recently Tandberg-Hanssen, 1995). As far as the cooler central part of prominences is concerned, an extensive NLTE radiative-transfer modelling has been performed recently by Gouttebroze, Heinzel, and Vial (1993 - GHV) and Heinzel, Gouttebroze, and Vial (1994 - HGV). Various laws that have been derived from the computation of the radiative outputs from a wide range of isobaric and isothermal models concern, for instance, the important correlation between the Hα emisssion and the emission measure *EM* of the cool plasma. Both models and the theoretical correlations constitute a powerful diagnostic tool provided that the hypotheses made in the computations are properly taken into account in the comparison with the observations.

In this respect the geometry is a key parameter to be considered. It is evident that most prominences, as observed with higher spatial resolution,

Solar Physics **164**: 211–222, 1996.
© 1996 *Kluwer Academic Publishers.*

do exhibit certain fine structure and all show a finite extent in the plane of sky (and in the perpendicular direction, as shown by filament observations). However, these *apparent* dimensions have little meaning as far as the physical conditions inside the prominences are concerned. On the contrary, we need to know the real volume of the emitting structures, i.e. to determine the parameter which is usually called the *effective* or *true* thickness. For example, this can be achieved using the above mentioned correlation between the total line intensity and EM which is the product of the square of the electron density and effective thickness. However, this method requires an independent determination of the electron density, at least a mean value over the prominence size as a first approximation.

Such an approach has been applied by Hirayama (1986) who used the Stark effect on Balmer lines for the determination of the electron density. From the integrated intensities of the Balmer line H_9 he has computed the effective thickness of several prominences or their parts, using simple analytical correlation between EM and the line intensity (this is possible for high Balmer lines which arise from hydrogen levels having almost equilibrium populations). The range of thicknesses obtained by Hirayama (1986) is between a few km and 10^4 km. He also discussed the role of the filamentary structure on the values of the filling factor. However, this analysis largely depends on the way how the Stark widths are evaluated. The ion-dynamics contribution to the Stark width could be important in this respect (Ch. Stehlé – private communication), leading to lower values of the electron density.

While the Stark-effect method can be used for the electron densities higher than, say, 2×10^{10} cm^{-3} (Hirayama 1986, 1990), there is another way of determining the electron density by using the collisional depolarization analysis (Bommier, Leroy, and Sahal-Bréchot, 1986). This technique works for lower densities, less than about 5×10^{10} cm^{-3}. Hirayama (1990) tried to use the results of the latter authors which concern the depolarization of the hydrogen $H\beta$ line, but he arrived at rather unrealistically large values of the effective thickness $4 - 7 \times 10^4$ km, which are about one order of magnitude higher then his results from H_9. We discuss this problem later on.

In the present paper, we combine the unique information obtained by polarimetric and intensity measurements in $H\alpha$ and He D_3 lines by Bommier *et al.* (1994-BLLS) with NLTE modelling of GHV, in order to derive the true geometrical thickness and to estimate other prominence parameters closely related to it (i.e. the line-of-sight filling factor and the total column mass). For this we use the theoretical GHV correlation between the integrated $H\alpha$ intensity and the plasma emission measure, in a similar way as discussed above. Sections 2 and 3 describe the measurements which lead to the determination of the electron density. Section 4 summarizes the re-

sults of the GHV modelling. In section 5 we compare the range of values
of the electron density as derived from modelling with that from polarimetric measurements. In section 6, we concentrate on the determination of
the emission measure from which we derive the true geometrical thickness
(Section 7). The different filling factors which can influence the emission
measure and consequently the determination of the electron density and the
geometrical thickness are considered in section 8. The implication in terms
of determining the total (column) mass is described in section 9. In section
10 we stress the importance of this combination of non-polarimetric and polarimetric measurements and NLTE modelling for the determination of the
true geometrical thickness of prominences and other parameters.

2. Linear polarization

Because of the low density, the basic mechanism of the prominence line
emission is the resonance scattering of the incident photospheric and chromospheric radiation. This irradiation of prominences is highly *anisotropic*
and the anisotropy increases with increasing prominence height above the
solar surface. Resonance scattering of anisotropic radiation then leads to *linear* polarization of the emitted photons. However, the presence of a (weak)
magnetic field generally decreases the amount of such linear polarization
and rotates the polarization direction with respect to the solar limb. This is
known as the Hanle effect.

The first detection of a small amount of linear polarization in prominences, particularly in the Hα line, was made by Lyot (1936). During the
last two decades, several new polarimetric observations have been carried
out and we shall concentrate here on the data from Pic-du-Midi Observatory. These data, obtained in the Hα and Hβ lines of hydrogen and in the
HeI D_3 line, have been extensively analysed in order to derive the prominence magnetic fields (see reviews by Leroy, 1989 and Landi Degl'Innocenti,
1990).

Quite recently, BLLS have published a detailed study of a set of 18 measurements in different prominences. Linear polarization and line integrated
intensities have been measured in the Hα and D_3 lines. These two lines allow
to resolve the ambiguity in the magnetic-field vector determination. However, the analysis of the Hα line is more complicated due to its non-negligible
optical thickness which causes a lower degree of polarization (a partial low-
ering of the radiation-field anisotropy). A typical value of the magnetic field
intensity is below 10 Gauss.

3. Collisional depolarization and electron densities

Besides the Hanle effect, another depolarizing mechanism is the so-called *collisional depolarization* caused by collisions of the emitting atoms with protons and free electrons. Having observed linear polarization in two different lines where one is sensitive to collisional depolarization (Hα in our case) and the other one almost insensitive (like the D_3 line of helium), one can disentangle between the depolarizing effect of the magnetic field and colliding particles. For the above mentioned set of observations in these two lines, a complete determination of the field vector and the electron density was made in all 18 observed prominence structures - for details see BLLS. Electron densities determined by this method range from 2.5×10^9 to 6.3×10^{10} cm^{-3}, for the set of studied prominences.

However, it should be noted that the polarimetric data used for the above analysis of the magnetic field and electron densities have been selected with respect to certain criteria and, thus, do not represent an arbitrary set of measurements. Particularly, low-brightness prominences have been excluded because the accuracy of the measurements of their linear polarization was insufficient. On the other hand, some of the brightest objects (or their parts) indicated a polarization degree too low to be accurately detected. This was ascribed to the higher electron density (collisional depolarization), and these measurements were also excluded because a joint electron-density and magnetic-field-vector diagnostic was impossible due to the lack of polarization. Nevertheless, higher brightness doesn't mean automatically high density since, as we shall demonstrate below, the low density may be compensated by a very large extension of the prominence.

4. Recent NLTE models

Another way of determining the electron density and other plasma parameters is to perform detailed NLTE computations for specific prominence models. Such models can be either in magnetohydrostatic equilibrium (Heasley and Mihalas, 1976) or simple isobaric-isothermal models (Heasley and Milkey, 1978, 1983; GHV); for a review see Tandberg-Hanssen (1995). However, until recently there has been no systematic study of relations between various prominence plasma parameters and the output radiation. Starting from schematic 1D isobaric-isothermal models, GHV have computed an extensive grid of 140 NLTE models covering most typical quiescent-prominence conditions. The models are parametrized by the gas pressure P, the prominence-slab thickness D, the kinetic temperature T, and the microturbulent velocity v_t. These computations have been done for hydrogen using a model atom with 20 levels plus continuum and taking into account partial-frequency-redistribution (PRD) effects in the Lyman-α and Lyman-β lines.

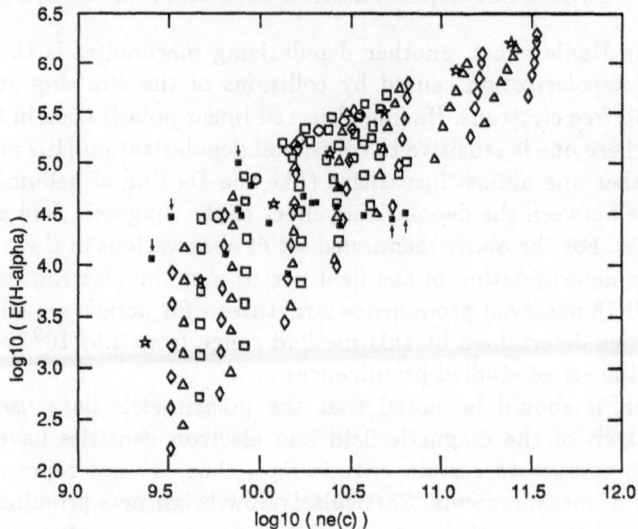

Fig. 1. $E(\mathrm{H}\alpha)$ versus n_e. E is in units of ergs cm^{-2} s^{-1} sr^{-1}, $n_e(c)$ means the electron density in the 1D-slab center. Full squares are from depolarization analysis. Other symbols correspond to different temperatures used in GHV models: circles - 4300 K, squares - 6000 K, triangles - 8000 K, losanges - 10000 K, stars - 15000 K. Arrows indicate the measurements which lie outside the range of GHV models.

As already demonstrated by Heinzel, Gouttebroze, and Vial (1987), PRD leads to higher electron densities inside the prominence body, as compared to the complete-redistribution (CRD) approach. Moreover, PRD changes significantly the Lyman-α and Lyman-β radiation fields inside the prominence and also the emergent intensities. As a consequence, the Hα *integrated* line intensity is also substantially changed.

In a subsequent paper, HGV used all these 140 NLTE models to study various correlations between prominence plasma parameters and the radiation emitted by hydrogen lines and Lyman continuum. We shall use some of these correlations in this paper. Of particular interest are the following two results of HGV: the correlation between the integrated Hα line intensity E and the electron density n_e, and an almost *unique* relation between E and the emission measure $EM = n_e^2 \times D$.

In Figure 1 we demonstrate the first correlation, i.e. between $E(\mathrm{H}\alpha)$ and n_e, where n_e corresponds to the values of the electron density in the central parts of the 1D prominence slab. Actually, as can be seen in GHV, the electron density is almost constant inside the 1D slab for typical conditions and only higher-pressure models exhibit a slow decrease of n_e towards both surfaces (these models are grouped around the right upward corner in Figure 1).

Concerning E, this corresponds to *normally* emergent Hα radiation. Different symbols in Figure 1 correspond to different kinetic temperatures considered by GHV (see figure captions). All 140 models are plotted, for gas pressures P ranging from 0.01 to 1.0 dyn cm^{-2} and the geometrical thickness D lying between 200 and 10000 km.

5. Electron densities - comparison with GHV models

It is interesting to see how the electron densities derived from polarimetric measurements fit these theoretical NLTE predictions. However, first we must convert the measured $E(\text{H}\alpha)$ (which corresponds to a particular direction of the output radiation) to the values which are consistent with Figure 1, i.e. to normally-emergent intensities. To achieve this, we have used the integrated-intensity conversion function and related formulas as published in HGV (note that only for an optically very thin medium a simple linear conversion can be used, which is not generally the case in our data). Converted integrated intensity E and the corresponding n_e from BLLS tables have been inserted into Figure 1 (see the full squares). As a result, we see a rather good agreement between the NLTE results of GHV and those obtained from the analysis of linear polarization by BLLS. A few points, denoted by arrows, which are just on the border of the range of GHV models, correspond in fact to geometrical thicknesses not considered by GHV.

6. Emission measure

The emission measure EM obtained from Hα integrated intensities can be used to derive n_e, *assuming* a reasonable range of D (HGV) and such a method was recently applied to the diagnostic of post-flare loops (Schmieder *et al.*, 1995). The aim of this paper is to use *independently* derived n_e from polarimetric data and then, from the EM, to evaluate D which is a parameter largely unknown in prominence studies. Correlation between $E(\text{H}\alpha)$ and the emission measure $EM = n_e^2 \times D$ is displayed in Figure 2, for different temperatures. The first part of this curve which is almost linear corresponds to a line center optical thickness less than unity. We also see a rather weak dependence on the kinetic temperature. Note that this correlation contains D implicitly in the definition of EM and, thus, is valid for a broader range of D than covered by GHV models.

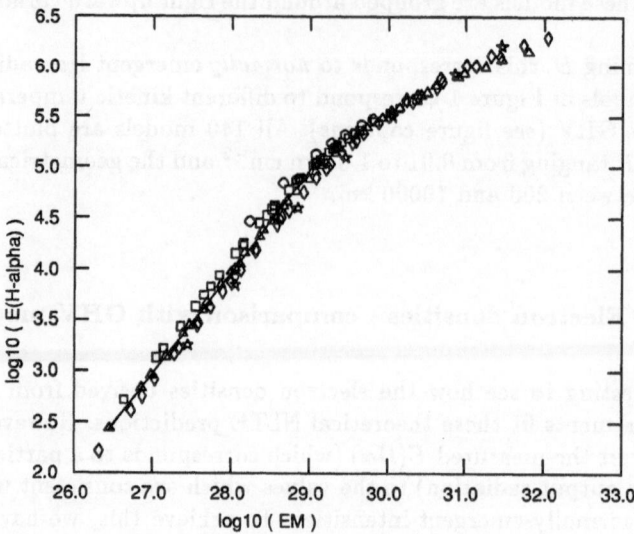

Fig. 2. $E(H\alpha)$ versus EM. The temperature symbols are same as in Figure 1. A full-line fit to $T = 8000$ K was used to derive EM form observed intensities.

7. True geometrical thickness

We have already seen that the prominence geometrical thickness D was one of the input parameters for GHV models and thus the output radiation fields strongly depend on it. Even if we can determine very well EM just from the observed Hα emission, from it we obtain the electron density to within the uncertainty of the unknown D (see the previous section). There is practically no way how to determine D from the prominence morphology. From the disk observations of the filaments we can measure the *apparent* thickness, which is related to D by the well-known *filling factor* f_D. Various authors report largely different filling factors, ranging from 1 to $10^{-2} - 10^{-3}$, or even lower.

The principal goal of this paper is to derive the prominence geometrical thickness D by using a very complex diagnostic based on polarimetric observations, measurements of unpolarized Hα radiation and using sophisticated modelling techniques. All the necessary tools have been described in previous sections. The basic approach is to use EM obtained from Hα data and to evaluate D from it by inserting the n_e derived from polarimetric analysis. To do this, we assume a typical prominence temperature around 8000 K, but the determination of EM is not very sensitive to T in the range 6500 K to 10000 K (Figure 2). Note that for E we use the converted values as discussed above. For all prominences analysed by BLLS we plot the relation between D and n_e in Figure 3. We see clearly a good correlation between these two parameters, showing that the most extended prominences have the lowest (electron) density and vice versa. Quantitatively, for reasonable values of n_e (discussed in previous sections), we obtain a relatively large scatter of D, ranging from about 100 km to a few 10^4 km. We shall discuss this behaviour later on. Unfortunately, we cannot estimate the filling factor f_D here because we do not know the apparent thicknesses of these prominences observed on the limb. Nevertheless, we have inserted into Figure 3 two dashed lines showing the typical range of apparent thicknesses, i.e. between the resolution limit (around 1 arcsec) and some 10 arcsec. The arrows indicate the same measurements as in Figure 1 and we see here that the corresponding D really lies outside the range of GHV models.

A correlation similar to Figure 2 was also plotted for the Hβ line, using again the full set of GHV models. For $T = 7000$ K we find the relation $E(\text{H}\beta) \simeq 10^{-25} EM$, while Hirayama (1990) gets $E(\text{H}\beta) = 5.3 \times 10^{-26} EM$ for the same kinetic temperature. This is probably due to the relatively high radiation temperature used by Hirayama to derive his relation. If our theoretical correlation is correct, then Hirayama's D based on the Hβ can be lowered by a factor of two. Moreover, in this paper we have converted all $E(\text{H}\alpha)$ to those corresponding to radiation normally emergent from the prominence sheet. As $E(\text{H}\beta)$ used by Hirayama is the line-of-sight value (J.L. Leroy – private communication), the derived effective thickness can be still somewhat higher as compared to our results.

8. Filling factors

We have already introduced a *line-of-sight* filling factor f_D. Its knowledge is important in those cases where we do not know D. In this paper we have determined D for a set of quiescent prominences, but there arises the question: to what extent the prominence porosity can affect the results so far obtained ? Prominence filamentary structure was reviewed by Heinzel and Vial (1992) and by Mein (1994). In fact, apart from f_D, we have to introduce another filling factor, say f_S, which characterizes the degree to which the entrance slit of the instrument is filled by the fine-structure elements. Let us now discuss the effect of both filling factors on our results.

First, as shown by HGV (see their Figure 21), the electron density is strongly dependent on the gas pressure, provided that we keep the incident solar radiation equal for all models. For typical temperatures and for $n_e \simeq 10^{10}$ cm^{-3}, the electron density exhibits a very weak dependence on D ranging from 200 to 10000 km (see our Figure 1). This suggests that even in highly inhomogeneous prominences, the individual threads will have the same ionization degree as if they would be isolated. The porosity can affect

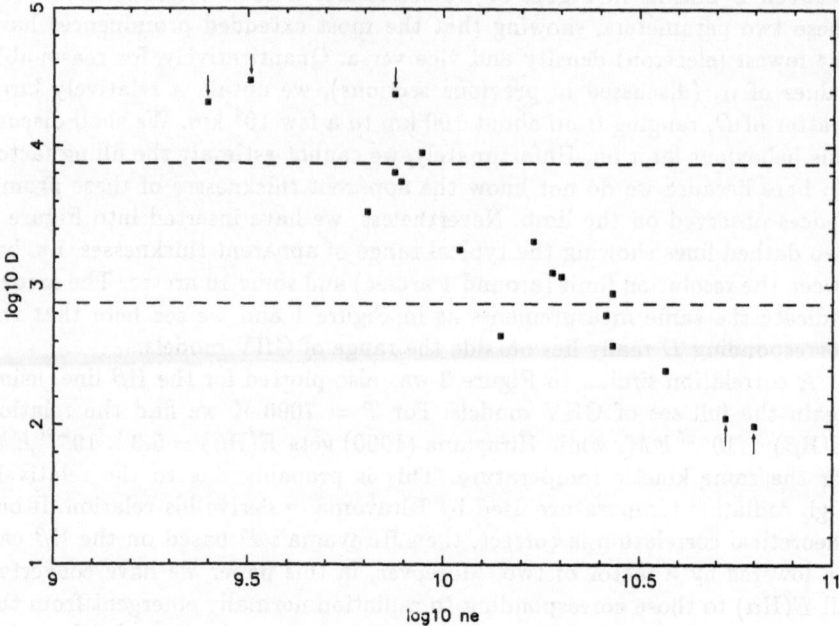

Fig. 3. True geometrical thickness D (in km) versus n_e, for the whole set of prominences studied by BLLS. Two dashed lines show the thickness of 1 and 10 arcsec, respectively. The arrows indicate the same squares as in Figure 1 (see the text).

the ionization in rather thin sheets around threads due to easier penetration of the incident Lyman continuum radiation into the prominence body (see also discussion in Milkey and Mihalas, 1976). Threads which are 500 km thick with a typical $P \simeq 0.1$ dyn cm^{-2} are already optically thick in Lyman continuum (GHV), so only the surface layers can be affected.

Second, it can be easily demonstrated that several fine-structure threads having the same constant Hα source function, will give exactly the same emergent line intensity as a 1D homogeneous slab with D equal to the sum of the thicknesses of all the threads and having the same source function (see Heinzel, 1989). Concerning the Hα source function, HGV again showed (see their Figures 18 and 19) that $S($H$\alpha)$ is almost independent of the thickness, unless the line-center optical thickness significantly exceeds unity.

The above discussion leads us to the conclusion that Figure 1 is not significantly affected by the prominence porosity. The same arguments can be also applied to Figure 2, provided that we consider a constant n_e in individual threads and much lower density in the interthreads medium. The degree of linear polarization in Hα is not influenced too much by the porosity.

So far, our conclusions seem to be unaffected by the presence of the prominence filamentary structure, assuming f_S to be unity. This condition requires high-resolution observations, where a typical thread fully fills the entrance slit of the spectrograph. If the threads are much thinner and sparsely distributed, then $f_S < 1$. If they are thinner but densely distributed (bright parts of prominences), f_S can be still around unity. For $f_S < 1$ the true $E(H\alpha)$ will be higher, leading to higher EM and thus D. Again, f_S will not substantially affect the degree of linear polarization, and thus n_e.

How higher values of $E(H\alpha)$ and thus of EM will affect our determination of D ? For example, assuming n_e to be correct, the points in Figure 3 with highest electron densities have a very small D. To see these prominences as filaments, the filling factor f_S must be much lower than unity. Then the real E will be higher, leading to higher EM and, keeping the same electron density, D will be higher (probably more realistic).

9. Implications for total mass

Our knowledge of the total mass contained in a quiescent prominence is critical for any considerations concerning the prominence formation and stability. This mass depends on the density and *true* volume. The prominence density is roughly equal to $\rho = 1.4\,m_H n_H$, where m_H is the mass of the hydrogen atom and n_H is the number density of hydrogen. For $log\,n_e \simeq 9.5$ we get also $log\,n_H \simeq 9.5$ (HGV, Figure 23) and thus $\rho \simeq 7.4 \times 10^{-15}$ g cm^{-3}. With the corresponding $D \simeq 2 \times 10^4$ km from Figure 3 we get a *total column mass* as $M = \rho D \simeq 1.5 \times 10^{-5}$ g cm^{-2}. For $log\,n_e \simeq 10.5$ we will have $log\,n_H \simeq 11$ for $T = 8000$ K. Then $\rho \simeq 1.5 \times 10^{-13}$ g cm^{-3} and $M \simeq 10^{-5}$ g cm^{-2} (with $D \simeq 400$ km). Apart from uncertainties in the ionization degree (which depends on T for higher densities), the total column mass seems to be fairly constant for the set of investigated prominences. However, ρ varies by almost two orders of magnitude.

From the Balmer H_9 line, Hirayama (1986) gets higher electron densities than BLLS (a selective property of the Stark-effect method he used) and thus his total plasma density is also somewhat higher than ours, around 10^{-12} g cm^{-3}. However, his values for the effective thickness D are lower (a few km to 10^4 km) which indicates a similar total column mass as we have found in this paper.

10. Discussion and conclusions

The main objective of this paper was to derive the true geometrical thickness D of quiescent prominences, in the direction normal to the prominence sheet.

For this purpose we used, for the first time, a combination of polarimetric data (degree of linear polarization) in Hα and HeI D_3 and the integrated Hα line intensities. Polarimetric data allow for an independent determination of electron densities (see BLLS) and these, together with the emission measure obtained from the integrated Hα emission, give us the information about the thickness D. BLLS electron densities have been found to be in good agreement with those derived from the NLTE models of GHV.

For 18 measurements, we obtained a rather large scatter in D, ranging from a few hundreds km up to a few 10^4 km. This confirms the large range of the line-of-sight filling factors, found by various authors. We have also estimated a total column mass along D and found that it has about the same value for all observed prominences, around 10^{-5} g cm^{-2}. On the other hand, the plasma density varies by almost two orders of magnitude, with typical values between $10^{-14} - 10^{-13}$ g cm^{-3}.

The plot in Figure 3 can be easily explained in terms of the relation $EM = n_e^2 \times D$, *assuming* that the range of EM (and thus of $E(H(\alpha))$ is rather narrow. For a broad range of E, as computed by GHV, the plot $log\,D$ vs. $log\,n_e$ is fully filled by points corresponding to individual models. However, in GHV models there is no relationship between the model parameters P, D and T which would lead us to prefer some models and thus to narrow the range of observable intensities.

Because the knowledge of D is critical for any spectral diagnostic of prominences and prominence-like objects (loops, spicules, mottles etc.), as well as for their MHD modelling, we suggest to carry out new polarimetric and intensity measurements, with higher spatial resolution and for a broader range of brightnesses. THEMIS will be well suited for such a task. Also by observing a large set of different prominences, one could reveal possible relationships between derived thickness D and other parameters like density, total column mass, etc. Finally, to properly interpret such new data, an improvement in NLTE modelling and treatment of polarized radiation transfer is necessary, taking into account prominence porosity and, therefore, 2D or even 3D effects.

Acknowledgements

We thank Drs. P. Démoulin, P. Mein and E. Tandberg-Hanssen for stimulating discussions and the referee Dr. J. Trujillo Bueno for his constructive comments. P.H. and V.B. acknowledge the support of GdR 'Magnétisme dans les étoiles de type solaire' which enabled them to participate at the *Solar Polarization Workshop* in St. Petersburg. P.H. appreciates the support of CNRS during the course of this work and kind hospitality of the Observatoire de Meudon.

References

Bommier, V., Leroy, J.-L., and Sahal-Bréchot, S.: 1986, *Astron. Astrophys.* **156**, 90.

Bommier, V., Landi Degl'Innocenti, E., Leroy, J.-L., Sahal-Bréchot, S.: 1994, *Solar Phys.* **154**, 231 (BLLS).

Gouttebroze, P., Heinzel, P., and Vial, J.-C.: 1993, *Astron. Astrophys. Suppl. Ser.* **99**, 513 (GHV).

Heasley, J.N. and Mihalas, D.: 1976, *Astrophys. J.* **205**, 273.

Heasley, J.N. and Milkey, R.W.: 1978, *Astrophys. J.* **210**, 827.

Heasley, J.N. and Milkey, R.W.: 1983, *Astrophys. J.* **268**, 398.

Heinzel, P.: 1989, *Hvar Obs. Bull.* **13**/1, No. 1, 317.

Heinzel, P. and Vial, J.-C.: 1992, ESA SP-344, 57.

Heinzel, P., Gouttebroze, P., and Vial, J.-C.: 1987, *Astron. Astrophys.* **183**, 351.

Heinzel, P., Gouttebroze, P., and Vial, J.-C.: 1994, *Astron. Astrophys.* **292**, 656 (HGV).

Hirayama, T.: 1986, in A.I. Poland (ed.), *Coronal and Prominence Plasmas*, NASA CP2442, 149.

Hirayama, T.: 1990, in V. Ruždjak and E. Tandberg-Hanssen (eds.), *Dynamics of Quiescent Prominences*, Lecture Notes in Physics 363, Springer-Verlag, Berlin, 187.

Landi Degl'Innocenti, E.: 1990, in V. Ruždjak and E. Tandberg-Hanssen (eds.), *Dynamics of Quiescent Prominences*, Lecture Notes in Physics 363, Springer-Verlag, Berlin, 206.

Leroy, J.-L.: 1989, in E.R. Priest (ed.), *Dynamics and Structure of Quiescent Solar Prominences*, Kluwer Acad. Publ., Dordrecht, 77.

Lyot, B.: 1936, *Compt. Rend. Acad. Sci.* **202**, 392.

Mein, P.: 1994, in V. Rušin, P. Heinzel, and J.-C. Vial (eds.), *Solar Coronal Structures*, Proc. IAU Coll. 144, Veda Publ. House, Bratislava, 289.

Priest, E.R. (ed.): 1989, *Dynamics and Structure of Quiescent Solar Prominences*, Kluwer Acad. Publ., Dordrecht.

Ruždjak, V. and Tandberg-Hanssen, E. (eds.): 1990, *Dynamics of Quiescent Prominences*, Lecture Notes in Physics 363, Springer-Verlag, Berlin.

Schmieder, B., Heinzel, P., Wiik, J.E., Lemen, J., Anwar, B., Kotrč, P., and Hiei, E.: 1995, *Solar Phys.* **156**, 337.

Tandberg-Hanssen, E.: 1995, *The Nature of Solar Prominences*, Kluwer Acad. Publ., Dordrecht.

POLARIZED RADIATION DIAGNOSTICS OF MAGNETOHYDRODYNAMIC MODELS OF THE SOLAR ATMOSPHERE

O. STEINER, U. GROSSMANN-DOERTH and M. SCHÜSSLER

Kiepenheuer-Institut für Sonnenphysik, Schöneckstrasse 6, D-79104 Freiburg, Germany

and

M. KNÖLKER

High Altitude Observatory, NCAR*, P.O. Box 30000, Boulder, CO 80307, USA

Abstract. Solar magnetic elements and their dynamical interaction with the convective surface layers of the Sun are numerically simulated. Radiation transfer in the photosphere is taken into account. A simulation run over 18.5 minutes real time shows that the granular flow is capable of moving and bending a magnetic flux sheet (the magnetic element). At times it becomes inclined by up to 30° with respect to the vertical around the level $\tau_{5000} = 1$ and it moves horizontally with a maximal velocity of 4 km/s. Shock waves form outside and within the magnetic flux sheet. The latter cause a distinctive signature in a time series of synthetic Stokes V-profiles. Such shock events occur with a mean frequency of about 2.5 minutes. A time resolution of at least 10 seconds in Stokes V recordings is needed to reveal an individual shock event by observation.

Key words: Magnetohydrodynamics – Polarization – Solar Atmosphere

1. Introduction

Magnetic elements are small scale magnetic flux concentrations in the solar photosphere. They contain a magnetic flux of approximately 10^{18} Mx with a field strength of the order of 1.5 kG. Their size of around 200 km is at the spatial resolution limit of present-day solar telescopes. The understanding of the nature of these magnetic fields is essential for a proper physical description of cool stellar atmospheres. These cannot be considered as plane parallel since the discrete nature of the magnetic field strongly structures the atmosphere in all of its different layers, convection zone, photosphere, chromosphere, transition zone, and corona.

Much of what we know today about magnetic elements has been derived from observations in the visible and infrared part of the spectrum, and mostly by analyzing spectral lines in polarized light. Since Stokes profiles (with the exception Stokes I) originate *only* in regions of magnetic fields they transmit information about the physical nature of magnetic elements even if these elements occupy only a small fraction of the observed part of

* The National Center for Atmospheric Research is sponsored by the National Science Foundation

the solar surface. Such observations have provided us with fairly accurate data about field strength, velocity, and temperature in the photospheric height range of magnetic elements. A comprehensive overview of these results has been given by Solanki (1993). Recently, speckle polarimetry together with speckle reconstructed white light images have provided a picture of the spatial distribution of magnetic small scale structures (Keller, 1992).

Despite these successes the intrinsic structure and the dynamical behaviour of magnetic elements remains obscure, mainly because of the limited spatial resolution. This applies not only in the direction parallel to the solar surface but also with respect to the height in the atmosphere. A second problem consists in the difficulty to obtain time sequences of high resolution images in order to properly investigate the dynamical behaviour of magnetic elements. This difficulty results from image distortion and smearing introduced by the Earth atmosphere. Moreover, the effect magnetic elements may have on the chromosphere and corona and their role in the heating mechanism is difficult to quantify since there the magnetic field escapes direct measurements completely. These shortcomings of observational methods can be partially compensated by theoretical considerations and numerical simulations. Observed quantities constrain the physics of the simulation which assures that we fairly realistically simulate solar conditions. On the other hand, model calculations are used as a tool in order to derive physical quantities by comparison with observed spectra and filtergrams. Considering the complexity of the magnetic field/plasma system of magnetic elements it is obvious that numerical simulations are indispensable for gaining a *physical understanding* of the nature of magnetic elements and the quiet solar atmosphere.

The standard model of magnetic elements is that they are magnetic flux tubes or magnetic flux sheets that extend more or less vertically from the convection zone into the photosphere, where they laterally expand due to the decreasing gas pressure. Neighbouring flux tubes merge in the lower chromosphere and completely fill the space in the higher chromosphere and corona (Solanki and Steiner, 1990). Magnetic elements of opposite polarity may be connected by small loops that reach into the lower corona, but they may also open up into the corona due to the streaming solar wind. In both cases the photospheric nature of the flux tube is expected to be similar since the magnetic field at coronal heights cannot act upon the much stronger flux tube fields in the photosphere.

In the past, hydrostatic or stationary model flux tubes have been constructed, mainly for studying the energy budget of the photospheric layers of flux tubes and for the interpretation of Stokes spectra (*e.g.*, Deinzer *et al.*, 1984a, b; Hasan, 1988; Knölker et al., 1988; Grossmann-Doerth et al., 1989; Steiner and Pizzo, 1989; Kalkofen *et al.*, 1989; Steiner and Stenflo, 1990; Fabiani Bendicho *et al.*, 1992; Pizzo *et al.*, 1993a, b). To progress further we have developed a MHD code for the time dependent simulation of magnetic flux tubes, including their interaction with the surrounding granular flow. With this code we are now, for the first time, able to investigate the dynamical behaviour of magnetic flux sheets in the photosphere.

2. The Equations

The physical processes which are relevant for the atmospheric layers under consideration (the uppermost part of the convection zone and the photosphere up to the temperature minimum) are described by the magnetohydrodynamic equations. Energy transfer by radiation is included together with an appropriate equation of state. The set of equations, written in conservation law form and in two-dimensional Cartesian coordinates consists of:

the continuity equation,

$$\frac{\partial \rho}{\partial t} + \frac{\partial}{\partial x}(\rho v_x) + \frac{\partial}{\partial y}(\rho v_y) = 0 , \tag{1}$$

the momentum equations,

$$\frac{\partial(\rho v_x)}{\partial t} + \frac{\partial}{\partial x}\left(\rho v_x^2 + p - \sigma_{xx}^t + \frac{1}{8\pi}\left(B_y^2 - B_x^2\right)\right)$$
$$+ \frac{\partial}{\partial y}\left(\rho v_x v_y - \sigma_{xy}^t - \frac{1}{4\pi}B_x B_y\right) = 0 , \tag{2}$$

$$\frac{\partial(\rho v_y)}{\partial t} + \frac{\partial}{\partial x}\left(\rho v_x v_y - \sigma_{yx}^t - \frac{1}{4\pi}B_x B_y\right)$$
$$+ \frac{\partial}{\partial y}\left(\rho v_y^2 + p - \sigma_{yy}^t + \frac{1}{8\pi}(B_x^2 - B_y^2)\right) + \rho g = 0 , \tag{3}$$

the entropy equation,

$$\frac{\partial(\rho s)}{\partial t} + \frac{\partial}{\partial x}\left(v_x \rho s\right) + \frac{\partial}{\partial y}\left(v_y \rho s\right)$$
$$- \frac{\Phi^t}{T} - \frac{\Phi^{mag}}{T} + \frac{1}{T}\left(\frac{\partial q_x^t}{\partial x} + \frac{\partial q_y^t}{\partial y}\right) - 4\pi\rho\frac{\kappa}{T}\left(J - \frac{\sigma_R}{\pi}T^4\right) = 0 , \tag{4}$$

and the induction equations,

$$\frac{\partial B_x}{\partial t} + \frac{\partial}{\partial y}\left(v_y B_x - v_x B_y + \eta_m\left(\frac{\partial B_y}{\partial x} - \frac{\partial B_x}{\partial y}\right)\right) = 0 , \tag{5}$$

$$\frac{\partial B_y}{\partial t} + \frac{\partial}{\partial x}\left(v_x B_y - v_y B_x - \eta_m\left(\frac{\partial B_y}{\partial x} - \frac{\partial B_x}{\partial y}\right)\right) = 0 . \tag{6}$$

ρ is the density, v_x and v_y are the velocity components, s is the entropy per unit mass, B_x and B_y are the components of the magnetic field strength, p the gas pressure, g the gravitational acceleration, and $\sigma_{\alpha\beta}^t$ are the components of the turbulent viscosity tensor. The entropy equation contains source terms with Φ^t and Φ^{mag}, the turbulent and magnetic dissipation function, q_α^t, the components of the turbulent heat flow, and the radiative source with J the frequency integrated mean intensity and κ Rosseland's mean opacity. T is the temperature, σ_R the Stefan-Boltzmann constant, and η_{m}, the coefficient of magnetic diffusion.

The Reynolds number of the fluid in the solar plasma is very high and flows are expected to be turbulent down to scales of a few cm. Since the mesh width of the computational grid for the simulation is of the order of 10 km the effect of turbulent eddys smaller than the grid resolution on the large scale flow is often modeled by adding special terms to the classical Navier Stokes equations. This approach to account for the sub-grid-scale (SGS) turbulence has been introduced by Smagorinsky (1965) and Deardorff (1971) and is used in the same form as in the present work and in the context of solar/stellar convection by Chan and Sofia (1986), Fox et al. (1991), Gigas (1990), and Steffen (1991).

Thus, $\sigma_{\alpha\beta}^t$ in Equations (2) and (3) correspond to the components of the usual viscosity tensor, but with the turbulent dynamic viscosity coefficient

$$\nu^t = c^2 \Delta^2 \sqrt{2\left(\frac{\partial v_x}{\partial x}\right)^2 + \left(\frac{\partial v_x}{\partial y} + \frac{\partial v_y}{\partial x}\right)^2 + 2\left(\frac{\partial v_y}{\partial y}\right)^2} . \tag{7}$$

Δ is the mesh width of the computational grid and c is a dimensionless scaling factor, set to 0.5 in our calculations.

According to Kolmogorov's law (e.g. Landau and Lifshitz, 1991) the velocity of the turbulent motion, v^t, is related to the scale length l of the corresponding eddy by

$$v^t \propto l^{1/3} . \tag{8}$$

With the granular flow having a velocity of the order of $v_0 = 10^5$ cm/s and a scale length of $L = 10^8$ cm we obtain for the turbulent velocity at the scale of our grid size, $l = 10^6$ cm, the relation

$$\frac{v^t}{v_0} = \left(\frac{l}{L}\right)^{1/3} \approx 0.2 , \tag{9}$$

thus, $v^t = 2 \cdot 10^5$ cm/s. It follows an equipartition field strength for the turbulent motion of the grid size scale of $B_{\mathrm{eq}} = v^t \sqrt{4\pi\rho} = 40$ G. Hence, in order to roughly account for the suppression of SGS-turbulence in the presence of magnetic fields we set $\nu^t = 0$ if the field strength surpasses a limit of 50 G.

The turbulent dissipation function in the entropy equation is given by

$$\Phi^t = \sigma^t_{\alpha\beta} \frac{\partial v_\alpha}{\partial x_\beta} , \tag{10}$$

where $\alpha, \beta = x, y$, and the turbulent heat flow is

$$q^t_x = -\frac{1}{\text{Pr}} \nu^t \rho T \frac{\partial s}{\partial x} , \tag{11}$$

with Pr the Prandtl number being 1 and s the entropy per unit mass.

To close the system of Equations (1)–(6) we require relations that specify the thermodynamic properties of the fluid. These relations are the (ideal gas) equation of state in which we allow for the mean molecular weight $\bar{\mu}$ to vary according to

$$\bar{\mu} = \frac{\bar{\mu}_0}{1 + n_e/n} , \tag{12}$$

where $\bar{\mu}_0 = 1.297$ is the mean molecular weight of the neutral gas and n_e/n is the ratio of the number densities of free electrons to the number densities of nuclei $n = \rho/(\bar{\mu}_0 A_0)$, with A_0 being the atomic mass unit. For the results presented here, we have only considered hydrogen ionization, the ionization of He being less than 0.1 % at $T \approx 13000$ K, the temperature close to the bottom of our computational box. Using Saha's equation one obtains an explicit formula for the electron density.

The temperature is computed from the specific entropy, which is the sum of the entropies of the different gas components, free electrons, H, H$^+$, He, and metals (Me) in our case. Using the expression given by Grosser (1991), for the entropy of a gas component, one obtains a transcendental algebraic equation for the temperature. This equation is solved by means of a Newton-Raphson iteration.

The radiative source term in Equation (4) is the grey approximation to the expression

$$4\pi\rho \int_0^\infty \kappa_\nu \left(J_\nu - B_\nu \right) d\nu , \tag{13}$$

where κ_ν is the specific opacity, J_ν the mean intensity, and B_ν the Planck function, all values evaluated at frequency ν. In Equation (4) the mean intensity is obtained by formal integration of the radiative transfer equation using $\int_0^\infty B_\nu d\nu = (\sigma/\pi)T^4$ as the source function. It is obvious that the grey approximation greatly reduces the computational cost of integrating the radiation transfer equation, since integration over the frequency spectrum is avoided, but it does not take into account the effects of line blanketing or backwarming. These effects would influence mainly the temperature structure of the uppermost layers in our model. However, the interplay between convective motions and radiative energy losses which determines the mor-phology of the layers near $\tau_c = 1$ is described reasonably well in a grey approximation of the radiation transfer.

3. Numerical Methods, Boundary Conditions, and Initial Values

We use a finite volume method for the discretization of Equations (1)–(6). In particular, this method keeps div $\mathbf{B} = 0$ exactly, during the integration. The numerical fluxes are computed according to the flux corrected transport (FCT) scheme as described by Oran and Boris (1987). This is a shock cap-turing scheme which automatically switches between second and first order accuracy near steep gradients in order to avoid spurious oscillations near discontinuities and to maintain "sharp" shock fronts.

The time integration is explicit and proceeds in three steps: In a first (half time) step all variables are calculated at the intermediate time level $t + \Delta t/2$. These values are used to compute the source terms and the convecting velocities which are introduced as time averaged quantities into the second (full time) step from time t to $t + \Delta t$. This last step is repeated in a third step using improved values for source terms and convecting velocities which are computed by averaging the solution at time t with the one just obtained and belonging to time $t + \Delta t$. This last step has first been used in a FCT-code by Schmitz and Fleck (1993) and it improves the scheme significantly.

The calculations presented in the following sections use a uniform grid with 240×120 grid cells.

Our present 2-D numerical model corresponds to a three-dimensional space in which variables do not depend on the third coordinate, z. The two-dimensional domain of integration represents a slab perpendicular to the direction of z. In order to numerically obtain the mean intensity $J(\mathbf{x})$, one has to compute the intensities $I(\mathbf{x}, \mathbf{n})$ by integration of the radiation transfer equation along a number of rays (of orientation $\mathbf{n}$) through a spatial point $\mathbf{x}$. We use three angles per quadrant for the angle quadrature in the x-y-plane and two angles for the azimuthal integration, which means that we have to carry out the integration in the x-y-plane only. The intensities are computed by integration of the radiation transfer equation along short characteristics as described by Kunasz and Auer (1988).

Since a large fraction of the computer time needed to solve the system (1)–(6) is spent for the integration of the radiative transfer equation, we switch to the diffusion approximation for computing the radiative source term at large optical depths ($\tau \geq 100$).

We have carried out a number of test calculations. Examples including a shock tube problem, a 1-D magnetic flux sheet, the magnetoconvection problem of Hurlburt and Toomre (1988), as well as the supersonic convection problem of Cattaneo *et al.* (1990) are described by Steiner *et al.* (1994).

3.1. BOUNDARY CONDITIONS

We use periodic side boundaries. Bottom and top are impermeable for the flow and the magnetic field is assumed to be vertical there. The temperature is fixed at 14000 K at the bottom boundary and we impose $\partial T/\partial y = 0$ at the top.

These boundary conditions are simple to implement and mathematically well posed. However, the closed bottom forces the descending gas to turn around when approaching the bottom, which may give rise to an artificial enhancement of turbulence (Nordlund *et al.*, 1994). Therefore, we are presently implementing an open lower boundary which allows for free out- and inflow, while keeping the gas pressure constant. At the same time, the entropy of inflowing material will be adjusted in order to obtain the correct value of the mean radiation output at the top (Steffen, 1991). This is not exactly satisfied with our present condition of a constant temperature, the mean radiation output in the present simulation being 5.4 % too small. A further improvement of the boundary condition, which is presently in the test stage, consist of a wave absorbing layer at the top.

3.2. INITIAL VALUES

The formation of the magnetic flux concentration by convective collapse is not included in these simulations since we assume a closed bottom of the computational box. Therefore, we start with a flux sheet whose internal density is a constant fraction, α, of the external density and which is in horizontal pressure equilibrium with a static, plane-parallel stratification. A smooth transition between flux sheet and the surrounding plasma is provided according to Equation (1.3) of Knölker *et al.* (1988). The parameters for the present simulation run (density reduction $\alpha = 0.5$ and a flux sheet width of 300 km at optical depth unity) are the same as those of model M350 of Grossmann-Doerth *et al.* (1994). The temperature is initially constant in horizontal planes across the whole domain but varies with height according to the model atmosphere of Spruit (1977). We perturb the initial configuration with a small non-symmetric velocity field. The magnetic diffusivity it set to zero in the equations, so that only the inherent numerical diffusion remains for the magnetic field.

During the first phase of the simulation, the energy equation is switched off and the flux sheet is allowed to relax and to approach a 2-D magnetostatic equilibrium while the temperature is kept equal to its (horizontally constant) initial value. After this relaxation phase, the energy equation is switched on and the convective instability leads to the development of a non-stationary pattern of "granules" (which are infinitely long rolls in our 2-D geometry) outside the flux sheet. The interior of the magnetic structure is cooled due to radiative losses and suppression of convection.

4. Dynamic Behaviour of Magnetic Elements

4.1. GENERAL DISCRIPTION

The two dimensional computational domain represents a section through the surface layers of the solar atmosphere. The abscissa (y-coordinate) is perpendicular to the solar surface with the coordinate $y = 0$ at optical depth $\tau_{5000} = 1$ of the initial atmosphere. It reaches 600 km above and 800 km below the level $\tau_{5000} = 1$. The width of the computational domain (x-coordinate) is 2400 km, which gives sufficient space for the evolution of normal granules. This plane section is thought to cut a magnetic element, which, in the present geometry, is a slab of magnetic flux infinitely extended in the horizontal direction perpendicular to the plane of the computational domain. The magnetic flux sheet is thought to represent a cross section of an elongated or thread-like magnetic element located in a intergranular lane.

Figure 1 shows a snapshot during the simulation which in total lasts for 18.5 minutes real time. The figure shows the temperature field (colour coded), the velocity field, and the magnetic field through indication of representative field lines. The horizontally running black curve indicates the optical depth $\tau_{5000} = 1$ for vertically incident lines of sight. Note that in the height scale scale we count positive with increasing height and negative into the convection zone.

Fig. 1.   (on the next page) Snapshot of a simulation which lasts over a period of 18.5 minutes real time. Representative magnetic field lines are shown in black, the velocity field is given by white arrows. The colours indicate the temperature field. The corresponding scaling is given in Kelvin at the top. The horizontally running black curve indicates the optical depth $\tau_{5000} = 1$ for vertically incident lines of sight. Two shock waves can be seen, one just above the downflow to the right ($x \approx 2000$ km) and one within the magnetic flux sheet at a height of around $y = 500$ km. The flux sheet is framed by two strong and narrow "downflow jets".

and the convective instability leads to the development of a non-stationary pattern of "granules," [within are initiable long rolls (near 2-D geometry) perpendicular to the sheet. The interior of the magnetic structure is convec-

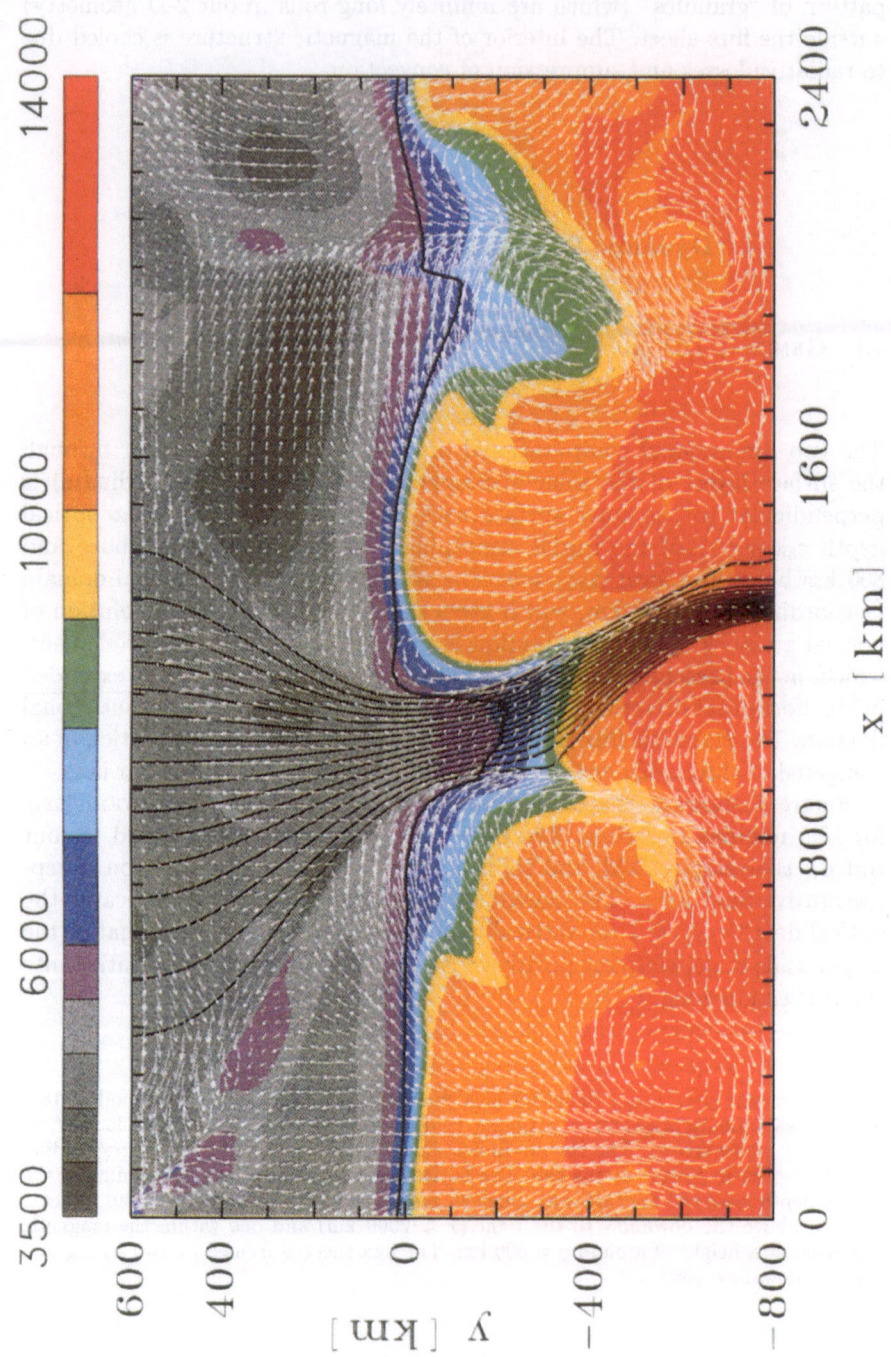

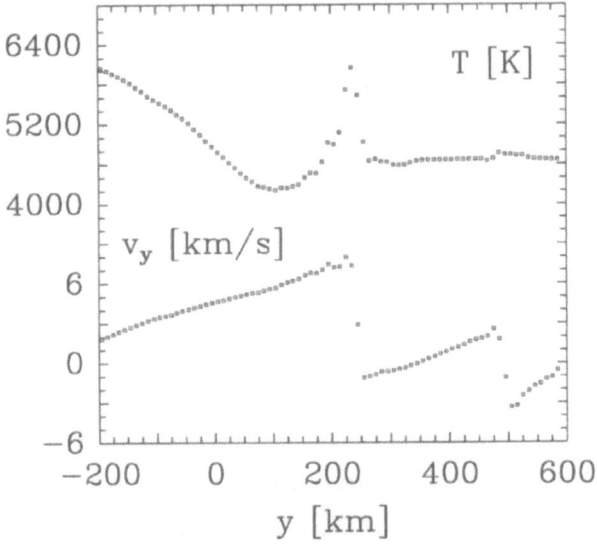

Fig. 2. Bottom: vertical component of the velocity along a vertical cross section through the photospheric layers of the magnetic flux sheet at the time of a shock wave occurrence. Two shock waves are visible at $y = 250$ km and $y = 500$ km. Top: Corresponding temperature as a function of height. The strong shock ($y = 250$ km) heats the post shock material by 1400 K. It rapidly cools off due to expansion and radiation.

A strong upwelling of hot material with a corresponding gas pressure enhancement to the right-hand side of the flux sheet appears. Since no comparable flow evolves on the left-hand side, the flux tube is pushed to the side by the upwelling fluid and bends to the left. This granular action drives and bends the magnetic field which itself reacts with magnetic tension forces. During the simulation run we obtain a maximum inclination angle of about 30° at $y = 0$ and horizontal displacements with a maximum velocity of about 4 km/s. This horizontal motion occasionally shows rapid variation in velocity and direction.

4.2. SHOCK WAVES

A conspicuous phenomenon occurring at several instances is the formation of shock fronts. They are visible as temperature peaks and discontinuities in the velocity field of Figure 1. Three kinds of shocks can be distinguished according to orientation and formation. Firstly, shock waves which form outside the magnetic flux sheet in the vicinity of downflows. They are vertically oriented and move (with about the local speed of sound) away from granular downflows. This can be seen in Figure 1 where a shock wave has formed

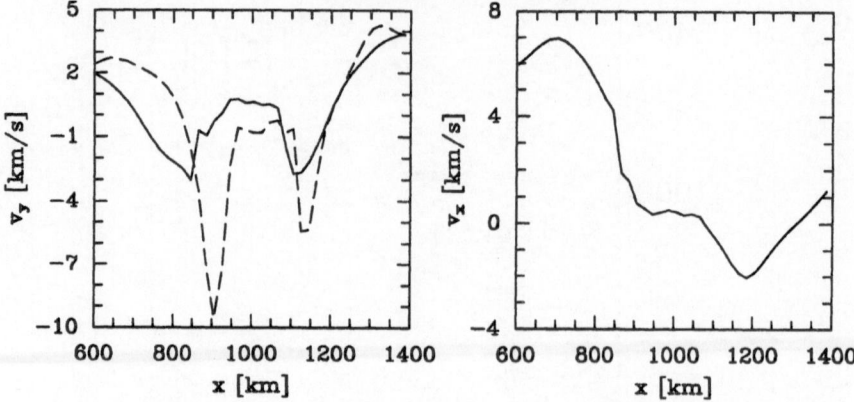

Fig. 3.   Left: Vertical velocity component in two heights (solid: $y = 0$ km, dashed: $y = -200$ km) as function of the horizontal coordinate (x) in the central part of the flux sheet shown in Figure 1. The strong jets adjacent to the flux sheet are clearly visible. Right: Horizontal velocity at height $y = 0$. The downflows are fed by horizontal flows directed towards the flux sheet.

near the downflow to the right ($x \approx 2000$ km). It moves to the left. Since the flux sheet is most of the time framed by narrow downflows, occasionally this type of shock forms close to this downflow and moves away from the flux sheet. The physics of this type of shock waves has been described by Cattaneo et al. (1990).

A second type of shock wave (not visible in Figure 1) is initiated by the magnetic flux sheet in the course of the relaxation of buoyancy and magnetic tension forces. The fast movement of the flux sheet "sweeps" material ahead of it. This "snow plough" effect leads to a shock front which again is vertically oriented and moves with a velocity of about 10 km/s (near the top boundary) away from the flux sheet.

Thirdly, shock waves frequently occur within the flux sheet. During the present simulation run of 18.5 minutes seven such shock events can be identified, an examples of which is visible in Figure 1. These shocks are more or less horizontal and move upward. The front is distorted when moving along the field because of the variation of density and temperature and, hence, of the local speed of sound across the flux sheet.

Figure 2 shows v_y, the vertical component of the velocity, and the temperature along a vertical line of sight through the photospheric layers of a magnetic flux sheet cutting through a strong shock. The shock has formed in deep photospheric layers and is now located at a height of $y = 250$ km. It moves upward with a velocity of about 10 km/s (positive y-direction).

The velocity of the fluid changes by 9 km/s across the discontinuity. The shocked material is heated up by about 1400 K, but rapidly cools off due to expansion and radiative losses. A weaker shock precursor is located at $y = 500$ km. The propagation of nonlinear, radiatively damped longitudinal waves along magnetic flux tubes has been discussed previously by Herbold et al. (1985).

4.3. DOWNFLOW JETS

During most of the time a persistent, strong, narrow downflow prevails in the close vicinity outside the flux sheet. It is only occasionally perturbed by the granular flow field. These downflows are "fed" by horizontal flows in the upper layers. Profiles of the vertical and horizontal flow velocities of the situation shown in Figure 1 are given in Figure 3. The left graph shows the *vertical* component of the velocity over a third of the computational domain as a function of the horizontal coordinate (600 km $< x <$ 1400 km) at heights $y = 0$ (solid line), which corresponds to continuum optical depth unity of an average atmosphere, and somewhat deeper, at $y = -200$ km (dashed line). The narrow downflow "jets" outside the flux sheet are clearly visible. They accelerate with depth and reach velocities of up to 9 km/s. The corresponding horizontal flows extend over a larger region. The right panel of the figure shows the profile of the horizontal velocity at $y = 0$, which reaches a maximum speed of about 7 km/s.

5. Polarization Diagnostics

Stokes polarimetry has proved to be the most accurate and productive diagnostic for magnetic elements. For the interpretation of Stokes profiles, careful analysis is required and realistic model atmospheres are an indispensable tool. So far all such analyses have been carried out on the basis of arbitrary, magnetohydrostatic, or stationary plasma/magnetic field configurations. With the present simulation code it has become feasible to compute time dependent synthetic Stokes profiles. Thus, for the simulation run discussed in Section 4 we have recorded on video disk Stokes I, V, and Q, simultaneously with a representation of the temperature, velocity, and magnetic field. This advance in theoretical analysis comes in time with newly developed polarimeters such as the Advanced Stokes Polarimeter, ASP, (Elmore et al., 1992) or the CCD-demodulator polarimeter of Povel et al. (1990).

5.1. STOKES PROFILES FROM VERTICAL LINES OF SIGHT

Many of the dynamic phenomena described in Section 4 cause specific signatures in the corresponding Stokes profiles. Figure 4 shows a sequence of

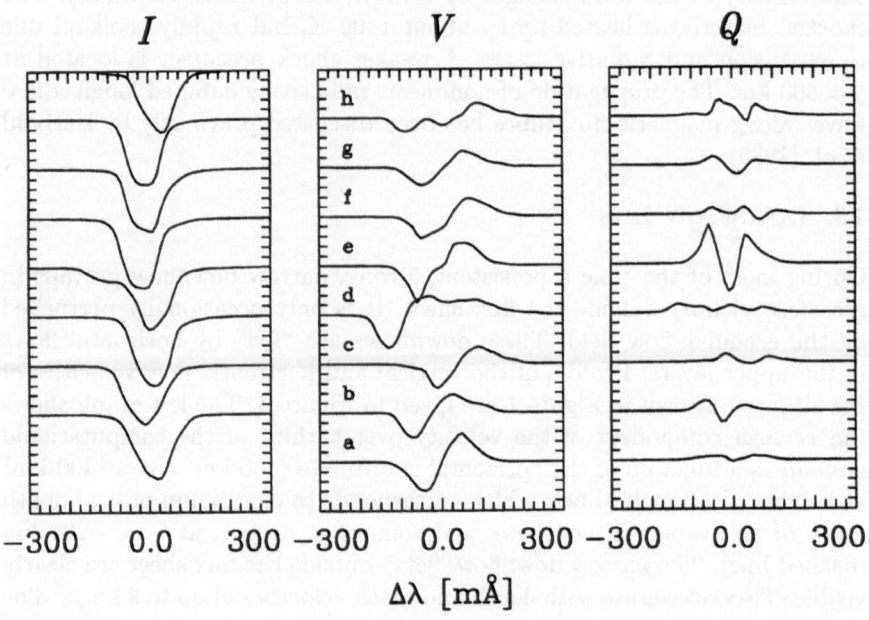

Fig. 4.   Stokes I, V, and Q normalized to the continuum intensity I_c of the spectral line
Fe I 5250.2. The eight profiles are computed from snapshots that are separated in time
by typically 100 seconds. Tick marks of the vertical axis indicate 10 % for the Stokes
I panel, 2 % for the Stokes V panel, and 0.6 % for the Stokes Q panel. The displayed
profiles are the average of profiles that have been computed for 50 vertically incident lines
of sight with a horizontally equidistant distribution across the simulation domain. The
strong Q-signal of profile (e) appears only over a period of 40 seconds. It is a consequence
of the strong inclination of the flux sheet. The strongly blueshifted V-lobes of profiles (d)
and (h) are each caused by a shock wave (see Figures 5 and 6, correspondingly).

Stokes I, V, and Q profiles for the spectral line Fe I 5250.2. They are an
average of profiles from 50 vertical lines of sight equidistantly distributed
over the computational box. They are separated in time by typically 100
seconds. The assumption of LTE is used for the line calculations.

The horizontal averaging over the computational domain of 2400 km is
thought to simulate low spatial resolution observations*. The swaying motion of the flux sheet can (for vertically incident lines of sight) most directly
be seen in Stokes Q, since this quantity is a direct measure of the transversal magnetic field component. The Q-signal becomes sizable only for a flux
sheet which is sufficiently inclined with respect to the vertical line of sight.
In our case a strong Stokes Q signal (above 2.5 %) occurs only during a pe-

* Note, however, that a resolution of about 200 km (the size of the flux sheet) would
only effect Stokes I since Stokes V and Q are only formed in the magnetic sheet.

riod of 40 seconds at the time of the extreme inclination to the right [profile
(e)]. This is a consequence of the quadratic dependence of Stokes Q on the
angle between the line of sight and the magnetic field. Hence, a high temporal resolution of at least 10 seconds is required in order to capture such an
event observationally. In the present simulation over 1110 seconds such an
event occurs only once so that the probability for detection may be small.
This may be a reason why Sánchez Almeida and Martínez Pillet (1994),
in a sample of 21 network magnetic elements, only find linear polarization
which corresponds to an inclination angle of at most 10°. A less extreme
inclination to the left leads to the Stokes Q of profile (b) with a maximal
amplitude of less than 1%. This instant corresponds to the situation shown
in Figure 1.

A small Stokes Q signal is always present. This is because even a perfectly
symmetric, vertical flux sheet yields a non-vanishing Q profile since the inclined field lines to the left and to the right of the symmetry axis positively
add to the profile and do not cancel. Cancellation would occur for a symmetric, vertical flux *tube* that lies completely within the spectrograph entrance
slit. This suggests that the field geometry of magnetic elements should be
taken into account when deriving inclination angles from Stokes Q and U.
Systematic errors may be introduced when measurements are interpreted on
the basis of a *homogeneous*, inclined magnetic field.

The Stokes profiles drastically change their shape when the line of sight
becomes inclined with respect to the vertical (observation off disk centre).
In this case the horizontal motion of the flux sheet is clearly visible in Stokes
V.

5.2. STOKES PROFILE SIGNATURE OF A SHOCK WAVE

The upward moving shock waves within the flux sheet leave distinctive "fin-
gerprints" in Stokes V. It also is discernible in Stokes I, but since the V-
profile originates in the magnetic region only, and since this type of shock
wave is confined to the magnetic sheet, the shock signature is more conspicuous in Stokes V than in Stokes I. With a shock front speed of about 10
km/s and a spectral line contribution width of about 450 km for Fe I 5250.2
(Grossmann-Doerth, 1994) one needs a temporal resolution of at least 10
seconds to properly observe this phenomenon. Figure 5 shows a sequence
(from bottom to top) of 20 Stokes V-profiles, equidistant in time and centred in time around the shock event which is shown Figure 2. The profiles
are labeled with the uppermost position (in km) of the shock front, when
existing. A superposition of two Stokes V-profiles originating in the pre-
and post-shock regimes can be seen. The post-shock material moves upward
with about the local speed of sound, leading to a strongly blueshifted profile, while the pre-shock medium moves downwards with an initial velocity

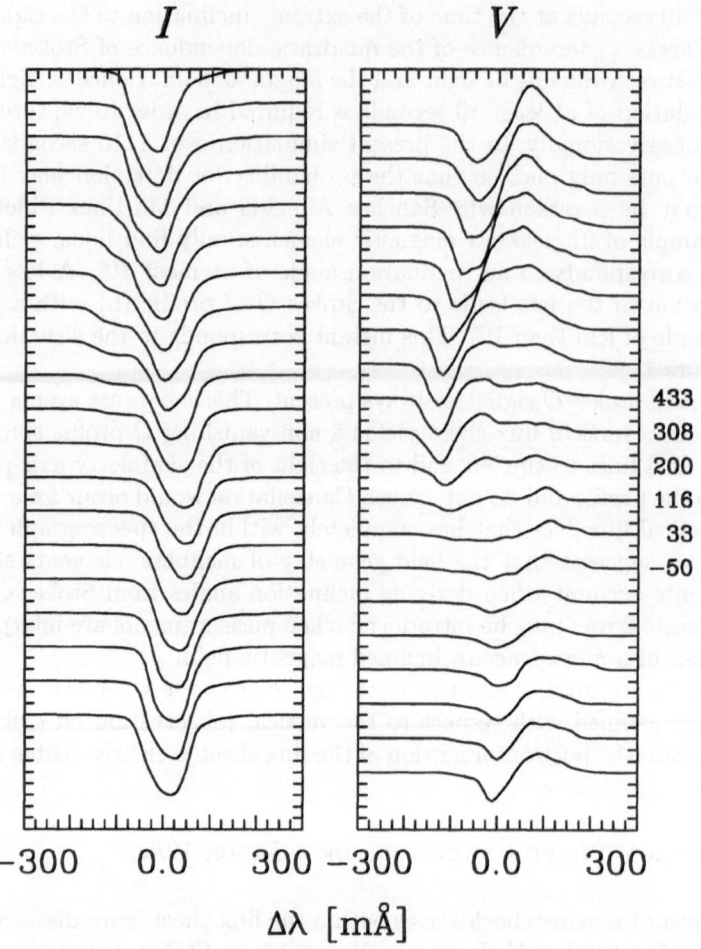

Fig. 5. Sequence (time increasing from bottom to top) of 20 Stokes I and V profiles of the spectral line Fe I 5250.2, centred in time around the shock event of panel (d). The profiles are separated in time by 10 seconds. Tick marks of the vertical axis indicate 10 % for the Stokes I panel and 2 % for the Stokes V panel, relative to the continuum intensity. Given in the right margin are the uppermost positions of the shock front in km. The superposition of a redshifted pre-shock and a blueshifted post-shock profile leads to the peculiar V-profiles around the middle of the sequence. Stokes I originates mainly in the field free region outside the flux sheet, so that the shock merely weakly effects the profile in the far blue wing.

which decreases from -4.6 km/s near $y = 0$ to 0.0 at a height of $y = 400$ km. In the first six profiles of Figure 5 the blueshifted profile cannot be seen since the upflow has not yet reached the spectral-line-forming layers of the atmosphere. Subsequently, the contribution of the blueshifted profile grows, while the portion of the downflowing fluid in the line formation region shrinks, thereby diminishing the contribution of the redshifted profile. After the shock has moved out of the line formation region, the speed of the upward flowing post-shock medium decreases and the Stokes V-profile shifts back towards a moderately blueshifted profile.

Stokes I mainly originates in the field free region around the flux sheet, so that the shock event only marginally effects the far blue wing of the profile. Most of the time, a strong downflow to the right of the flux sheet dominates since no comparable upflow is present. This leads to a redshifted Stokes I which only toward the end of the sequence shifts back and becomes blueshifted because of increasing upflow.

The upflow within the flux sheet which later leads to the shock initially causes a flat temperature profile. This is the reason for the initially weak Stokes V-profile. Later on, a steep temperature gradient builds up in the post shock regime due to the cooling by expansion and radiation. This leads to the strong V-profiles in the upper part of Figure 5.

Not all shock waves that form within the magnetic flux sheet lead to the characteristic pattern in Stokes V that can be seen so clearly in Figure 5. Sometimes shocks form only above the formation height of photospheric spectral lines. This, for example, applies for the shock wave that can be seen in Figure 1 within the flux sheet at around a height of 500 km.

Figure 6 shows a time sequence of Stokes I and V profiles similar to Figure 5 but 7.6 minutes later. Again, this sequence features a shock wave that originates in deep photospheric layers. Profiles (h) to (m) show the superposition of the redshifted pre-shock and the blueshifted post-shock contribution to Stokes V. The separated blue lobes of the two superpositioned profiles are best seen in profile (l). Although the exact location of their maximum amplitude may be influenced by superpositional effects, their separation in wavelength yields an estimate of the difference in velocity of the two media giving rise to the profile. In the case of profile (l) this separation is 100 mÅ corresponding to a velocity of 6 km/s. This value is in good agreement with what we would expect for a shock wave as the one shown in Figure 2, taking the finite width of the line formation into account.

Recalling the geometry of the flux sheet, the Stokes profiles of Figures 5 and 6 of course are not just a superposition of two profiles originating in two plane-parallel pre- and post-shock media. They also are a spatial average across the magnetic flux sheet. Since the shock front deviates from being plane-parallel, the individual profiles which form along vertical lines of sight at different horizontal locations across the flux sheet vary strongly. This can be seen in Figure 7, which shows the Stokes V-profiles of 6 lines of sight equidistantly distributed across the magnetic flux sheet. The average of these profiles corresponds to profile (l) of Figure 6. Because of the different

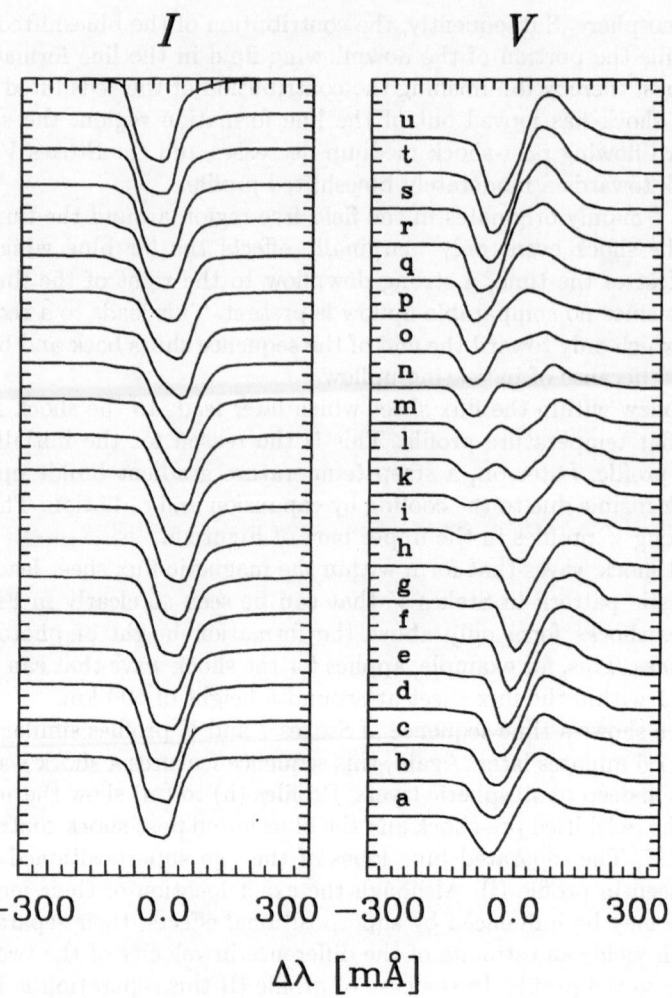

Fig. 6. Sequence of 20 Stokes I and V profiles as in Figure 5 but for a shock event that took place 7.6 minutes later. Tick marks of the vertical axis indicate 10 % for the Stokes I panel and 2.5 % for the Stokes V panel.

strength, not each of the profiles of Figure 7 equally contributes to the average profile. Although the individual profiles strongly differ from each other, the average profile still conveys the presence of the shock wave as is seen in Figure 6 and discussed above.

From the observation of an event as shown for the case of synthetic profiles in Figures 5 and 6, one cannot unambiguously imply the actual existence of a shock front. Such a pattern could be caused by an upward moving front

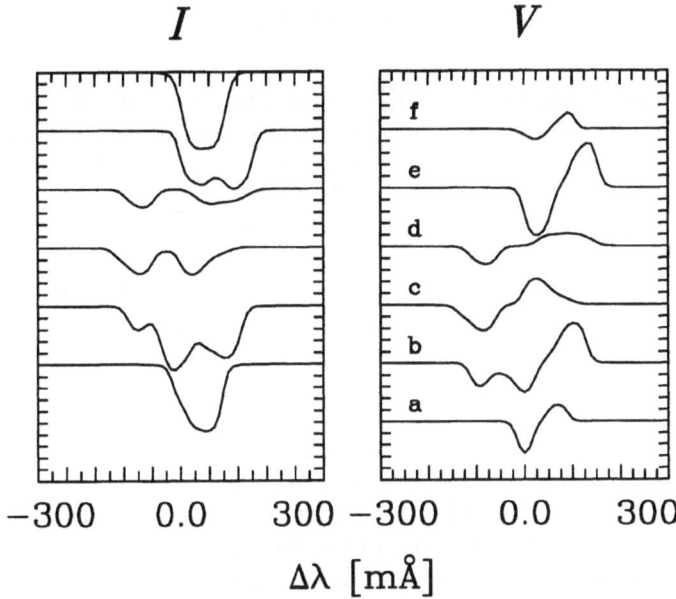

Fig. 7. Stokes I and V profiles originating from 6 lines of sight equidistantly distributed across the magnetic flux sheet at a time of shock incidence. Tick marks of the vertical axis indicate 10 % of I_c in both panels. The average of these profiles yield profile (1) of Figure 6. Only profiles b, c, and d have contributions from the upward moving post-shock medium.

between an up flow and a down flow, which, however, would sooner or later give rise to a shock, because of the strongly stratified atmosphere.

Shocks outside the magnetic flux sheet are less conspicuous in spectral diagnostic since only Stokes I can be used. These shocks are vertically oriented so that the fluid flow across the shock front is more or less perpendicular to the line of sight when observing at disk centre. Hence, the only effect these shocks may have on Stokes I is that of the temperature peak in the post-shock fluid which leads to a temperature weakening of the absorption line core. This temperature peak however, is quite narrow, much smaller than the resolution capabilities of present solar telescopes so that their weakening effect on Stokes I would be barely detectable as shown by Steffen *et al.* (1994). However, with space based observations with a spatial resolution of $0.2''$ the detection of such shocks should be possible as synthetic spectral line calculations have shown. From this we conclude that time sequence recordings of Stokes V of network magnetic elements are probably the best method for detecting photospheric shock waves with ground based instruments.

6. Conclusions

We solve numerically the set of MHD-equations taking radiative transfer (in the grey approximation) and hydrogen ionization into account. Initial and boundary conditions that comply with realistic solar conditions are specified. Thus, we simulate a magnetic flux sheet that dynamically interacts with a non-stationary, convectively unstable subsurface layer and the overlying photosphere.

The simulation, on the one hand, confirms results obtained on the basis of previous simulations. This includes the continuum contrast signature across these structures (Steiner *et al.*, 1994) as well as the flow pattern in the surrounding of the flux sheet. On the other hand, it has revealed a variety of new features. We observe in our simulation strong bending of the magnetic flux sheet by asymmetric convective flow, upward propagating shock waves within the magnetic field, and horizontally propagating shock waves outside it, which are partly caused by the motion of the magnetic flux sheet itself. The magnetic flux sheet is most of the time framed by narrow downflow "jets", which reach velocities close to the speed of sound. Horizontal displacement of the flux sheet takes place with a velocity of up to 4 km/s near $\tau_{5000} = 1$.

Shocks within the magnetic flux concentrations lead to a distinctive pattern in Stokes V and their evolution can directly be traced in a time series of Stokes V-profiles, provided the profiles are recorded with a time resolution of at least 10 seconds. From this pattern one cannot unambiguously infer the actual presence of a shock wave, but one can directly estimate the velocity difference of the bulk pre- and post-wave medium. In cases when this velocity is close to the local speed of sound a shock front has already formed or is about to evolve.

Upward propagating shock waves within the magnetic field concentration occur quite frequently (7 events are identified in the present simulation of 18.5 minutes real time). They certainly propagate into the chromosphere where they are expected to produce strong emission in chromospheric spectral lines. Thus, the Ca II emission of the chromospheric network, which are known to coincide with magnetic flux, may be a direct consequence of these shocks.

Acknowledgements

This work has been supported by the Deutsche Forschungsgemeinschaft (DFG) under grant SCHU 500/6-1.

References

Cattaneo, F.C., Hurlburt, N.E., and Toomre, J.: 1990, *Astrophys. J.* **349**, L63

Chan, K.L. and Sofia, S.: 1986, *Astrophys. J.* **307**, 222

Deardorff, J.W.: 1971, *J. Comput. Phys.* **7**, 120

Deinzer, W., Hensler, G., Schüssler, M., and Weisshaar, E.: 1984a, *Astron. Astrophys.* **139**, 426

Deinzer, W., Hensler, G., Schüssler, M., and Weisshaar, E.: 1984b, *Astron. Astrophys.* **139**, 435

Elmore, D.F., Lites, B.W., Tomczyk, S., Skumanich, A.P., Dunn, R.B., Schuenke, J.A., Streander, K.V., Leach, T.W., Chambellan, C.W., Hull, H.K., and Lacey, L.B.: 1992, SPIE (Society of Photo-Optical Instrumentation Engeneers) **1746**, 22

Fabiani Bendicho, P., Kneer, F., and Trujillo Bueno, J.: 1992, *Astron. Astrophys.* **264**, 229

Fox, P.A., Theobald, M.L., and Sofia, S.: 1991, *Astrophys. J.* **383**, 860

Gigas, D.: 1990, *Ph.D. thesis*, Christian-Albrechts-Universität, Kiel

Grosser, H.: 1991, *Ph.D. thesis*, Georg-August-Universität, Göttingen

Grossmann-Doerth, U.: 1994, *Astron. Astrophys.* **285**, 1012

Grossmann-Doerth, U., Knölker, M., Schüssler, M., and Solanki, S.K.: 1994, *Astron. Astrophys.* **285**, 648

Grossmann-Doerth, U., Knölker, M., Schüssler, and E. Weisshaar, M.: 1989, in R. Rutten and G. Severino (eds.), *Solar and Stellar Granulation*, NATO ASI Series C-263, Kluwer, Dordrecht, p. 481

Hasan, S.S.: 1988, *Astrophys. J.* **332**, 499

Herbold, G., Ulmschneider, P., Spruit, H.C., and Rosner, B.: 1985, *Astron. Astrophys.* **145**, 157

Hurlburt, N.E. and Toomre, J.: 1988, *Astrophys. J.* **327**, 920

Kalkofen, W., Bodo, G., Massaglia, S., and Rossi, P.: 1989, in R. Rutten and G. Severino (eds.), *Solar and Stellar Granulation*, NATO ASI, C-263, Kluwer, Dordrecht, p. 571

Keller, C.U.: 1992, *Nature* **359**, 307

Knölker, M., Schüssler, M., and Weisshaar, E.: 1988, *Astron. Astrophys.* **194**, 257

Kunasz, P.B. and Auer, L.H.: 1988, *J. Quant. Spectrosc. Radiat. Transfer* **39**, 67

Landau, L.D. and Lifschitz, E.M.: 1991, *Hydrodynamik*, Akademie Verlag, Berlin, 167

Nordlund, Å, Galsgaard, K., and Stein, R.F.: 1994 in R.J. Rutten and C.J. Schrijver (eds.), *Solar Surface Magnetism*, NATO ASI Series C-433, Kluwer, Dordrecht, p. 471

Oran, E.S., Boris, J.P.: 1987, *Numerical Simulation of Reactive Flow*, Elsevier, New York

Pizzo, V.J., MacGregor, K.B., and Kunasz, P.B.: 1993a, *Astrophys. J.* **404**, 788

Pizzo, V.J., MacGregor, K.B., and Kunasz, P.B.: 1993b, *Astrophys. J.* **413**, 764

Povel, H., Aebersold, H., and Stenflo, J.O.: 1990, *Applied Optics* **29**, 1186

Sánchez Almeida, J. and Martínez Pillet, V.: 1994, *Astrophys. J.* **424**, 1014

Schmitz, F. and Fleck, B.: 1993, *Astron. Astrophys.* **279**, 499

Smagorinsky, J.S., Manabe, S., and Holloway, S.: 1965, *Monthly Weather Rev.* **93**, 727

Solanki, S.K.: 1993, *Space Science Reviews* **63**, 1

Solanki, S.K. and Steiner, O.: 1990, *Astron. Astrophys.* **234**, 519

Spruit, H.C.: 1977, *Ph.D. thesis* University Utrecht

Steffen, M.: 1991, in L. Crivellari, I. Hubeny, and D.G. Hummer (eds.), *Stellar Atmospheres: Beyond Classical Models*, NATO ASI Series C-341, Kluwer, Dordrecht, p. 247

Steffen, M., Freytag, B., and Holweger, H.: 1994, in M. Schüssler and W. Schmidt (eds.), *Solar Magnetic Fields*, Cambridge University Press, p. 298

Steiner, O. and Pizzo V.J.: 1989, *Astron. Astrophys.* **211**, 447

Steiner, O. and Stenflo, J.O.: 1990, in J. O. Stenflo (ed.), *Solar Photosphere: Structure, Convection and Magnetic Fields*, IAU-Symp. No. 138, Kluwer, Dordrecht, p. 181

Steiner, O., Knölker, M., and Schüssler, M.: 1994 in R.J. Rutten and C.J. Schrijver (eds.), *Solar Surface Magnetism*, NATO ASI Series C-433, Kluwer, Dordrecht, p. 441

Steiner, O., Grossmann-Doerth, U., Knölker, M., and Schüssler, M.: 1994 in G. Klare (ed.), *Reviews in Modern Astronomy* Vol. 8, Astron. Gesellschaft, Hamburg, p. 81

RECENT PROGRESS IN IMAGING POLARIMETRY

C.U. KELLER

National Solar Observatory, P.O.Box 26732, Tucson, AZ 85726-6732, USA

Abstract. Recent instrumental developments in imaging polarimetry allow array detectors to reach a polarimetric sensitivity of 1×10^{-4} of the intensity. New instrumental effects appear at these levels of sensitivity and generate spurious polarization signals with amplitudes of up to 5×10^{-4}. Here I discuss these effects and present methods to avoid them. Polarized spectra with an rms noise of 6×10^{-6} may then be obtained. Furthermore a method is brought to the reader's attention that allows polarization measurements at the 1×10^{-4} level with regular array detectors, e.g. in the near-infrared.

Key words: Polarimetry – Spectroscopy – Array detectors

1. Introduction

Highly sensitive polarimetric measurements are needed in many parts of solar physics. The measurement of the transverse magnetic field component in the quiet sun and the spectroscopy of intrinsically weak magnetic fields require sensitivities of about 1×10^{-4} of the intensity. Investigations of resonance and coherence effects in spectral lines are best assessed by looking at weak linear polarization signals close to the solar limb (see Stenflo, 1996).

Array detectors provide an inherently high efficiency of measurements by taking 10^4–10^6 simultaneous measurements, which reduce the influence of seeing and allow the application of image reconstruction techniques. The main disadvantages of array detectors in terms of very precise differential measurements are variations of pixel properties, the relatively slow read-out process (< 100 Hz frame rate), the large amount of data to store and process, the sensitivity to optical aberrations due to the extended field of view, and the limited signal-to-noise ratio in a single exposure.

The near-infrared spectral region has the important advantage of increased Zeeman-splitting. However, array detectors in the near-infrared tend to be small, have bad flat-field properties, and require sophisticated cooling equipment.

Recent technical advances in imaging polarimetry provide much improved sensitivities in the visible as well as in the near-infrared. When trying to reach polarization sensitivities at 1×10^{-5} of the intensity, new observational problems appear that have not been noticed before, such as modulated fringes and coupling of instrumental polarization with detector non-linearities. In the following I will discuss these problems and ways to circumvent or calibrate them.

Solar Physics **164**: 243–252, 1996.

All visible data shown in the following sections have been obtained with ZIMPOL I, the first generation Zurich Imaging Stokes Polarimeter. This instrument is insensitive to seeing, differential gain effects, and optical aberrations. It is based on fast polarization modulation at 42 kHz for circular polarization and 84 kHz for linear polarization. The demodulation is performed with special CCDs that have every other pixel row covered with an opaque mask. The charges below the mask are shifted forth and back by one pixel in synchrony with the modulation for the particular, polarized Stokes parameter. Detailed information can be found in Keller et al. (1992), Povel et al. (1994), and Povel (1995). Nevertheless, all effects discussed in the following apply to any type of polarimeter that is capable of achieving a high enough sensitivity.

2. Reaching Beyond 1×10^{-4}

Although array detectors are most often used in polarimetry to obtain two-dimensional images of the sun in a narrow spectral bandpass or spectra along a slit, they can also be used for very sensitive one-dimensional measurements. Typical CCDs have a maximum capacity of about 3×10^5 electrons per pixel. Therefore, the maximum signal-to-noise ratio (SNR) that can be achieved by considering the photon noise in a single exposure is about 500. To attain a higher SNR, many individual exposures need to be added up. Considering only photon noise, 3.3×10^4 exposures would be needed to achieve a SNR of 1×10^5. Even at a frame rate of 10 exposures per second, this would take almost one hour. However, if an integration along the CCD rows or columns can be performed, the same SNR can be achieved within about 10 seconds, assuming an integration over 300 pixels.

Figure 1 shows an example of how the SNR is improved by averaging along the spectral lines. If spectra are considered and the spectral domain is oversampled, then some smoothing in the spectral direction can be applied without distorting the line profiles. Figure 2 shows an example of very sensitive polarimetry where the rms noise in the continuum is 6×10^{-6}.

3. Instrumental Effects Producing Spurious Polarization Signals

Any real detector system has non-linearities, which are mostly due to the analog read-out electronics. When trying to look for very small polarization signals (e.g. 1×10^{-5} of the intensity) on top of a small polarization signal of 1% due to polarization introduced by the telescope, non-linearities become important, as is shown in the following.

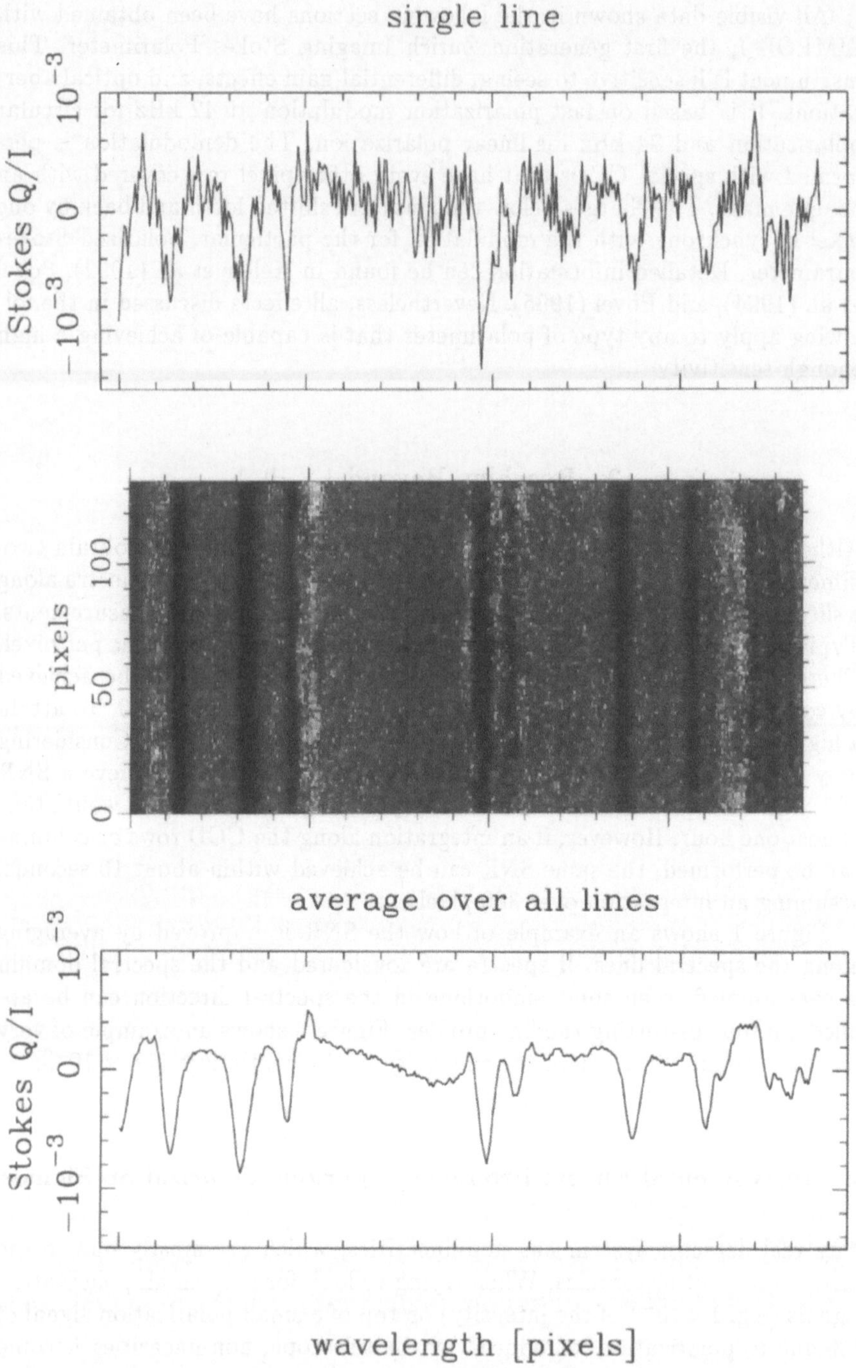

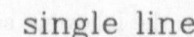

Fig. 1. Improvement of the signal-to-noise ratio by integrating along the spectral lines. The slit of the spectrograph was parallel to the solar limb at $\mu = 0.1$ and shows the amount of linear polarization around the MgI 5167.3 Å line (see Stenflo, 1996 for a discussion of the spectral features).

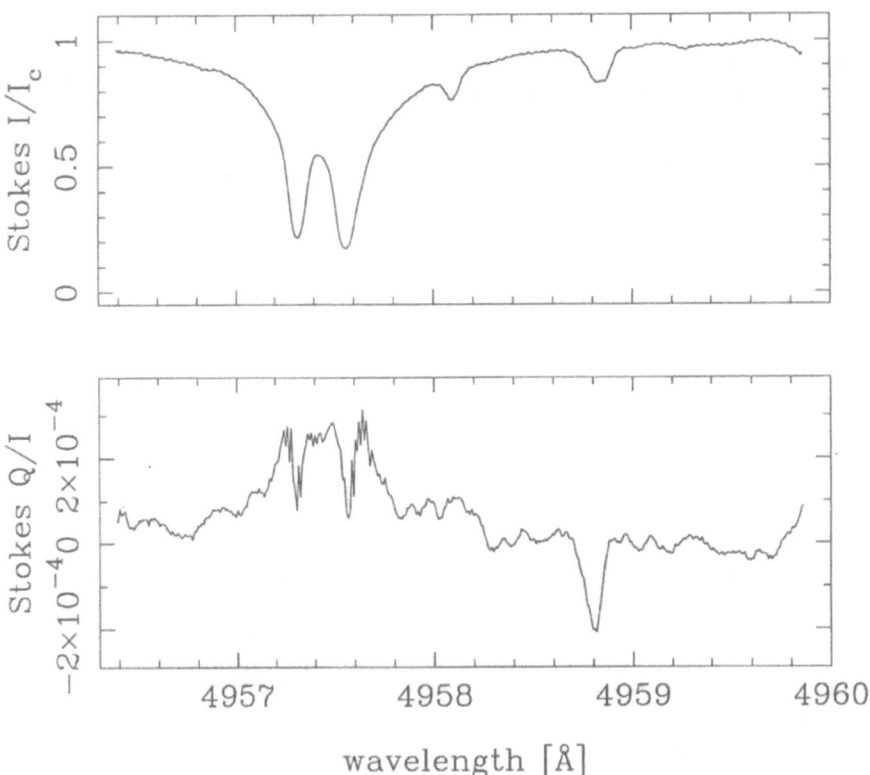

Fig. 2. An example of highly sensitive linear polarization measurements close to the solar limb ($\mu = 0.2$) showing features in the FeI 4957 Å doublet and a NdII line. The rms noise in the continuum is 6×10^{-6}. The instrumental polarization has been minimized by tilting a glass plate in front of the polarimeter.

Let the measured signal S be a quadratic function of the incoming intensity I

$$S = aI^2 + bI + c, \tag{1}$$

which is the most simple form for a non-linear behavior. The coefficient of the linear term represents the (arbitrary) gain, the offset any remaining influence of bias or dark current that has not been correctly removed, and the quadratic term models the non-linearity. The constant term also models effects of stray-light in a filter or a spectrograph.

For unpolarized light, the light corresponding to the two opposite polarization states of the polarimeter can be described by

$$I^+ = I + \delta I, \quad I^- = I - \delta I, \tag{2}$$

where δI is the additional polarization signal introduced by the telescope or the polarimeter.

The measured amount of polarization is determined from

$$P_m = \frac{S^+ - S^-}{S^+ + S^-} , \tag{3}$$

which in the ideal case of $a = 0$, $c = 0$ corresponds to the instrumentally introduced polarization

$$P = \frac{I^+ - I^-}{I^+ + I^-} = \delta . \tag{4}$$

In the case of non-linearities and an offset error (see Eq. 1), the apparent, measured polarization signal becomes

$$P_m = \frac{2a\delta I^2 + b\delta I}{aI^2 + a\delta^2 I^2 + bI + c} . \tag{5}$$

As expected, in the absence of instrumental polarization, i.e. $\delta = 0$, no polarization would be measured.

In the following we will only keep terms up to second order in a, δ, and their cross-products. Let us look at two cases. First we assume that $c = 0$. The measured polarization then becomes

$$P_m = \delta \left(1 + \frac{a}{b} I \right) . \tag{6}$$

The observed polarization has therefore an additional component that is essentially proportional to aI, the coefficient of the non-linear term times the intensity. If the detected signal corresponds to a spectrum, the measured polarization is not constant anymore, but has a Stokes I-like additive component.

To measure very small polarization signals in the presence of non-linearities in the detector system (which are present in any real detector system), it is necessary to minimize the instrumental polarization. After normalizing Stokes I with the continuum intensity I_c, variations of Stokes I are of order unity. A typical value for the non-linear term is $a/b = 0.01$. To achieve a sensitivity of 1×10^{-5}, the instrumental polarization δ must be smaller than 1×10^{-3}. This is particularly hard to achieve for linear polarization where a typical value for existing telescopes is $\delta = 0.05$. Therefore, a typical magnitude of the coupling of instrumental polarization with non-linearities in the detector system is 5×10^{-4}.

Figure 3 compares two linear polarization measurements of the same spectral region with and without compensation of the instrumental polarization. The ZIMPOL I camera used for these measurements did show a non-linearity

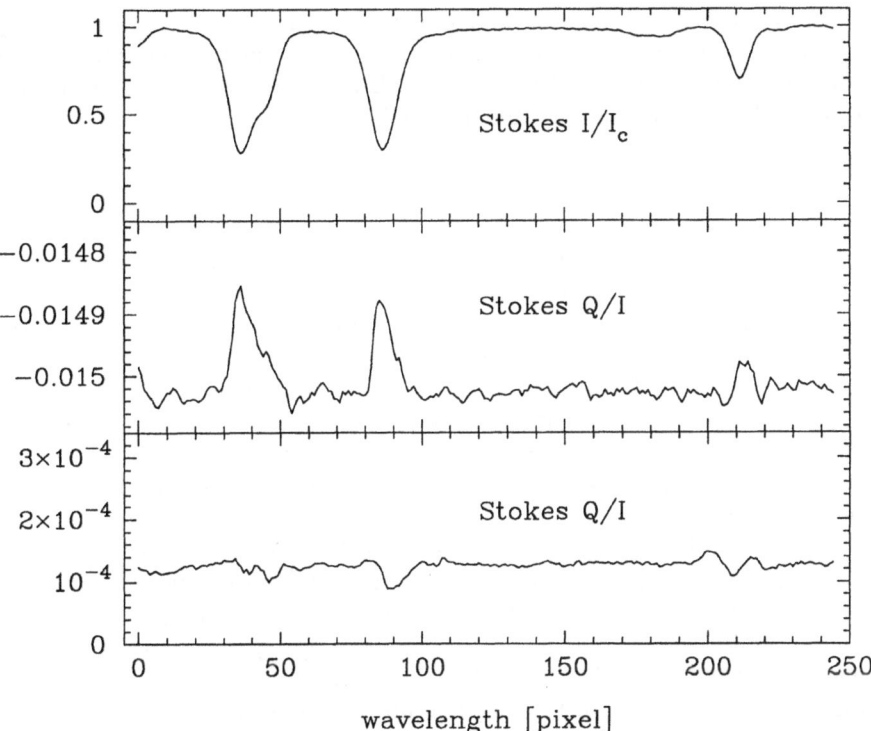

Fig. 3. Artificial polarization signal created by the coupling of non-linearities in the CCD readout electronics and instrumental polarization. These measurements were performed at the McMath-Pierce main telescope and the vertical spectrograph with one channel of ZIMPOL I in the green part of the spectrum. The top panel shows the Stokes I profile normalized with the continuum intensity I_c. The center panel shows the amount of linear polarization or Stokes Q/I in the case of -1.5% linear polarization induced by oblique reflections in the telescope. The lower panel shows the same linear polarization after the polarimeter has been rotated to minimize the linear polarization induced by the telescope. This procedure, however, maximizes the instrumental Stokes V to Q crosstalk in this particular setup. Indeed, the remaining structure seen in the lower panel is only due to magnetic fields in the quiet sun.

of about $a/b = 1.4\%$. The simple model developed above explains the sign as well as the magnitude of the effect.

The other case is $a = 0$. The measured polarization then becomes

$$P_m = \delta \left(1 - \frac{c}{bI}\right) . \tag{7}$$

Since I shows structure, e.g. spectral lines, the observed polarization will show signatures proportional to $1/I$. Any offset such as bias, dark current, stray-light from the observing room etc. must be removed to a high accuracy. However, due to minuscule changes in the observing conditions (e.g.

variation of the bias in time), there is always an offset error of approximately 1×10^{-3}. Therefore, this effect is about one order of magnitude smaller than the influence of non-linearities under realistic circumstances. It disappears completely if there is no instrumental polarization.

4. Modulated Interference Fringes

Many modern solar polarimeters work with one or several variable retarders and a linear polarizer. When the retardation is varied, the optical path length within the retarder also changes. Multiple reflections between the surfaces of the retarder lead to spectral fringes, whose pattern changes when the retardation is changed. A polarized spectrum (difference between two measurements with different retardations) will show a fringe pattern. This effect has been described by Oakberg (1995) for piezo-elastic modulators as used in ZIMPOL. I have also observed such fringe patterns in liquid crystal retarders.

Tilting, wedging, and coating the variable retarders reduce the fringe amplitude considerably. However, at the 1×10^{-5} level, fringes are always present. Often the fringes can be removed by an appropriate filtering in the Fourier domain. However, if the fringes are almost parallel to the spectral lines or if the line profiles need to be very accurate, another approach is required that does not distort the line profiles. Here, the parts of the spectrum that do not contain spectral lines are used to fit a model of the fringes, which is then used to predict the fringes at the location of the spectral lines. Experience has shown that it is best to fit the fringes with an autoregressive model (also called linear prediction) where the value y_n of the pixel at position n is predicted by

$$y_n = \sum_{j=1}^{N} c_j y_{n-j} . \tag{8}$$

The c_j are the coefficients of the autoregressive model. Each spatial location is treated separately. Since the model implies a direction, two models are constructed, one from left to right and one from right to left. Both model predictions for the fringes at the location of spectral lines are combined with a weighted average. The further the extrapolation, the smaller the weight. This method is fast and successfully predicts the fringes over gaps as wide as 5 fringe cycles (e.g. Rüedi et al., 1996).

5. Achieving 10^{-4} with Regular Array Detectors

A sensitive method for stellar circular polarimetry was developed by Semel et al. (1993). It is based on a rotatable quarter-wave plate and a double-calcite beam-splitter, which produce two beams corresponding to opposite circularly polarized light. The quarter-wave plate may be rotated to +45° and −45° with respect to the polarization axes of the beam-splitter. Both beams are recorded simultaneously. The quarter-wave plate is then rotated by 90° and another image is exposed. The four measurements of the same object are then combined to obtain an estimate of the Stokes V/I ratio that is largely free of effects from seeing and gain variations between different detector areas if the polarization signal is small. This approach has been used in stellar polarimetry with great success by Donati et al. (1990). It can be applied to any polarized Stokes parameter.

This approach also works very well for solar applications where the spectrum in the first and the second exposures are different. Consider the measured intensities in the two beams in the first exposure after subtraction of the dark current (for the case of Stokes V)

$$S_1^l = g_l \alpha_1 (I_1 + V_1), \quad S_1^r = g_r \alpha_1 (I_1 - V_1). \tag{9}$$

The subscript 1 indicates the first exposure, the subscripts l and r indicate the left and the right beams of the polarizing beam-splitter. S describes the measured signal, g the gain in a particular beam, and α the average transmission of the atmosphere and the instrument for a given exposure.

In the second exposure, the measured signals are given by

$$S_2^l = g_l \alpha_2 (I_2 - V_2), \quad S_2^r = g_r \alpha_2 (I_2 + V_2). \tag{10}$$

Note that the incoming intensity in the second exposure may be completely different from the first exposure. Such changes may be due to seeing as well as instrumental changes. This also includes a shift of the two beams between exposures due to beam-wobble induced by rotation of a wave plate.

The following combination of these four measured intensities removes the effect of transmission changes and differential gain variations of different detector areas:

$$\frac{1}{4} \left(\frac{S_1^l S_2^r}{S_2^l S_1^r} - 1 \right) = \frac{1}{2} \frac{I_2 V_1 + I_1 V_2}{I_1 I_2 - I_2 V_1 - I_1 V_2 + V_1 V_2}. \tag{11}$$

In the case of $V \ll I$, this is equivalent to

$$\frac{1}{2} \left(\frac{V_1}{I_1} + \frac{V_2}{I_2} \right). \tag{12}$$

Therefore we obtain the average V/I signal of the two exposures. No spurious polarization signals are introduced. If V is comparable to I, i.e. in the case of large polarization signals, the flat-field is accurate enough to directly subtract the simultaneously measured beams.

This data reduction method for images taken with a polarizing beam-splitter is not well known in the solar community. Kuhn (1995) has recently

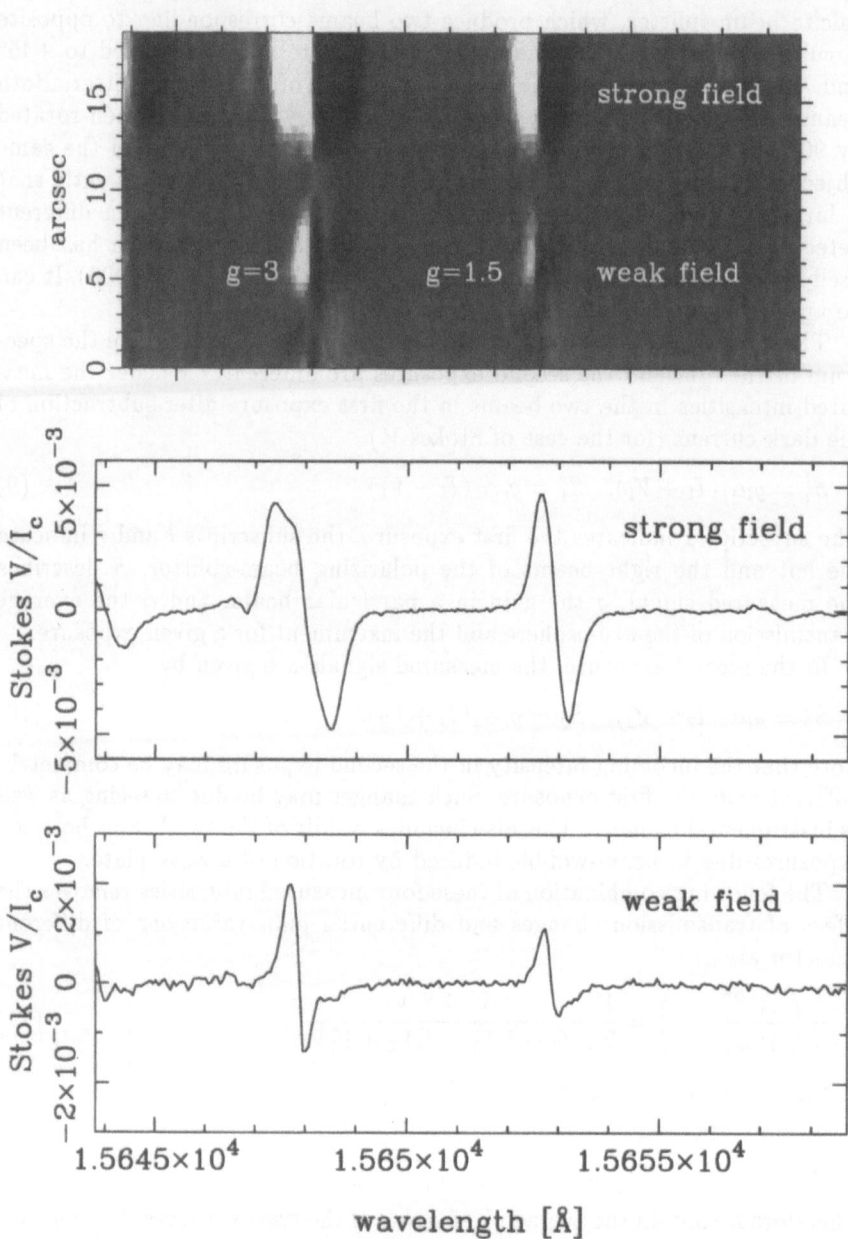

Fig. 4.   Strong and weak fields in the quiet sun close to disk center. The upper panel shows a two-dimensional spectrum, while the center and lower panels show cuts through the strong and the weak field regions. The splitting and the ratio of the amplitudes of the Stokes V profiles is an indication of the field strength. The data were obtained with the Near-Infrared Magnetograph (Rabin et al. 1991), which was enhanced with a polarizing beam-splitter.

used a similar method in an attempt to detect magnetic fields in the corona. Figure 4 shows an example of very sensitive polarimetry in the near-infrared using liquid crystal retarders and a double-calcite beam-splitter to obtain scans in the quiet sun using lines around 1565 nm.

6. Conclusions

Today we have technologies and methods at our disposal to reach a polari-metric sensitivity of 10^{-4} between 0.3 and 2.5 μm with array detectors. In the visible, spectra with a sensitivity of better than 10^{-5} have been obtained by integrating along the CCD columns. To avoid spurious polarization sig-nals due to non-linearities and offset changes in the detector system, either telescopes without intrinsic instrumental polarization must be used, or the instrumental polarization must be compensated. Instrumental linear polar-ization can be minimized efficiently by tilting and rotating a glass plate in front of the polarimeter.

Acknowledgements

I thank Douglas Rabin for collaborating on the near-infrared polarimetry and Jan Olof Stenflo for collaborating on the visible polarimetry. The Na-tional Solar Observatory is one of the National Optical Astronomy Obser-vatories, which are operated by the Association of Universities for Research in Astronomy, Inc. (AURA) under cooperative agreement with the National Science Foundation.

References

Donati, J.-F.,Semel, M., Rees, D.E., Taylor, K., Robinson, R.D.: 1990, *A&A* **232**, L1
Keller, C.U., Aebersold, F., Egger, U., Povel, H.P., Steiner, P., Stenflo, J.O.: 1992, *LEST Technical Report* **53**,
Kuhn, J.R.: 1995, in Kuhn, J.R., Penn, M. (eds.), IR Tools for Solar Astrophysics, World Scientific, Singapore, in press
Oakberg, T.C.: 1995, *Opt. Eng.* **34**, 1545
Povel, H.P.: 1995, *Opt. Eng.* **34**, 1870
Povel, H.P., Keller, C.U., Yadigaroglu, I.-A.: 1994, *Applied Optics* **33**, 4254
Rabin, D.M., Jaksha, D., Plymate, C., Wagner, J., Iwata, K.: 1991, in November, L.J. (ed.), Proc. of the 11th Sacramento Peak Summer Workshop on Solar Polarimetry, Sunspot, New Mexico, 361
Rüedi, I., Keller, C.U., Solanki, S.K.: 1996, these proceedings
Semel, M., Donati, J.-F., Rees D.E.: 1993, *A&A* **278**, 231
Stenflo, J.O.: 1996, these proceedings

THE INFLUENCE OF SUNSPOT CANOPIES ON MAGNETIC INCLINATION MEASUREMENTS IN SOLAR PLAGES

SAMI K. SOLANKI, WOLFGANG FINSTERLE and ISABELLE RÜEDI

Institute of Astronomy, ETH Zentrum, CH-8092 Zurich, Switzerland

Abstract. Sunspots are known to have large, low-lying magnetic canopies, i.e. horizontal magnetic fields overlying a field-free medium, that cover substantial fractions of active region plage. In this paper we consider the influence of such canopies on the inclination of plage magnetic fields. We find that for observations in spectral lines like 5250.2Å the neglect of a sunspot canopy when determining magnetic inclination angles of plage fields can introduce errors exceeding 5–10°. This is particularly true if the observations do not have high spatial resolution. Thus this effect may explain some of the measurements of substantially inclined fields in solar plages. Furthermore we find that the Fe I 15648 Å line is far superior in giving correct flux-tube inclinations in the presence of a sunspot magnetic canopy. Finally, the inversion of full Stokes profiles is shown to produce more reliable results than results obtained by considering only ratios of individual Stokes profile parameters.

Key words: Magnetic fields – active regions – plage – sunspots – flux tubes

1. Introduction

Reliable observations of solar vector magnetic fields are central to the understanding of solar MHD processes, as testified by the increasing efforts being undertaken to derive magnetic vectors with high precision from measured Stokes profiles.

One of the important parameters one wishes to determine is the inclination angle γ' (relative to the surface normal) of the magnetic elements composing solar plages.* Initial theoretical estimates (e.g. Schüssler, 1986) predicted that kG magnetic elements should not deviate from the vertical direction by more than 1°. Later, Schüssler (1990) allowed for the possibility that some magnetic elements, those with lower field strengths, may be more strongly inclined. More recently these estimates have had to be revised in the light of the discovery of supersonic flows in solar granulation (e.g. Cattaneo *et al.*, 1990, Nesis *et al.*, 1992, Solanki *et al.*, 1995a). The increased strength of the granular buffeting can tilt individual flux tubes by as much as 15–25° (Solanki, 1993). Finally, 2-D MHD simulations (Steiner *et al.*, 1995) have demonstrated directly that flux tubes become periodically

---

* Primed quantities, inclination angle γ' and azimuth χ', refer to local solar coordinates, unprimed ones to inclinations and azimuths relative to the line of sight.

Solar Physics **164**: 253–264, 1996.
© 1996 *Kluwer Academic Publishers.*

inclined by large amounts due to granular buffeting. This mechanism is expected to act on each flux tube individually, so that at any given moment the average inclination of a sufficiently large ensemble of flux tubes is expected to be small. In addition, further simulations by Steiner, Grossmann-Doerth, Knölker and Schüssler (private communication) show that only small flux tubes (diameter $\lesssim$ 400 km in the continuum forming layer) react strongly to granular buffeting. Larger flux tubes remain basically vertical throughout their simulation.

In the quiet magnetic network only an upper limit of 10° on the *average* inclination of magnetic elements is known from direct measurement (Sánchez Almeida & Martínez Pillet, 1994), although Murray (1992) finds evidence for on average highly inclined weak magnetic field from a statistical analysis of full-disk magnetograms, in particular of the centre-to-limb variation of the magnetogram signal. How far his results near the limb are affected by horizontal canopy fields due to the quiet network (Jones & Giovanelli, 1983) is an open question.

In active region plages recent observations, even when averaged over many magnetic elements, give inclinations to the vertical that often are larger than 10° (Solanki *et al.*, 1987; Lites & Skumanich, 1990; Bernasconi *et al.*, 1995). A statistical study of magnetograms (Howard, 1991) also indicates large inclinations, with the opposite polarity fields in active regions being inclined by 16° towards each other.

Although theory and quiet-sun observations are mutually consistent, they disagree with active region observations. In the present paper we investigate a possible explanation for this discrepancy. In addition to small-scale magnetic features active regions contain sunspots which possess large-scale, low-lying magnetic canopies that cover a significant fraction of active region plages (e.g. Giovanelli, 1980; Giovanelli & Jones, 1982; Solanki *et al.*, 1992b, 1994; Bruls *et al.*, 1995). We investigate how such a low-lying magnetic canopy influences and possibly falsifies magnetic element inclinations derived from the Stokes vector.

2. Technique

We have carried out a large set of calculations of Stokes parameters in a simple 2-component model of a horizontal superpenumbral magnetic canopy (1st component) and vertical magnetic elements, modelled as flux tubes (2nd component). Each magnetic component is described by a simple 1-D (i.e. plane-parallel) atmosphere. For the canopy this is clearly a reasonable assumption, while for the flux tubes its validity, as far as the inclination of the magnetic field is concerned, has been shown to hold using 2-D models (Solanki *et al.*, 1995b).

The magnetic field of the canopy, with strength B_c and an inclination to the surface normal $\gamma'_c = 90°$, is homogeneous above the canopy base-height Z_c and disappears below Z_c. The magnetic field of the flux tube has $\gamma'_{FT} = 0°$ and a strength B_{FT} whose vertical variation is given by the thin-tube approximation (Defouw, 1976). Thus in this model there is a magnetic field everywhere above the height Z_c, while below this height only the fraction of the solar surface covered by flux tubes, α_{FT}, has a magnetic field. Consequently the filling factor of the canopy component $\alpha_c = 1 - \alpha_{FT}$. In accordance with previous work the thermal structure of the canopy component is the same as that of the quiet sun (Maltby et al., 1986), while the flux-tube atmosphere is described by the plage flux-tube model of Solanki & Brigljević (1992). In the present simple model pressure balance is not enforced across the canopy boundary.

At a given heliocentric angle θ we calculate the Stokes profiles for each component separately (obtaining S_{FT} and S_c, where $S = I, Q, U, V$), weight them by the surface filling factor of the respective component and add them together, forming

$$S_{tot} = S_{FT}\, \alpha_{FT} + S_c(1 - \alpha_{FT}) .$$

In order to determine the inclination γ with respect to the line-of-sight we take two approaches. In the first the ratio $R = V/\sqrt{Q^2 + U^2}$ is used as a measure of γ. The difference between

$$R_{tot} = \frac{V_{tot}}{\sqrt{Q_{tot}^2 + U_{tot}^2}}$$

and

$$R_{FT} = \frac{V_{FT}}{\sqrt{Q_{FT}^2 + U_{FT}^2}}$$

is an indication of the error, $\Delta\gamma_{FT}$, introduced into γ measurements of the flux-tube field by neglecting the magnetic canopy. To estimate $\Delta\gamma_{FT}$ we have calculated R_{FT} for a number of γ'_{FT} values for each θ (i.e. flux tubes with different inclinations to the vertical; primed quantities refer to the local solar coordinate system). If γ_1 is the inclination with respect to the line of sight of the original flux tube (whose Stokes spectrum is added to that of the canopy) and γ_2 is the inclination of the flux tube that gives the same R value as R_{tot}, then $\Delta\gamma_{FT} = |\gamma_2 - \gamma_1|$.

In the second approach we have used the inversion code described by Solanki et al. (1992b, 1994) to fit the combined Stokes profiles S_{tot} with a pure flux-tube model. The code then gives us γ_{FT} (fit) and χ_{FT} (fit), which can be compared with the true values to yield $\Delta\gamma_{FT}$ and $\Delta\chi_{FT}$. Here χ is the magnetic azimuth. We define $\chi = \chi' = 0$ in the direction of solar disk centre.

We have calculated the Stokes profiles of four Fe I spectral lines using the Stokes transfer code of Solanki et al. (1992a) based on the Diagonal Element Lambda Operator (DELO) technique (Rees et al., 1989; Murphy, 1990). The four lines are: 5247.1Å (excitation potential $\chi_e = 0.09$ eV, Landé factor $g_{eff} = 2$), 5250.2Å ($\chi_e = 0.12$ eV, $g = 3$), 5250.6Å ($\chi_e = 2.20$ eV, $g_{eff} = 1.5$) and 15648Å ($\chi_e = 5.43$ eV, $g = 3$). The three lines in the visible have been shown to be very useful diagnostics of the field strength and the temperature in small-scale magnetic features by Stenflo et al. (1987) and Zayer et al. (1990).

The free parameters of the model calculations are: Z_c, α_{FT}, B_c, B_{FT}, ξ_{mac}, χ_c', where ξ_{mac} is the macroturbulence velocity and χ_c' is the azimuthal angle of the canopy magnetic field in local (solar) coordinates. We have carried out model calculations for $Z_c = 300$, 400 and 500 km, with the corresponding B_c values 500, 300 and 200 G. This Z_c–B_c combination corresponds roughly to values derived from infrared lines (Solanki et al., 1992b, 1994; Hewagama et al., 1993 and Bruls et al., 1994). Following the constraints imposed by observations (e.g. Rüedi et al., 1992), only a single value of B_{FT}, namely B_{FT} $(z = 0) = 1500$ G is chosen, where $z = 0$ corresponds to continuum optical depth $\tau_c = 1$ in the quiet sun. χ_c' values of $0°$, $180°$ and $90°$ have been considered, i.e. canopies looking away from the limb, towards the limb, and parallel to the limb as seen by the observer (recall that by considering horizontal canopies we have chosen $\gamma_c' = 90°$). 0 and 2 km/s have been employed for ξ_{mac}, and 0.1, 0.3 and 0.5 for α_{FT}. 0.1 is typical of α_{FT} values obtained with low spatial-resolution data, while 0.5 can only be achieved with rather high spatial resolution data. A microturbulence of 0.8 km/s is applied in the non-magnetic portions of the atmospheres and in the canopy, while 1.0 km/s is used within the flux tubes.

3. Results

3.1. AMPLITUDE RATIOS

We first discuss a selection of the results obtained with the simpler technique, i.e. from the ratio $V/\sqrt{Q^2 + U^2}$.

Figure 1 shows $\Delta\gamma_{FT}$ as a function of θ for $Z_c = 500$, 400 and 300 km. Plotted are results for Fe I 5250.2Å, $\alpha_{FT} = 0.1$ and $\xi_{mac} = 0$. We take $0 < \gamma_c < 90°$ and $\chi_c = 0$, i.e. we consider positive polarity vertical flux tubes and the canopy on the diskward side of a positive polarity sunspot. Note that for $\chi_c = 0°$ we have $\Delta\gamma_{FT} = \Delta\gamma_{FT}'$, i.e. the errors plotted in Fig. 1 and subsequent figures have exactly the same magnitude as the errors in the

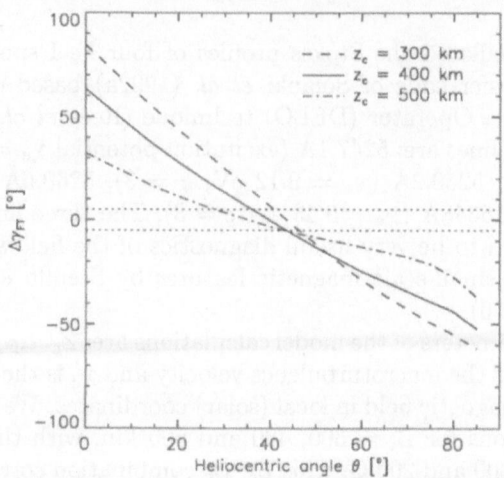

Fig. 1.   Error in flux-tube inclination $\Delta\gamma_{FT}$ vs. heliocentric angle θ at which the flux tube is observed. See text for a description of $\Delta\gamma_{FT}$. Each curve represents $\Delta\gamma_{FT}$ for a different value of the sunspot canopy base-height, Z_c. Other parameters: Flux-tube magnetic filling factor $\alpha_{FT} = 0.1$, spectral line Fe I 5250.2Å, flux-tube and canopy magnetic inclinations relative to line-of-sight $\gamma_{FT} < 90°$, $\gamma_c < 90°$, while the azimuths for the flux-tube and canopy fields are $\chi_{FT} = 180°$ and $\chi_c = 0$. The horizontal dotted lines bound the interval $-5° \leq \Delta\gamma_{FT} \leq 5°$. The plotted results were obtained by comparing ratios of σ-amplitudes.

inclination angle relative to the surface normal. As expected $\Delta\gamma_{FT}$ decreases with increasing Z_c, i.e. the higher the canopy base the smaller its influence on the Stokes profiles. The qualitative dependence of $\Delta\gamma_{FT}$ on θ can be understood relatively easily. For $\theta < 45°$ we have $\gamma_c > 45° > \gamma_{FT}$. Thus the combined flux-tube and canopy Stokes profiles mimic a flux tube with a larger inclination to the line-of-sight, if interpreted in terms of a flux-tube component only ($\gamma_{tot} > \gamma_{FT}$, i.e $\Delta\gamma_{FT} > 0$). For $\theta > 45°$, on the other hand, $\gamma_c < 45° < \gamma_{FT}$, so that the combined profiles mimic flux tubes inclined towards the observer ($\gamma_{tot} < \gamma_{FT}$). If $B_c = B_{FT}$ then $\Delta\gamma_{FT}$ must disappear at $\theta = 45°$. Since $B_c \neq B_{FT}$ and the ratio $V/\sqrt{Q^2 + U^2}$ depends on the field strength (Stenflo, 1985), $\Delta\gamma_{FT} = 0$ near, but not exactly at $\theta = 45°$.

The most striking feature of Fig. 1 is the magnitude of the effect. The two horizontal dotted lines lie at $|\Delta\gamma_{FT}| = 5°$. We consider $\Delta\gamma_{FT}$ values lying outside the area enclosed by these two lines to be significant in the sense that they are larger than the usual errors in measured flux-tube inclinations. In Fig. 1 this is the case over most of the θ range.

Figure 2 compares the results for three different values of α_{FT}: 0.1, 0.3 and 0.5 ($Z_c = 400$ km, and the rest of the model parameters are the same as for Fig. 1). As expected $|\Delta\gamma_{FT}|$ decreases strongly with increasing α_{FT}, so

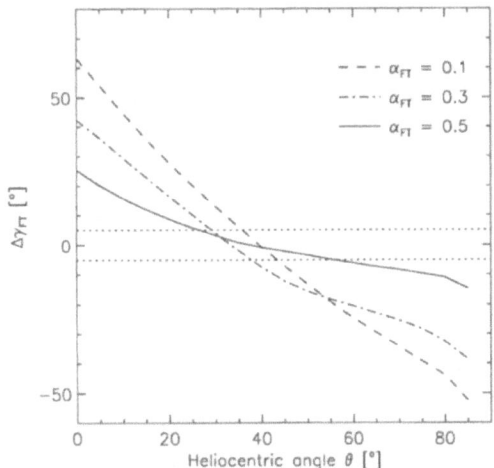

Fig. 2.   Same as Fig. 1, except that each curve now represents $\Delta\gamma_{FT}$ for a different α_{FT}. The height of the canopy base $Z_c = 400$ km.

that for $\alpha_{FT} = 0.5$ $|\Delta\gamma_{FT}|$ remains below 5° over a θ interval of 30°. Figure 2 thus illustrates the value of high spatial resolution observations.

A non-zero ξ_{mac} reduces $\Delta\gamma_{FT}$, since the lobes of the narrower and generally weaker canopy V and Q profiles are cancelled more readily by turbulent broadening than the flux-tube profiles. However, reasonable values of ξ_{mac} reduce $|\Delta\gamma_{FT}|$ by only 10–20% and do not significantly modify conclusions based on calculations with $\xi_{mac} = 0$.

In Fig. 3 we plot $\Delta\gamma_{FT}$, determined individually from each of the three lines $\lambda5250.2$Å $(g = 3)$, $\lambda5247.1$Å $(g_{eff} = 2)$ and $\lambda5250.6$Å $(g_{eff} = 1.5)$ $(Z_c = 400$ km, $\alpha_{FT} = 0.5$, $\chi_c = 0°$ and $\gamma_c < 90°)$. $|\Delta\gamma_{FT}|$ increases in the order of increasing g_{eff} since the strength of the canopy Q and V profiles increases relative to the flux-tube profiles for increasing g_{eff} (the flux-tube Q and V profiles are increasingly Zeeman saturated for increasing g_{eff}). This implies that it is of advantage to use lines with lower g_{eff} to measure flux-tube inclinations.

This result may not be generalized, however, since by far the most reliable results were obtained with the most Zeeman-sensitive line in our sample, Fe I 15648Å. For the plotted case $|\Delta\gamma_{FT}| < 1°$ over almost the whole θ range. This result is to a large part due to the infrared line's low height of formation (it obtains its main contribution at $\log \tau_{5000} \gtrsim -1$ in the quiet sun), which makes it rather insensitive to canopy fields with $Z_c \gtrsim 300$ km. In addition, due to its large Zeeman sensitivity the canopy and flux-tube contributions

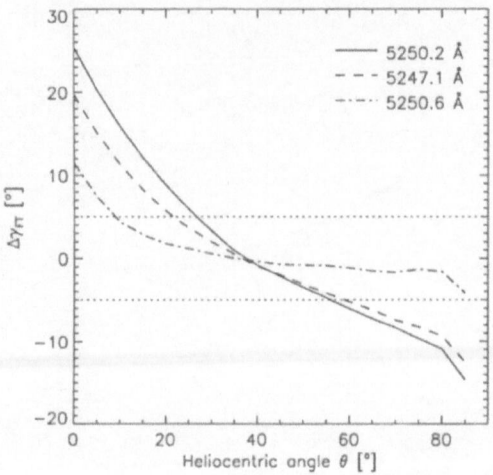

Fig. 3. Same as Fig. 1, except that each curve now represents the $\Delta\gamma_{FT}$ derived from a different spectral line. We have chosen $Z_c = 400$ km and $\alpha_{FT} = 0.5$ (instead of $\alpha_{FT} = 0.1$ as in Figs. 1 and 2).

to its V and Q profiles are clearly separated as a result of the different field strengths in these components (Solanki *et al.*, 1992b, 1994).

3.2. INVERSIONS

Due to the substantially larger computing demands of the inversion technique we have applied it to a smaller model grid. We discuss only the case $Z_c = 400$ km with $\alpha_{FT} = 0.5$ and $\alpha_{FT} = 0.1$.

In the upper panel of Fig. 4 we plot $\Delta\gamma_{FT}$ resulting from the inversion of the combined flux-tube and canopy Stokes profiles for $\alpha_{FT} = 0.5$, $Z_c = 400$ km and three different canopy azimuths: $\chi'_c = 0°$ (i.e. $0° < \gamma_c \leq 90°$ and $\chi_c = 0°$), $\chi'_c = 90°$ (i.e. $\gamma_c = 90°$ and $\chi_c = 90°$) and $\chi'_c = 180°$ (i.e. $180° > \gamma_c \geq 90°$ and $\chi_c = 180°$). The first of these cases corresponds, e.g., to observations on the diskward side of a sunspot with the same polarity as the flux tubes. The second case corresponds to canopy field lines lying parallel to the solar limb, as seen, e.g. when observing plage at the same θ as a nearby sunspot. The third case corresponds, e.g., to observations on the limbward side of a sunspot with the same polarity as the flux tubes. The difference to the first case ($\gamma_c < 90°$) is that now the V profiles due to the canopy and flux-tube components have opposite signs. Each curve in Fig. 4 is based on the combined inversion of all three lines. We allowed α_{FT}, B, γ, χ and ξ_{mac} to be determined by the code.

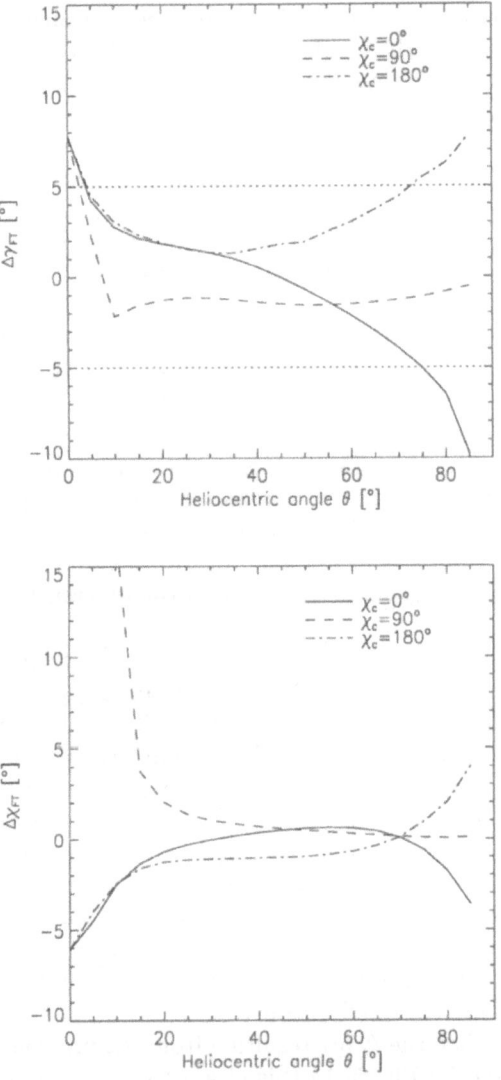

Fig. 4.   Upper panel: $\Delta\gamma_{FT}$ determined from multi-line inversions for three different directions of the sunspot canopy vs. θ. Solid curve: $\gamma_c \leq 90°$, $\chi_c = 0°$ (canopy field oriented away from the limb), dot-dashed curve: $\gamma_c \geq 90°$, $\chi_c = 180°$ (canopy field oriented towards the limb), and dashed curve: $\gamma_c = 90°$, $\chi_c = 90°$ (canopy field parallel to the limb). $\gamma_{FT} < 90°$, $\chi_{FT} = 180°$ in all cases. Other model parameters are $Z_c = 400$ km and $\alpha_{FT} = 0.5$.   Lower panel: Error in the azimuth $\Delta\chi_{FT}$ vs. θ for the same fits as in the upper panel.

A comparison between the solid curves in Figs. 2 and 4 (upper panel) indicates that the inversion technique gives on average a factor of 2–3 smaller $|\Delta\gamma_{FT}|$. One reason is that the inversion code considers not only a single wavelength point but the whole line profile. It also combines fits to all three lines. Finally, the fact that the inversion determines a field strength somewhat lower than that of the flux tube (due to the canopy contribution) also plays a significant role in lowering $\Delta\gamma_{FT}$. The field strength returned by the inversion code at the limb is approximately 500 G lower than the field strength of the original model flux tube. The relative strengths of the V and Q profiles of a vertical flux tube with lower B are more similar to those of the combined canopy and flux-tube profiles than the relative strengths for a strong-field flux tube.

The inversion also provides $\Delta\chi_{FT}$, which is plotted in the lower panel of Fig. 4. Except near the centre of the solar disk, where $\Delta\chi_{FT}$ due to the best fit tends to 90° for $\chi_c = 90°$, $\Delta\chi_{FT}$ is rather small and should not affect results significantly.

In the case plotted in Fig. 4, corresponding to very high spatial resolution measurements, sunspot canopies can for most purposes effectively be neglected, except at disk centre and at the limb. In addition, the accuracy in γ_{FT} for the parts of the canopy with field lines parallel to the limb ($\chi_c = 90°$) is on the whole better than on the diskward and limbward sides of the sunspot.

If all parameters excluding γ and χ are fixed during the inversion, the combined inversion of all three lines results in $\Delta\gamma_{FT}$ values that are intermediate between those obtained by fitting 5250.6Å alone (which gives somewhat smaller $|\Delta\gamma_{FT}|$ values) and those from fitting 5250.2Å and 5247.1Å (larger $|\Delta\gamma_{FT}|$ values). A simultaneous fit to all three lines is nevertheless to be preferred over fits to $\lambda5250.6$Å alone. Firstly, it gives more reliable χ_{FT} values than any of the lines individually. Secondly, by fitting all three lines simultaneously we obtain in addition to γ_{FT} and χ_{FT} also parameters such as field strength, macroturbulence velocity and temperature. Since each of these parameters influences the derived γ_{FT} and χ_{FT}, and as they are usually not known beforehand, it is important to fit multiple lines and derive all the free parameters simultaneously.

If $\alpha_{FT} = 0.1$, then the $\Delta\gamma_{FT}$ resulting from the inversion is quite substantial, as illustrated by Fig. 5. In this case $|\Delta\gamma_{FT}| > 5°$ over a large fraction of the solar disk. The inversion nevertheless gives on average an almost two times smaller $\Delta\gamma_{FT}$ than that produced by the amplitude ratio (compare with Fig. 1). Once again the inversion code underestimates the field strength of the flux-tube, which contributes to reducing $|\Delta\gamma_{FT}|$ relative to the simple $V/\sqrt{Q^2 + U^2}$-ratio method. The quality of the fits is not too good very close to the limb, mainly because a single magnetic component cannot properly fit the combined Stokes profiles S_{tot}, which begin to exhibit the influence

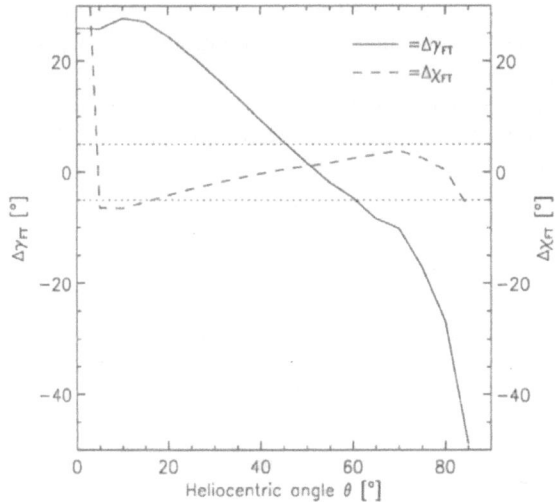

Fig. 5. $\Delta\gamma_{FT}$ (solid) and $\Delta\chi_{FT}$ (dashed) vs. θ for the following choice of parameters: $Z_c = 400$ km, $\alpha_{FT} = 0.1$, $\gamma_c \leq 90°$. The diagram has been obtained by simultaneous inversion of all three visible lines.

of the second magnetic component quite clearly there. Thus, by considering only fits which reproduce the observations with high quality, it should be possible to restrict errors to less than approximately 20° in $\Delta\gamma_{FT}$ due to the sunspot magnetic canopies. The error in γ_{FT} (dashed curve in Fig. 5) is once again relatively small.

4. Discussion and Conclusion

We have tested how the presence of a sunspot magnetic canopy affects measurements of the direction of the magnetic vector. The conclusions from the present work are:

1. For a line like 5250.2Å the presence of a low-lying sunspot canopy can significantly affect the flux-tube inclination derived from the observed Stokes parameters, in particular if the spatial resolution is not very high.
2. We find that an inversion approach regains the direction of the flux-tube magnetic vector more reliably than the modelling of simple line ratios.
3. The height of formation of a spectral line is important for its sensitivity to the canopy field. The lower the height of formation of a spectral line, the better it provides estimates of the flux-tube magnetic vector. In particular the Fe I 15648Å line turns out to give flux-tube magnetic

inclinations that are relatively unaffected by the canopy. For lines in the visible the error in the flux-tube inclination due to the canopy increases with increasing Zeeman sensitivity. If, however, the Zeeman sensitivity is large enough to separate the flux-tube and canopy contributions to the observed line profile, then the error in the derived inclination introduced by a sunspot canopy is also reduced.

4. Since the Stokes profiles produced by the canopy component are narrower than the flux-tube Stokes profiles (e.g. Rüedi *et al.*, 1995), we propose that it is better to observe or analyse the outer parts of the Stokes profiles in order to derive the true inclination of flux tubes, particularly when observing with an imaging instrument that has a fixed spectral passband. This choice of wavelength also possesses other advantages (cf. Jefferies & Mickey, 1991).

5. The current investigation demonstrates yet again that no diagnostic or inversion is free from model dependence and that it pays to carefully select the spectral line(s) and to consider the various possible magnetic field configurations. In other words, even the best inversion technique is no better than the underlying model and spectral diagnostics.

6. The smallest errors are achieved in the interval $\mu = \cos\theta \approx 0.5$–$0.8$. It may be worthwhile to give priority to magnetic inclination measurements in plages within roughly this range of μ values.

7. It is possible to extend the Stokes fitting technique to encompass two magnetic components (Bernasconi & Solanki, 1995) and thus model the effect of the canopy directly during the inversion. This may be a promising approach for the future.

The relevance of the current investigation to the real sun depends on the fraction of plage area covered by low-lying sunspot canopies. Giovanelli & Jones (1982) found for two active regions that approximately half of the surface area free from magnetic fields in the low photosphere is covered by nearly horizontal canopies with a base height below 500 km. Although it is not obvious how closely their canopy base height corresponds to ours, it is nevertheless clear that the influence of sunspot canopies cannot be neglected in active regions.

Sunspot magnetic canopies may thus affect magnetic inclination measurements of the type presented by Solanki *et al.* (1987), Lites & Skumanich (1990) and Bernasconi *et al.* (1995), all of which have low spatial resolution.* We suggest that in order to deliver more reliable values of γ_{FT} and χ_{FT} future Stokes vector measurements in plages should either make use of high spatial resolution, employ lines formed low in the atmosphere (e.g. Fe I 15648Å),

---

* The results of Lites & Skumanich (1990), being based on Fe I 6302.5Å, which has a somewhat lower height of formation than the Fe I lines around 5250Å favoured by the other investigators, probably are somewhat less affected by sunspot canopies.

or include the influence of a possible sunspot canopy directly in the model on which the inversion is based.

References

Bernasconi P.N., Solanki S.K.: 1995, *Solar Phys.* in press
Bernasconi P.N., Keller C.U., Povel H.P., Stenflo J.O.: 1995, *Astron. Astrophys.* in press
Bruls J.H.M.J., Solanki S.K., Carlsson M., Rutten R.J.: 1995, *Astron. Astrophys.* **293**, 225
Cattaneo F., Hurlburt N.E., Toomre J.: 1990, *Astrophys. J.* **349**, L63
Defouw R.J., 1976, *Astrophys. J.* **209**, 266
Finsterle, W.: 1995, *Bestimmung der Magnetischen Struktur von Sonnenflecken mit Infrarot und Optischen Spektren (in German)*, Diplomarbeit, Institute of Astronomy, ETH, Zürich
Giovanelli R.G.: 1980, *Solar Phys.* **68**, 49
Giovanelli R.G., Jones H.P.: 1982, *Solar Phys.* **79**, 267
Hewagama T., Deming D., Jennings D.E., Osherovich V., Wiedemann G., Zipoy D., Mickey D.L., Garcia H., 1993, *Astrophys. J. Suppl. Ser.* **86**, 313
Howard R.F.: 1991, *Solar Phys.* **134**, 233
Jefferies J.T., Mickey D.L.: 1991, *Astrophys. J.* **372**, 694
Jones H.P., Giovanelli R.G., 1983, *Solar Phys.* **87**, 37
Lites B.W., Skumanich A.: 1990, *Astrophys. J.* **348**, 747
Maltby P., Avrett E.H., Carlsson M., Kjeldseth-Moe O., Kurucz R.L., Loeser R.: 1986, *Astrophys. J.* **306**, 284
Murphy G.A.: 1990, *The Synthesis and Inversion of Stokes Spectral Profiles*, NCAR Cooperative Thesis No. 124
Murray N.: 1992, *Astrophys. J.* **401**, 386
Nesis A., Bogdan T.J., Cattaneo F., Hanslmeier A., Knölker M., Malagoli A., 1992, *Astrophys. J.* **399**, L99
Rees D.E., Murphy G.A., Durrant C.J.: 1989, *Astrophys. J.* **339**, 1093
Rüedi I., Solanki S.K., Livingston W., Stenflo, J.O.: 1992, *Astron. Astrophys.* **263**, 323
Rüedi I., Solanki S.K., Livingston W., 1995, *Astron. Astrophys.* **293**, 252
Sánchez Almeida J., Martínez Pillet V.: 1994, *Astrophys. J.* **424**, 1014
Schüssler M.: 1986, in *Small Scale Magnetic Flux Concentrations in the Solar Photosphere*, W. Deinzer, M. Knölker, H.H. Voigt (Eds.), Vandenhoeck & Ruprecht, Göttingen, p. 103
Schüssler M.: 1990, in *Solar Photosphere: Structure, Convection and Magnetic Fields*, J.O. Stenflo (Ed.), Kluwer, Dordrecht, *IAU Symp.* **138**, 161
Solanki S.K.: 1993, *Space Science Rev.* **61**, 1
Solanki S.K., Brigljević V.: 1992, *Astron. Astrophys.* **262**, L29
Solanki S.K., Keller C., Stenflo J.O.: 1987, *Astron. Astrophys.* **188**, 183
Solanki S.K., Rüedi I., Livingston W.: 1992a, *Astron. Astrophys.* **263**, 312
Solanki S.K., Rüedi I., Livingston W.: 1992b, *Astron. Astrophys.* **263**, 339
Solanki S.K., Montavon C.A.P., Livingston W.: 1994, *Astron. Astrophys.* **283**, 221
Solanki S.K., Rüedi I., Bianda M., Steffen M.: 1995a, *Astron. Astrophys.* submitted
Solanki S.K., Steiner O., Bünte M., Murphy G.: 1995b, *Astron. Astrophys.* to be submitted
Steiner O., Grossmann-Doerth U., Knölker M., Schüssler M.: 1995, *Rev. Mod. Astron.* **8**, p. 81
Stenflo J.O.: 1985, in *Measurements of Solar Vector Magnetic Fields*, M.J. Hagyard (Ed.), NASA Conf. Publ. 2374, p. 263
Stenflo J.O., Solanki S.K., Harvey J.W., 1987, *Astron. Astrophys.* **171**, 305
Zayer I., Solanki S.K., Stenflo J.O., Keller C.U., 1990, *Astron. Astrophys.* **239**, 356

MEASUREMENT OF THE FULL STOKES VECTOR OF HE I 10830 Å

I. RÜEDI

Institute of Astronomy, ETH Zentrum, CH-8092 Zürich, Switzerland

C.U. KELLER

National Solar Observatory, NOAO*, P.O.Box 26732, Tucson, AZ 85726, USA

and

S.K. SOLANKI

Institute of Astronomy, ETH Zentrum, CH-8092 Zürich, Switzerland

Abstract. First observations of the full Stokes vector in the upper chromosphere are presented. The He I 10830 Å line, which has been shown to give reliable measurements of the line-of-sight component of the magnetic field vector, has been used for this purpose. It is shown that the difference between the appearance of chromospheric and photospheric magnetic structures observed close to the solar limb is largely due to the difference in height to which they refer and projection effects. The observations do suggest, however, that the magnetic field above sunspot penumbrae is somewhat more vertical in the chromosphere than in the photosphere.

Key words: Chromosphere – Polarization – Infrared – Magnetic Field

1. Introduction

Only few measurements of the upper chromospheric magnetic field exist. All of them are restricted to a single magnetic vector component, i.e. they are based on only a single polarized Stokes parameter. Simultaneous and cospatial measurements of all the components of the magnetic vector would be of great interest. In the upper chromosphere the ratio of magnetic to gas pressure is much larger than in the photosphere and consequently more closely representative of coronal conditions. Measurements of the magnetic vector in the upper chromosphere consequently are better starting values for the extrapolation of the magnetic configuration into the corona, or can at least be used to test the predictions of extrapolations that start in the photosphere.

The He I 10830 Å line has been shown to be a convenient and useful diagnostic of the line of sight component of the magnetic field vector (Penn *et al.*, 1995; Rüedi *et al.*, 1995a). Due to its narrow range of formation in the upper chromosphere it gives direct information on that layer, in contrast to

* The National Optical Astronomy Obervatories are operated by the Association of Universities for Research in Astronomy, Inc. (AURA) under cooperative agreement with the National Science Foundation

Solar Physics **164**: 265–275, 1996.

lines such as, for example, Hα, which are formed over a much larger height range.

In the present paper we point out that He I 10830 Å is also suited for the measurement of the other components of the magnetic vector and demonstrate that simultaneous and cospatial measurements of all the components of the Stokes vector can be carried out using current detectors and polarimeters.

2. Observations

The observations presented here were carried out on 9 Feb. 1995 at Kitt Peak. The McMath-Pierce telescope was used with the main spectrograph and the visible grating in combination with ZIMPOL I (Povel, 1995; Povel et al., 1990, 1994; Keller et al., 1992), a vector polarimeter consisting of three separate CCD cameras that simultaneously records $I \pm V$, $I \pm Q$, and $I \pm U$.

The observations covered the wavelength range from 10823 Å to 10836 Å. The pixel size was $3.23 \cdot 10^{-2}$ Å in the spectral direction and $0.36''$ in the spatial direction. The resolving power was 85 000.

The slit was positioned along the solar N-S direction. Stokes Q and U each made an angle of $22.5°$ to the slit. The observed feature was a sunspot (NOAA 7838) located close to the eastern solar limb at $\mu = \cos\theta = 0.3$ with the coordinates N 09 E 71. The slit was first placed through the umbra. Then it was moved to the outer edge of the eastern (limb-side) penumbra and stepped through the penumbra in the direction of the umbra in steps of roughly $3''$. Spectra were recorded at a total of 6 slit positions. The integration time was 10 minutes at each position. The long integration time was needed because of the high S/N ratio required to measure linear polarization signals in the He I line and the low quantum efficiency of the CCD at this wavelength.

3. Data Reduction

The reduction procedure included the removal of the dark current, flat-fielding, correction for the difference of illumination of the CCD rows for each pair of opposite polarization and calibration of the relative positioning of the cameras. The modulation efficiency along the slit was determined and taken into account.

Every second row of the CCD is covered with an opaque mask (see Povel et al., 1994). While this mask is very effective in providing a storage area for charges in the visible, the scattering of photons in the near-infrared in the silicon substrate leads to an apparent leakage under the mask. This reduction in the efficiency of the polarization measurements was calibrated and removed during the data reduction.

Strong *modulated* interference fringes affect our data. They are produced by the piezo-elastic modulators. The appearance of modulated interference fringes is a problem intrinsic to all variable retarders (Oakberg, 1995). Multiple reflections between the two surfaces, between which the optical path length is varied, lead to spectral fringes. The optical path length changes in synchrony with the polarization modulation, and this change results in a change of the fringe pattern. Even when the incoming light is unpolarized there will be a fringe-like polarization pattern in the polarized spectrum due to the fringe pattern varying in synchrony with the polarization modulation. Anti-reflective coatings and tilting of the modulators strongly reduce their amplitude. Generally these fringes are of no importance, since their amplitude is below the noise level when the coating is appropriate for the observed wavelength range. ZIMPOL is optimized for the visible, so that the anti-reflective coatings are not very efficient at 1.1 μm. In addition, the instrumental set-up did not allow us to tilt the modulators significantly.

Due to their polarized character these fringes cannot be removed by simple flat fielding. We removed (or reduced) them in the following way. First of all, those fringes that were neither parallel nor perpendicular to the slit direction were removed by 2D-Fourier filtering. If the spectrum is not known in advance, fringes parallel to spectral lines are hard to remove in this manner. We then employed the following approach: Since the polarization signal in the continuum is only due to the modulated fringes, we could fit an autoregressive model to the fringe pattern in the continuum and use this model to predict the fringe pattern at the location of spectral lines. This method is similar to linear predictive coding and represents a maximum entropy interpolation. We were not able to fit simple periodic functions because of optical distortions in the spectra. Using this procedure, the amplitude of the fringes could be reduced by at least a factor of 10.

4. Results

Sample CCD frames corresponding, from top to bottom, to the four Stokes parameters I, V/I_c, Q/I_c, and U/I_c are plotted in Fig. 1. They were recorded in the limbside penumbra of the sunspot ($\mu = 0.3$). The strong line in the left half of the plots is the Si I 10827.14 Å line, a photospheric line with a Landé factor of 1.5. The upper chromospheric He I line, which actually consists of three components, is located at 10830 Å. Two of these components are blended together at 10830.3 Å, the third is the very weak line that appears at 10829.0 Å. Some fringes are still present in Stokes Q at the position of these

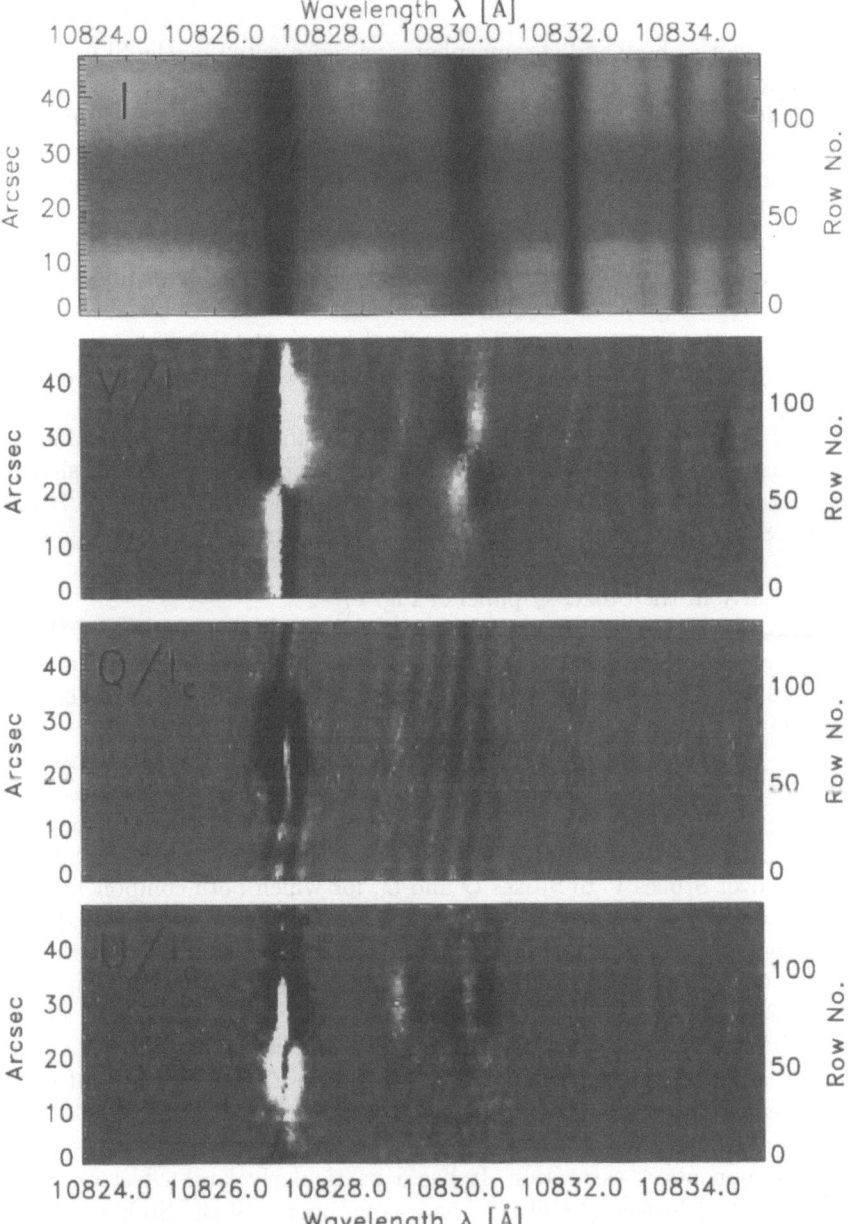

Fig. 1.   CCD frames of Stokes I, V/I_c, Q/I_c, and U/I_c recorded simultaneously in a limbside sunspot penumbra at $\mu = \cos\theta = 0.3$. The vertical scale on the left of the plots gives the position along the slit in arcsec. The scale on the right indicates the row numbers of the CCD.

spectral lines, while they have been removed almost completely in Stokes V and U.

The photosphere and chromosphere (as represented by the Si I and He I lines, respectively) possess on the whole similar polarization signals. For example both lines show a sign reversal in Stokes V. In Stokes Q only little signal can be seen above the noise (or rather fringe) level in the helium line, while in Stokes U a distinct signal is seen in the upper part of the panel. In the lower part, a very weak signal can be distinguished in the helium line at a few places. Its weakness may be due to the reduced field strength expected in the chromosphere and the small Landé factor of this line. The He I Stokes U signal has the same sign as that of the silicon line, i.e. reversed relative to the upper part of the panel, although the S/N ratio is very low in the lower part (however, averaged profiles distinctly show the reversal of the sign of Stokes U).

In Fig. 2 we show a Stokes vector spectrum. The plotted profiles are averages over rows number 81–90 of Fig. 1. A distinct Stokes U signal is present in both apparent line components of the helium line (at 10829.0 and 10830.3 Å). The signal seen in Stokes Q is a mixture of true signal and fringes that could not be removed completely (the remaining fringes can be seen clearly in the Stokes Q panel of Fig. 1).

The plotted He I Stokes profiles represent the first detections of the full Stokes vector coming from the upper chromosphere and demonstrate that He I 10830 Å may be used to measure the full magnetic vector in the chromosphere.

One striking feature of Fig. 2 is the difference in relative strength between the weak and strong He I spectral components in the various Stokes parameters. The ratio of the 10829 Å component's amplitude (respectively line depth) to that of the 10830 Å component increases from the Stokes I profile, over Stokes V to Stokes Q and U, for which both components show almost equally strong signals. The main cause for this difference is the fact that the strength of the Stokes V profile is proportional to g_{eff}, while for Stokes Q and U it is proportional to g_{eff}^2. Since the 10829 Å component has $g_{\text{eff}} = 2$, while the 10830 Å component has $\langle g_{\text{eff}} \rangle = 1.22$ (weighted average over the two sub-components) it is clear that the increasingly strong weighting by g_{eff} when going from Stokes V to Stokes Q and U will lead to the observed effect.

Another factor which influences the relative strength of the two separate line components is saturation. It has a particularly large effect on the π-component of Stokes Q and U. The σ-components of the Stokes U profile at 10830 Å are considerably stronger than those of the 10829 Å line. The π-components are, however, almost equal in strength. This and the spectral shape of the Stokes U profiles suggest that the He I line in this sunspot is

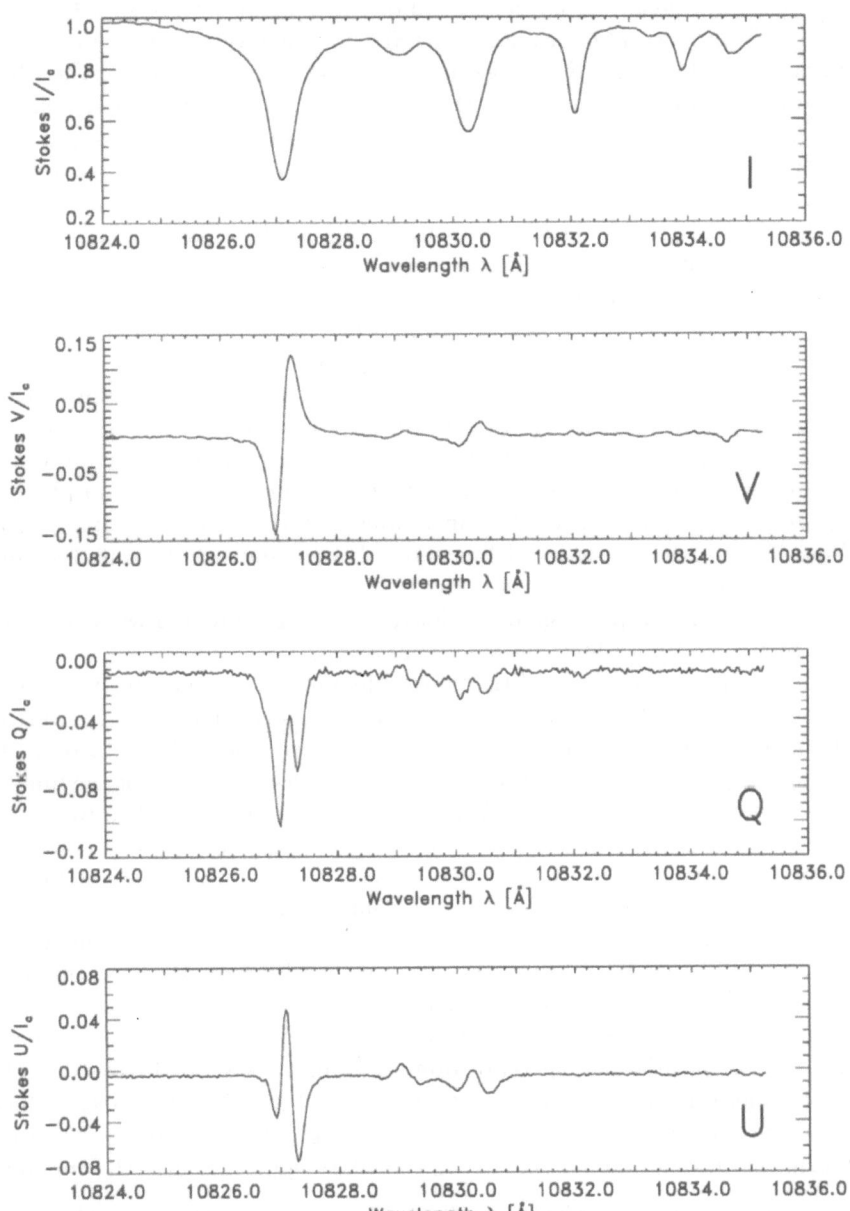

Fig. 2.   From top to bottom: examples of Stokes I, V, Q, and U profiles extracted from Fig 1. They correspond to an average of spectra 81–90 of Fig. 1.

significantly saturated, confirming the suggestion by Rüedi *et al.* (1995b) that this may generally be the case above sunspots near the solar limb.

The Stokes Q and U profiles of the Si I line show very large asymmetries. In Stokes U, for which the fringes could be properly removed over the whole spectrum, both components of the He I line show an asymmetry similar to that of the Si I line. These asymmetries are produced by cross-talk from Stokes V into Stokes Q and U and have not been removed in the preliminary data reduction. *

In spite of the similarities the photospheric and chromospheric Stokes profiles exhibit significant differences. In Stokes V, for example, the polarity reverses at row number 57 in the photosphere but at row 75 in the chromosphere. Fig. 3 shows Stokes V profiles corresponding to different spatial positions in Fig. 1. The top panel corresponds to row numbers 84–88. At that position both lines show the same magnetic polarity. In the second panel from the top, corresponding to row Nos. 73–77, the Si I line still shows a strong signal of the same polarity as above, while the He I line Stokes V profile has disappeared. The apparent neutral line in the chromosphere is located at this position. In the third panel of Fig. 3 (corresponding to row Nos. 55–59 in Fig. 1) the Si I line shows hardly any signal, while the He I Stokes V profile exhibits the opposite polarity to that in the first panel. The neutral line in the photosphere is located here. Finally at row Nos. 46–50 (bottom panel) both lines show a signal again.

Is the different position of the neutral line in the chromosphere and photosphere produced by a strong height evolution of the magnetic structure of the sunspot penumbra or is it just due to a projection effect? In order to settle this point we consider the whole Stokes V scan through the penumbra. Four adjacent Stokes V CCD frames are plotted in Fig. 4. The top panel was recorded closest to the solar limb at the outer edge of the penumbra, while the lowest panel was recorded closest to the sunspot umbra. Firstly, closer to the umbra the position of the neutral line in the Si I line is shifted upwards suggesting that the neutral line runs at an angle to the slit, which cuts it at different latitudes as the slit is moved in the E-W direction. This phenomenon can also be seen in the helium line in the two uppermost frames where the neutral line position is clearly defined. In effect, the chromospheric neutral line positions in the two topmost frames correspond closely to the photospheric neutral line positions in the third and fourth frames (which have been recorded further from the solar limb). In summary, the neutral line in the chromosphere lies closer to the white-light limb than the apparent

---

* In principle this cross-talk can be calibrated by observing the nearby phosphorus line at 10597 Å. This line has a Landé factor $g_{eff} = 1.333$. Due to the configurations of its atomic levels, this transition produces no linear polarization (Vela Villahoz *et al.*, 1994). Consequently, any signal observed in its Stokes Q or U spectrum has to be produced by cross-talk from Stokes I or V and can thus be used to directly determine the amount of cross-talk also affecting the Si I and He I lines.

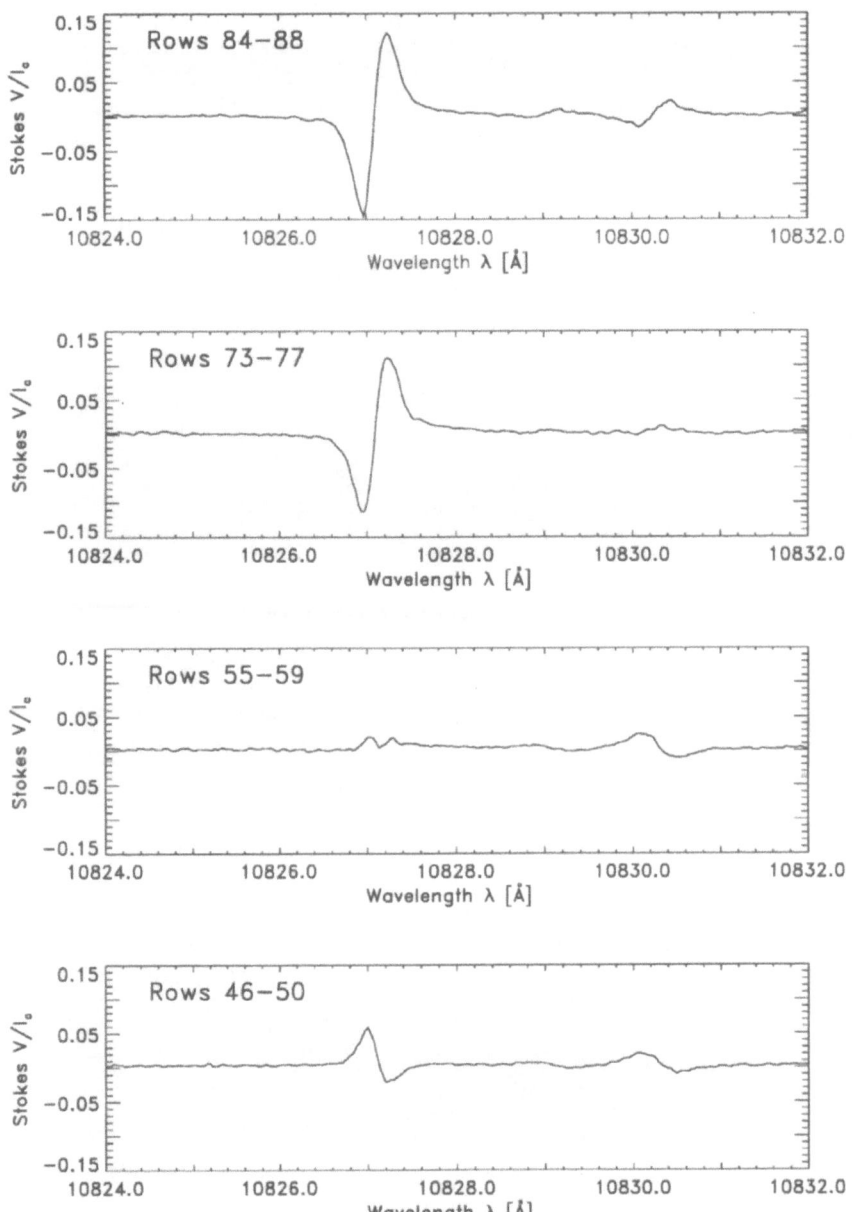

Fig. 3. Sample of Stokes V profiles extracted from Fig. 1. From top to bottom the plotted spectra correspond to averages over rows 84–88, 73–77, 55–59, and 46–50, respectively.

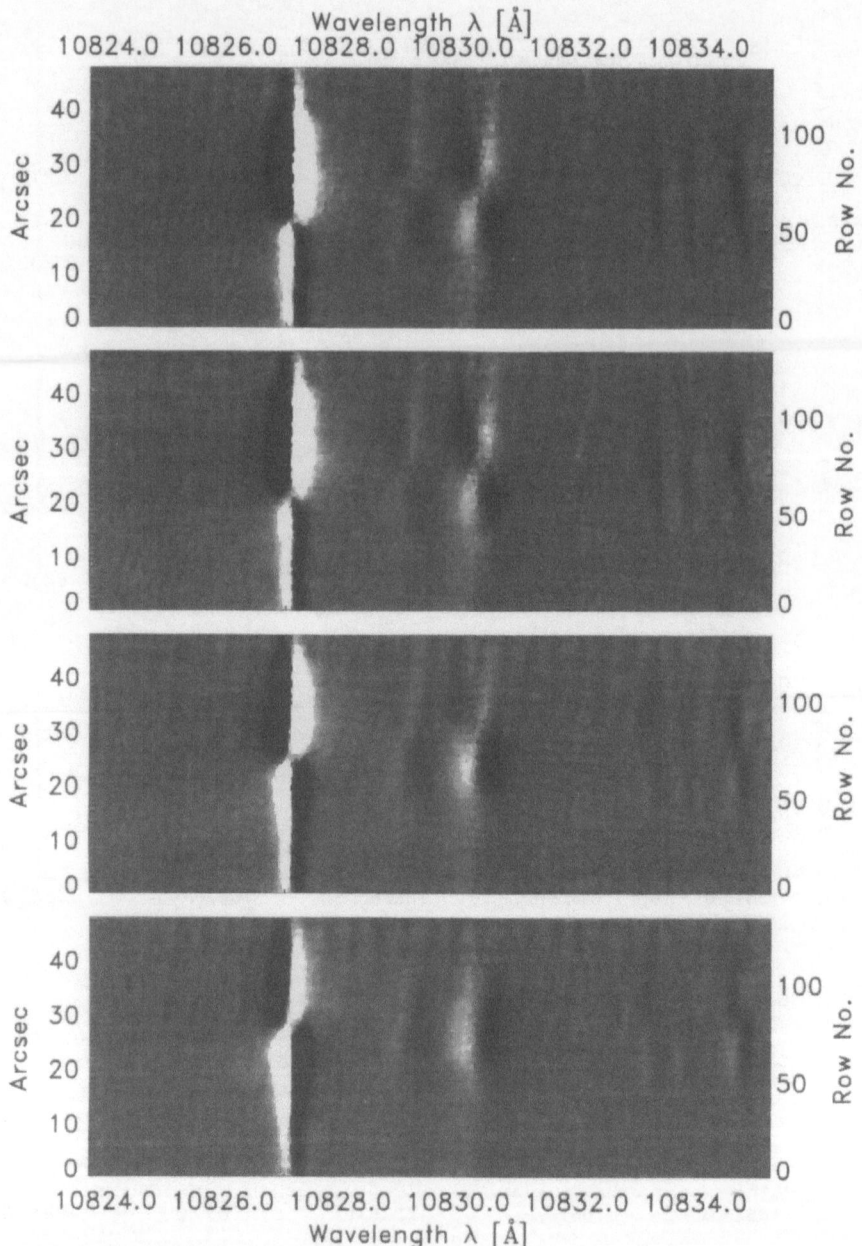

Fig. 4. Stokes V CCD frames recorded at different positions in the limb-side penumbra of the sunspot. The top frame was recorded closest to the limb. Equidistant steps of approximately $3''$ separate the frames.

neutral line in the photosphere. The shift between the neutral line position in the chromosphere and the photosphere can be explained by a projection effect. Consider the simple case of a neutral line lying at the same horizontal position at all heights in the atmosphere. Further, consider it to be located near the solar limb, so that the line of sight makes only a small angle to the solar surface. If we now observe this neutral line in two spectral lines formed at different heights, then it will be seen along two different rays, with the one passing through the neutral line in the upper atmosphere intersecting the solar surface closer to the solar limb. Thus the neutral line *apparently* lies closer to the limb in the upper chromosphere than in the photosphere.

Assuming that the difference in location of the neutral line in the two spectral lines is due only to projection, we can estimate the difference in height of formation of the two spectral lines. This difference, determined in the described manner from the current data set is 4500 km, which is significantly larger than theoretical predictions (Avrett *et al.*, 1994; Fontenla *et al.*, 1993) and off-limb observations (Schmidt *et al.*, 1994). This suggests that, although the magnetic structure in the upper chromosphere is similar to that in the photosphere, the two are not identical. For example, previous investigations have suggested that the chromospheric structures appear diffuser than the photospheric structures (Harvey & Hall, 1971; Giovanelli, 1980; Giovanelli & Jones, 1982; Solanki& Steiner, 1990; Rüedi *et al.*, 1995a). In the case of the observed sunspot the chromospheric neutral line appears to be located closer to the limb (even after removal of projection effects) by approximately 2000 km if we employ the He I 10830 Å height of formation quoted by Schmidt *et al.* (1994). This implies that the magnetic field is somewhat more vertical in the chromosphere than in the underlying photosphere, in agreement with the general picture of a sunspot as an expanding flux tube, e.g. with a potential field.

5. Conclusions

The Helium 10830 Å line can be used for complete magnetic field vector measurements in the upper chromosphere.

ZIMPOL I has been successfully used to measure the complete Stokes vector in the He I 10830 Å line in a sunspot region. Observations in a filament, not discussed here, also show linear polarization signatures. The modulated fringes will be strongly reduced in future observations by tilting the piezo-elastic modulators by a considerable amount. The low quantum efficiency of the CCD at these wavelengths remains a problem, however. Future efforts will address the accurate determination of chromospheric vector magnetic fields from observed Stokes spectra.

Acknowledgements

We thank Pietro Bernasconi, whose help during the data reduction was invaluable.

References

Avrett E.H., Fontenla J.M., Loeser R.: 1994 in D. Rabin, J. Jefferies, C.A. Lindsey, (Eds.), 'Infrared Solar Physics', *IAU Symp.* **154**, 35

Fontenla J.M., Avrett E.H., Loeser R.: 1993, *Astrophys. J.* **406**, 319

Giovanelli R.G.: 1980, *Solar Phys.* **68**, 49

Giovanelli R.G., Jones H.P.: 1982, *Solar Phys.* **79**, 267

Harvey J.W., Hall D.N.B.: 1971, in R.F. Howard, (Ed.), 'Solar Magnetic Fields', *IAU Symp.* **43**, 279

Keller C.U., Aebersold F., Egger U., Povel H.P., Steiner P., Stenflo J.O.: 1992, LEST Techn. Report No. 53

Oakberg T.C.: 1995, *Optical Engineering* **34**, 1545

Penn M.J., Kuhn J.R.: 1995, *Astrophys. J.* **441**, L51

Povel H.P.: 1995, *Optical Engineering* **34**, 1870

Povel H.P., Aebersold H., Stenflo J.O.: 1990, *Applied Optics* **29**, 1186

Povel H.P., Keller C.U., Yadigaroglu I.-A.: 1994, *Applied Optics* **33**, 4254

Rüedi I., Solanki S.K., Livingston W.: 1995a, *Astron. Astrophys.* **293**, 252

Rüedi I., Solanki S.K., Balthasar H., Livingston W., Schmidt W.: 1995b in J. Kuhn, M. Penn, (Eds.), 'Infrared Tools for Solar Astrophysics', *Proceedings of the 20th NSO Workshop*, in press

Schmidt W., Knölker M., Westendorp Plaza C.: 1994, *Astron. Astrophys.* **287**, 229

Solanki S.K., Steiner O.: 1990, *Astron. Astrophys.* **234**, 519

Vela Villahoz E., Sánchez Almeida J., Wittmann A.D.: 1994, *Astron. Astrophys. Suppl.* **103**, 293

INVERSION OF STOKES VECTOR PROFILES IN TERMS
OF A 3-COMPONENT MODEL

P.N. BERNASCONI and S.K. SOLANKI

Institute of Astronomy, ETH Zentrum, CH-8092 Zurich, Switzerland

Abstract. Various spectropolarimetric observations show peculiar Stokes profiles that reveal the coexistence of at least two magnetic components in the same resolution element. An example is given by observations of the full Stokes vector in a complex active region performed with the ZIMPOL I Stokes polarimeter. In order to deduce the physical parameters of the observed regions from such measured profiles, we have extended an existing inversion code, so that it can now fit the data with models composed of up to three different atmospheric components. Two of these components are magnetic and may possess different field strengths, field geometries, temperature stratifications, and velocity fields. The third component describes the field free atmosphere surrounding the magnetic features.

The so extended inversion code has then been applied to the ZIMPOL I data. In this paper we present and discuss sample fits. The code is able to reproduce the observed complex Stokes profiles with good accuracy and provides physical parameters for both of the coexisting magnetic atmospheres. Inversion tests with a 2-component model (with one magnetic and one non-magnetic component) applied to the same profiles do not reproduce the measurements sufficiently well.

Key words: Sun – Polarization – Magnetic fields – Diagnostic techniques

1. Introduction

With the instrumentation currently available it is generally impossible to fully resolve small solar magnetic features. Consequently both magnetic and non-magnetic atmospheric components are often present in the same resolution element. Using polarimetric techniques it is possible to overcome this problem, since only the magnetic component produces polarized line profiles, Stokes Q, U, and V. The non-magnetic component contributes only to the intensity, Stokes I.

The interpretation of observed Stokes profiles is much more complex if more than one magnetic component is present in the same resolution element. The polarized profiles, in particular Stokes V, do not look "simple" any more and it is not so straightforward to determine the atmospheric parameters, such as magnetic field strength B, field geometry (angles γ and χ), and temperature stratification $T(\tau)$, for each component. Although complex Stokes profiles have been successfully interpreted in the past, they always had to be fit manually (i.e., without an inversion approach). Stenflo (1968) first introduced the concept of multicomponent models. Skumanich and Lites (1991) observed complex Stokes V profiles in the line Fe I 6302.5 Å

Solar Physics **164**: 277–290, 1996.
© 1996 *Kluwer Academic Publishers.*

recorded at the neutral line between two sunspots with different polarities. They interpreted the measured profiles by manually fitting them with a model possessing two magnetic components. Martínez Pillet *et al.* (1994) found strongly distorted Stokes V profiles that they interpreted as the superposition of Stokes V spectra belonging to two different magnetic components with opposite polarities. One of the two components was highly red-shifted with respect to the other. Rüedi *et al.* (1992a,b) manually analyzed numerous Stokes V profiles of 1.56 μm lines exhibiting clear signatures of two magnetic components. In this case the analysis is simplified by the larger Zeeman splitting of infrared magnetic lines.

Inversion techniques have so far not been applied to the analysis of complex Stokes profiles and Stokes inversion codes (e.g., Skumanich and Lites, 1987; Ruiz Cobo and Del Toro Iniesta, 1992; Solanki *et al.*, 1992a,b) have previously been incapable of inverting such profiles. Here we present an inversion code capable of reproducing complex Stokes profiles. As an illustration we apply it to anomalous Stokes profiles of spectral lines in the visible, which need a model with two magnetic and one non-magnetic component to be correctly reproduced. The main aim of this work is to demonstrate the feasibility of applying an automated fit procedure with a 3-component model to complex Stokes profiles in the visible.

2. Observational data

The present paper is based on observations performed at the Swedish Vacuum Solar Telescope on La Palma (Canary Islands) on May 13, 1993 with the Zürich Imaging Stokes Polarimeter (ZIMPOL I) together with U. Egger, C.U. Keller, H.P. Povel and P. Steiner (see Stenflo, 1991; Povel *et al.*, 1990, 1994, and Povel, 1995 for an overview of the polarimeter). We recorded the complete Stokes vector of the three well known lines Fe I 5247.1 Å ($g_{\text{eff}} = 2.0$), Fe I 5250.2 Å ($g = 3.0$) and Fe I 5250.7 Å ($g_{\text{eff}} = 1.5$) in an active region belonging to NOAA group No. 7500 located at 18°N, 33°W. The angle between the normal to the solar surface and the line of sight was $\theta = 40°$, which corresponds to $\mu = \cos\theta = 0.76$.

At the time of the observations the instrument was working with a single CCD camera placed in the focus of the spectrograph, so that it was only possible to record Stokes I plus one other Stokes parameter (Q, U or V) in a single exposure. We obtained the full Stokes vector sequentially by measuring each polarized Stokes parameter in turn, with an integration time of 1 s per measurement and an interval of $\sim$ 150 ms between two consecutive measurements. The sequence of three exposures was repeated seven times, after which the seven images for each Stokes parameter were co-added, so that the net integration time for each parameter was 7 s. This procedure ensures quasi-simultaneity, even if strictly simultaneous images of all four Stokes parameters presently cannot be obtained with a single CCD (but should become possible with the next generation of polarimeters, e.g., ZIM-POL II, see Keller *et al.*, 1995). For one slit position the total integration time per Stokes image was 21 s, and the noise level was approximately 0.25% rms in units of the continuum intensity. Each set of four reduced images contains 120 spectra each in Stokes I, Q, U and V, ranging from 5246.3 Å to 5251.5 Å, with a spectral resolving power of 175 000 and a pixel size in the spatial direction of 0.4″. More information on the observations and analysis is given by Bernasconi *et al.* (in preparation).

The presence of three oblique reflections in the optical train of the Swedish Vacuum Solar Telescope introduces a large amount of instrumental polarization and cross-talk into the Stokes measurements. To free the data from this unwanted effect, we developed a theoretical model for the instrumental polarization of the telescope. The complex refractive indices of the three mirrors were considered as free parameters. They were determined by fitting the theoretical curves describing the daily behavior of the instrumentally produced continuum polarization to the observed values recorded at different times of the day. With this procedure we were able to directly fit only the first column of the telescope Mueller matrix, i.e. we could correctly determine only three elements of the matrix. The remaining elements were, however, consistently calculated by means of the theoretical Mueller matrix for reflections by mirrors given by Stenflo (1994) using the refractive indices derived from the observations. The so deduced telescope Mueller matrix, transformed to the time at which the solar Stokes spectrum we analyze was recorded, was applied to the Stokes measurements. The resulting Stokes images are free from the effects of instrumental polarization with an accuracy better than 0.5%, which corresponds to twice the rms noise level.

The remainder of the data reduction follows procedures standard for ZIM-POL. The noise level was additionally reduced by a factor of 1.5 using a Fourier low-pass filter in the spectral direction.

Although most of the Stokes profiles along the slit are "normal", in shape a number of them are "complex", i.e. they show signs of two magnetic components. The solid lines in Figs. 1, 2 and 3 show three examples of reduced Stokes profiles requiring two magnetic components to be fit. In Figs. 1 and 2 the presence of a second component is clearly visible in the profile shape of Stokes V (it is distorted relative to the usual almost antisymmetric profile shape well beyond the possible influence of remaining uncompensated cross-talk). The influence of a possible second magnetic component is more subtle on the profiles plotted in Fig. 3: it mainly exhibits itself through the wavelength shift of V relative to Q and U.

3. Inversions

To derive the magnetic field strength, field orientation, and temperature stratification we used an inversion code that fits observed Stokes profiles using a non-linear least-squares technique (Levenberg-Marquardt, see Press *et al.*, 1990) with synthetic profiles that are solutions of the radiative transfer equations for polarized light. The inversion code is basically the same as that used by Bernasconi *et al.* (1995) and described and applied by Solanki *et al.* (1992a,b) and Emonet (1992). It has, however, been modified to allow complex Stokes profiles such as those plotted in Figs. 1–3 to be fit. The main features of the modifications are discussed below.

3.1. 3-COMPONENT ATMOSPHERIC MODEL

To allow the inversion code to fit complex Stokes profiles of the type illustrated in Figs. 1, 2, and 3, we modified the code by introducing the possibility of inverting Stokes profiles based on a 3-component model. The first two components are magnetic. They may possess different field strength, magnetic field geometries, temperature stratifications and velocity fields. The third component is field free and usually describes the quiet sun.

Two filling factors α_1 and α_2 describe the fraction of the spatial resolution element covered by the first and the second magnetic components, respectively. The remaining portion of the resolution element, i.e. $1 - \alpha_1 - \alpha_2$, is covered by the third, field-free component. α_1 and α_2 are free parameters that have to be determined during the fit procedure. First the Stokes vector resulting from each component is calculated, then a linear combination of the three Stokes vectors weighted by their respective filling factors is determined:

$$
\begin{aligned}
I_{res} &= I_1\,\alpha_1 \;+\; I_2\,\alpha_2 \;+\; I_3\,(1 - \alpha_1 - \alpha_2)\,, \\
Q_{res} &= Q_1\,\alpha_1 \;+\; Q_2\,\alpha_2\,, \\
U_{res} &= U_1\,\alpha_1 \;+\; U_2\,\alpha_2\,, \\
V_{res} &= V_1\,\alpha_1 \;+\; V_2\,\alpha_2\,,
\end{aligned}
\tag{1}
$$

where the indices refer to the contributions of the components 1, 2, and 3, respectively, and $(I_{res}, Q_{res}, U_{res}, V_{res})^{T}$ is the transpose of the resulting Stokes vector which is compared to the observations.

3.2. FREE PARAMETERS AND MODEL ATMOSPHERES

Each component is described by a model atmosphere. Key parameters describing each atmosphere are allowed to vary and be determined by the fitting procedure. The remainder of the parameters are calculated self-consistently therefrom. We distinguish between two types of free parameters, those which depend on the spectral lines and those which do not. We first discuss the latter.

The spectral line-independent free parameters in the magnetic model atmospheres are the temperature T, the field strength B, the filling factor α, the inclination γ of the field lines with respect to the line of sight, and the field azimuth χ counted counterclockwise starting from the geocentric W direction. With the exception of the temperature, all these parameters are depth independent. In some cases the field strength is also allowed to vary with depth, i.e. a field gradient $\nabla B = \partial B / \partial \log \tau$ is introduced as a further free parameter when it reproduces the real magnetic field structure in the solar atmosphere better. A magnetic gradient was used only when the measured Stokes profiles appeared "simple", i.e. a single magnetic component was sufficient to reproduce the observed spectra. In the cases where two magnetic atmospheres were needed we preferred to have a simple model for the magnetic structure of each component, in order to keep the number of free parameters as small as possible. Note that the inversion code allows $B(\tau)$ to be determined consistently from horizontal pressure balance if B is specified at a single height, but we have chosen not to use this option (which corresponds to the thin-tube approximation), since many of the inverted spectra lie in penumbra-like structures or are affected by pore magnetic fields, to which the thin-tube approximation does not apply.

Although the temperature T is a function of the continuum optical depth τ (or height z) we prefer to determine it only at a single τ value in the case of magnetic atmospheres. $T(\log \tau_{5000} = -1.5)$ is allowed to vary, while T at other τ values is determined by linearly interpolating between the two nearest temperature stratifications of the empirical plage flux tube model of Solanki and Brigljevíc (1992), the quiet-sun atmosphere of Maltby *et al.* (1986), and the umbral model M of Maltby *et al.* (1986).

For the quiet-sun component the entire temperature stratification of the Maltby *et al.* (1986) quiet-sun model atmosphere is simply translated until it passes through the prescribed temperature at $\log \tau_{5000} = -1.5$.

The parameters calculated consistently from $T(\tau)$ are the geometrical depth $z(\tau)$, the electron pressure $P_{el}(\tau)$, the gas pressure $P_{gas}(\tau)$, and the gas density $\rho_{gas}(\tau)$.

For the Doppler width of the microturbulent velocity distribution ξ_{mic} we chose a constant value of 1.0 km/s for all the atmospheric components.

In addition, for each magnetic component we assigned a different Doppler shift $\Delta \lambda_l$ and macroturbulence $\xi_{mac,l}$ to each spectral line (the index l labels the three observed spectral lines: $l = 5247.1$ Å, 5250.2 Å, 5250.7 Å). These parameters were kept free. In this way we could take into account also the effect of different velocities in the various atmospheric components. To reduce the number of free parameters in our model, we decided to force ξ_{mac}

for the two lines Fe I 5247.1 Åand 5250.2 Å to be the same, since they have almost the same atomic parameters (Solanki, 1987).

3.3. Tests

In order to test the extension of the inversion code to three components we fit a number of infrared Stokes I and V spectra of λ 15648 Å and λ 15652 Å that obviously required two magnetic components. These data had earlier been analyzed by Rüedi *et al.* (1992). Our results, based on the automated inversion procedure, were in good agreement with their findings which are based on manual fits. The large Zeeman splitting of the infrared lines made the fitting procedure easier and faster.

Once the inversion code had passed these tests, we applied it to the observed Stokes V, Q, and U spectra of the visible lines that are discussed in Section 2.

4. Examples of inversions with a 3-component model

We now present and discuss three examples of fits to complex Stokes spectra. We aim to show that for these profiles a model with two magnetic components in addition to the quiet-sun component reproduces the data much better than a model with a single magnetic component, at least as long as extremely strong velocity-magnetic gradient combinations along the line-of-sight (LOS) are not allowed.

We decided not to fit the Stokes I profiles because they are heavily influenced by the non-magnetic atmosphere surrounding the magnetic features and are not required to determine the strength and orientation of the magnetic field on which our major interest was focused. Consequently, in the following figures we do not plot the synthetic Stokes I profiles.

Table I lists the most important atmospheric parameters referring to the best fit curves shown in the following Figures. γ is the field inclination with respect to the LOS (angles larger than 90° signify field lines oriented away from the observer), χ the field azimuth counted counterclockwise from the heliocentric W direction, and T the temperature at continuum optical depth $\log \tau_{5000} = -1.5$. In the last column the relative flow velocities $\Delta\lambda$ between the two magnetic components along the LOS are listed. They have been derived by computing the difference of the Doppler shifts for the two components (since the spectrograph was not absolutely calibrated, only relative Doppler shifts are meaningful). $\Delta\lambda = 1$ km/s means that the velocity flow in the first magnetic component is 1 km/s faster towards the observer than the flow in the second component. To determine the errors in the best-fit parameters we repeated the fits several times with different initial parameters and compared the various resulting fits with each other. This procedure

TABLE I

Parameters derived from the inversion.

Fig. No.	Comp. No.	B [G]	α	γ	χ	T [K]	Δλ [km/s]
1	1	1450	0.30	133	33	4680	1.0
	2	1830	0.24	31	70	4430	
2	1	781	0.64	118	24	5340	3.3
	2	2187	0.23	54	42	4640	
3	1	1300	0.38	117	44	4460	1.1
	2	1580	0.41	43	87	4370	
4		1290	0.36	60	49	4380	—

allowed us to test the uniqueness of the results, which turned out to be quite good, especially for the simpler profiles. We estimated an accuracy of 100 G for the field strength, 0.05 for the filling factors, 5° for γ and χ, 100 K for the temperature, and 0.2 km/s for $\Delta\lambda$.

The first example we wish to discuss is shown in Figure 1. The observed profiles refer to a slit position between two regions with reversed magnetic polarity (relative to the LOS). The Stokes V profiles look unusual, having a profile shape more typical of Q or U. This is however common for V spectra along apparent neutral lines (see for example Skumanich and Lites, 1991; Sánchez Almeida and Lites, 1992; Rüedi et al., 1992b). Since the observed region lies outside sunspots (at the edge of a pore without a visible penumbra) it is likely that two magnetic components with similar field strengths but with opposite magnetic polarities and different flow velocities coexist in the same resolution element (within penumbrae V profiles of the same general shape can be produced by a single magnetic component: Sánchez Almeida and Lites, 1992; Solanki et al., 1994). The discontinuity in the synthetic profiles around $\lambda = 5250.5$ Å is due to an overlap of the selected spectral wavelength ranges over which these two lines are synthesized. Note how well the V and Q profiles are reproduced. The fit of Stokes U is of lower quality, possibly due to the presence of a third, weak magnetic component almost perpendicular to the LOS, which was not taken into account in our model (the wavelength shift between U and V supports this interpretation, see discussion of Fig. 3 below). Another possibility is the presence of significant gradients of magnetic and velocity fields in one or both of the components (this is supported by the asymmetry of the Stokes U profiles).

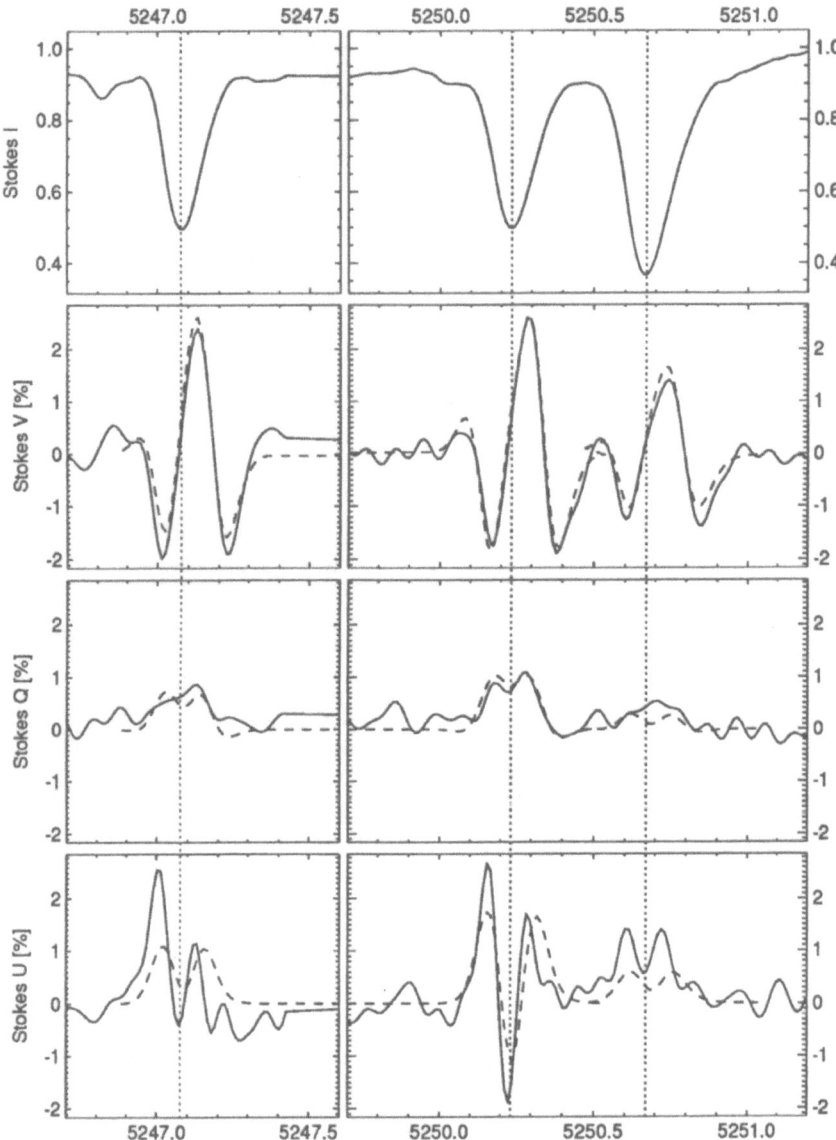

Fig. 1. Observed (solid) and synthetic (dashed) Stokes profiles. A model with two magnetic components of opposite polarity, $B_1 = 1450G$, $\gamma_1 = 133°$, and $B_2 = 1830G$, $\gamma_2 = 31°$,
and a Doppler shift between the two components of 1.0 km/s was used to fit the data.
The vertical dotted lines mark the wavelengths of the Stokes I minima.

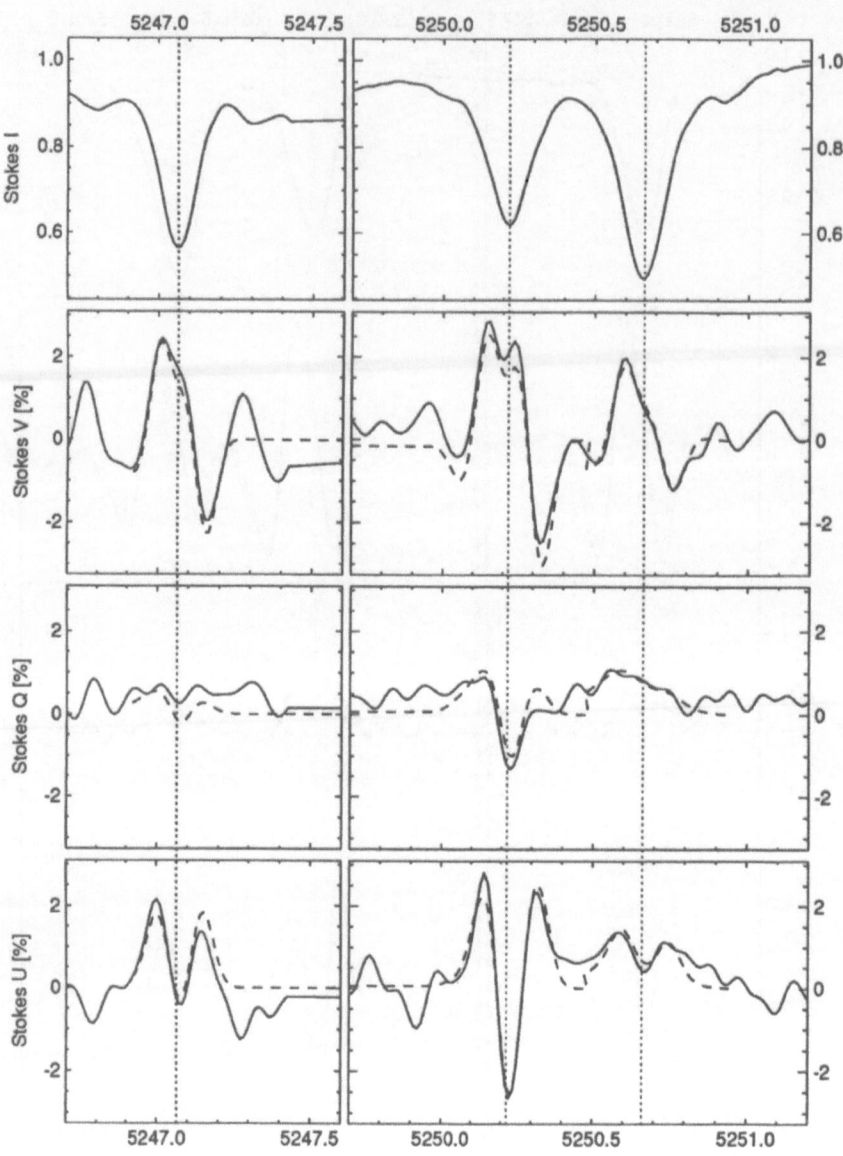

Fig. 2. Observed (solid) and synthetic (dashed) Stokes profiles. A model with two magnetic components of opposite polarity, $B_1 = 780\,\mathrm{G}$, $\gamma_1 = 118°$, and $B_2 = 2190\mathrm{G}$, $\gamma_2 = 54°$, and a Doppler shift between the two components of 3.3 km/s was used to fit the data. The vertical dotted lines mark the wavelengths of the Stokes I minima.

As a second example we plot in Fig. 2 another set of complex Stokes profiles. The blue wing of Stokes V is much broadener than the red wing, and the Fe I 5250.2 Å Stokes V profiles exhibit a double peak that is not due to noise. It is suggestive of the presence of (at least) two magnetic components. Once more the 3-component model inversion finds synthetic profiles that are in good agreement with the observations. These data were recorded very close to a pore, which contributes to the strong component deduced from the fit (Component No. 2 in Tab. I), while the weaker component (Component No. 1) probably corresponds to the canopy of another pore located in the vicinity. This interpretation also explains the large temperature difference between the two components.

The evidence for two magnetic components is less obvious in Fig. 3. In this case the Stokes V profiles look almost "normal", but for all the lines we notice a strong wavelength shift between the zero crossing of V, the minimum of Stokes I, and the core of the π components for Stokes Q and U. We can explain these shifts by considering two magnetic components with slightly different velocity fields, having different inclinations relative to the LOS. The component more aligned with the LOS (Component No. 1) contributes to the Stokes V profiles much more than the more transverse one (Component No. 2). The latter, however, dominates the Stokes Q and U signals. Finally, the difference between the velocity fields in the two atmospheres gives rise to the observed shifts between the various Stokes profiles.

Our inversion code with the 3-component model easily separated the two magnetic components, and provided synthetic Stokes profiles that match the observations relatively satisfactorily. For comparison purposes we inverted the same set of profiles by applying a model with just a single magnetic component. The result is plotted in Fig. 4. The Stokes V profiles are reasonably well fit, but the synthetic Stokes U profiles are too strongly shifted towards the red. The atmospheric parameters obtained from the fit (last line in Tab. I) turn out to be a mixture of the parameters determined from the 3-component model. The field strength, filling factor and the field azimuth (χ) are similar to those of component number 1, while field inclination (γ) and temperature are more like those of component number 2. Note that the sum of the filling factors deduced with the 3-component model is approximately twice the filling factor obtained with the 2-component model (see Tab. I). This is probably due to the fact that since the two magnetic components have opposite polarities, their Stokes V contributions partially cancel out, reducing the effective observed Stokes V amplitude. Consequently, a large filling factor is required to reproduce the amplitudes of the observed profiles.

We were unable to obtain a reasonably accurate fit to these Stokes profiles within the confines of a 2-component model. Note, however, that since the Stokes profiles have not been observed *strictly* simultaneously we cannot

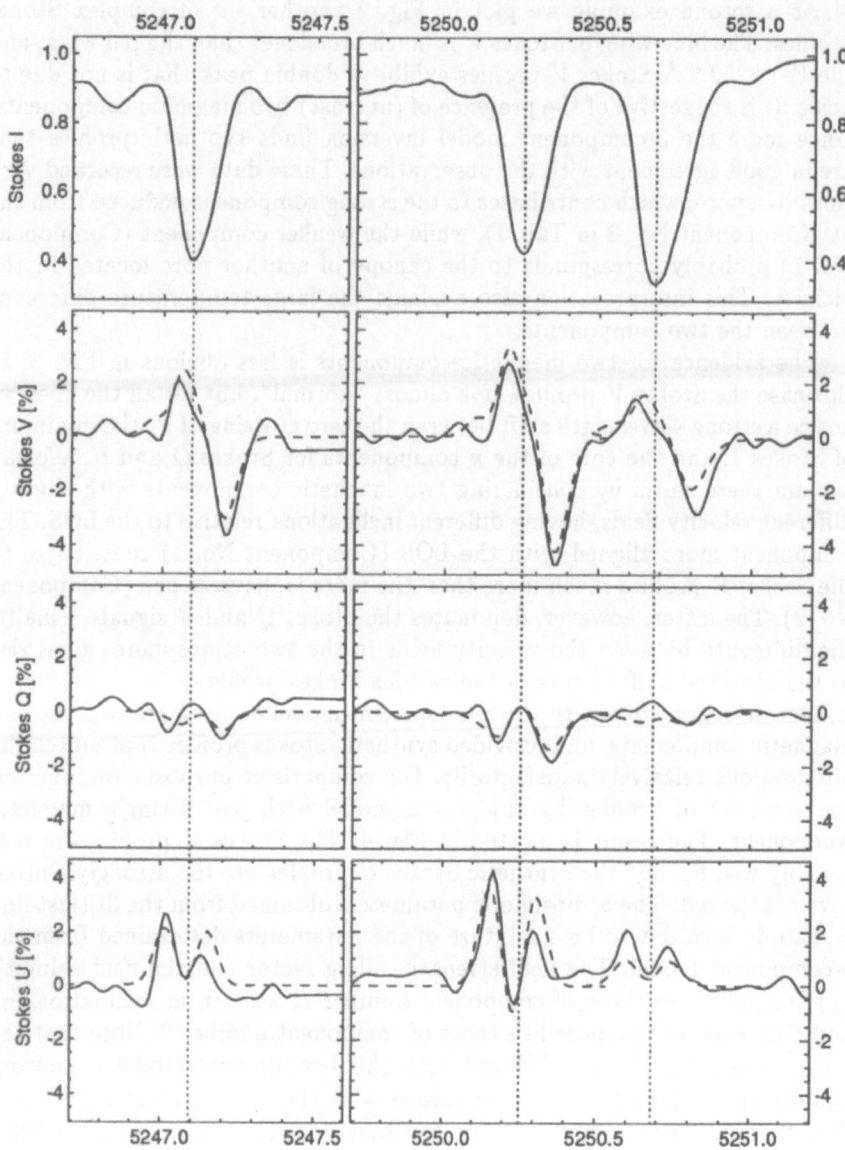

Fig. 3.   Same as Fig. 1, but for another slit position. The fit parameters are: $B_1 = 1300\,\text{G}$, $\gamma_1 = 117°$, $B_2 = 1580\,\text{G}$, $\gamma_2 = 43°$ and $\Delta\lambda_{5250.2} = 3.5\,\text{km/s}$. Note the difference between the zero crossing wavelength of the V profiles (of all the lines) and the wavelengths of the Stokes I minima. See the text for details.

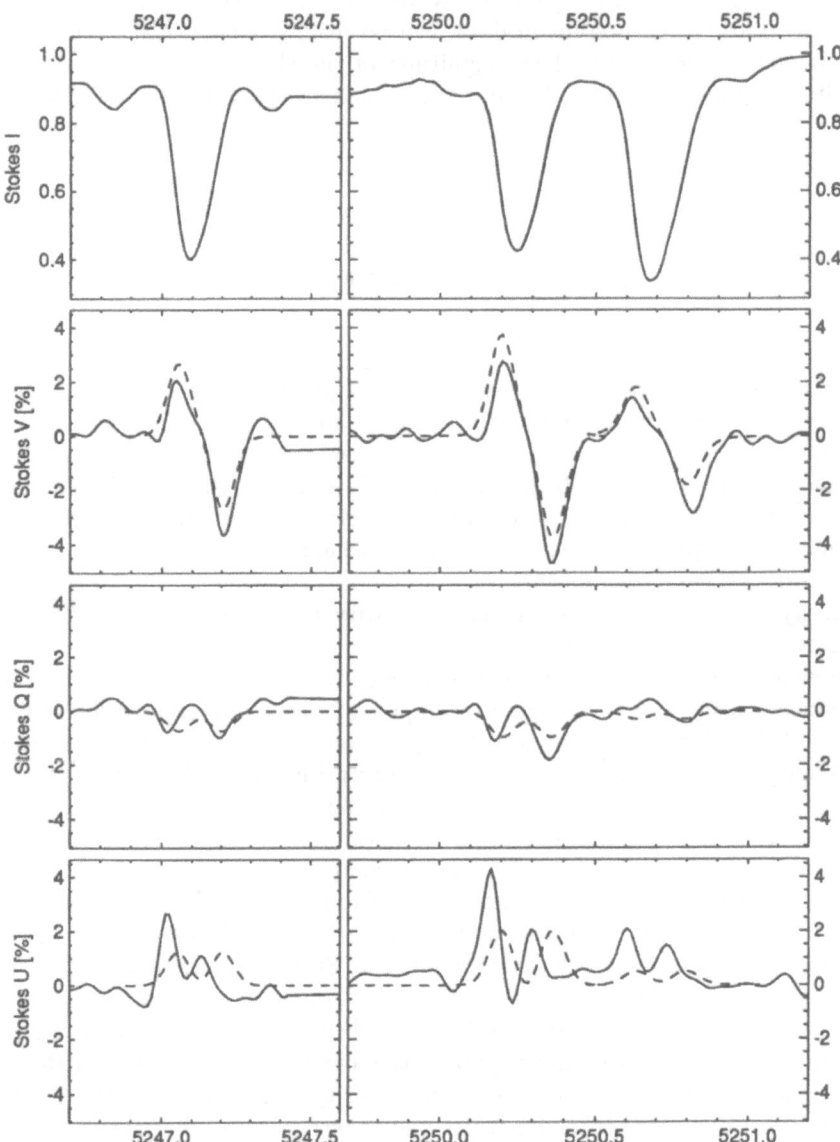

Fig. 4. Same as Fig. 3, but for a fit performed by adopting a model with just one magnetic component.

definitely rule out the possibility that the wavelength shift between V and U is produced by seeing effects. Nevertheless, we have demonstrated that such shifts, when present in data, can be readily explained in terms of two magnetic components. The magnitude of the shift between Stokes V and Stokes Q and U (2 km/s) also makes it unlikely that it is an artifact.

5. Conclusions and final remarks

In the present work we have presented some preliminary results on the inversion of complex Stokes profiles based on a 3-component model atmosphere. Our automated approach is new for Stokes profiles produced by multiple magnetic components. In general we find that the infrared is more suited to identify the presence of multiple magnetic components than the visible, mainly due to the comparatively large Zeeman splitting in the infrared. It allows magnetic components with different field strengths to be separated easily, even if they have the same polarity and there is no velocity shift between them. In our observations the evidence for multiple magnetic components within the same resolution element is in general restricted to those cases in which the two magnetic components posses different velocity fields coupled with opposite magnetic field polarities. An exception is when the different Stokes parameters of the same spectral line exhibit different Doppler shifts, which can be explained quite naturally by a second magnetic component. In such cases the measurement of the full Stokes vector is crucial for the identification of a second magnetic component.

In general our inversion code turns out to be rather efficient in separating the two coexisting magnetic components for even relatively complex Stokes profiles. Nevertheless, for a small number of Stokes profiles it was impossible to obtain an acceptable fit even with the 3-component model. In particular V profiles for which the total area under their positive part is very different from the total area of the negative part cannot be reproduced without a coupled variation of the magnetic and velocity field along the LOS. Even in the more numerous cases for which the Stokes V area asymmetry is not so large we cannot rule out that part of the effect that we successfully model using two discrete magnetic components is in fact due to coupled longitudinal gradients of the magnetic and velocity fields within a single magnetic component. This basic ambiguity can probably only be successfully resolved by simultaneously considering spectral lines with widely different line strengths. (i.e. different levels of saturation).

Acknowledgements

We are grateful to H.P. Povel and U. Egger for developing and constructing the electronics of the vector polarimeter, P. Steiner for developing the software to control it, and F. Aebersold for constructing the mechanical parts. We also thank C.U. Keller for his help during the observations and together with J.O. Stenflo for the interesting discussions on topics concerning this paper. This work was partly supported by the Swiss National Science Foundation, grant No. 20-37323.93, which is gratefully acknowledged.

References

Bernasconi, P.N., Keller, C.U., Povel, H.P., Stenflo, J.O.: 1995, *Astron. Astrophys.* , in press

Bernasconi, P.N., Aebersold, F., Egger, U., Keller, C.U., Povel, H.P., Solanki, S.K., Steiner, P., Stenflo, J.O.: *Astron. Astrophys.*, in preparation

Emonet, T.: 1992, *Diploma Thesis*, ETH, Zürich

Keller, C.U., Bernasconi, P.N., Egger, U., Povel, H.P., Steiner, P., Stenflo, J.O.: 1995, LEST Technical Report No. 59

Maltby, P., Avrett, E.H., Carlsson, M., Kjeldseth-Moe, O., Kurucz, R.L., Loeser, R.: 1986, *Astrophys. J.* **306**, 284

Martínez Pillet, V., Lites, B.W., Skumanich, A., Degenhardt, D.: 1994, *Astrophys. J.* **425**, L113

Povel, H.P.: 1995, *Optical Engineering* **34**(7), 1870

Povel, H.P., Aebersold, H., Stenflo J.O.: 1990, *Applied Optics* **29**, 1186

Povel, H.P., Keller, C.U., Yadigaroglu, I.-A.: 1994, *Applied Optics* **33**, 4254

Press, W.H., Flannery, B.P., Teukolsky, S.A., Vetterling, V.T.: 1990, *Numerical Recipes. The Art of Scientific Computing*, Cambridge University Press, Cambridge

Rüedi, I., Solanki, S.K., Rabin, D.: 1992a, *Astron. Astrophys.* **261**, L21

Rüedi, I., Solanki, S.K., Livingston, W., Stenflo, J.O.: 1992b, *Astron. Astrophys.* **263**, 323

Ruiz Cobo, B., Del Toro Iniesta, J.C.: 1992, *Astrophys. J.* **398**, 375

Sánchez Almeida, J., Lites, B.W.: 1992, *Astrophys. J.* **398**, 359

Skumanich, A., Lites, B.W.: 1991, in *Solar Polarimetry*, L. November (Ed.), National Solar Observatory, Sunspot, NM, p. 307

Skumanich, A., Lites, B.W.: 1987, *Astrophys. J.* **322**, 473

Solanki, S.K.: 1987, *Ph.D. Thesis*, ETH Zürich, p. 66

Solanki, S.K., Brigljevíc: 1992, *Astron. Astrophys.* **262**, L29

Solanki, S.K., Rüedi, I., Livingston, W.: 1992a, *Astron. Astrophys.* **263**, 312

Solanki, S.K., Rüedi, I., Livingston, W.: 1992b, *Astron. Astrophys.* **263**, 339

Solanki, S.K., Montavon, C.A.P.: 1994, *Astron. Astrophys.* **283**, 221

Stenflo, J.O.: 1968, *Acta Univ. Lund* II, No. 2 = *Medd. Lunds Astronom. Obs.* II, No. 153

Stenflo, J.O.: 1991, LEST Technical Report No. 44

Stenflo, J.O.: 1994, *Solar Magnetic Fields*, Kluwer Academic Publishers, Dordrecht, p. 320

FIELD AZIMUTH DISAMBIGUATION USING AMBIGUITY-FREE CURRENTS

A. SKUMANICH

High Altitude Observatory, NCAR, Boulder, CO 80302, U.S.A.*

and

M. SEMEL

Observatoire de Paris, URA 326 (CNRS), F 92195, Meudon, Cedex, France

Abstract. Using the ambiguity-free vertical current defined by Semel and Skumanich (1995) we derive a minimum-current azimuth disambiguation for the observed magnetic field in the active region NOAA 7201. A comparison of such a minimum-current azimuth resolution with those from other extant methods indicates that the resulting resolution, even though found to be limited by noise, is a useful first approximation. A comparison of our minimum current distribution with the currents we derive from an extant disambiguation (Lites *et al.*, 1995) indicates the presence of current discontinuities in the form of linear features near the magnetic neutral line of the associated δ-spot.

Key words: magnetic azimuth resolution, vertical currents, δ-spots

1. Introduction

Vertical photospheric current densities, i.e. currents along the solar radius, derived from Stokes polarization observations depend on the choice of the ambiguous azimuth of the plane-of-sky magnetic field component. However, a reference current density (hereafter called current) may be derived, as shown by Semel and Skumanich (1995), which is free of the ambiguity in the observed field. In this paper we use this reference current to derive a minimum-current azimuth resolution for the active region NOAA 7201. The Stokes polarimetric measurements for this region and their interpretation are to be found in Lites *et al.* (1995), hereafter called L95.

2. Construction of the Ambiguity-free Vertical Current

We consider an abbreviated description of the Semel-Skumanich invariant current. Let $\mathbf{B_p}$ be the "true" plane-of-sky field (POS) component, i.e. with the correct solar azimuth assignment, mapped into the *local* heliocentric coordinate frame (HCF). The total magnetic field may then be written as

* The National Center for Atmospheric Research is funded by the National Science Foundation

Solar Physics **164**: 291–302, 1996.

$$\mathbf{B}(\mathbf{x}) = \mathbf{B}_l(\mathbf{x}) + \mathbf{B}_p(\mathbf{x}) , \tag{1}$$

where $\mathbf{B}_l$ is the azimuth-unambiguous line-of-sight (LOS) field component mapped to the local HCF, and $\mathbf{x} = (x, y)$, the local Solar West and North directed coordinates in the HCF, i.e., in the local plane tangent to the solar surface. Any given selection of azimuth angles for the POS-field may be considered as a stage in the disambiguation process. Consider that one is at some ϵ-th stage, the resultant field resolution may be written

$$\mathbf{B}^{(\epsilon)}(\mathbf{x}) = \mathbf{B}_l(\mathbf{x}) + S^{(\epsilon)}(\mathbf{x})\mathbf{B}_p(\mathbf{x}) \equiv \mathbf{B}_l + \mathbf{B}_p^{(\epsilon)} , \tag{2}$$

where $S^{(\epsilon)}$, the (unknown) disambiguation factor, is either $+1$ or -1 at each $\mathbf{x}$. A "true" disambiguation would yield $S^{(\epsilon)} \equiv 1$. Note that the *alternate* disambiguation may be written as

$$\mathbf{B}^{(-\epsilon)} = \mathbf{B}_l - S^{(\epsilon)}\mathbf{B}_p \equiv \mathbf{B}_l - \mathbf{B}_p^{(\epsilon)} . \tag{3}$$

Equations (2) and (3) allow one to determine $\mathbf{B}_l$, $\mathbf{B}_p^{(\epsilon)}$. Note also that $B_{ip}^{(\epsilon)} B_{jp}^{(\epsilon)} = B_{ip}B_{jp} = |B_{ip}||B_{jp}|$, where (i, j) represent any of the (x, y) components. The current associated with the ϵ-th resolution may be calculated from

$$\mathbf{j}^{(\epsilon)} = \nabla \times \mathbf{B}^{(\epsilon)} = \nabla \times \mathbf{B}_l + \nabla \times \mathbf{B}_p^{(\epsilon)} = \mathbf{j}_l + \mathbf{j}_p^{(\epsilon)} . \tag{4}$$

Note that the radial or vertical current j_{zl} is ambiguity-free. Consider

$$j_{zp}^{(\epsilon)} = \frac{\partial B_{yp}^{(\epsilon)}}{\partial x} - \frac{\partial B_{xp}^{(\epsilon)}}{\partial y} , \tag{5}$$

and multiply it separately by each of the three factors $\left(B_{xp}^{(\epsilon)4}, 2B_{xp}^{(\epsilon)2} B_{yp}^{(\epsilon)2}, B_{yp}^{(\epsilon)4} \right)$ and sum the three expressions. One obtains, if one drops the p-subscript from B_{ip},

$$
\begin{aligned}
j_z^{(\epsilon)} B_h^4 = B_x^{(\epsilon)} &\left\{ +B_x^{(\epsilon)2} \frac{\partial}{\partial x} B_x^{(\epsilon)} B_y^{(\epsilon)} - B_y^{(\epsilon)2} \frac{\partial}{\partial y} B_x^{(\epsilon)2} + \frac{B_y^{(\epsilon)2}}{2} \frac{\partial}{\partial y} B_y^{(\epsilon)2} - \frac{B_x^{(\epsilon)2}}{2} \frac{\partial}{\partial y} B_x^{(\epsilon)2} \right\} \\
+ B_y^{(\epsilon)} &\left\{ -B_y^{(\epsilon)2} \frac{\partial}{\partial y} B_x^{(\epsilon)} B_y^{(\epsilon)} + B_x^{(\epsilon)2} \frac{\partial}{\partial x} B_y^{(\epsilon)2} - \frac{B_x^{(\epsilon)2}}{2} \frac{\partial}{\partial x} B_x^{(\epsilon)2} + \frac{B_y^{(\epsilon)2}}{2} \frac{\partial}{\partial x} B_y^{(\epsilon)2} \right\} \\
\equiv B_x^{(\epsilon)} f &+ B_y^{(\epsilon)} g ,
\end{aligned} \tag{6}
$$

where

$$B_h^2 = \left(B_x^2 + B_y^2 \right) = \left(B_x^{(\epsilon)2} + B_y^{(\epsilon)2} \right) . \tag{7}$$

The factors f and g are ambiguity-free. Since $B_{ip}^{(\epsilon)} = S^{(\epsilon)} B_{ip}$ we have, after adding the missing p-subscript,

$$j_{zp}^{(\epsilon)} B_{hp}^4 = S^{(\epsilon)}[B_{xp}f + B_{yp}g] \equiv S^{(\epsilon)} j_{zp} B_{hp}^4 , \qquad (8)$$

where j_{zp} is the signed "true" vertical current. Note that as given by our construction it does not have all the possible discontinuities that might occur from the use of Eq. (5) when $S^{(\epsilon)} = 1$. We refer to this version of the j_{zp} current then as being quasi-continuous. Finally we may define an *invariant* current,

$$\tilde{j}_{zp} \equiv |j_{zp}^{(\epsilon)}| = |j_{zp}| , \qquad (9)$$

where $\tilde{j}_{zp}$ is independent of the *unknown* sign of j_{zp} and is a quasi-continuous representation of the "true" POS contribution to the vertical current. As Semel and Skumanich have noted, each component part of f and g are treated symmetrically from the point of view of measurement errors in the field components and their contribution to the current.

3. Construction of Minimum Current Disambiguation

Consider a *sign* selection $S_b^{(MIN)}$ such that the absolute value of $j_z = j_{zl} + S_b^{(MIN)} \tilde{j}_{zp}$ is a minimum. This sign factor is *not* the ambiguity factor. Since $B_{xp}^{(MIN)} = B_{hp} \cos \Phi_p^{(MIN)}$ and $B_{yp}^{(MIN)} = B_{hp} \sin \Phi_p^{(MIN)}$, where Φ_p is the field azimuth with respect to the x-axis (Solar West) in the local HCF, we have

$$\begin{aligned} j_{zp}^{(MIN)} \equiv S_b^{(MIN)} \tilde{j}_{zp} &\propto \left[\cos \Phi_p^{(MIN)} f + \sin \Phi_p^{(MIN)} g\right] \\ &\propto \left[\cos \Phi_p^{(MIN)} \cos \tilde{\Phi} + \sin \Phi_p^{(MIN)} \sin \tilde{\Phi}\right] \\ &\propto \cos(\Phi_p^{(MIN)} - \tilde{\Phi}) , \end{aligned} \qquad (10)$$

where we have taken $\cos \tilde{\Phi} = f/(\sqrt{f^2 + g^2})$ and $\sin \tilde{\Phi} = g/(\sqrt{f^2 + g^2})$. We note that $\tilde{\Phi}$ is ambiguity-free. Let ζ_p be the zenith angle of $\mathbf{B}_p$ [the angle from the outward radial (z-axis) in the HCF] then the sets $\{\zeta_p, \Phi_p\}$ and $\{\pi - \zeta_p, \Phi_p + \pi\}$ represent the two sign choices $+\mathbf{B}_p$ and $-\mathbf{B}_p$, respectively. The set $\{\zeta_p^{(\epsilon)}, \Phi_p^{(\epsilon)}\}$ for a particular disambiguation contains elements from either sign choice (selected by $S^{(\epsilon)}$). We select a value from either set for $\Phi_p^{(MIN)}$ such that $sign\left(\cos(\Phi_p^{(MIN)} - \tilde{\Phi})\right) = S_b^{(MIN)}$. This yields $\left\{\zeta_p^{(MIN)}, \Phi_p^{(MIN)}\right\}$, or $\mathbf{B}_p^{(MIN)}$, and hence $\mathbf{B}^{(MIN)} = \mathbf{B}_l + \mathbf{B}_p^{(MIN)}$.

4. Analysis of Active Region NOAA 7201

4.1. DECOMPOSITION OF VERTICAL CURRENT SYSTEMS

The active region NOAA 7201 was observed at $\mu = 0.77$ $(24N, 32E)$ with the Advanced Stokes Polarimeter on 1992 June 17. A discussion of the data, the inferred magnetic field and its azimuth resolution is found in L95. On the basis of this resolution, labeled "AZAM" here, we have calculated, for illustration, the vertical currents (in kG/Mm; note 1 $kG/Mm = 80\ mA/m^2$) in the local HCF, associated with the LOS-, the POS-, and total-field as well as the invariant current. Numerical derivatives were calculated in the POS reference frame (i.e., the observers frame) at each observed point using a three-point Lagrangian scheme and were then transformed to the HCF using appropriate Jacobians. Figure 1 illustrates the spatial distribution of j_{zl}, j_{zp} and $j_z = (j_{zl} + j_{zp})$. (The upper left image of Figure 5 shows $\tilde{j}_{zp}$). We also show the total-field magnitude for comparison in the lower left panel of Figure 1 with the field neutral line, $B_z^{(AZAM)} = 0$, indicated by the black curve, and umbral boundaries ($I = 0.7I_c$) by the closed white contours. The large spot on the (solar) Western side is negative polarity, while the plage on the (solar) Eastern side is positive. Note the δ-spot to the left of center. For a discussion of the evolution of the δ-spot refer to L95.

The frequency distribution of the j_{zl} and $j_{zp}^{(AZAM)}$ currents in Figure 1 is illustrated in Figure 2a where we have plotted the natural logarithm of the number in 0.1 kG/Mm bins. The solid curve represents the symmetrized (i.e. reflected about the origin and reduced by two to keep the same normalization) values of the invariant current $\tilde{j}_{zp}$. In Figure 2b, we consider the inner core of the current distribution function for a bin size of 0.01 kG/Mm. The j_{zl} current is shown by the asterisk ($*$) symbols while the symmetrized $\tilde{j}_{zp}$ current, by the plus ($+$) symbols. A Gaussian fit to the central portion is plotted as a dashed curve and yields a standard deviation of $\sigma_{ASP} = \pm 80\ G/Mm\ (\pm 6.4\ mA/m^2)$. We believe that this represents the system noise in the ASP currents for step sizes of 0.37 arcsecs along and 0.75 arcsecs perpendicular to the slit. The seeing was 1 - 2 arcsecs. This is consistent with the values of $\pm 25\ G/Mm$ quoted for the Haleakala Stokes polarimeter for a step size of either 2.8 arcsecs or 5.6 arcsecs and a 6.0 arcsecs pinhole, refer to Leka et al. (1993) and de la Beaujardière et al. (1993).

The maximum current values for NOAA 7201 appear to be $\simeq \pm 25\ \sigma_{ASP}$ and are sited around the magnetic neutral line in and near the δ-spot. Oppositely signed current regions are associated with the δ-spot-umbrae. Such a current distribution but with a helicity signature (i.e. j_z/B_z) that is opposite to ours was found by Hagyard (1988) for AR 2372. A result similar to ours was reported by Leka et al. (1994). Helicity of either sign is naturally associated with δ-spots, according to the model of L95. Outside the δ-spot the strong currents appear as narrow linearly extended features near the neutral line. Further discussion of the nature of the vertical currents is not pursued here. We do note, however, that the positive side of the j_{zl} distribution appears to exceed the symmetrized $\tilde{j}_{zp}$ distribution while it is the negative side for the $j_{zp}^{(AZAM)}$ currents. Recall that since $\tilde{j}_{zp}$ is a quasi-continuous represen-

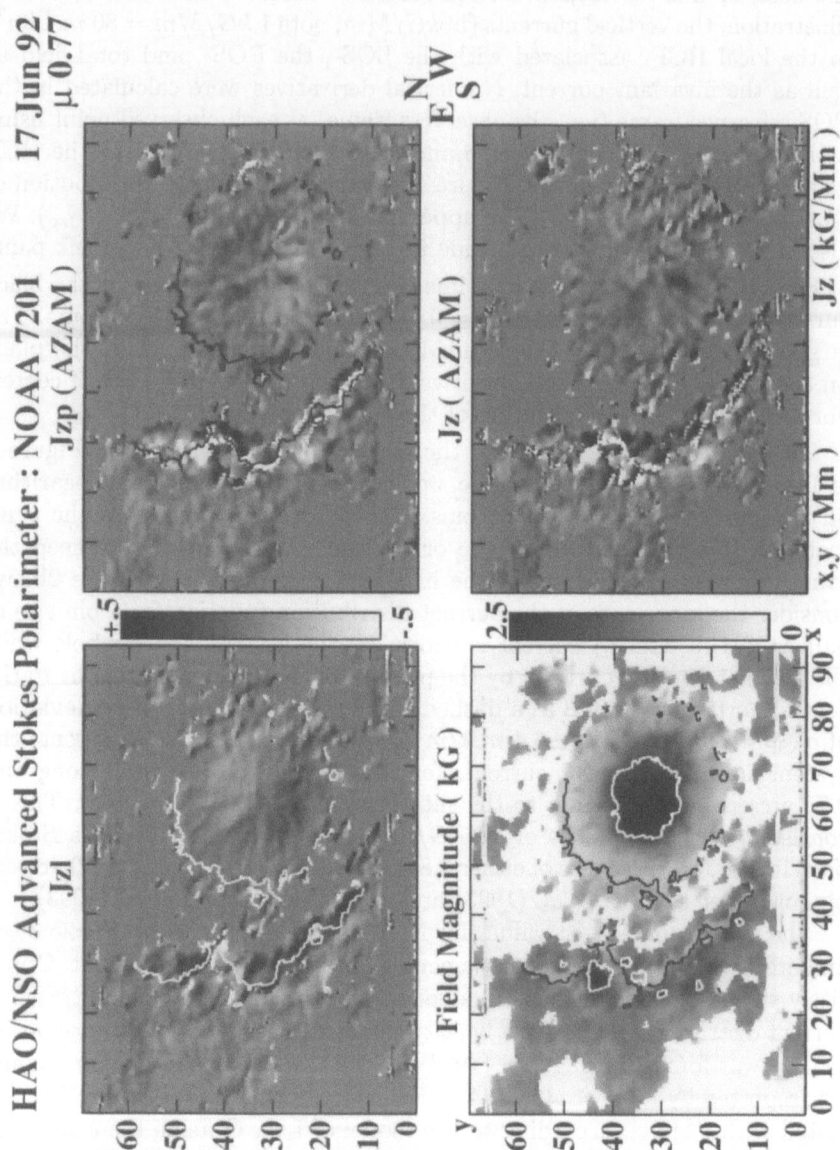

Fig. 1.   The spatial distribution in local heliocentric coordinates of the vertical (radial) current, j_z (kG/Mm, lower right) and its decomposition into the contribution from the LOS-field, j_{zl} (upper left) and POS-field, j_{zp} (upper right), after the "AZAM" resolution (see text). The total-field (kG) with umbral boundaries in closed white contours is at the lower right. The neutral line, $B_z = 0$, is shown as either a black or white curve. Solar West (x-axis) is to the right, North (y-axis) is to the top.

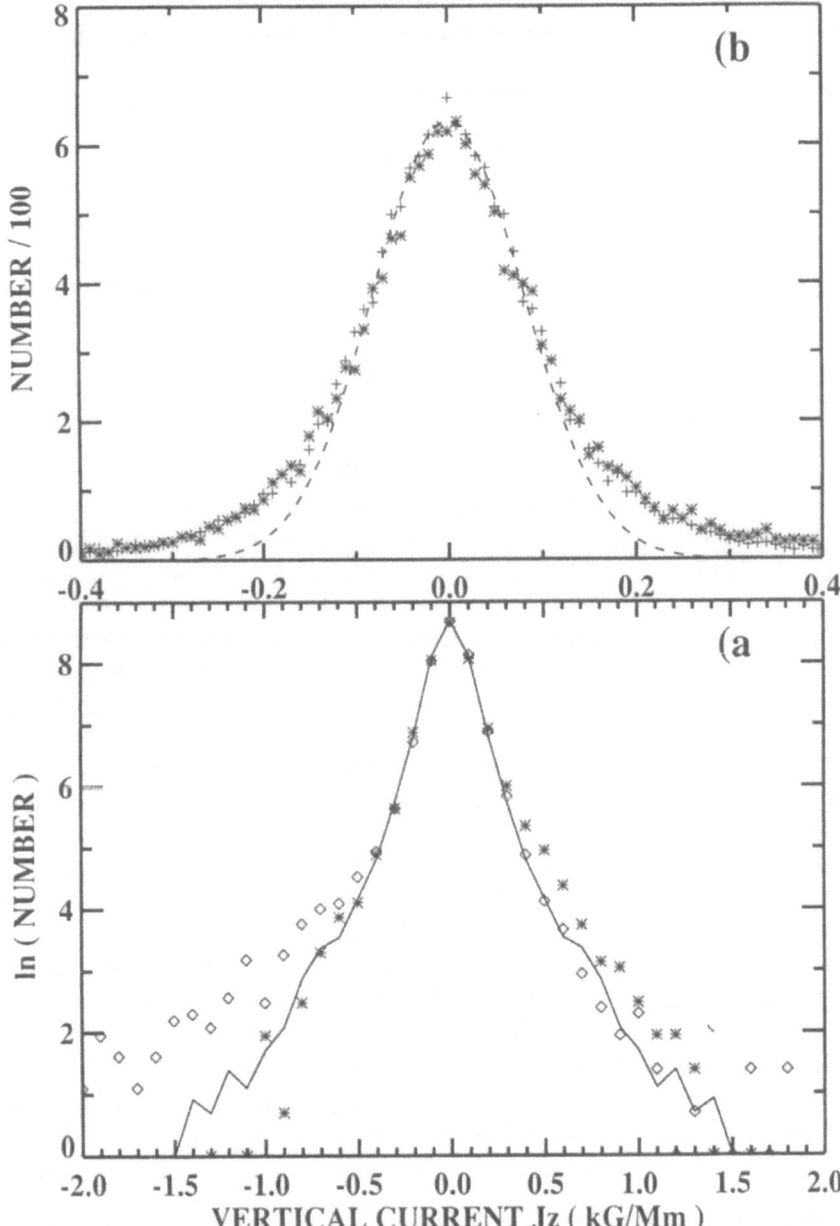

Fig. 2.   The frequency distribution is given in panel (a) of the vertical POS-current, j_{zp} ($\Diamond$), the LOS-current, j_{zl} (∗), and the symmetrized invariant current $\tilde{j}_{zp}$ (solid curve). The upper panel (b) shows a Gaussian fit (dashed curve) to the j_{zl} (∗) and symmetrized $\tilde{j}_{zp}$ (+) distributions for lower current values.

tation of the current distribution this excess must represent regions where the currents are discontinuous due to jumps in either sign or magnitude.

4.2. MINIMUM CURRENT DISAMBIGUATION

Of the two possible azimuth solutions for the magnetic field available to the non-linear least squares (LSQ) inversion of the Stokes profiles (Skumanich and Lites, 1987; Lites and Skumanich, 1990) the one closest to the initial guess is usually found. The initial guess is an arbitrary choice of one of the two available from the observed U to Q ratio. If available the azimuth at the preceding point along the slit is then used to select whichever of the two possible azimuths is closer. The upper left grey scale images in Figures 3 and 4 illustrate the LSQ solutions for the azimuth Φ (labeled "AZM"), and zenith angle ζ (labeled "ZA") of the total-field in the local HCF. The invariant current

$\tilde{j}_{zp}$ for this "LSQ" azimuth resolution differs fractionally from that for the "AZAM" resolution by less than 1×10^{-4}.

Applying the minimum current sign assignment $S_b^{(MIN)}$ to the LSQ solution, we find the resolution illustrated in the upper right images in Figures 3 and 4 (labeled "MIN"). It is apparent that this resolution has general features similar to the "AZAM" resolution but for the occurrence of noise. The use of the interactive AZAM-algorithm (refer to L95) to force continuity on such noise regions would proceed more quickly than starting with the "LSQ" resolution.

In the lower left panel we illustrate the resolution, labeled "POT", that one finds using a potential field comparison method (Semel, 1967; Fan et al., 1990). This method uses a "closest direction" criterion when comparing the two possible $\mathbf{B}_h$ fields to the $\mathbf{B}_h^{(POT)}$ field. The dark (white) lines represent the neutral line, $\zeta = 90° \equiv ZA$, for the "POT" ("AZAM") resolution, refer to Figure 4. Except for some differences in the neutral line regions near the δ-spot, the "POT" and "AZAM" resolutions appear quite similar. It would seem that the "POT" method yields a more continuous and less noisy resolution than the "MIN" method.

It is of interest to compare the resulting spatial structure of the POS-currents, $j_{zp}^{(\epsilon)}$, which we illustrate in Figure 5. The upper left panel presents the invariant current, $\tilde{j}_{zp}$, with a superimposed "AZAM" neutral line in white for reference. The lower left (right) shows $j_{zp}^{(POT)}$ $\left(j_{zp}^{(AZAM)} \right)$. Finally the upper right panel shows the *signed* invariant current, $S_b^{(MIN)} \tilde{j}_{zp}$, rather than $j_{zp}^{(MIN)}$ calculated from the curl of $\mathbf{B}_p^{(MIN)}$ which is significantly noisier (point-like discontinuities). Note the noise, fine scale black/white discontinuities, on the "POT"-map (e.g. in the region $x = 15$ to 25, $y = 30$ to 55 Mm) as compared to the "AZAM"-map. All four maps show narrow

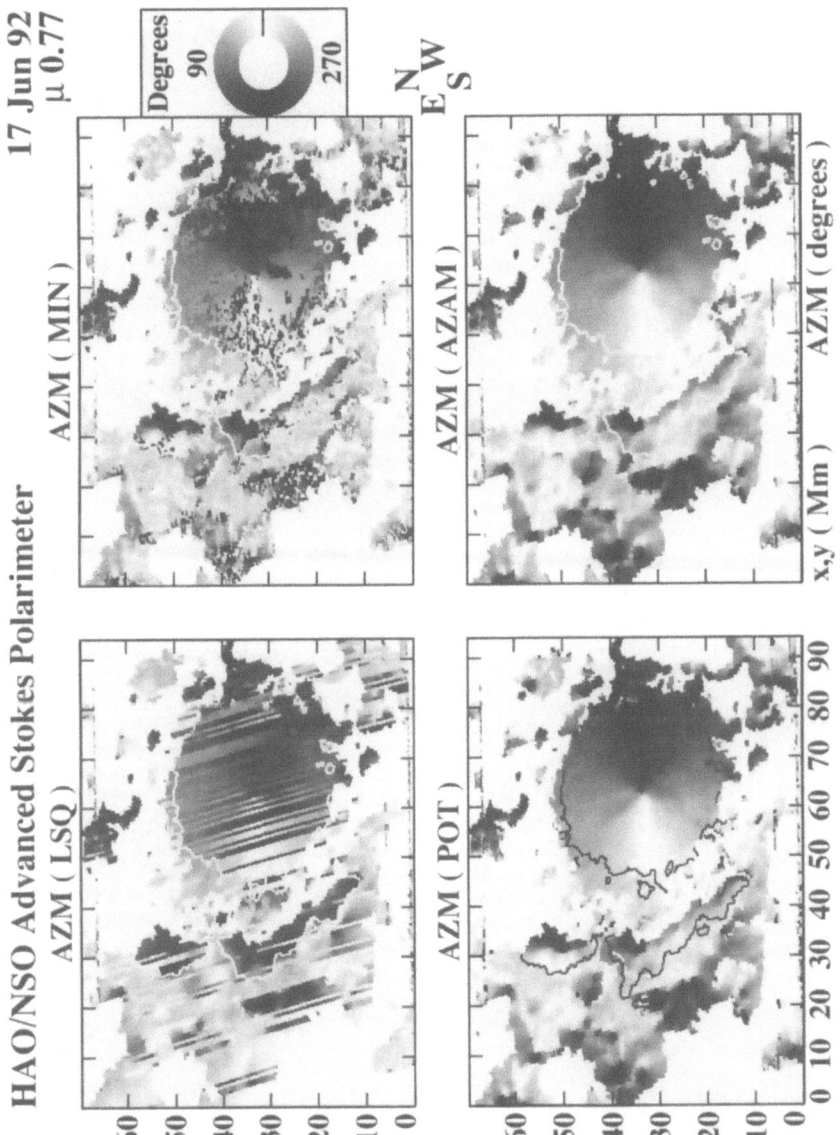

Fig. 3.   Images are shown of the spatial distribution of the total-field azimuth $\Phi(\equiv \text{AZM})$ in the local HCF ($\Phi = 0°$ is Solar West, $90°$ is Solar North) for different disambiguations, for details see text. The magnetic neutral line for the "AZAM" resolution is shown in white, while that for "POT" in black.

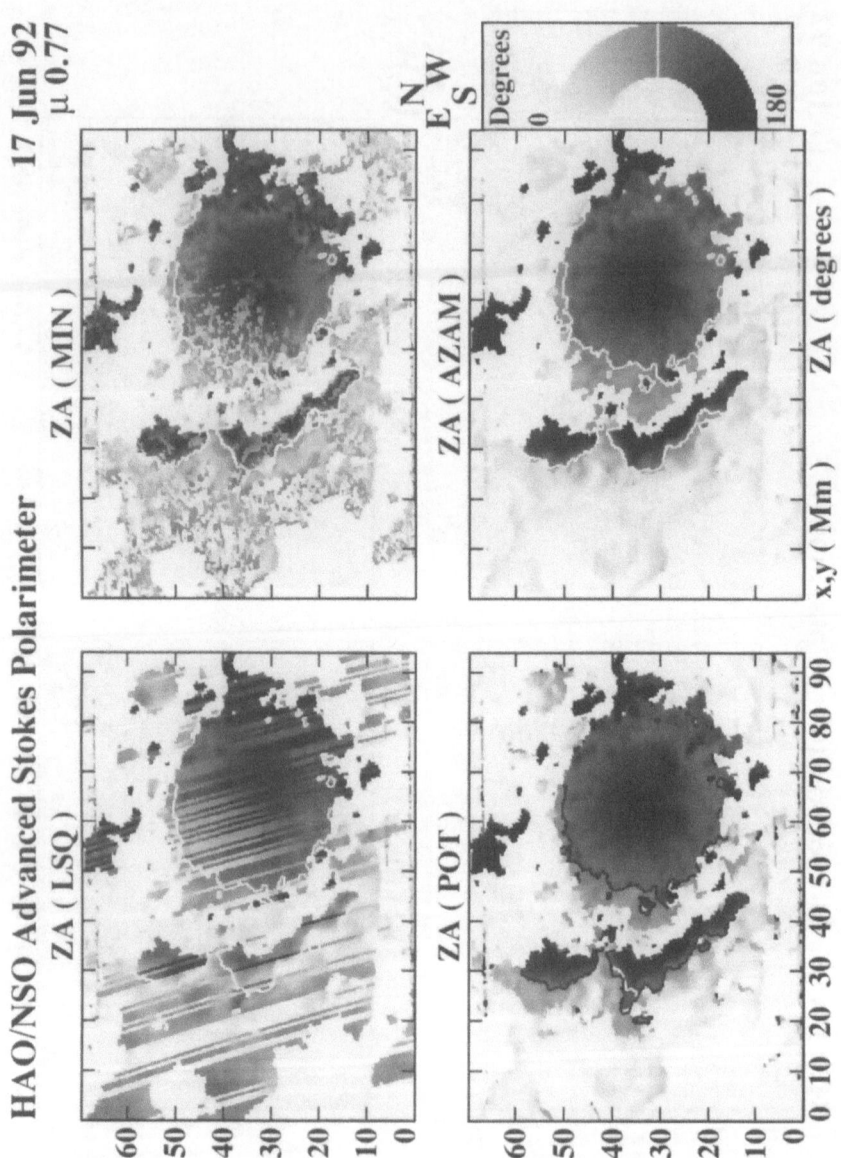

Fig. 4. Same as Fig. 3 except for total-field zenith angle from the outward radius, ζ ($\equiv$ ZA).

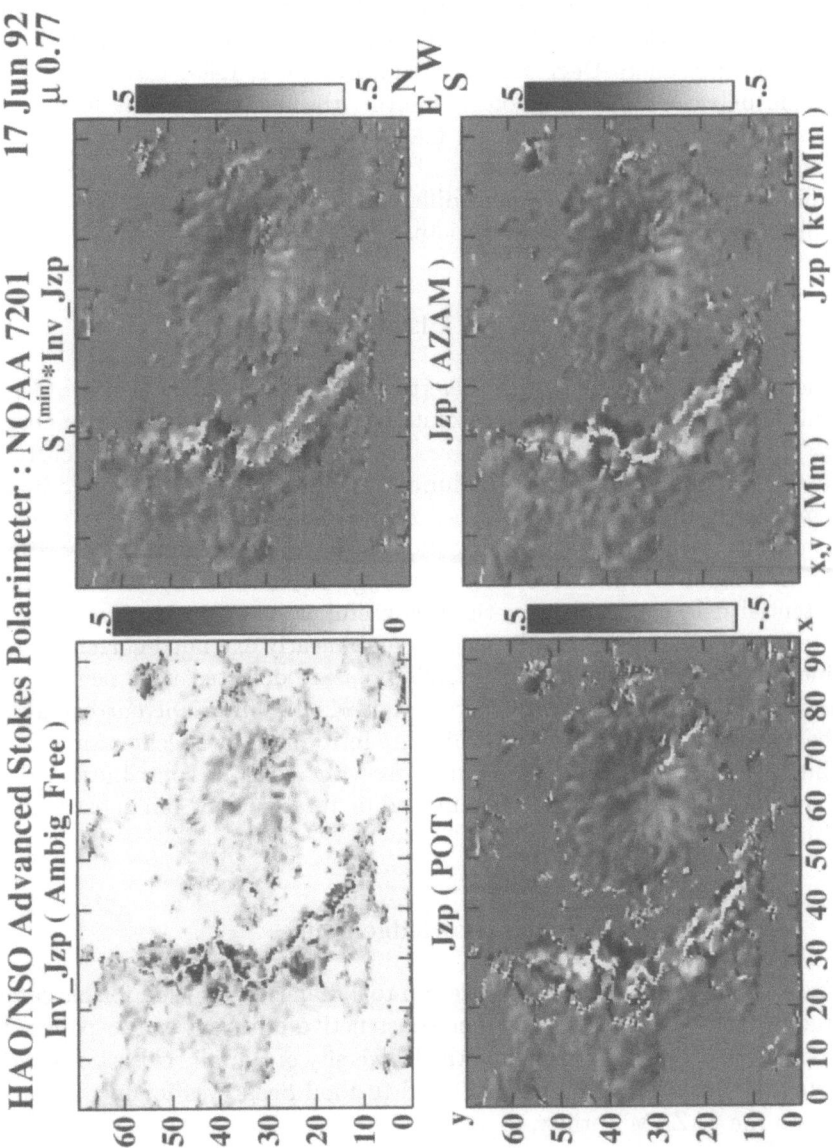

Fig. 5. Shown is the vertical POS-current, j_{zp}, derived from the "AZAM" ("POT") disambiguation, lower right (lower left). The upper left panel presents the invariant current, $\tilde{j}_{zp}$, while the *signed*, according to the "MIN" disambiguation, invariant current is in the upper right.

linearly extended discontinuities. However, the location and length of such linear "discontinuities" near the neutral line in the neighborhood of the δ-spot are significantly different. All four maps show strong current regions of opposite signs associated with the umbræ of the δ-spot. In the absence of any detectable physical signature (e.g. second order linear polarization asymmetries, Landi Degl' Innocenti and Bommier, 1993) that allow one to decide between the ambiguous azimuths, the reality of those linear "discontinuities" not seen in the $\tilde{j}_{zp}$ map (refer Semel and Skumanich, 1995) remains in question. Observations of the same region, if slowly evolving, at a different disk position, i.e. at a different $\mu = \cos\theta$, may allow one to verify which resolution is more appropriate.

5. Discussion

The introduction of an invariant vertical (radial) current which is independent of any assumed azimuth resolution provides a useful reference current for the study of various azimuth-resolutions and their associated currents. Its use in a minimum current resolution scheme is found to be limited by system noise in the calculated currents. However, it must be noted that there is no physical reason why solar currents must be at a minimum. In any respect the minimum azimuth resolution, although not as smooth as that using a potential field comparison method, is useful as an alternate first guess for interactive methods. Spatially narrow and linearly extended current discontinuities with magnitudes up to $\pm 25\ \sigma_{ASP}$ are detected near neutral lines for NOAA 7201 in the POS-current, both in the invariant current and in the additional current features which are introduced by specific azimuth resolutions, as well as in the ambiguity-free LOS-current contribution. Some sign cancellation occurs in the summed or total current but most features remain.

Acknowledgements

The authors express their appreciation to J. Sánchez Almeida for his careful reading and to K.D. Leka for her constructive criticism of the paper. We thank T.R. Metcalf for providing the University of Hawaii "closest direction" potential-resolution code to HAO and to Paul Seagraves for incorporating it into the "AZAM" utility.

References

de la Beaujardière, J.-F., Canfield, R. C., and Leka, K. D.: 1993, *Astrophys. J.* **411**, 378

Fan, Y., Canfield, R. C., and McClymont, A. N.: 1990, *B.A.A.S.* **22**, 827

Hagyard, M. J.: 1988, *Solar Phys.* **115**, 107

Landi Degl'Innocenti, E. and Bommier, V.: 1993, *Astrophys. J. Lett.* **411**, L49

Leka, K. D., Canfield, R. C., McClymont, A. N., de la Beaujardière, J.-F., Fan, Y., and Tang, F.: 1993, *Astrophys. J.* **411**, 370

Leka, K. D., van Driel-Gesztelyi, L., Nitta, N., Canfield, R. C., Mickey, D. L., Sakurai, T., and Ichimoto, K.: 1994, *Solar Phys.* **155**, 301

Lites, B. W., Low, B. C., Martínez Pillet, V., Seagraves, P., Skumanich, A., Frank, Z. A., Shine, R. A., and Tsuneta, S.: 1995, *Astrophys. J.* **446**, 877

Lites, B. W. and Skumanich, A.: 1990, *Astrophys. J.* **348**, 747

Semel, M.: 1967, *Ann. d'Astrophys.* **30**, 513

Semel, M. and Skumanich, A.: 1995 (in preparation)

Skumanich, A. and Lites, B. W.: 1987, *Astrophys. J.* **322**, 473

HIGH RESOLUTION OBSERVATIONS OF SMALL-SCALE MAGNETIC ELEMENTS AND INTERPRETATION

F. KNEER and F. STOLPE

Universitäts-Sternwarte, Geismarlandstr. 11, D-37083 Göttingen, Germany

Abstract. This contribution deals with the properties of small-scale magnetic elements in plages. Spectro-polarimetric observations, obtained with the highest possible spatial resolution with the German solar telescopes at the Observatorio del Teide on Tenerife, were analysed. We conclude from the spread of line parameters measured in the Stokes I and V profiles of Fe I and Fe II lines that a wide range of magnetic properties is realised in the solar atmosphere. The flow velocities in small-scale magnetic flux tubes, deduced from the zero-crossing of the V profiles at high spatial resolution, show a fluctuation of $v_{Doppler}$ = 580 m s^{-1}. This is substantially smaller than the "turbulent" broadening velocities of $v_{Doppler} = 2 - 3$ km s^{-1} commonly derived by fitting V profiles from flux tube models to low spatial resolution data, e.g. from a Fourier Transform Spectrometer. Attempts to explain the high resolution I and V profiles by models of hydrostatic flux tubes are discussed. It appears impossible to accomplish agreement between the modeled and observed radiation of lines with strong and weak magnetic sensitivity at the same time. We suggest a scenario in which small-scale magnetic elements possess substructure and are dynamic, with gas flows and magnetic field strengths varying in space and time.

Key words: Small-scale magnetic fields – High spatial resolution observation – Dynamic properties

1. Introduction

Small-scale magnetic elements on the Sun are under intense investigation since many years. They occur predominantly in active regions, in plages with their bright chromospheric emission, but also in the quiet Sun, especially at the borders of the supergranulation cells where also the chromospheric network is seen in filtergrams of chromospheric lines. These fields and their dynamic behaviour are an essential ingredient for many solar atmospheric properties and processes such as strongly enhanced transfer and deposition of energy into the chromosphere and the corona, strong radiative emission from plages and the chromospheric network, spicular dynamics with their mass input to the corona, and many more phenomena of the outer solar atmosphere. In addition to the important role in producing and maintaining the chromosphere, the corona, and the solar wind, small-scale magnetic fields take part in the budget of the global magnetic flux and its variation during the solar cycle. Thus, apart from being an interesting phenomenon of plasma physics by themselves, these magnetic features and their dynamics represent an important key to the understanding of the Sun's atmosphere and its interaction with the heliosphere.

Solar Physics **164**: 303–310, 1996.
© 1996 *Kluwer Academic Publishers.*

In the mid sixties, by means of high spatial resolution polarimetry, the existence of small-scale magnetic fields was shown e.g. by Howard and Harvey (1966, quoted by Sheeley, 1967) and Sheeley (1967). Such fields are co-spatial with positions of "line gaps", as the strong weakening and broadening of spectral lines were coined. Outside sunspots and pores, Beckers and Schröter (1968) found magnetic knots with kilogauss field strength.

Starting in the seventies, high precision polarimetry with high signal to noise ratio (though with low spatial resolution) brought to light rich information about the structure of small-scale fields. The kilogauss field strength was confirmed by Stenflo (1973) and later by Wiehr (1978) and others. By means of polarimetry with the Fourier Transform Spectrometer at the National Solar Observatory at Kitt Peak/Arizona, by means of polarimetry in the infrared, and together with sophisticated modeling, the Zürich group and its collaborators could establish detailed pictures of small-scale magnetic flux tubes in plages and the network of the quiet Sun. The models, based on hydrostatic equilibrium, include the height variation of the magnetic field, the temperature and density structure, and the velocity inside the tubes and in the medium surrounding them. Extensive reviews on these results can be found in Solanki (1993) and Stenflo (1994).

One might argue that there remains little to learn about small-scale magnetic fields, and close the act. However, some discomfort comes with the above models. Firstly, the findings from low spatial resolution data are very indirect. For instance, the intensity signal, i.e. the Stokes I component, is of no use to determine the brightness and the temperature structure. It stems mainly from the non-magnetic atmosphere, which is believed to be known. Thus, true continuum intensities of the radiation from the fluxtube proper are very uncertain. Secondly, to explain the widths of the V profiles, large "turbulent" velocities of $2 - 3$ km s^{-1} had to be adopted. This amounts to as much as $30 - 50$ percent of the tube speed $v_T = v_s v_A/(v_s^2 + v_A^2)^{1/2}$, which is $5 - 6$ km s^{-1} under the conditions adopted in tube models. v_s and v_A are the sound velocity and the Alfvén velocity, respectively. Such high turbulent velocities render the hydrostatic assumption questionable. It is thus worthwhile to attempt a closer inspection also of the intensities and to try to actually see the high velocities. So we resume observations with spatial resolution as high as possible, though sacrificing on polarimetric accuracy. We intend to obtain direct information on the brightness, the magnetic field, and the dynamics of small-scale magnetic elements.

2. Observations, Analysis, and Results

The observations discussed here were obtained in 1991 and 1992 with the Gregory Coudé Telescope (GCT) at the Observatorio del Teide/Tenerife

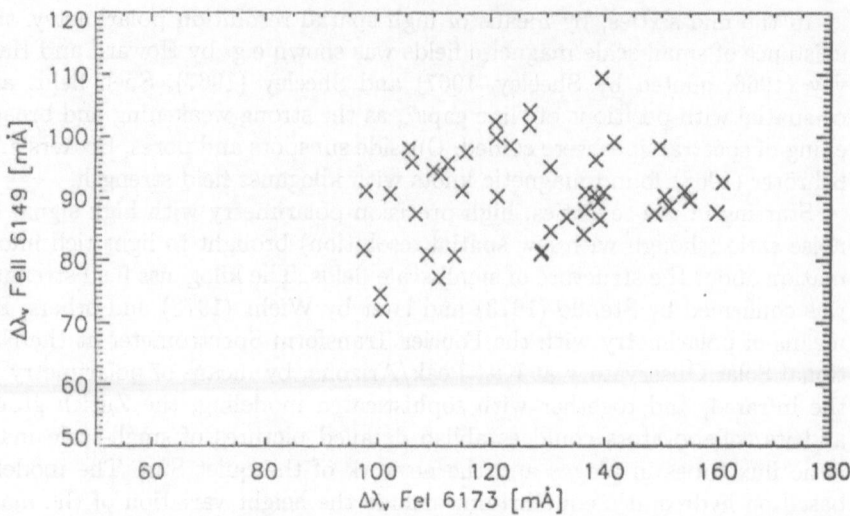

Fig. 1.   Separation of the Stokes V extrema $\Delta\lambda_V$ of Fe II 6149 versus $\Delta\lambda_V$ of Fe I 6173.

from plage regions near disc centre of the Sun ($\sin\theta \leq 0.3$). They have been
discussed (Amer and Kneer, 1993) or will be discussed in detail elsewhere.
So only a short description, with few results and examples, is given here. (We
shall come back to additional efforts to scrutinize small-scale magnetic field
dynamics with the Vacuum Tower Telescope (VTT) on Tenerife, in Section
4.) The spectrum cutter of the GCT allows simultaneous recording of two
spectral regions. For the present study alternate pairs of the lines Fe II 6149
(Landé factor $g_{eff} = 1.33$), Fe I 6151 ($g_{eff} = 1.83$), and Fe I 6173 ($g = 2.5$)
were selected. These lines span a wide range of magnetic sensitivity and, due
to the high ionization degree of iron, of temperature sensitivity. By means
of a $\lambda/4$ retarder plate and two crossed calcites behind the entrance slit of
the spectrograph, spectrograms of $\frac{1}{2}(I+V)$ and $\frac{1}{2}(I-V)$ were exposed onto
different parts of a CCD chip. The exposure time was 0.2 – 0.3 sec. From the
fine structure visible in the spectrograms, we estimate the spatial resolution
to 0.6 – 0.8 arcsec.

The data analysis yields profiles of I and V along the slit position. From
these, continuum intensities, half widths and residual intensities of I_λ, amplitudes of V, the separation $\Delta\lambda_V$ of the V extrema, velocities v_I deduced
from the position of the minima of I_λ, velocities v_{zc} of the magnetized gas
deduced from the zero-crossings of the V profiles, and other parameters were
determined. We explicitly exclude data from pores.

Figure 1 shows, as one example, $\Delta\lambda_V$ values of the Fe II 6149 line versus
those of the 6173 line. We note several points. Firstly, there exists certainly
a correlation between the two data sets. The slope of the regression line is

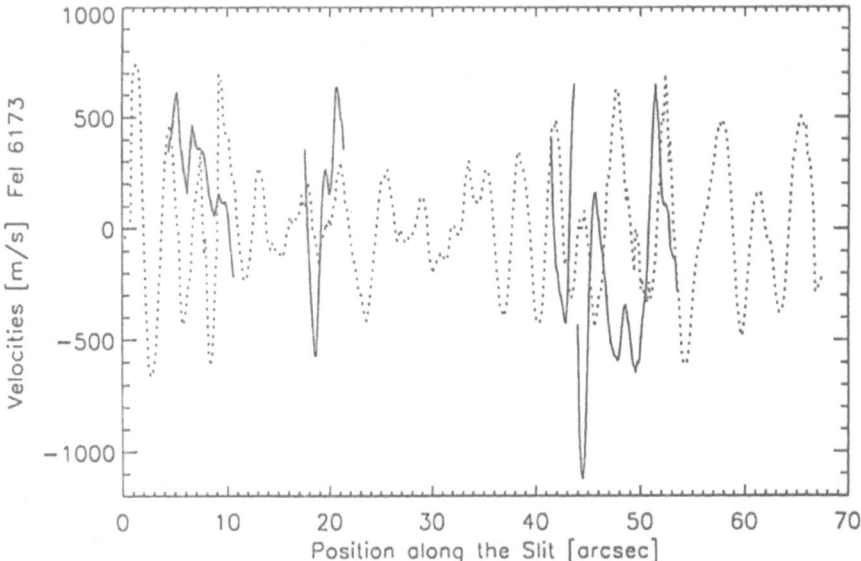

Fig. 2. Velocities in the Fe I 6173 line. v_{zc} (solid lines) is measured from the position of the zero-crossings of the Stokes V profiles and v_I (dotted line) from the position of the line centres of the Stokes I profiles.

about 0.3, while the ratio of the Landé factors is 0.53. Secondly, the separations spread over a rather large range, from about 70 to 110 mÅ for Fe II 6149 with an average of 90 mÅ, and from 95 to 160 mÅ for Fe I 6173 with an average of 125 mÅ. Thus, at high spatial resolution one detects many possible physical manifestations, i.e. thermal, dynamic, and magnetic states, of small-scale elements. Thirdly, the correlation is not perfect. The data show a substantial scatter, much larger than the accuracy of the determination of about 7 mÅ. This demonstrates again that a multitude of configurations and dynamic properties of magnetic features is realized. It appears questionable that their structure may be described by few parameters such as the flux tube radius and the plasma β (the ratio of gas pressure to magnetic pressure) at some height. We rather suspect an element of stochasticity as an essential physical contribution, which is produced by the complex interaction of the magnetic field with the turbulently convecting gas and which causes fluctuations of the (thermo-)dynamic state including magnetic field strength and velocity, and of their variation with height.

Figure 2 depicts, as a further example, the velocities v_{zc} (solid lines) and v_I (dotted line) measured in Fe I 6173 from one spectrogram. Our convention here is that positive velocities are away from the observer. The statistical error of the determination amounts to $50 - 80$ m s^{-1}. The systematic effects due to limited spatial resolution may be larger. The rms velocity of v_{zc} is 410 $m s^{-1}$, that of v_I 280 $m s^{-1}$. To be comparable with commonly used macro-turbulent velocities for line broadening the rms values have to be multiplied with a factor $2^{1/2}$, assuming a Gaussian distribution. This gives for the V zero-crossing velocity a macroturbulence parameter of $v_{Doppler} = 580$ $m s^{-1}$ which falls short by a factor $3 - 5$ of the value needed to fit the Stokes V profiles measured with low spatial resolution by the profiles calculated from static models. Thus, the high velocities in the magnetized plasma are not (yet) seen. We do see, however, strong horizontal changes of the line of sight velocity in magnetic elements on short distances, up to 1 $km s^{-1}$ per arc-sec. This indicates again the highly dynamic nature of small-scale magnetic fields and the deviation from vertical and horizontal pressure balance.

We mention, without illustrations, few more results of the data analysis: The structures with V signal, irrespective of the strength of V, do not show up as special features in the continuum intensities. Typical variations are ±5 percent about average. This is a known fact (see e.g. Kneer and von Uexküll, 1991, and references there). Furthermore, the line width of the I profiles, their line centre intensity, i.e. the "strength" of the line gap phenomenon, and the amplitude of the V profiles are well correlated. We may understand this by the notion that these parameters all depend on the total magnetic flux contained in the resolution element considered. And finally, the separation of the V extrema, which according to Figure 1 varies strongly, is not correlated with the former parameters. It does not depend on the observed flux in the resolution element.

3. Modeling

The observational results of the last Section deserve thorough investigation of dynamic models, starting with exploratory simulations of the various effects of velocity and magnetic field fluctuations. Here we still restrict ourselves to static fluxtube models and pose the question whether they are compatible with high spatial resolution data. The results of our study are published elsewhere (Kneer *et al.*, 1995). So we give here only a summary.

Radiation in the above three Fe lines from small-scale flux tube models is calculated and compared with the observations. The slender tube approximation and horizontal and vertical pressure balance are adopted. The run of temperature with geometric height inside the tube is closely tied to that of the ambient medium by irradiation from the outside gas. For the surrounding atmosphere a quiet Sun model is adopted.

The information contained in the Stokes I profiles in addition to the V data is very useful. Satisfactory models require that their overall thermal as well as dynamic and magnetic structure be compatible also with the Stokes I

data. The plasma β is found in the range of 0.3 – 0.5 in agreement with Rüedi *et al.* (1992). Too low values of β give very high flux tube temperatures at the line forming layers with too strong an ionization of iron, consequently very weak V signal in the Fe I lines. In the models of the ambient medium and the flux tube proper, the temperature as well as the temperature gradient appear to be too high. This is concluded from the notion that the continuum intensity, averaged over an area containing the flux tube and the surrounding photosphere with appropriate contributions, are much too high and the I and V profiles of the Fe II 6149 line are rather strong compared with observation. Thus, with respect to the temperature structure, the tube models can be improved.

As to the modeling of the width of the I profiles and of the separation of the V extrema, an additional broadening mechanism is still needed to fit the observed strongly split Fe I 6173 line, even when observed with high spatial resolution. We may accept the possibility that the observations have still not sufficient spatial resolution and test the effect of a high, unresolved turbulent velocity. This may explain the width of Fe I 6173. But then one faces the problem that the separation of the V extrema of Fe II 6149 becomes too large. A broadening that is proportional to the Landé factor, i.e. a magnetic broadening, appears more appropriate. This requires that the observed features possess not resolved substructures in which the magnetic field is not ordered as in a static fluxtube, but in which horizontally varying conditions prevail and which are dynamic. We take the observed strong fluctuations of the resulting parameters such as the separation of the V extrema, Figure 1, and the velocity of the gas in the magnetic elements, Figure 2, as hints to the property of "disorder" on not yet resolved spatial scales.

4. Discussion and Conclusion

Our observations and modeling attempts emphasize the essentially dynamic nature of small-scale magnetic elements. Recent observations with the VTT on Tenerife support our view. Denker *et al.* (1995) have applied speckle methods to obtain images of the moustache phenomenon in the wing of Hα with a spatial resolution of 0.2 arcsec. (We may assume that moustaches are linked to small-scale magnetic fields.) It is shown that these features possess sub-structure with spatial and temporal intensity variations at scales below the resolution of the observations discussed above in Section 2. Furthermore, with the Fabry-Perot interferometer installed in the VTT, Volkmer *et al.* (1995) have obtained time sequences of two-dimensional spectropolarimetric images from a plage region. There one sees small-scale magnetic features being pushed about, changing shape, exhibiting strong horizontal

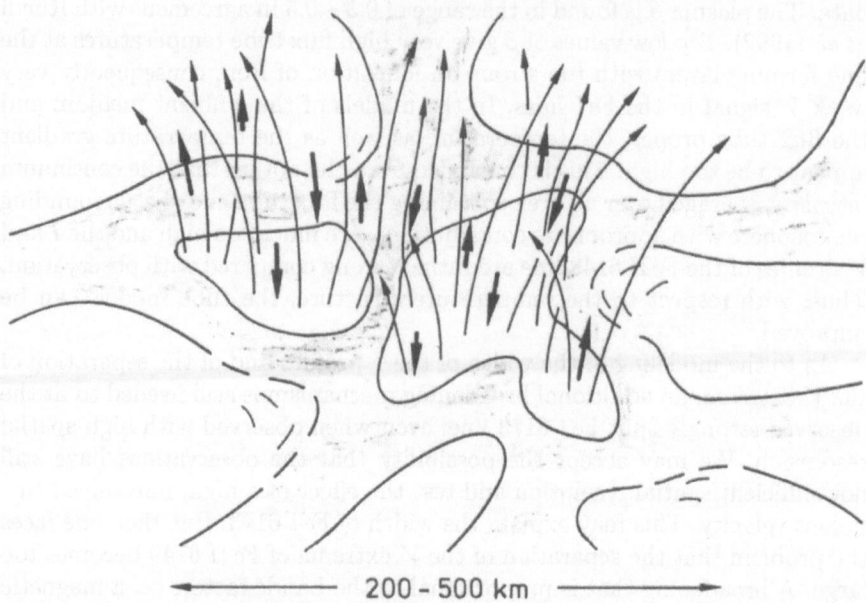

Fig. 3. Scenario of magnetic elements embedded in the granular pattern. The magnetic features possess substructure with magnetic field strength fluctuating stochastically in space and time. Accordingly, the pressure imbalance sets up gas flows (thick arrows).

variations of the line of sight velocity as in Figure 2 above, and oscillating with periods of about 100 sec and an amplitude of 500 m s^{-1}.

In conclusion, we propose the scenario of small-scale magnetic features embedded in the granular pattern as depicted in Figure 3. It differs from the models of static, axially symmetric tubes. Bundles of magnetic field lines emerge in irregular patches from the solar interior. Most of the field lines connect to areas of opposite flux at close or more distant locations. Few of them may go out to the heliosphere and enter the interstellar medium. The flux bundles are continuously distorted and displaced under the action of convection and of their own buoyancy and stresses in subphotospheric layers. This causes a permanent imbalance of magnetic and gas pressure, thus gas flows inside and outside the magnetic elements are driven.

There is certainly a need for modeling of magnetic fields, guided by this dynamic and stochastic picture, and to study its consequences on the emergent polarized radiation in spectral lines. Even more important are observations at very high spatial resolution to have a better insight into the processes in small-scale magnetic elements, which are important ingredients to the physics of the dynamic Sun. We need LEST, the Large Earth-based Solar Telescope.

Acknowledgements

The Gregory Coudé Telescope is operated by the Universitäts-Sternwarte, Göttingen, and the Vacuum Tower Telescope by the Kiepenheuer-Institut für Sonnenphysik, Freiburg, at the Spanish Observatorio del Teide of the Instituto de Astrofísica de Canarias. F. K. wishes to thank the staff of the Pulkovo Observatory for the hospitality he had received.

References

Amer, M.A., Kneer, F.: 1993, *Astron. Astrophys.* **273**, 304
Beckers, J.M., Schröter, E.H.: 1968, *Solar Phys.* **4**, 142
Denker, C., de Boer, C.R., Volkmer, R., Kneer, F.: 1995, *Astron. Astrophys.* **296**, 567
Kneer F., von Uexküll, M.: 1991, *Astron. Astrophys.* **247**, 556
Kneer, F., Hasan, S.S., Kalkofen, W.: 1995, *Astron. Astrophys.* , in press
Rüedi, I., Solanki, S.K., Livingson, W., Stenflo, J.O.: 1992, *Astron. Astrophys.* **263**, 323
Sheeley, N.R., Jr.: 1967, *Solar Phys.* **1**, 171
Solanki, S.K.: 1993, *Space Science Reviews* **63**, 1
Stenflo, J.O.: 1973, *Solar Phys.* **32**, 41
Stenflo, J.O.: 1994, *Solar Magnetic Fields — Polarized Radiation Diagnostics*, Kluwer, Dordrecht
Volkmer, R., Kneer, F., Bendlin, C.: 1995, *Astron. Astrophys. Letters* , submitted
Wiehr, E.: 1978, *Astron. Astrophys.* **69**, 279

POLARIMETRY OF SOLAR PORES

P. SÜTTERLIN, E.H. SCHRÖTER AND K. MUGLACH

Kiepenheuer Institut für Sonnenphysik, Freiburg, Germany

Abstract. We address the magnetic field structure of solar pores. The data were obtained at the Gregory Coudè telescope at Izaña using the AT1 CCD camera system to observe pores with three spectral lines: one magnetically sensitive line, recording all 4 Stokes profiles, and two $g = 0$ lines where only the intensity profiles were measured. The data reduction included the standard procedure (removing dark current and flatfielding) as well as destretching of the polarimetric spectra and removing the non–magnetic straylight by means of a 2–d deconvolution of the observed intensity variation using a Lucy–Richardson restoration algorithm. In the following analysis we first determined the temperature- and pressure stratification of the pore using the $g = 0$ lines and then applied an inversion of the Stokes profiles to get the parameters of the magnetic field. Across the pore we find a strong variation of the resulting field strength as well as of the inclination and the azimuth, consistent with the assumption of a canopy forming in the higher atmosphere.

Key words: Solar magnetic fields – Pores – Polarimetry

1. Introduction

We present an investigation of the internal properties of solar pores using polarimetric spectra. The usual definition of a pore is that it is a sunspot lacking a penumbra but recent high resolution speckle observations (Keller, 1992) as well as the magnetic field variation across the object (Muglach *et al.*, 1994) indicate that this is a rather arbitrary definition. Its brightness should be less than about 80% of the surrounding atmosphere (Brants and Steenbeck, 1985; Muller, 1992). Many contributions have used photometric information only (Bumba, 1967; Beckers and Schröter, 1968; Brants and Zwaan, 1982; Bonet *et al.*, 1994), most of them are hampered by the limited size of the pores and thus the influence of seeing except Bonet *et al.* who use the moon limb for seeing correction.

When interested in their internal properties the use of polarized light has the advantage that it originates in the magnetic part of the observed resolution element and is not mixed with non–magnetic straylight. Information derived from polarimetric measurements can be found in Sütterlin (1991), Thim (1993), Muglach *et al.* (1994) and Sütterlin *et al.* (1994).

We analyse intensity spectra taken in two magnetically insensitive lines in addition to spectra of the complete Stokes vector in a g=2.5 line. An inversion technique is used to determine the thermodynamic and magnetic structure of a pore.

Solar Physics **164**: 311–320, 1996.

2. Instrumental Setup and Observations

The spectra were taken on May 30th, 1993, at the 45-cm Gregory-Coudé-Telescope of the Observatorio del Teide at Izaña, Tenerife. This telescope is well suited for polarimetric observations due to the low instrumental polarization which only depends on the observed declination δ (and is even zero for $\delta = 0$). The focal length of 25 m yields a spatial scale of $8.25''$/mm in the plane of the spectrograph entrance slit. As detector we used the AT1 CCD camera system with a chip of 1024×1024 pixels, each having a size of $19\mu^2$. The linear dispersion of the GCT spectrograph is 1.79Å/cm at 630.2 nm, resulting in a scale of 3.41mÅ $\times 0.157''$ on the chip.

The observed object was a pore in the near surroundings of the complex B-group NOAA 7515 and located at $\cos\theta = 0.87$. Under very good seeing conditions, the best spectra show structures down to approximately $0.6''$. Due to this as well as to increase the signal to noise ratio we averaged four spatial rows (giving an image resolution of $0.63''$).

The spectra of the magnetically insensitive lines Fe I 512.3 nm and Fe I 709.0 nm were recorded simultaneously using the so-called spectrum cutter. The Stokes I, Q, U and V spectra of Fe I 630.2 nm (g=2.5) had to be recorded consecutively due to instrumental constraints. For the polarimetric observations we used a device (Soltau, 1993) consisting of two birefringent calcites orientated at $\pm 45°$ relative to the entrance slit of the spectrograph, thereby also setting the local reference frame for the azimuth measurement. Together with a quarter-wave retarder at $0°$ this device splits the beam into two parts, each of which carrying the information of $\frac{1}{2}(I - V)$ and $\frac{1}{2}(I + V)$, respectively. By projecting the two beams onto the same CCD camera it is possible to simultaneously observe the two polarization states (e.g. $I \pm V$). Stokes Q and U are measured by rotating the quarter-wave plate by $45°$ (U) and adding a half-wave retarder at $22.5°$ (Q).

3. Data Reduction

The data reduction consisted of a number of steps, including the standard procedure of subtracting the dark current and flatfielding the image.

The correction for instrumental scattered light introduced by the polarization optics and the spectrum cutter (Thim, 1993) was done by measuring the equivalent widths of solar spectral lines and comparing them to those in a quiet sun FTS atlas compiled by Neckel. Atmospheric scattered light has also to be taken into account. We used the continuum intensity fluctuations along the slit to produce a pseudo-two dimensional intensity image by rotating them around the pore center. This image was then de-convoluted with a seeing function using an iterative Lucy-Richardson restoration algorithm.

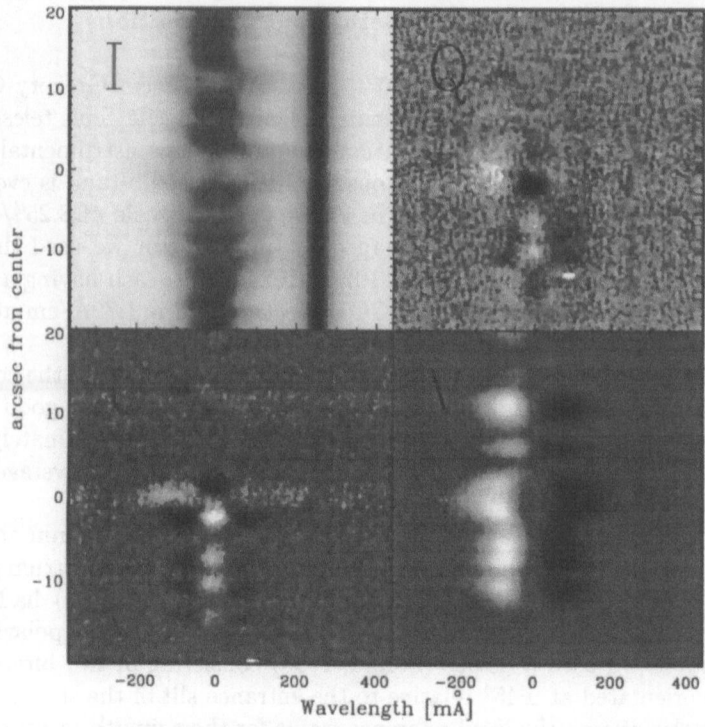

Fig. 1. Reduced and normalized Stokes I, Q, U and V spectra of Fe I 6302.5.

As point spread function we chose an analytical one (Rossbach and Schröter, 1970) which consists of two Gaussians taking into account the influence of image motion and background scattered light. One of them, representing image motion, has a half-width of 0.6″ which was derived from the size of the smallest visible structure in the spectrum.

This algorithm can strictly only be applied if the *complete* information of the observed object is included in the image, which is obviously not the case for the (spatially) one–dimensional spectrum (see Keller and Johannesson, 1995, for a method to circumvent this by means of a rapid scanning across the observed structure). However, there are two points to justify this procedure: First, the surroundings of the pore is granulation, which has a self-similar structure and the pore itself has also an axisymmetric appearance. Therefore the error introduced by taking the same spatial structure in all directions is likely to be small. And second, assuming of a point spread function in itself is quite a constraint. Our resulting continuum intensity at the center of the pore is 28% of the continuum intensity averaged along the slit, which

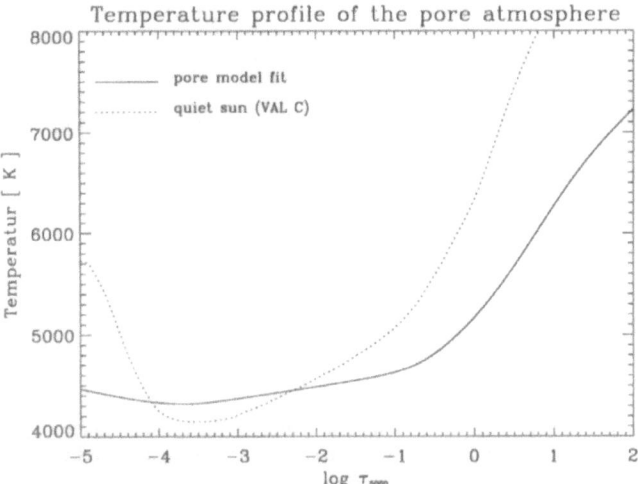

Fig. 2.   Temperature as a function of optical depth τ (at $\lambda = 5000$ Å) of the model atmosphere of the pore which we derived from a fit to the two $g = 0$ lines. The solid line represents the pore atmosphere, the dotted one is a model of the quiet sun.

is in good agreement with similar investigations (Sütterlin (1991) gives 31%, Thim (1993) gives 28% and Bonet *et al.* (1994) give a value of 30%).

Then each spectrum was normalized to the local continuum and the best one of each series (determined by the largest continuum rms value and the lowest continuum intensity) was selected for further evaluation.

One additional reduction step was necessary for the polarimetric data: The two polarization states of the image show some spatial degradation caused by the different light paths in the calcites. Therefore the spatial scale had to be adjusted to allow the subtraction of the two parts without introducing artefacts. For this purpose the granular continuum intensity variations were used as tracers for a 1-D local correlation tracking routine.

The reduced Stokes I, Q, U and V spectra of the pore can be seen in Figure 1 as a gray–scale plot.

4.   Analysis and Results

Our least square fitting routine applies a Levenberg-Marquardt algorithm to find the minimum in parameter space. To describe the atmosphere we choose a set of 8 parameters: four of them determine its thermodynamic structure (minimum temperature, temperature range, a parameter for the temperature gradient and the gas pressure at the deepest point), the other four the magnetic field (the field strength at $\tau = 1$, the field gradient, linear

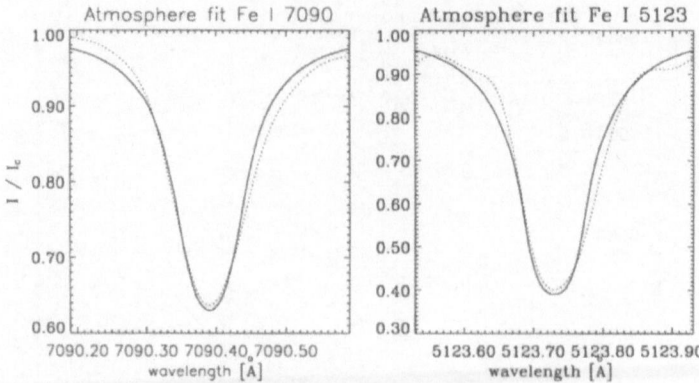

Fig. 3. Fits of the two $g = 0$ lines using the model atmosphere shown above in Figure 2. The dotted line is the measured profile, the solid one represents the best fit.

in $\log \tau$, the inclination and azimuth angle of the field, which are constant with depth in the atmosphere). For a given set of parameters the atmosphere is computed and the radiative transfer equation of the Stokes vector is solved using a code kindly provided by U. Grossmann–Doerth.

The applied inversion of the reduced spectra consists of three steps: First we determine the temperature- and pressure stratification of the pore by fitting the two non-splitting lines. Figure 2 shows the resulting model atmosphere together with a model of the quiet sun (VAL C, Vernazza *et al.*, 1976) and Figure 3 gives the resulting fit of the lines.

In the next step we applied this thermodynamic model to determine the parameters of the magnetic field from the polarization spectra of Fe I 630.2 nm. After test calculations we decided not to try a simultaneous fit of all 4 Stokes parameters. They were not recorded simultaneously and therefore do not show the same orientation of the field vector. Furthermore the accuracy of the azimuth measurement (derived from the linear polarization states) is much less than that of the inclination: The signal to noise ratio of Q and U is worse due to the small signal. In addition the effects of smearing and averaging over a range of magnetic field vectors (introduced both by the finite width of the spectrograph slit and image motion during the exposure) is larger for the azimuth than for the inclination.

It turned out that the fit of the gradient of the magnetic field is not possible when using only one spectral line. Although the gradient tends to broaden the wings of I and V and therefore should be easily detectable, the data are too noisy to give a reliable fit. Therefore we used a fixed increment of 400 Gauss per decade in the logarithmic optical depth scale as field gradient. For our atmosphere, this corresponds to approximately 3.5 Gauss/km.

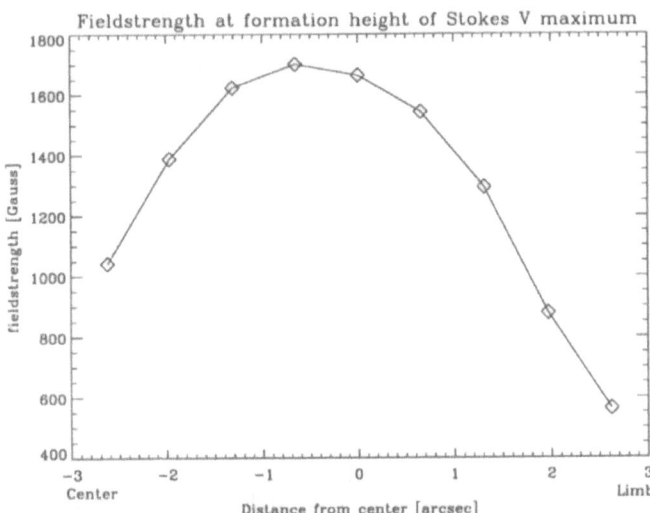

Fig. 4.   Field strength B (in Gauss) across the pore. The values were determined from a fit of Stokes I and V of the Fe I 6302.5 Å line. They refer to the height of formation of the Stokes V maximum, which lies around $\log\tau = -2$. The direction of disk center and limb are indicated at the bottom of the plot. A gradient of the field of 3.5 G/km was also included in the model. The size of the pore seen in Stokes V is larger than the size of the visible pore (defined by the points where 90 % of the photospheric intensity is reached) which extends $\pm 2''$ from the central position.

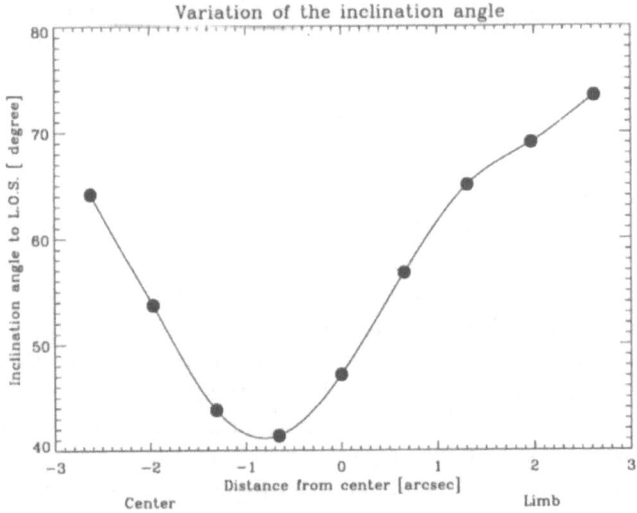

Fig. 5.   Inclination angle of the field (with respect to the line of sight) across the pore also determined from a fit to Stokes I and V.

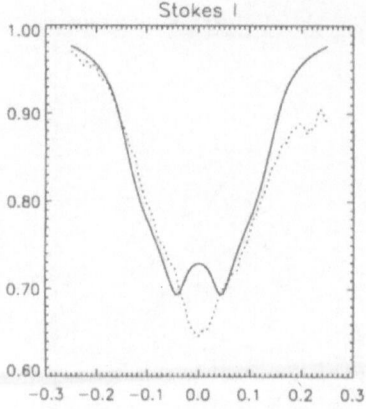
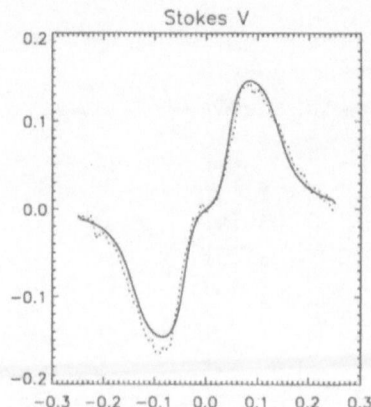

Fig. 6.   This figure shows a representative example of the fits to Stokes I and V (the dotted line is the measured profile, the solid one is the fit). The temperature stratification of Figure 2 was used. The field strength and inclination angle of the field were varied to give the best fit.

Fitting Stokes I and V together gives strength and inclination of the field: Figure 4 shows the results for the variation of the field strength across the pore, extrapolated to the height of formation of the 630.2 nm line (at about $\log \tau = -2$ for our atmosphere). The field strength at $\tau = 1$ is about 800 Gauss larger. Figure 5 shows the inversion results for the inclination of the field. The values are not corrected for the position of the pore on the disk, 30°. The fact that $\gamma = 0$ is reached nowhere in the pore is a hint that we did not cut through the center of the pore. An example of the line profiles is shown in Figure 6.

The determination of the azimuth angle of the field also suffers from the fact that the Q and U spectra could not be recorded simultaneously and therefore show different orientations. We decided to drop the Stokes Q measurement and to use only the U spectrum for the fit. In general it is not possible to measure the field azimuth using only one of the linear polarization signals. However, as we fixed all other parameters in earlier stages, there was only one parameter to fit. Adopting a smooth variation of the azimuth across the pore, we used for each spatial point the result of the preceeding fit as the initial value, starting the whole process at the point with maximum U signal. The results given in Figure 7 show a large scatter, the solid line represents a weighted linear regression. If only those points are used where a clear signal had been detected (marked by black dots) one can see the expected smooth variation of the azimuth of nearly 180°. Finally in Figure 8 one can find the complete set of Stokes U profiles and their fits. At

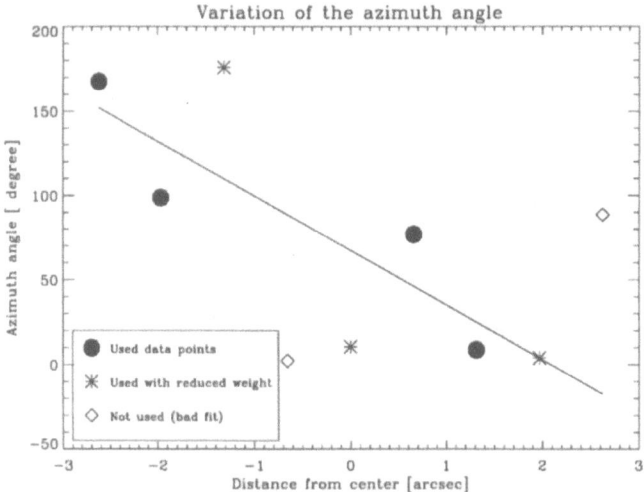

Fig. 7. The azimuth angle ϕ of the magnetic field (in degrees) across the pore is shown which is derived from a fit of Stokes U profiles. The straight line is a weighted linear regression through the points, the weight was selected according to the quality of the fit (see Figure 8).

two positions the U signal almost vanishes, indicating an azimuth of 0 or 90°. Due to the lack of a simultaneous measurement of Stokes Q the code can not distinguish between these two cases.

5. Conclusions and Outlook

This investigation shows that high spatial and spectral resolution polarimetry of small elements is possible and leads to good and consistent results. Additional efforts are required for the determination of the atmospheric straylight, which we plan to do by introducing a filling factor which is then also a parameter to be fitted, and using absolute intensity profiles. There is also a clear need to observe magnetically sensitive lines with different heights of formation simultaneously. In this case the gradient of the field strength can be determined with $< 20\%$ error, as tests with synthetic data have shown.

Acknowledgements

We wish to thank U. Grossmann-Doerth, who kindly supplied his radiative transfer code for the computation of the full Stokes vector, and S.K. Solanki

Stokes U amplitude

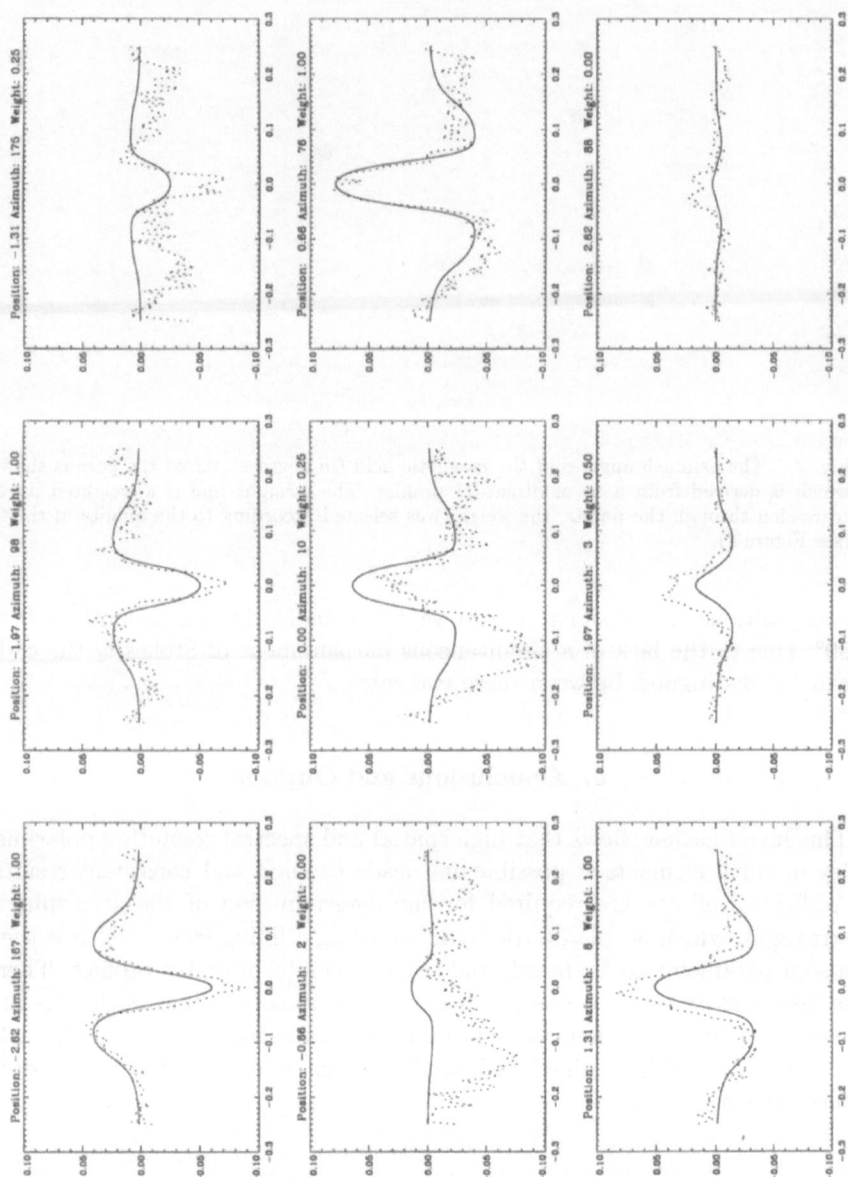

Fig. 8.   The complete set of Stokes U profiles of the pore. The dotted lines are the
measured spectra, the solid ones are calculated. The weight we used for the regression in
Figure 7 is indicated on each plot.

for providing his inversion code which inspired us with certain improvements on our code. And finally, cordial thanks go to the organizers of the St. Petersburg workshop for their help and support.

References

Beckers, J.M., Schröter, E.H.: 1968, *Sol. Phys.* **4**, 142
Bonet, J.A., Sobotka, M., Vásquez, M.: 1994, *Astron. Astrophys.* **296**, 241
Brants, J.J., Steenbeck, J.C.M.: 1985, *Sol. Phys.* **96**, 229
Brants, J.J., Zwaan, C.: 1982, *Sol. Phys.*80, 251
Bumba, V.: 1967, *Sol. Phys.* **1**, 371
Keller, C.U.: 1995, *Nature* **359** 307
Keller, C.U., Johannesson, A.: 1995, *Astron. Astrophys. Suppl.* **110**, 565
Muller, R.: 1992, in J.H.Thomas, N.O.Weiss (eds), *Sunspots: Theory and Observations*, Kluwer, p. 175
Muglach, K., Solanki, S.K., Livingston, W.C.: 1994, in R.J.Rutten, C.J.Schrijver (eds), *Solar Surface Magnetism*, Kluwer, p. 127
Rossbach, M, Schröter, E.H.: 1970, *Sol. Phys.* **12**, 95
Soltau, D.: 1993, Dissertation Universität Freiburg
Sütterlin, P.: 1991, Diplomarbeit Universität Freiburg
Sütterlin, P., Thim, F., Schröter, E.H.: 1994, in M.Schüssler, W.Schmidt (eds), *Solar Magnetic Fields*, Cambridge university Press, p. 213
Thim, F.: 1993, Diplomarbeit Universität Freiburg
Vernazza, J.E., Avrett, E.H., Loeser, R.: 1976, *Astrophys. J. Suppl.* **30**, 1

HIGH RESOLUTION POLARIMETRIC MEASUREMENTS
IN A SUNSPOT

T. HORN, A. HOFMANN

*Astrophysikalisches Institut Potsdam, Sonnenobservatorium Einsteinturm,
Telegraphenberg, D-14473 Potsdam, F.R.G.*

and

H. BALTHASAR

Kiepenheuer Institut für Sonnenphysik, Schöneckstrasse 6, D-79104 Freiburg, F.R.G.

Abstract. A Fabry-Perot interferometer is being used for two-dimensional spectropolarimetric measurements. We demonstrate the suitability of the setup for the measurement of the magnetic field and present some preliminary results from first observations.

Key words: Polarimetry – Sunspots

1. Introduction

Two methods are usually applied to obtain information on vector magnetic fields in the solar photosphere:

(i) magnetography: spectral pass-band averages of the Stokes parameters I, Q, U, and V are measured. Alternate I±V, I±Q, and I±U images are recorded by imaging filter systems and analyzed. This method allows the measurement of fast changes of the magnetic field very well, but consideration of the filling factor and the thermal and dynamic states of the solar atmosphere are insufficient. The advantages of this method were well demonstrated by the use of tunable filters at different telescopes (Tarbell et al., 1988; Tarbell et al., 1990; Keller et al., 1992). During two observational campaigns at the German Vacuum Tower Telescope (VTT) on Tenerife, the Lockheed tunable filter was used in combination with a beam-splitting polarimeter (Schmidt et al., 1992, Hofmann et al., 1994).

(ii) spectropolarimetry uses spectrally resolved observations of Stokes parameters and yields more accurate quantitative field information based on an analysis including the thermodynamic properties of the magnetized plasma. But, compared to magnetography, there are some limits concerning the recording of fast field changes and the uniform spatial imaging. When a spectrograph is used, only one spatial dimension can be recorded and the spectrograph slit has to be scanned over the active region. Substantial progress has been achieved with the Advanced Stokes Polarimeter (Lites et al., 1991). This polarimeter operates at the National Solar Observatory's Vacuum Tower Telescope at Sacramento Peak, New Mexico. The main parameters of this system are a spatial pixel sampling of about 0.4 arcsec and a

Solar Physics **164**: 321–332, 1996.

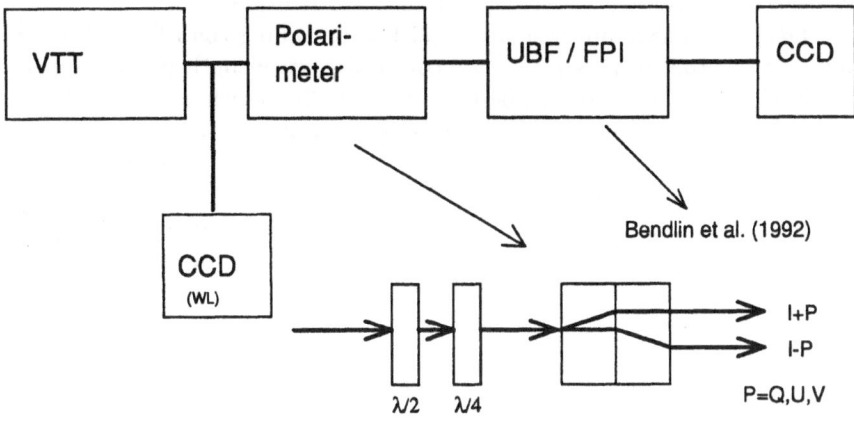

Fig. 1.   Schematic of the optical configuration. Abbreviations: VTT - Vacuum Tower Telescope, UBF - Universal Birefringent Filter (FWHM = 460 mÅ), FPI - Fabry Perot Interferometer, CCD (WL) - white light image CCD camera , CCD - narrow band filter image CCD camera

scan-time of about 10 min for a region of 100 x 100 $arcsec^2$. The polarimetric sensitivity lies between 10^{-4} and 10^{-3} of the continuum intensity. The magnetic field parameters are determined with a multi-parameter nonlinear least-squares fit to all four Stokes profiles.

In 1994 we performed observations at the German Vacuum Tower Telescope at the Observatorio del Teide (Tenerife) by using a full Stokes polarimeter in front of a two-dimensional imaging spectrometer with a narrow-band Fabry-Perot interferometer (FPI). Our goal was to perform magnetic field measurements of high spatial as well as spectral resolution and moderate temporal resolution. This paper deals with the basic setup, the data reduction, and some first observations using the system.

2. The Instrumentation and Observations

Figure 1 shows the basic optical setup of the observations. The spectrometer has been described by Bendlin et al. (1992) and Bendlin and Volkmer (1995). Its spectral resolution is about 3×10^5 (about 20 mÅ at FeI 6173.4 Å). The narrow-band filtergrams were recorded on a 286 x 384 pixel CCD camera with 0.2 arcsec pixels. The polarimeter consisted of rotating quarter and half wave plates in front of a polarizing beam-splitter (two crossed calcites). The polarimeter produced sequences of pairs of Stokes I±V, I±Q, and I±U images. This avoids the generation of artificial polarization signals within the individual Stokes pairs due to variable seeing. The resulting field of view for each of the two subimages was about 55 x 37 arcsec.

The spectral scanning was accomplished by tuning the FPI with a sampling rate of 10.9 mÅ. A total scan covered 45 wavelength points and took about 54 s, so that the recording time for a full Stokes set was around 160 s. Integral white-light images from the same field of view were taken strictly simultaneously to all narrow-band images by a secondary beam separated in front of the polarimeter.

The observations of the spectral line FeI 6173.4 Å presented in this paper were taken at the Vacuum Tower Telescope at Tenerife during an observing run on July 20, 1994. The particular spot NOAA 7757 was 30° outside of disc center ($\mu =0.89$).

3. Data Reduction and Calibration

3.1. DARK CURRENT, GAIN TABLE CORRECTION

Each observation consists of a series of scans through a Fraunhofer line obtained by tuning the FPI, i.e. the maximum transmission of the FPI is shifted to another wavelength for each image of a scan. We therefore obtain a data cube (a line scan) with 2 spatial dimensions and a spectral profile for each spatial pixel. For the data reduction, measurements of dark counts and a scan of flat-field images are needed, too. The flat-field scan was obtained by defocusing the telescope at disk center (besides the spot). The observed narrow-band filter profiles shows a combination of the observed Fraunhofer line and the passband of the UBF. To reconstruct the line profile, it is necessary to take a sample of the passband shape of the UBF too. This was done by sampling a continuum scan (a lamp with collimator optics).

The correction of the white-light images is straightforward, simply by subtracting an averaged dark image from each picture and dividing the result by the gain table obtained from the flat field images.

To correct the narrow-band filter images, the profile of each pixel must be divided by the passband shape of the UBF and all images of a scan have to be corrected for the flat-field. To do this, an artificial flat-field scan was generated. This is necessary because the observed flat-field scan contains both, the profiles of the UBF-passband and the Fraunhofer line. Therefore this flat-field scan is not suitable to reconstruct the spectra.

The reconstruction of the spectral data is shown in Figure 2. After the dark current reduction (step 1), 5 images of the continuum scan and the flat field scan with equal wavelength of the FPI transmission maximum were averaged and divided by each other. The obtained 'normalized' image is equivalent to a gain table (step 2). The artificial flat-field scan is the

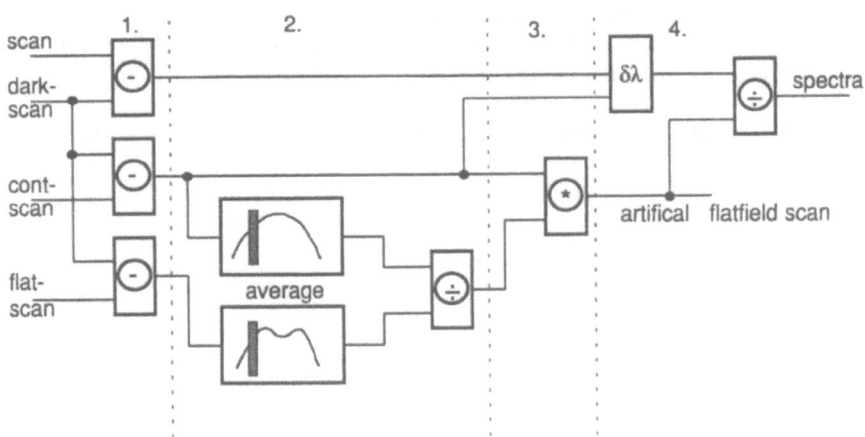

Fig. 2. Scheme for narrow-band data correction. The numbers at the top correspond to the different steps of the data reduction: 1. dark current correction, 2. calculation of a normalized picture by averaging 5 scan positions with equal λ, 3. generation of an artificial flat field scan, 4. reconstruction of spectra

product of the images of the continuum scan with the normalized image (step 3). The profiles of all pixels of the artificial scan are combined in an averaged profile. This has to be done with the profiles of the line scan, too. The ratio of the resulting averaged profiles shows an averaged profile of the Fraunhofer line in the observed solar region. The averaged artificial and the averaged observed line profiles often need to be shifted with respect to each other to correct for drifts of the passband position of the UBF. The criterion for the correct fitting of these two profiles were equal intensities of the continua near the red and blue wings of the observed Fraunhofer line in the reconstructed averaged line profile.

One feature of the FPI in the pupil plane is that the maximum transmission wavelength depends also on the angle between the optical axis and the picture point:

$$\Delta\lambda = \lambda - \lambda_0 = -\frac{\lambda_0}{2}\left(\Theta\frac{f_{VTT}}{f_3}\right)^2$$

(λ_0 = unshifted wavelength, beam perpendicular to FPI plates; Θ = angle of pixel to optical axis; f_3 = focus length of collimator lens before FPI = 2.25 m.; f_{VTT} = focus length of telescope = 46 m). The consequence is, that pixels far from the optical axis have a maximum transmission which is blue shifted relatively to pixels near the optical axis. To correct for this effect, the function $\Delta\lambda = f(x,y)$ was obtained by analyzing the shifts of the passband shape of the UBF (continuum scan) in the individual pixels. These individual shifts, together with a wavelength shift of the whole scan obtained by analyzing the averaged profiles (see above), are applied to the line scans, and then the scans are divided by the artificial scan to reconstruct the Fraunhofer line of each individual spatial pixel (step 4).

3.2. OPTICAL DISTORTIONS AND SEEING INDUCED DISPLACEMENT

The two images of the polarizing beam-splitter are transmitted through different optical paths behind the polarimeter. Therefore optical distortions might be different and generate artificial polarization signals. We used measurements with a high contrast grid target to eliminate these residual distortions by cross correlation techniques.

In a similar way the displacements and distortions between the single images of a spectral scan caused by seeing were eliminated by using the series of simultaneously recorded white light pictures as references.

3.3. INSTRUMENTAL POLARIZATION

The oblique reflections at various mirrors, the windows of the telescope, and the polarimeter itself cause a crosstalk between different Stokes parameters. To eliminate such distortions we placed large linear or circular polarizers in front of the first coelostat and measured the Müller matrix (instrumental response) of the whole equipment.

A typical example of such a matrix at the time of our observation is

$$M = \begin{bmatrix} 1.000 & -.517 & .035 & *** \\ -.375 & 1.030 & .041 & -.029 \\ .044 & -.036 & 1.015 & -.386 \\ .047 & .003 & .230 & .893 \end{bmatrix}$$

This example shows large crosstalk between Stokes U and V (M_{43} and M_{34}), which can lead to substantial distortions of the V or U spectra in the umbra and penumbra, respectively. We made these measurements over a whole day to obtain the temporal variations of each matrix element as a function of the hour angle of the telescope (see Figure 3). In this paper we analyze Stokes I±V frames only, so that the correction was restricted to the use of the first and last elements of the last line (M_{41} and M_{44}).

4. First Analysis of Stokes V

During the observational program, we took series of full Stokes frames as well as of Stokes I±V frames, only. We present some results of the first analysis of the Stokes V recordings.

Figure 4 shows cospatial images of continuum, Stokes I, and Stokes V. In the umbra and the center side (left-hand top) of the penumbra, Stokes V

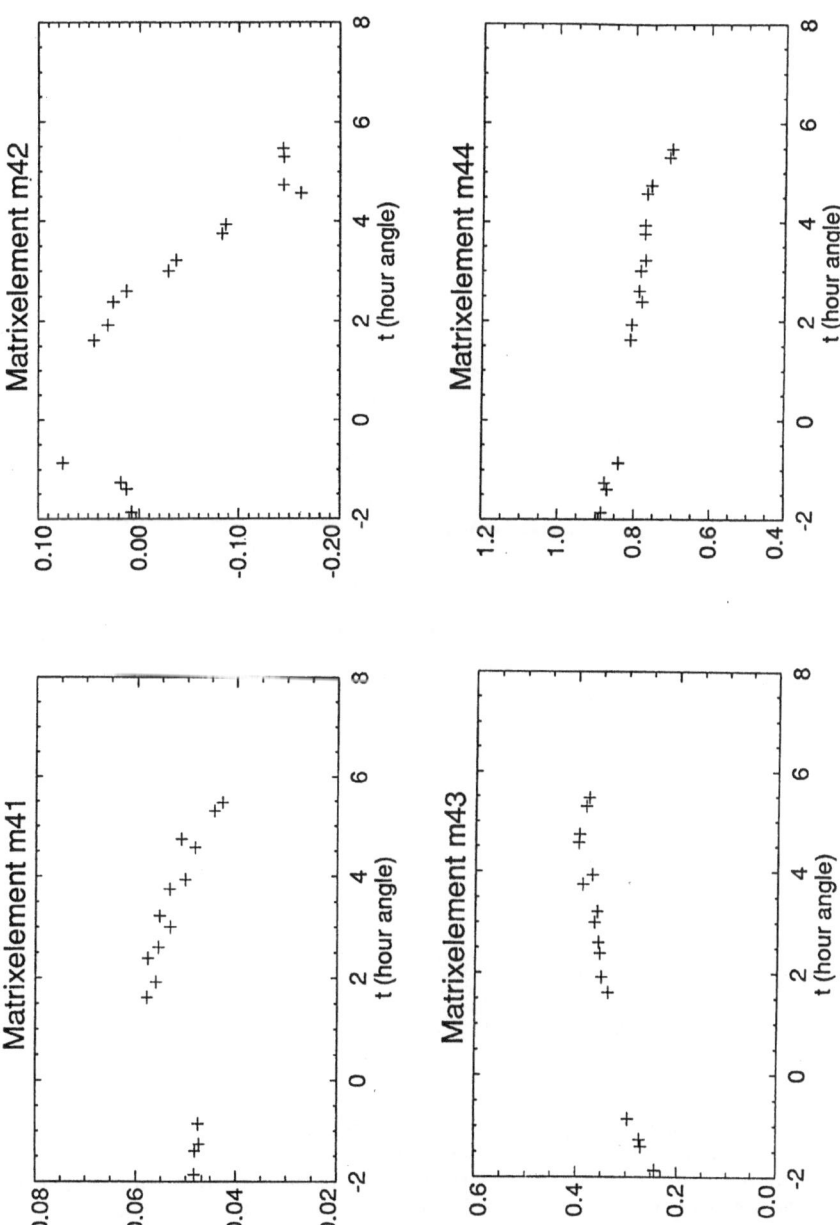

Fig. 3.   Variation of the V-related elements of the Müller matrix of the telescope and the polarimeter.

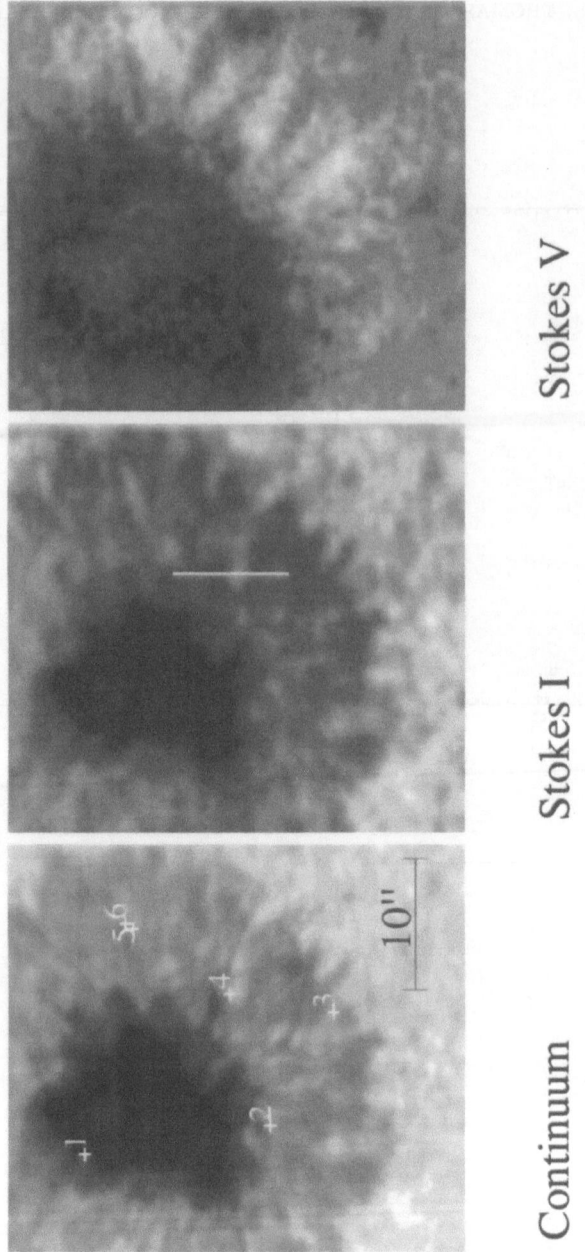

Fig. 4.   Part of the sunspot NOAA 7757 on July 20, 1994. The I and V images are narrow-band filtergrams (30 mÅ) in the red wing at about 130 mÅ from the center of the Fe I 6173 Å line. The numbers mark the positions of the spectra shown in Figure 5. The bright bar marks the path of the profiles of Fig. 7. In the Stokes V image: dark= + and bright= - .

shows a positive sign or a left-hand circular polarization (dark in the image). This means that the magnetic field vectors are inclined by more than 90° to the spot normal. The most striking features of Stokes V are the radial filamentary structures best visible at the limb side (right-hand bottom) of the spot. The bright filaments are negative, i.e. of opposite sign to their dark neighbors and to the umbra. This indicates that the inclination of the bright (dark) V structures is smaller (larger) than 90° to the line-of-sight and less (more) than 60° to the spot normal. This confirms the results of numerous authors (Beckers and Schröter, 1969, Title et al., 1993, Degenhardt and Wiehr, 1991, Kalman, 1991, and Schmidt et al., 1992) who found hints of differently inclined structures in the penumbra.

At first view there seems to be a correlation between dark and bright penumbral intensity features and variations of V in some cases, but in other cases it cannot be clearly seen. Variations of the circular polarization at a fixed wavelength position can also be produced by changes of the spectral line profile in the filaments due to Doppler shifts and their variations with temperature and height of line formation. It requires a more thorough analysis to separate these influences from those of the magnetic field.

In Figure 5 we show some Stokes I and Stokes V spectra at different spatial positions as constructed from a spectral scan. The positions are marked in the continuum image. At the center-side umbra (Pos.1) the line is clearly split and a pure V profile is observed. From this splitting we obtain a field strength of about 2075±240 G. The line is still split (B=1590±185G) at the umbral-penumbral boundary at the limb side (Pos.2) .

An influence of cross-talk between linear (Stokes Q and U) and circular (Stokes V) polarization is evident at Pos.5 and 6. There the penumbral filaments are nearly perpendicular to the line-of-sight, and the spectra show distinct signatures of the transverse Zeeman effect (symmetric Q and U profiles) and represent more or less the strong cross-talk (25 percent) from U into V. This effect is more obvious in Pos.6 (two minima symmetrical to the maximum) than at Pos.5 (non-symmetric maximum shift to the I minimum) where we find some V influence. The two positions have a separation of only 1.4", but Pos.5 is located in a dark and Pos.6 in a bright filament, respectively. This variation of Stokes V indicates changes of the field inclination over small distances, similar to those found by Degenhardt and Wiehr (1991).

Spectra 3 and 4 are penumbral positions close to the center direction and show distinct V signatures. At Pos. 3, Stokes V already reverses its sign due to projection effects.

Figure 6 shows a profile of the relative intensity (quiet sun = 1) and the field strengths for CCD column 55 (close to Pos.2). The field strengths were determined by fitting synthetic line profiles to the observed profiles as described in Balthasar and Schmidt (1993). The right-hand sides of the profiles

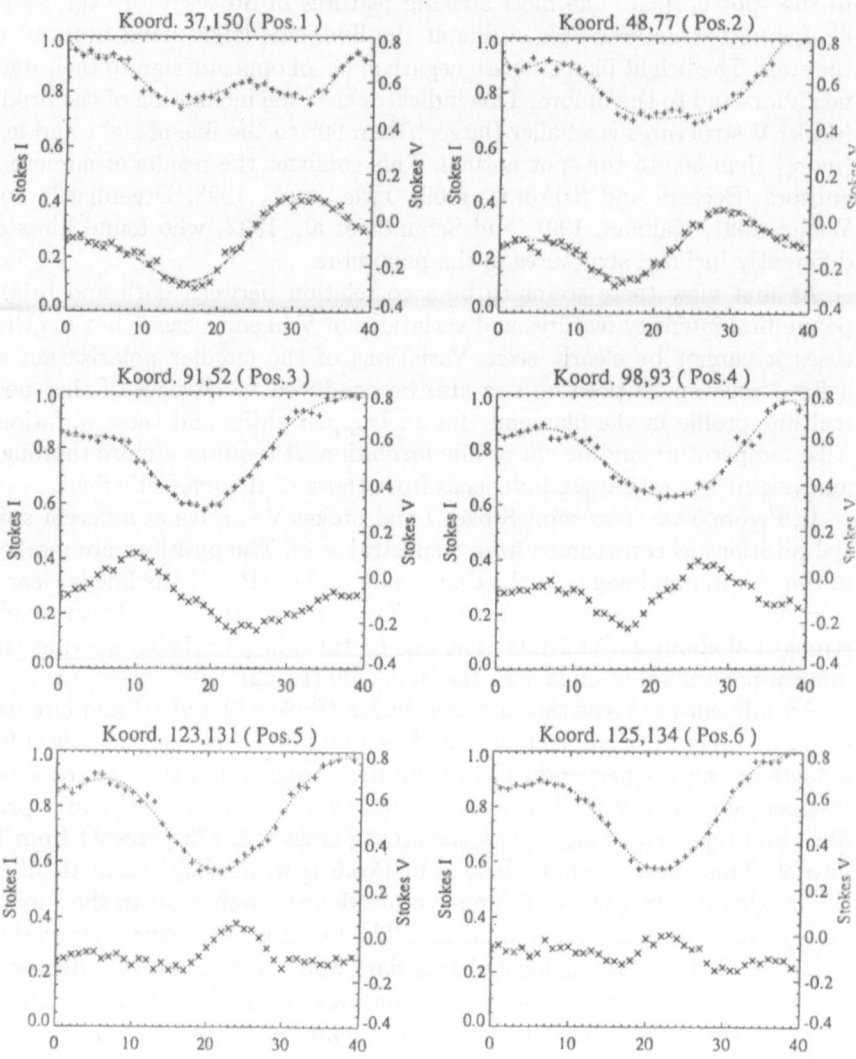

Fig. 5. Spectra for selected points. The symbols: x = Stokes V, + = Stokes I. The abscissae are expressed in wavelength steps of 10.9mÅ

include parts of the penumbra that are not shown at the top of the images in Figure 4. Figure 7 displays a vertical cut through a large bright structure in the penumbra at position 4. The cut is indicated as a bright bar in the

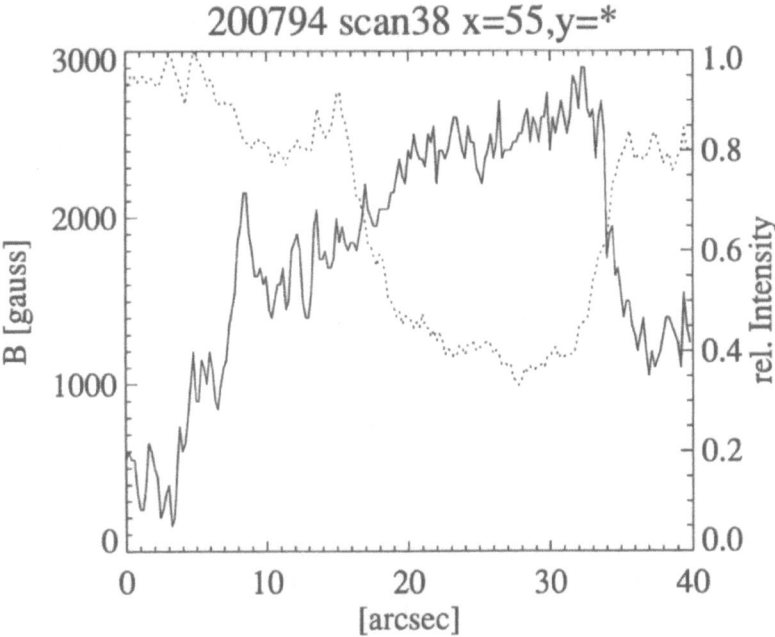

Fig. 6.   Profiles of field strength (solid line) and relative intensity (dotted line).

Stokes I image in Figure 4. In the bright structure the field strength drops by about 500 G. Assuming that bright and dark penumbral filaments differ more or less only by their inclination as argued by Schmidt et al. (1992), this structure should not be a 'normal' bright penumbral filament. This might be the signature of underlying photospheric matter visible through a gap in the penumbra.

5. Remarks and Future Goals

This first data analysis shows that the FPI setup is well suited for spectropolarimetric observations with high spectral, spatial, and moderate temporal resolution. The next step will be the analysis of full Stokes frames to enable the full correction for instrumental polarization. This will form the basis for a more detailed quantitative analysis of the spectra by fitting synthetic line profiles to the observed ones.

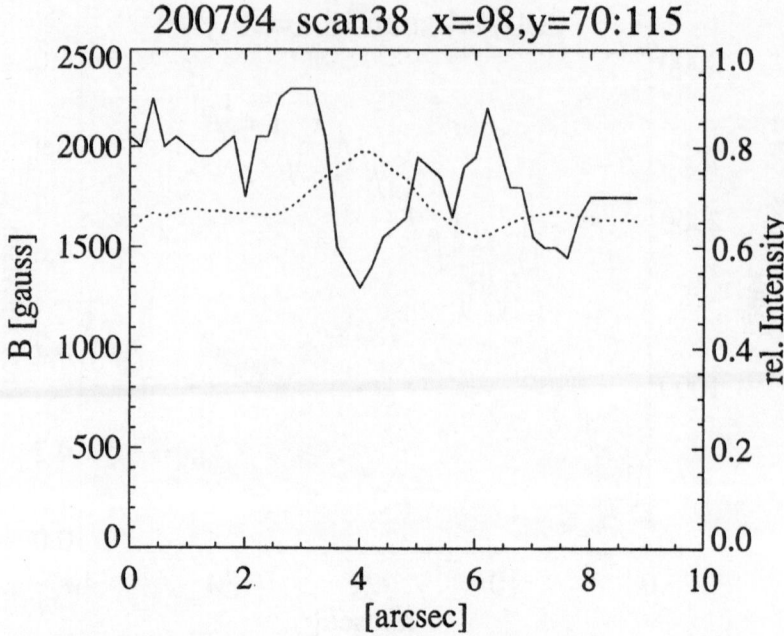

Fig. 7. Profiles of field strength (solid line) and relative intensity (dotted line) across a bright structure in the penumbra

Acknowledgements

The spectrometer was built and tested by Dr. C.Bendlin and Dr. R.Volkmer from the Universitätssternwarte Göttingen under support of the DFG (Grant Nos. Kn 152/4-1, Kn 152/4-2, Kn 152/9-1, Kn 152/9-2). We are greatly indebted to them for their generous introduction in operating the spectrometer. We thank Dr. Ch.Keller for a lot of helpful comments and suggestions which have considerably improved the paper. T.H. acknowledges travel support from the DFG (DFG-Grant No. Sta-351/3-1). The Vacuum Tower Telescope is operated by the Kiepenheuer-Institut für Sonnenphysik Freiburg at the Observatorio del Teide/Tenerife of the Instituto de Astrofísica de Canarias.

References

Balthasar, H. and Schmidt, W.: 1993, *Astron. Astrophys.* **279**, 243
Beckers, J.M. and Schröter, E.H.: 1969, *Solar Phys.* **10**, 384
Bendlin, C. and Volkmer, R.: 1995, *Astron. Astrophys. Suppl.* **112**, 371
Bendlin, C. Volkmer, R., and Kneer, F.: 1992, *Astron. Astrophys.* **257**, 817
Degenhardt, D. and Wiehr, E.: 1991, *Astron.Astrophys.* **252**, 821

Hofmann, A., Shine, R., Frank, Z., Schmidt, W., and Balthasar, H.: 1994 in M. Schüssler and W. Schmidt (eds.), *Solar Magnetic Fields*, Cambridge U. Press, Cambridge, p. 204

Kalman,B.: 1991, *Solar Phys.* **135**, 299

Keller, C.U., Stenflo, J.O., and von der Lühe, O.: 1992, *Astron. Astrophys.* **254**, 355

Lites, B.,Elmore, D.,Murphy, G.,Skumanich, A.,Tomczyk, S., and Dunn, R.: 1991 in L.November (ed.), *Solar Polarimetry*, 11th NSO/Sacramento Peak Summer Workshop, Sunspot, New Mexico, p.3

Schmidt, W., Hofmann, A., Balthasar, H., Tarbell, T., and Frank, Z.: 1992, *Astron. Astrophys.* **264**, L27

Tarbell, T., Topka, K., Ferguson, S., Frank, Z., and Title, A.: 1988 in O.von der Lühe (ed.), *High Spatial Resolution Solar Observations*, 10th Sacramento Peak Summer Workshop, Sunspot, New Mexico, p.506

Tarbell, T., Ferguson, S., Frank, Z., Shine, R., Title, A., Topka, K., and Scharmer, G.: 1990, in J.O.Stenflo (ed.), *Solar Photosphere: Structure, Convection and Magnetic Fields*, IAU-Symp. 138, 147

Title, A.M., Frank, Z.A., Shine, R.A., Tarbell, T.D., Topka, K.P., Scharmer, G., and Schmidt, W.: 1993, *Ap.J.* **403**, 780

MAGNETIC FIELDS OF SUNSPOTS
BASED ON COMBINED OPTICAL AND RADIO
OBSERVATIONS

V.E. ABRAMOV-MAKSIMOV, G.F. VYALSHIN and G.B. GELFREIKH

Main (Pulkovo) Astronomical Observatory RAS
St.-Petersburg 196140, Russia

and

V.I. SHATILOV

Special Astrophysical Observatory RAS
St.-Petersburg 196140, Russia

Abstract. In the present paper we present the results of measurement of magnetic fields in some sunspots at different heights in the solar atmosphere, based on simultaneous optical and radio measurements. The optical measurements were made by traditional photographic spectral observations of Zeeman splitting in a number of spectral lines originating at different heights in the solar photosphere and chromosphere. Radio observations of the spectra and polarization of the sunspot - associated sources were made in the wavelength range of 2–4 cm using large reflector-type radio telescope RATAN-600. The magnetic field penetrating the hot regions of the solar atmosphere were found from the shortest wavelength of generation of thermal cyclotron emission (presumably in the third harmonic of electron gyrofrequency). For all the eight cases under consideration we have found that magnetic field first drops with height, increases from the photosphere to lower chromosphere, and then decreases again as we proceed to higher chromosphere and chromosphere-corona transition region. Radio measurements were found to be well correlated with optical measurements of magnetic fields for the same sunspot. An alternative interpretation implies that different lines used for magnetic field measurements refer to different locations on the solar surface. If this is the case, then the inversion in vertical gradients of magnetic fields may not exist above the sunspots. Possible sources of systematic and random errors are also discussed.

Key words: Sunspots – Magnetic fields – Optical and Radio Observations

1. Introduction

Detection of strongly polarized sources observed above sunspots opened the way for experimental study of their magnetic fields at different levels of the corona and corona-chromosphere transition region. The first solar eclipse observations have shown that, this type of sources are situated at lower heights (not more than say 3000 km above the photosphere), and have sizes comparable with that of umbra at a wavelength of around 3 cm. That implies penetration of strong magnetic fields into the solar corona without large widening of the magnetic tube of a sunspot. In other words, it may be interpreted as "descending" of the corona above a sunspot (see Livshits *et al.*, 1967).

Solar Physics **164**: 333–343, 1996.

In the early sixties Zheleznyakov and Kakinuma proposed the interpretation of these sources as thermal electron emission of the corona predominantly in the second and third harmonics of their gyrofrequency (see Zheleznyakov, 1970). Model computations confirmed agreement of observed parameters of the sources with those expected from the model computations. It was found that emission at the second and the third harmonics of gyrofrequency could be distinguished if we have got polarization measurements, because extraordinary mode is normally dominant at the third harmonic, while the ordinary mode is dominant at the second harmonic (both the modes are optically thick in this case). The two modes differ by the sign of circular polarization.

The results above allow to develop the methods of measuring the magnetic fields in the region of generation of cyclotron emission above sunspots. It was found first that the magnetic field in the corona above sunspots exceeds usually 1000 G (unexpectedly high value) and the corona begins at the height of few thousands km above sunspots (unexpectedly low value). Further development in the techniques of observations with a combination of high spatial resolution, and detailed spectral analysis at cm-wavelengths showed that in fact the coronal magnetic fields are still higher and normally amount to about 80% of the maximum photospheric values. This result was obtained from observations using the radio telescope RATAN-600 (see Akhmedov et al., 1982). Magnetic fields in the corona, exceeding 2000 G are found to be quite normal.

Magnetic field mapping in three dimensions have also been done by Lee et al. (1993) and by Gary and Hurford (1994), using methods that are similar to those of the present paper.

Of special interest were cooperative programs, where magnetic fields of some chosen sunspots were measured using optical methods, based on Zeeman splitting in several lines originating at different heights of the solar atmosphere. Two experiments based on the RATAN-600 observations (see Gelfreikh et al., 1981; Bogod et al., 1982) showed rather complicated distribution of magnetic field with height as follows from optical observations. At the same time good correlation of magnetic field strength found from optical and radio methods was confirmed. It was clear that further experiments of this kind are highly desirable both to improve models of magnetic field structure above sunspots, and to find the nature of discrepancies between different cases. Such a program was performed in August 1990 using coordinated observations with the RATAN-600 and optical data obtained with the horizontal optical telescope in Pulkovo observatory. The results are presented below.

TABLE I
The chosen sunspot groups

TABLE I

The chosen sunspot groups

Date	Number of the group	Sunspot area (m.h.)	Area of the main spot
14.08.1990	298	318	92
14.08.1990	303	225	171
15.08.1990	297	206	152
15.08.1990	298	664	230
15.08.1990	303	260	183
16.08.1990	297	153	128
16.08.1990	303	273	169
17.08.1990	303	250	179

2. Sunspots investigated

For study we have chosen three sunspot groups which were observed in the period from August 14 to 17, 1990. For these we had the optical and radio observations of good quality made on the same dates. So the time difference between optical and radio measurements did not exceed three hours. Table I contains the list of chosen sunspot groups and their parameters. For optical measurements we preferably choose the sunspots of simple structure with one umbra, or two umbrae comparable in size.

3. Optical Measurements of the Magnetic Fields

The measurements of magnetic fields of sunspots were carried out by photographic method from the observations made with horizontal solar telescope ACU-5 in Pulkovo observatory. To measure magnetic fields of the solar atmosphere at different heights — from lower photosphere to middle chromosphere we have used eight different spectral lines. The basic parameters of these lines are given in Table II with the effective heights of their formation according Koval and Stepanyan (1983). In Table II the following quantities are tabulated: the wavelength in Angstrom units, effective height of formation in km, the splitting factor (which is the product of Lande factor and the square of wavelength in cm multiplied by 10), and the intensity in relative units.

For each sunspot and each date we constructed dependence of the magnetic field strength on height. Then cross-correlation between different curves for the same sunspot was calculated. All the curves were compared with one which showed the best correlation with the others and in this way the errors

TABLE II

Spectral lines used in measuring magnetic fields of sunspots

Element	Wavelength	Effective height	Factor of splitting	Intensity
Ca I	6102.727	1200	74.0	25
Ni I	6108.125	400	40.4	8
Ni I	6111.078	60	46.6	1
V I	6119.532	400	41.6	8
Ni I	6128.984	120	56.5	2
Fe I	6137.002	200	75.3	4
Ba II	6141.727	630	41.4	12
Fe I	6151.623	290	69.2	6

were estimated. The errors are in the range of 60 to 240 G for an accuracy of 90%, and are in the range 70 to 300 G for an accuracy of 95%.

The correctness in the estimation of line formation heights employed is important for our studies. The estimates from model computations are dependent on the model atmospheres used. So, we prefer to employ the heights presented in Krat and Vyalshin (1965), Mattig (1969), Huseynov (1978), and Koval and Stepanyan (1983). In these papers, the heights were found from observations of the line shifts with respect to continuum, when a sunspot is near the limb (the lines in sunspots are formed in a layer above the continuum forming layers).

Intensity of the line depends on the height of line formation. To estimate errors in the height of line formation, we have used Figures 4 and 5 of Mattig (1969), where the measured heights for 60 different lines of sunspot spectra are presented as a function of Roland's height. These heights are shown in the present paper in Figure 3. Numbers of the sunspot groups used in this work are taken from Bogod et al. (1982).

4. Radio Measurements of Magnetic Fields

In the present paper we use the method of measuring the magnetic field above a sunspot from the shortest wavelength where one can trace the presence of the thermal gyroresonance (cyclotron) emission, generated by electrons of the corona, or corona-chromosphere transition region (see Akhmedov et al., 1982; Zheleznyakov, 1970). The separation of the sunspot-associated component generated by cyclotron mechanism was based on the following characteristics: (i) high degree of circular polarization; (ii) agreement of co-ordinates with sunspot positions; and (iii) small angular size – of the order of corresponding sunspot.

The radio method of measuring the magnetic field of sunspots (see Akhmedov et al., 1982), is based on the interpretation of radio emission at short centimeter wavelengths as cyclotron emission in the first harmonic of the electron gyrofrequency, generated by hot ($T \geq 10^5$ K) thermal plasma in strong ($B \geq 10^3$ G) magnetic field of a sunspot. According to the model computations and numerous observations, the radio sources connected with sunspots are associated with steep (growing with λ) spectral emission in the wavelength range of 2 - 4 cm, and high degree of polarization. Generally the positive sign of circular polarization corresponds to an excess of e-mode over the o-mode. This property can be employed in accurate measurement of the shortest wavelength λ' where this type of emission occurs.

From the theory of generation of cyclotron emission and using the known parameters of solar atmosphere, we conclude that high optical thickness in the third harmonic of gyrofrequency is essential for generation of this strongly polarized emission. This condition may be represented as

$$B(G) = \frac{3570}{\lambda'(cm)}. \tag{1}$$

To model the observed emission, we need strong magnetic field as well as high temperature of the plasma. For such a choice of parameters, Equation (1) gives an estimate of the highest value of the magnetic field penetrating the hot layer of the solar atmosphere above the spot. We consider it as the magnetic field at the "base" of corona, probably in the transition region.

In this study we have used observations of solar spectral-polarization complex made with the radio telescope RATAN-600 (see Bogod et al., 1985). The antenna was used in a standard mode of the transit instrument: south sector of the main mirror plus a flat periscopic mirror. Both intensity (I) and circular polarization (V-Stokes parameter) were registered. Table III presents wavelength dependence of the beam sizes of instrument under consideration (see Bogod et al., 1992). For calibration of the scans, we have assumed typical values of the antenna temperature for the quiet sun.

There were many active regions on the sun during the period of observations. Thus in the fan-beam diagram, sometimes several sources were registered simultaneously. Nevertheless, sunspot-associated sources were not very difficult to distinguish due to their typical spectra and high degree of polarization. So we do not expect significant errors caused by this interference, with the results of the spectrum measurements. Figure 1 illustrates an example of the scan of the solar disk with radio telescope RATAN-600 at several wavelengths, and identification of the sources with the sunspots.

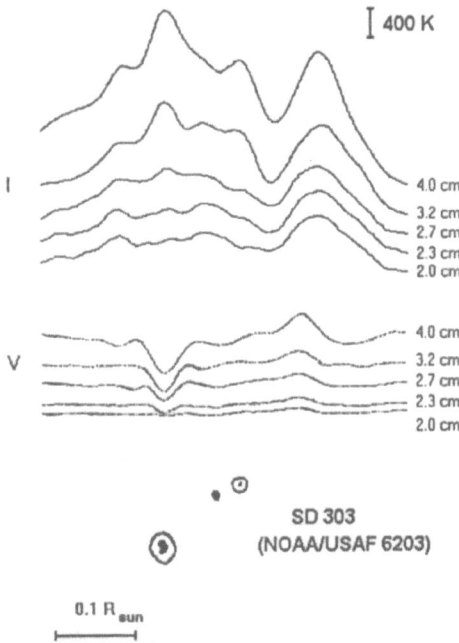

Fig. 1.   Scans of the solar disk obtained with RATAN-600 in intensity (I) and circular polarization (V). The photospheric map with a picture of sunspots is also shown for the date of observations – August 15, 1990.

TABLE III

The beam sizes of RATAN-600

Wavelength (cm)	E-W (arcsec)	N-S (arcmin)
2.0	18	12.5
2.3	21	14.4
2.7	25	16.9
3.2	29	20.0
4.0	36	25.0

Determination of the magnetic field strength is based on the curves in Figure 2, which are actually used to find the shortest wavelength of emission, generated by cyclotron mechanism.

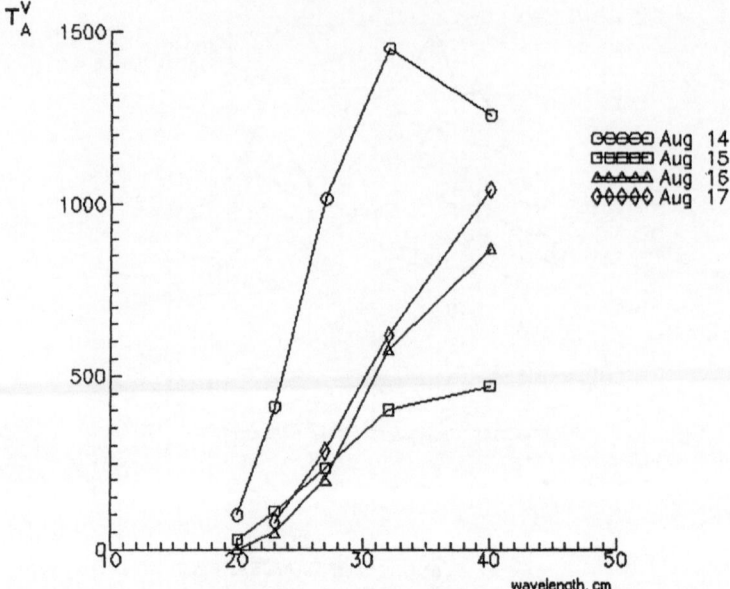

Fig. 2.   Dependence of the intensity of emission of the sunspot-associated sources on the wavelength used to find the maximum magnetic field strength at the base of solar corona.

5.   The Height Distribution of Magnetic Fields

Combined data of optical and radio observations, presented a unique opportunity to follow the variation of magnetic field strength above the sunspots, within a wide range of temperatures of the solar atmosphere – from the photosphere upto corona. The results of such a study for a typical case are shown in Table IV and in the Figure 3. This Figure shows dependence of the maximum value of magnetic field strength on the height of formation of a particular line. For magnetic fields responsible for radio emission, we have used a rather typical value of 2000 km –representing an estimate from both telescopic and eclipse observations. Any way it represents the magnetic field of hot plasma at the base of corona, as follows from brightness temperatures of the radio sources.

When we constructed these functions using the data of the Table II and I, we have found very similar distributions of the magnetic fields with height, for all sunspots under consideration. In all the cases the maximum strength was found at heights of 400 to 700 km, while at heights of 300-400 km, in most cases one could notice a drop in the magnetic field strength.

An important feature of all the cases was a good correlation of magnetic fields found from optical and radio measurements. This situation is rather remarkable, if one takes into account the principal difference in methods

TABLE IV

Combined data of optical and radio observations of magnetic fields

Group	Date	60	120	200	290	400 Ni I	400 V I	630	1200	2000
297	15.08	2200	1800	2200	1900	2900	2600	3100	2400	1900
	16.08	1900	1600	1800	1900	2300	2000	3400	2400	2000
298	14.08	1800	1800	1700	900	2300	1900	2500	1900	1900
	15.08	2800	2200	2400	2500	3200	2800	3800	2500	1700
303	14.08	2900	2200	2300	2500	3700	3600	3400	2600	1800
	15.08	2300	2200	2200	2400	3400	3500	3400	2500	1900
	16.08	2500	2000	2300	2300	3300	3100	3500	2600	1600
	17.08	2700	2100	2200	2200	2300	2600	3400	2500	1600

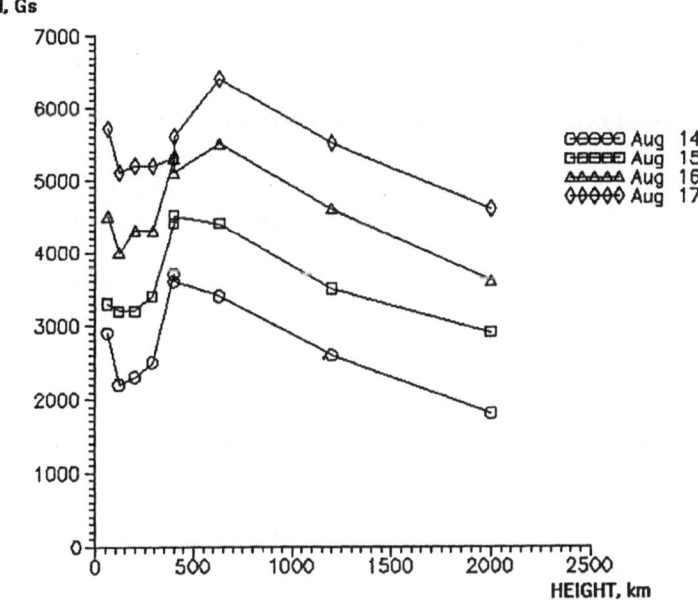

Fig. 3. The maximum value of magnetic field strength as a function of height above the photosphere.

used in the two types of observations (the Zeeman splitting and gyroresonance emission). This essentially is an argument in favour of the simultaneous observations using these two independent methods. At the same time, the dependence of the magnetic field strength on height, is adopted from independent investigations, and should be chosen with some care.

6. Discussion

Detection and analysis of the cyclotron sources above sunspots, opened new vistas to study magnetic fields of sunspots in the hot layers of the solar atmosphere – in the corona and chromosphere-corona transition regions. An unexpected result of such an analysis was detection of very strong coronal magnetic fields, comparable with the field strengths at the photospheric level. In most cases, strength of the magnetic field at the base of corona was only 15-20% lower than those at the photosphere, the latter found by routine methods from Zeeman splitting of optical spectral lines. An interpretation of this effect was given by Livshits *et al.* (1967) in terms of a "descending" corona above a sunspot, due to additional flux of energy in some types of MHD-waves.

Cooperative observational programs of radio and optical measurements of sunspot magnetic fields appears promising. The magnetic fields are measured at several heights using optical observations in different spectral lines. The radio observations give a new basis to study peculiarities of the solar atmosphere above sunspots, especially the magnetic field distribution. A limited number of such programs have been conducted so far, but satisfactory model interpretation of the results is lacking.

Earlier, a comparison of the radio measurements of magnetic fields in some sunspots was made, with EUV and optical line measuremsnts (see Abramov-Maksimov *et al.*, 1983; Gelfreikh *et al.*, 1988). In that case, optical and radio data were found to be in good agreement as usual, but the EUV data (from SMM - satellite observations) gave much lower values for the magnetic field strengths. An interpretation of this difference was given by Obridko (1979), through model computations on the basis of an inhomogenious solar atmosphere above a sunspot. In this case, the radio and EUV measurements in fact correspond to different parts of the solar atmosphere above the spot, with different heights of generation of the observed emission. This resulted in differences with measured magnetic fields. These models also implied that, radio measurements correspond to lower and hotter layers of the solar atmosphere. Besides, a couple of observational programs similar to the one above, included magnetic field measurements at different heights, using several optical lines, and comparing them with radio data. For instance, Bogod *et al.* (1982) derive quite different height distributions for three sunspots. Gelfreikh *et al.* (1981) present results that have many features similar to the present work.

If we assume that the curves obtained by us reflect the actual distribution of the magnetic field with height, we come to the conclusion that somewhere in the upper photosphere the magnetic field strength first declines and then increases with height leading to a local maximum in the lower chromosphere. This implies the presence of local currents in the observed layers of the solar atmosphere with rather fine structure (few hundreds of km). This characteristic length scale is comparable with characteristic heights in the photosphere and chromosphere. So the cross-section of the magnetic tube of the sunspot may be affected by differences in gas pressure distribution inside and outside the magnetic structure of the sunspot. The necessary explanation for this behaviour in terms of local currents is yet to be found.

The small-scale structures of the solar atmosphere above sunspots, affect the heights of line formation. Thus different lines may in fact be formed over different areas of the solar surface. To examine this alternative interpretation, an analysis similar to that of the EUV line of C IV needs to be performed.

7. Conclusions

We would like to draw attention to the problem of structure of the magnetic fields and other parameters of the solar atmosphere above sunspots, as they can be derived from combined optical and radio observations. Of great interest, are simultaneous observational data in short wavelength bands – EUV and possibly X-rays. As we have illustrated above, combination of even radio and optical measurements lead to problems which are difficult to resolve without consideration of other parameters of the solar atmosphere. Further progress in this direction needs cooperative programs involving radio observations (of spectral and polarization characteristics of sunspot-associated sources with high spatial resolution, using large radio telescopes), and optical observations (in a set of different spectral lines).

From the point of view of model computations, it appears that further development of inhomogenious models of the solar atmosphere above sunspots is necessary to account for the observed data. One should keep in mind the essential differences in the structure of the solar atmosphere above a sunspot and the surrounding plasma. Note that it is possible to resolve the discrepancy found earlier between radio and EUV measurements of the magnetic fields at the base of corona, in terms of a multi-component model of the solar atmosphere (cf. Obridko, 1979).

8. Acknowledgments

The authors feel happy to express their thanks to a number of Russian and international grants which made this work possible. Partial support for this investigation was provided by the state scientific program of Russia ("Astronomy" 4-24 and 4-110); Russian foundation of fundamental studies (94-02-06510 and 94-02-06508); International science foundation (NS800 and NVH000); and also the Russian Government (NS8300 and NVH300).

References

Abramov-Maksimov, V.E., Gelfreikh, G.B.: 1983, *Sov. Astron. Lett.* **9**, 132 (=*Pisma Astron. Zh.* **9**, No. 4, 244)

Akhmedov, Sh.B., Gelfreikh, G.B., Bogod, V.M., Korzhavin, A.N.: 1982, *Solar Phys.* **79**, 41

Bogod, V.M., Boldyrev, S.I., Zueva, V.A., Korzhavin, A.N., Petrov, Z.E., Plotnikov, V.M., Shatilov, V.A.: 1992, *Materials of the World Data Center B. Solar radio observations with the RATAN-600 radiotelescope in the wavelength range of 0.8–31.6 cm during the year 1984* , Moscow

Bogod, V.M., Gelfreikh, G.B., Petrov, Z.E.: 1985, *Astrophys.Issled. (Izv. SAO.)* **20**, 102

Bogod, V.M., Vyalshin, G.F., Gelfreikh, G.B., Petrova, N.S.: 1982, *Solnechnie Dannie.* **1**, 104

Gary, D.F., Hurford, G.J.: 1994, *Astrophys. J.* **420**, 903

Gelfreikh, G.B., Abramov-Maksimov, V.E., Akhmedov, Sh.B., Bogod, V.M.: 1988, in V.E. Stepanov, V.N. Obridko, G.Ya. Smolkov, (eds.), *Solar Maximum Analysis. Proceedings of the International Workshop held at Irkutsk, USSR*, "Nauka" (Siberian branch), Novosibirsk, p. 71

Gelfreikh, G.B., Koval, A.N., Stepanyan, N.N.: 1981, in V.N. Obridko, (ed.), *Solar Maximum Year. Proceedings of International Workshop*, IZMIRAN, Moscow, p. 193

Huseynov, M.J.: 1978, *Izv. Krymsk. Astrofiz. Obs.* **58**, 31

Koval, A.N., Stepanyan, N.N.: 1983, *Izv. Krymsk. Astrofiz. Obs.* **68**, 3

Krat, V.A., Vyalshin, G.F.: 1965, *Izv. Glav. Astron. Obs. Pulkovo* **178**, 26

Lee, J.W., Hurford, G.J., Gary, D.E.: 1993, *Solar Phys.* **144**, 45

Livshits, M.A., Obridko, V.N., Pikel'ner, S.B.: 1967, *Sov. Astron.* **10**, 909 (=*Astron.Zh.* **43**, 1135)

Mattig, W.: 1969, *Solar phys.* **8**, 291

Obridko, V.N.: 1979, *Sov. Astron.* **23**, 38 (=*Astron.Zh.* **56**, 67)

Zheleznyakov, V.V.: 1970, *Radio Emission of the Sun and Planets*, Pergamon Press, Oxford

POLARIMETRIC STUDY OF SOLAR FLARES

E. VOGT and J.-C. HÉNOUX
Observatoire de Paris, DASOP, URA326,
92195 Meudon, France

Abstract. The theory of impact polarization is briefly reviewed. Spectropolarimetry provides a tool to derive the nature, the number flux, and the main characteristics of the angular velocity distribution function of energetic particles accelerated in solar flares. As an exemple of application of polarimetry the spatial and temporal characteristics of the linear polarization of the hydrogen Hα line observed in a solar flare is presented.

Key words: Impact Polarization – Solar Flares

1. Introduction

Particles are accelerated in solar flares in a wide range covering keV up to GeV energies. Electrons are detected through the radio, X-ray and γ-ray emission they produce, or directly when they move in the interplanetary space. They also cause enhancements of the visible and UV emission continua by interacting with the solar atmosphere. Protons above energies of 1 MeV produce line emission in the γ-ray line spectrum by bombarding the solar atmosphere. They are also detected in the interplanetary space. Low energy protons below 1 MeV do not produce γ-ray lines and are more difficult to detect.

The detection of low energy protons is important since high energy protons may be accelerated by stochastic Fermi acceleration and this process requires a preacceleration of the protons to energies in the 10-100 keV range. Moreover, low energy protons contribute possibly to the formation of the hot soft- X-ray emitting plasma observed in solar flares. Unfortunately, the detection of low energy protons is difficult. Till recently the only method of detection was the observation of the enhancement of the red wings of lines in the Lyman series, emitted by neutral hydrogen atoms. Such enhancement is expected to result from charge exchange between the incident protons and the ambient neutral hydrogen atoms. Since these lines are in the UV part of the radiation spectrum, their observation requires the use of specially dedicated UV space telescopes. The wings of lines from the Balmer series are more easily observed from the ground. However, the emision due to charge exchange in these lines is expected to be too weak to be detected (Canfield and Chang, 1985; Fang, Feautrier and Hénoux, 1995).

Solar Physics **164**: 345–359, 1996.
© 1996 *Kluwer Academic Publishers. Printed in Belgium.*

Another possible diagnostic for proton beams is the observation of impact linear polarization in atomic lines. This linear polarization arises when atoms are collisionally excited by anisotropic beams of particles. As shown below, the resulting polarization provides information on the nature of the incident particles, their number flux and their direction of propagation. In section 2, the mechanism of impact polarization is presented. In section 3, the instrumentation used for polarization measurements is briefly described and recent observations of Hα polarization are presented. In section 4 the origin of the observed polarization is discussed.

2. Impact polarization as a diagnostic for beams of energetic particles

The monochromatic radiation emitted by an atom collisionally excited by beams of particles may be linearly polarized (Skinner, 1926; Percival and Seaton, 1959; Kleinpoppen, 1969; Heddle 1979, Aboudarham et al. , 1992a; Kazantsev et al. , 1994; Kazantsev and Hénoux, 1995). For a monoenergetic beam of particles of individual energy E, the polarization is defined as

$$P(\beta, E) = (I_{\parallel} - I_{\perp})/(I_{\parallel} + I_{\perp}), \tag{1}$$

where β is the angle between the line of sight and the beam propagation direction, and $I_{\parallel}$ and $I_{\perp}$ are the intensities of the vibrations parallel and perpendicular to the plane defined by the beam and the line of sight.

The degree of polarization depends on the particle energy and on the angle of observation. The maximum polarization $P(90^0, E)$ is observed at an angle 90^0 of the beam propagation direction. For dipole radiation, the angular dependence of the degree of linear polarization $P(\beta, E)$ (Hénoux et al. 1983) is given by

$$P(\beta, E) = P(90^0, E)\frac{\sin^2 \beta}{1 - P(90^0, E)\cos^2 \beta}. \tag{2}$$

The polarization degree is energy dependent. Particles of threshold excitation energy lead to a polarization degree opposite to the one produced by particles of very high energy. Before collision, every particle in an incoming beam has no component of orbital angular momentum along the beam propagation direction and its quantum number m_0 is null. At the excitation threshold, the particle is left after collision with zero velocity and zero angular momentum ($m_1 = 0$). From angular momentum conservation law, we can write

$$M_0 + m_0 = M_1 + m_1, \tag{3}$$

POLARISATION (%)

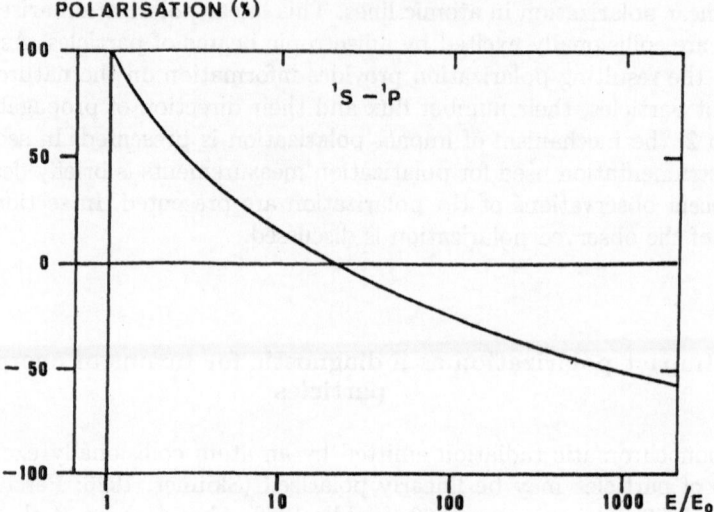

Fig. 1. Schematic representation of the energy dependence of the linear polarization of a $^1S - {}^1P$ transition. E_0 is the threshold excitation energy.

where M_0 and M_1 are the magnetic quantum numbers of the atom before and after the collision. Since $\Delta m = 0$, only substates such that $\Delta M = 0$ are excited. For excitation from a ground state with $L_0 = 0$, only the substate with $M_1 = 0$ is excited at threshold. At very high energy the relative velocity variation of the incoming particle is dominant perpendicular to the beam and quantization axis. Therefore, the momentum transferred to the atom is also perpendicular to the quantization axis, and substates corresponding to $\Delta M = \pm 1$ are excited. A schematic representation of the energy dependence of the linear polarization of a $^1P \rightarrow {}^1S$ transition collisionally excited from level ^1S is shown in Figure 1. Since the magnetic quantum number M_2 of the final level after deexcitation is zero, the changes in quantum number $M_1 - M_2$ at threshold and at high energy are respectively 0 and ± 1. Consequently they are associated with a linear polarization of 100% perpendicular and parallel to the beam.

Fine and hyperfine structure splittings reduce the degree of linear polarization. At threshold, even if the only quantum number excited is $M_L = 0$, different values of the electron and nuclear spin quantum numbers M_S and M_I and therefore of M_J are possible, leading to the emission of both π ($\Delta M_J = 0$) and σ ($\Delta M_J = \pm 1$) transitions, and to a reduction of the polarization degree. However fine and hyperfine structures play a role only if the level natural and/or collisional width of the excited state is very small com-

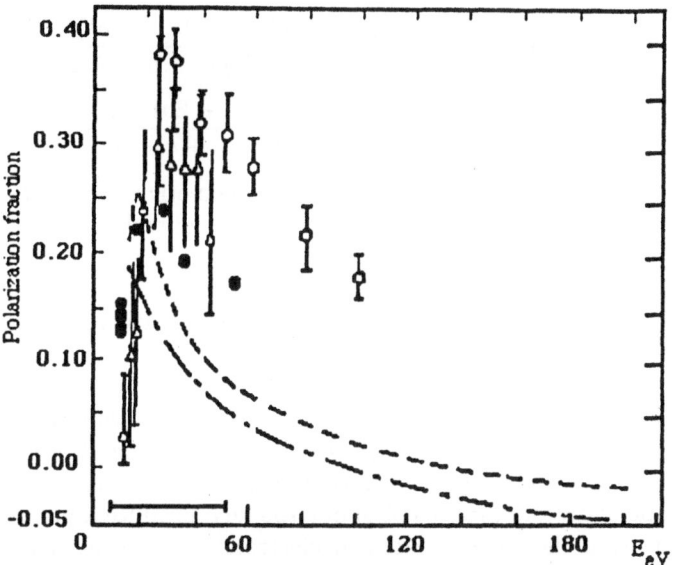

Fig. 2.   Polarization fraction of Hα radiation $P(90^0, E)$ observed at 90^0 of an electron beam. Dashed line: Syms *et al.* (1975); dotted-dashed line: Born; circles: Kleinpoppen *et al.* (1962); triangles: Kleinpoppen and Kraiss (1968); black squares: Aboudarham *et al.* (1992b)

pared to the fine and hyperfine structure separation. Bommier and Sahal-Bréchot (1982) showed that the computed linear polarization of Lyα due to scattering is practically unchanged, wether the hyperfine structure is taken into acount or not, and the case of Hα is not very differnet as for the relative orders of magnitude of the hyperfine structure and natural width. The $3l$ levels of hydrogen are degenerate and the Hα polarization is obtained by summing the contributions from the $3s$, $3p$ and $3d$ levels. Syms *et al.* (1975) give the following expression for the Hα polarization

$$P(90^0, E) = \frac{1}{\sigma_{90}(\mathrm{H}\alpha)}[\sigma_{90}(3s)P_{90}(3s) + 0.12\sigma_{90}(3p)P_{90}(3p) + \sigma_{90}(3d)P_{90}(3d)],$$

(4)

where

$$\sigma_{90}(\mathrm{H}\alpha) = \sigma_{90}(3s) + 0.12\sigma_{90}(3p) + \sigma_{90}(3d), \qquad (5)$$

$$\sigma_{90}(3s) = \sigma(3s), \quad \sigma_{90}(3p) = \frac{h_0\sigma(3p0) + h_1\sigma(3p1)}{6},$$

$$\sigma_{90}(3d) = \frac{h_0\sigma(3d0) + h_1\sigma(3d1) + h_2\sigma(3d2)}{100}, \qquad (6)$$

and

$$P_{90}(3s) = 0, \quad P_{90}(3p) = \frac{G(\sigma(3p0) - \sigma(3p1))}{h_0\sigma(3p0) + h_1\sigma(3p1)},$$

$$P_{90}(3d) = \frac{G(\sigma(3d0) + \sigma(3d1) - 2\sigma(3d2))}{h_0\sigma(3d0) + h_1\sigma(3d1) + h_2\sigma(3d2)}. \tag{7}$$

$\sigma(3lm_l)$ is the cross section for exciting the magnetic sublevel $3lm_l$. The coefficients G and h depend on the magnetic sublevel and are given in Percival and Seaton (1958) and in Syms et al. (1975).

Observed and theoretical values of the polarization fraction at an angle of 90^0 to the beam, as a function of impact electron energy are presented in Figure 2. Recently Werner and Schartner (1995) measured perpendicular impact polarization of $H\alpha$ excited by a beam of protons of energy 40 keV to 1 Mev and found that if at high velocities electrons and protons of same velocity (protons of energy higher than 800 keV) lead to the same degree of linear polarization, protons of about 160 keV produce a polarization five time lower that the one generated by electrons of same velocity.

3. Measurement of impact linear polarization

3.1. PARIS, A PATROL HELIO-POLARIMETER

A few solar flares have been observed with the Meudon $H\alpha$ polarimeter. These observations have been reported in Hénoux and Chambe (1990), Hénoux et al. (1990) and Hénoux (1991). All these flares have been observed with a flare patrol heliograph equiped with a Lyot filter with a band-pass of 0.75 Å. A rotating half-wave plate was inserted in front of the linear polarizer at the entrance of the Lyot filter. During the observations this plate is rotated through 22.5^0 per step. $H\alpha$ images are taken every four secondes and give respectively the spatial variation of the parameters $I+Q, I+U, I-Q, I-U$. Then, the Stokes parameters are obtained by adding and substracting appropriate digitized images after compensating for image motions by cross-correlation technics. The polarization degree P and the azimuth Φ of the linear polarization are given respectively by:

$$P = \sqrt{(U^2 + Q^2)}/I, \qquad \tan 2\Phi = U/Q. \tag{8}$$

The observations presented below have been obtained with this instrument equipped with a photographic camera. A new instrument called PARIS* is

* Polarimeter for Active Region Instabilities Studies

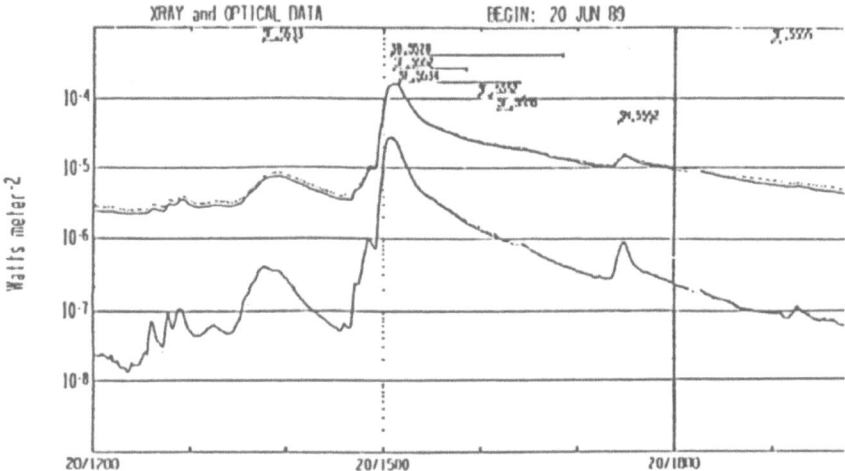

Fig. 3.   Time evolution of the soft X-ray emission (0.5-4.0 and 1.0-8.0 Å) on June 20, 1989.

expected to be operating by fall 1995. This instrument is equipped with a 288 × 384 pixels CCD camera similar to the ones used in THEMIS. The field of view is 4. × 5.5 arc minutes. Compared to the photographic polarimeter, the time resolution is improved by a factor five.

3.2. MAIN CHARACTERISTICS OF THE JUNE 20, 1989 SOLAR FLARE Hα POLARIZATION

A solar flare was observed on June 20, 1989 at 24^0N and 68^0W. The figures 3 and 4 show the time evolution of the soft (0.5-4.0 and 1.0-8.0 Å) and hard (50-270 keV) X-ray emission and of the radio emission at 1.4 GHz. The hard X-ray and the microwave emission are impulsive. The hard X-ray emission last about 15 minutes. It starts at 14:55 UT, reaches a peak at 14:57:30 UT and decreases. The microwave emission shows two impulsive peaks and last also about 15 minutes. The soft X-ray emission extends over nearly four hours and shows a maximum at 15:08 UT.

The high energy electrons that radiate in hard X-rays are loosing most of their energy in Coulomb collisions with the cold electrons of the solar atmosphere. As a result the atmosphere is heated and starts emitting in soft X-rays. Additional possible bombardment by hecta keV protons would contribute to the heating and enhance the soft X-ray emission.

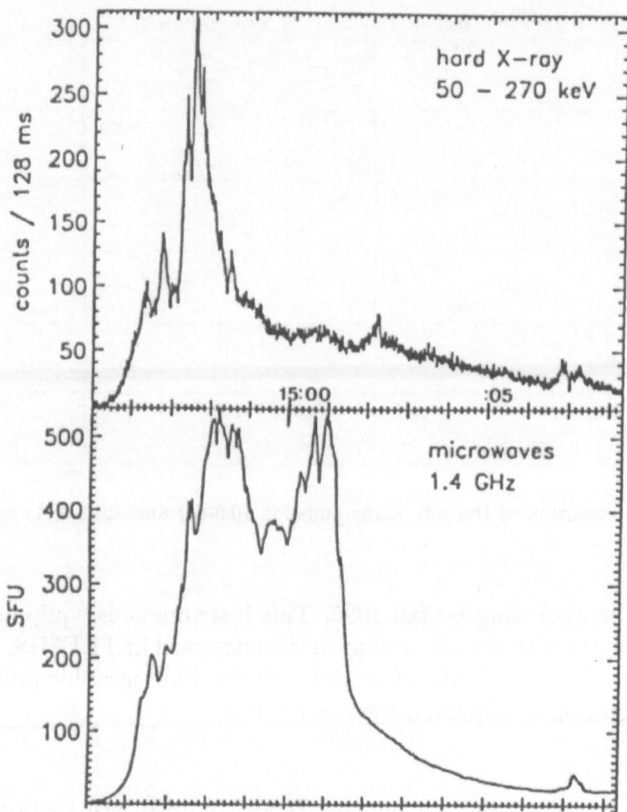

Fig. 4.   Time evolution of the hard X-ray (50 - 270 keV) and microwave (1.4 GHz) emission on June 20, 1989.

The polarization data were analyzed over 17 minutes. They cover the time of maximum hard and soft X-ray and radio emission. Linear polarization is observed all along the observed event. The degree of linear polarization and its azimuth vary with time and space. However, there are times - at 15:00 and 15:08 UT - where a dominant radial direction of polarization is observed as shown in Figures 5 and 6. The polarization integrated over 17 minutes of observation has also a dominant radial direction, as can be seen on the histogram of Figure 7. The main features of the observed polarization can be described as the following:

1 - Near the time of maximum of hard X-ray emission the angular distribution of the azimuth of polarization shows a maximum for an angle that differs by 20^0 to 30^0 from the radial direction.

2 - At 15:00 UT the hard X-ray emission has decreased by nearly an order

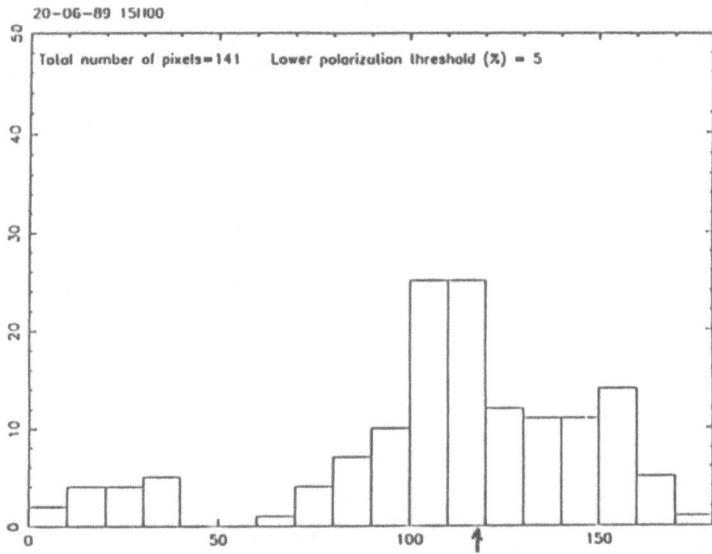

Fig. 5. Azimuthal distribution of the one minute time integrated polarization exceeding 5% at 15:00 TU on 20 June 1989. An arrow indicates the flare to disk-center direction.

of magnitude and the Hα intensity peaks. This peak of emission coincides with a dominant radial direction of the linear polarization and with a second peak of microwave emission that is indication for the upward acceleration towards the corona of electrons. We will suggest later that low energy protons are presumably impacting on the chromosphere at that time. The spatial distribution of the Hα polarization is shown in Figure 8.

3 - The observation at 15:08 UT coincides with the maximum of soft X-ray emission and may also be associated with an enhanced heating by proton bombardment.

4. Origin of the polarization observed in the June 20, 1989 solar flare

The linear polarization that is observed only during solar flares is interpreted as resulting from the impact of energetic particles on the solar atmosphere. Assuming that the velocity distribution of these particle is symmetrical around the solar vertical, the linear polarization in the line relative to the radial and tangential directions on the solar disk at a position of

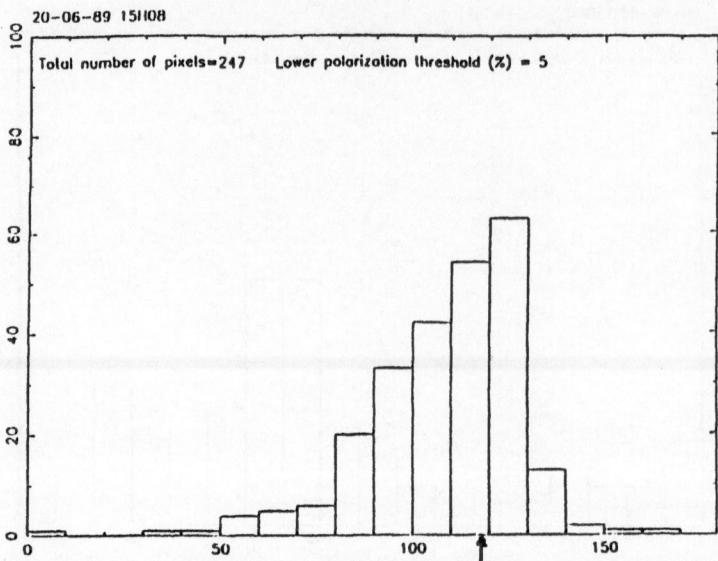

Fig. 6. Azimuthal distribution of the one minute time integrated polarization exceeding 5% at 15:08 TU on 20 June 1989. An arrow indicates the flare to disk-center direction.

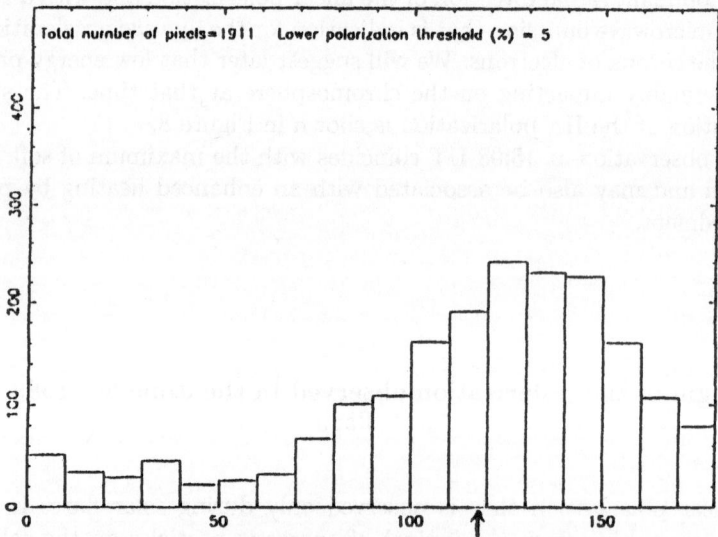

Fig. 7. Azimuthal distribution of the polarization integrated from 14:55 to 15:12 UT on 20 June 1989 and exceeding 5%. An arrow indicates the flare to disk-center direction.

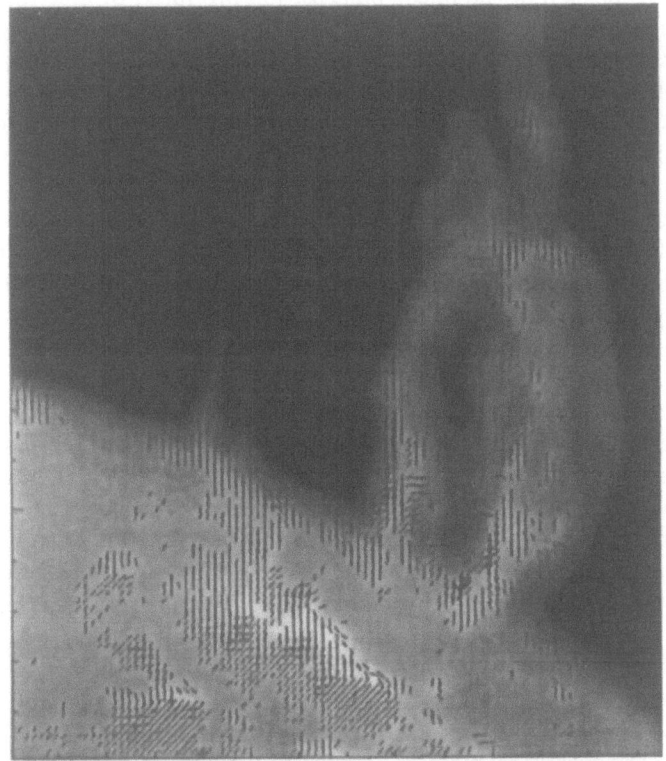

Fig. 8. Spatial repartition of the one minute time integrated polarization exceeding 5% at 15:00 UT on 20 June 1989

heliocentric angle θ is given by (Hénoux *et al.* 1983)

$$\mathcal{P}(\theta) \simeq P(90^0, E) \sin^2 \theta \frac{3J_2(E) - J_0(E)}{2J_0(E)}, \tag{9}$$

where

$$J_n(E) = \int_0^\pi f(E, \alpha) \cos^n \alpha \sin \alpha d\alpha, \tag{10}$$

and α is the particle pitch angle. This leads to define an anisotropy factor b(E) such that

$$b(E) = \frac{3J_2(E) - J_0(E)}{2J_0(E)}. \tag{11}$$

So the observed linear polarization gives information on the anisotropy of the energetic particles through the ratio $J_2(E)/J_0(E)$.

Since the sign of the linear polarization produced by electrons or protons changes with their energy, these particles may have a low energy and a vertical direction of propagation or be at high energy and circling horizontally. In the limited number of observations we obtained, there is no evidence of correlation between a radial direction of propagation and a strong hard X-ray emission, which rules out energetic electrons as the main source of the observed polarization (see Fletcher and Brown, 1995). High energy protons would lead to γ-ray emission that is not present in the observed flares. So the set of observations available shows Hα linear polarization that is not associated with highly energetic particles. As a matter of fact, due to the decrease with energy of the excitation cross section, the hypothesis of excitation by low energy particles leads to the minimum of energy requirement.

Assuming that the particles generating the observed polarization are accelerated somewhere in the solar corona, we interpret the observed polarization as due to low energy protons with an energy when they reach the chromosphere that has fallen below 200 keV (energy at which the perpendicular polarization changes of sign). Taking into account the energy losses in Coulomb collisions from the corona down to the chromosphere for empirical values of the coronal density in medium and strong flares, we are left with an upper value of their initial energy between about 250 and 350 keV at the coronal acceleration site. Due to their large mass, protons are not significantly deviated from the corona to the chromosphere by Coulomb collisions. This property makes the protons to keep or even to increase their directivity when the horizontal stratification in density is taken into account. Coulomb collisions in the transition region generate naturally a significant anisotropy in velocity (Hénoux *et al.* 1990). For protons in the energy ranges we consider, Coulomb collisions are efficient at transition region level. Low energy protons with high pitch angle cover an increased distance compared to the ones that travel vertically. Consequently, for a given proton energy at chromospheric level, the initial energy of the proton at the place of its origin in the corona must be higher for higher pitch angle. Since the proton number density in a beam is expected to decrease with energy, an energy dependent velocity distribution leads to an angular dependence of this distribution at chromospheric level. The resulting angular velocity distribution of protons at chromospheric level at energies close to threshold, and for an isotropic distribution in the corona is given by:

$$f(\cos \theta) \propto \cos^{\gamma/2}(\theta), \tag{12}$$

where θ is the heliocentric angle and γ is the power index of the proton energy distribution function at the coronal site.

In the absence of a significant effect of Alfven waves and assuming an energy distribution of the particles given by a power law $n(E) \propto E^{-4}$ (Hénoux *et al.* 1990), the required particle and energy fluxes required to explain the observed Hα polarization are respectively:

- Particle flux: $\Phi(E > 200\text{keV}) > 5.10^{14}\text{cm}^{-2}\text{s}^{-1}$
- Energy flux: $F(E > 200\text{keV}) > 3.10^{8}$ ergs cm^{-2}s^{-1}.

Indeed proton beams can generate Alfven waves and be scattered by these waves. For a very high level of Alfvén wave energy density, the protons are highly scattered. A resulting very strong diffusion in the corona enhances strongly the requirement on the individual energy that a proton must have to reach the chromosphere - up to 2-20 MeV (Smith *et al.* 1990). However, for an intermediate regime of Alfvén wave generation, where the isotropisation time is of the same order as the mean free propagation time from the top of the loop to the transition region (energy density $W_a^{tot} = 10^{-1}$ ergs cm^{-3}), a significant anisotropy is present at chromospheric levels at the feet of a magnetic loop. Karlicky, Hénoux and Smith (1995) have shown using Monte-Carlo computations that, even if Alfvén wave produce some isotropisation, the effect of Coulomb collisions dominates for protons of energy less than about 500 keV. A beam of proton accelerated at the top of a magnetic loop leads to a velocity distribution function that peaks along the vertical and therefore produces a radial direction of polarization.

Low energy electrons loose very quickly their energy and their directivity in Coulomb collisions. Anisotropy of the electron distribution function at the base of the transition zone associated with heat flux has been first suggested to explain the observation of linear polarization in a Sulfur UV line with the UVSP instrument on SMM (Hénoux *et al.* 1983). However, the computation of the velocity distribution of electrons in the part of the solar atmosphere requires solving the Fokker-Planck equation in the solar corona to chromosphere transition region. Existing calculations are valid mainly in the transition zone outside the region where the Hα line is formed. The particle density at chromospheric level is high enough for the low energy electrons to be quickly isotropized before reaching the Hα line forming level where the expected anisotropy is presumably very low. So, in the hypothesis where the anisotropy is not a local phenomena, protons are the best candidates to explain the observed polarization. Neutral beams, where electrons and protons are moving with the same velocity are also good candidates. As a matter of fact, they may be more efficient than beams carrying only protons, since both electrons and protons would contribute to the polarization. However, it is still difficult to derive from polarization observations if the proton beam is neutralized or not.

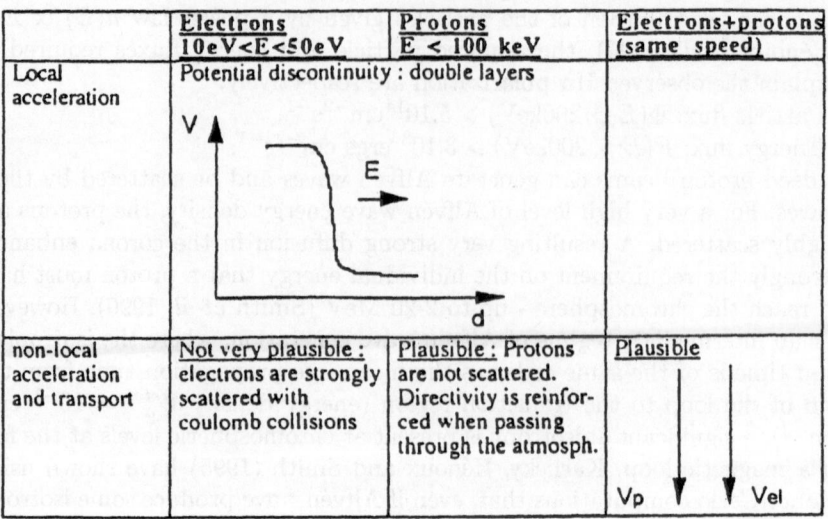

Fig. 9. Summary of the various processes that may produce impact polarization in Hα.

In the case where some mechanism creates a local electric field, like in the formation of electric potential discontinuities called double-layers, a local anisotropy of electrons and protons could be generated. So the formation of double-layers at chromospheric level could also explain the polarization observations. However, chromospheric bombardment by a reasonable flux of protons or by a neutral beam can easily generate significant linear polarization. Therefore, the observation of linear polarization does not establish the existence of chromospheric double-layers. The Figure 9 taken from Aboudarham *et al.* (1992a) shows the few possible scenarios that could explain the observed polarization. The double layers that could lead to anisotropic velocity distribution of electrons and protons would most probably be a succession of weak double layers with a potential energy of each layer not greatly exceeding the local thermal energy.

The results presented here come from a first analysis that stresses the main observed features of the June 20, 1989 flare, that may be interpreted as resulting directly through Hα impact excitation from the hydrogen ground level. Bommier *et al.* (1992) pointed out that at the high electron number densities of 10^{12} cm^{-3} that could be present in a flaring chromosphere, exchange of population between fine structure levels of different orbital quantum number l due to collisions with ambient electrons and protons could destroy the polarization. A quantitative study of this effect has to be done since it is a fact that, despite a probably high local electron and proton densities, a significant polarization $\mathcal{P}(\theta)$ that could exceed 10% is observed even if it seems to be fluctuating with time. A detailed study of the relative importance of all the processes that populate and depopulate all sublevels is under way (Sahal-Bréchot *et al.* 1995).

5. Conclusion

Contrary to Zeeman and Hanle spectroscopy, impact polarization as a diagnostic tool began to be applied in Solar Physics only recently during 1980-1990. With the development of fast data acquisition systems and storage capability, it may replace in future the classical chromospheric flare patrol observations. The investigation of the particle velocity anisotropies opens a new source of information on the nature of the energetic particles at the origin of chromospheric flares, on the energy flux associated with them, and presumably on their acceleration mechanism. Since the physical conditions in the chromospheric target may also play a role in the resulting polarization, there is presumably an opportunity to learn more about the evolution of the local physical conditions at chromospheric level that result from the flare energy downward propagation. Progress must still be made in the interpretation of the measurements. They require to solve the transfer and statistical equations describing the transfer of Hα radiation and the equilibrium between the various processes that control the level populations and the polarization of the emitted radiation. Complementary observations of the spatial distribution of the polarization integrated over the line profile, made by PARIS, and of the spectral variation of the polarization along the profile, the way it could be done in Irkuskt, as by the DPSM on THEMIS, would be of mutual benefit.

Acknowledgements

Thanks go to Drs. Sylvie Sahal-Bréchot, S. Kazantsev, N. Feautrier and V. Bommier for their contributions.

References

Aboudarham, J., Chambe, G., Feautrier, N., Hénoux, J.-C.:1992a, *Méthodes de détermination des champs magnétiques solaires et stellaires*, Paris Observatory Publication, Faurobert-Scholl, M., Frisch, H. and Mein, N. eds., pp.163-172
Aboudarham, J., Berrington, K., Callaway, J., Feautrier, N., Hénoux, J.-C., Peach, G. and Saraph, H.E.: 1992b, *Astron. Astrophys.* **262**, 302

Bommier, V. and Sahal-Bréchot, S.: 1982, *Solar Physics* **78**, 157
Bommier, V., Chambe, G., Feautrier, N., Hénoux, J.-C., Sahal-Bréchot, S., Saraph, H.:1994, Paris Observatory Publication, Frisch, H. ed., p.152
Canfield, R.C. and Chang, C.R.: 1985, *Astrophys. J.* **295**, 275
Fang, C., Feautrier, N. and Hénoux, J.-C.: 1995, *Astron. Astrophys.* **297**, 854
Fletcher, L. and Brown, J.-C.: 1992, *Astron. Astrophys.* **294**, 260
Heddle, D.W.O.: 1979, *Advances in Atomic and Molecular Physics*, D.R. Bates and B. Bederson eds., Academic Press, N.Y., 321
Hénoux, J.-C., Heristchi, D., Chambe, G., Machado, M., Woodgate, B., Shine, R., and Beckers, J.: 1983, *Astron. Astrophys.* **119**, 233
Hénoux, J.-C, Chambe, G., Smith, D., Tamres, D., Feautrier, N., Rovira, M. and Sahal-Bréchot, S.: 1990, *Astrophys. J. Suplt. Ser.* **73**, 303
Hénoux, J.-C. and Chambe, G.: 1990, *J.Q.S.R.T.* **44**, 193
Hénoux, J.-C.: 1991, *Solar Polarimetry*, Proceedings of the Eleventh National Solar Observatory/Sacramento Peak Summer Workshop, p.285
Karlicky, M., Hénoux, J.-C. and Smith, D.: 1995, *Astron. Astrophys.* , (submitted)
Kazantsev, S.A. and Hénoux, J.-C.: 1995, *Polarization Spectroscopy of Ionized Gases*, Kluwer Academic Publisher
Kazantsev, S.A., Feautrier, N., Hénoux, J.-C., Liaptsev, A.V. and Luchinkina, V.V.: 1994, *Astron. Astrophys. Rev.* **6**, 1
Kleinpoppen, H.: 1969, *Advances in Atomic and Molecular Physics*, F. Bopp and H.Kleinpoppen eds, North Holland, Amsterdam, 612
Kleinpoppen, H., Krueger, H. and Ulmer, R.: 1962, *Phys. Let.* **2**, 78
Kleinpoppen, H. and Kraiss, E.: 1968, *Phys. Rev. Let.* **20**, 36
Percival, I.C. and Seaton, M.J.: 1959, *Phil. Trans. R. Soc. London A* **251**, 113
Skinner, H.W.B.: 1926, *Proc. R. Soc. London A* **112**, 642
Smith, D.F., Chambe,G., Hénoux, J.-C. and Tamres, D.: 1990, *Astrophys. J.* **358**, 674
Syms, R.F., McDowell, M.R.C., Morgan, L.A. and Myerscough, V.P.: 1975, *J. Phys. B. Atom. Molec. Phys.* **8**, 2817
Sahal-Bréchot S., Vogt E., Thoraval S. and Diedhiou, L.: 1995, *Astron. Astrophys.* , (submitted)
Werner, A and Schartner, K.H.: 1995, (*Private communication*)

CHARACTERISTICS OF THE LINEAR POLARIZATION
OBSERVED IN THE 16 MAY 1991 SOLAR FLARE

N.M. FIRSTOVA and A.V. BOULATOV

Institute of Solar-Terrestrial Physics (ISTP), P.O.Box 4026, Irkutsk, 664033, Russia
Internet: bulat@sitmis.irkutsk.su

Abstract. In a search for linear polarization effects, 37 profiles of the H_α line emitted in the 16 May 1991 flare have been analyzed. Linear polarization is clearly present in the central part of line. On average, the degree of polarization is 7 % , but it reaches 20 % in regions with lower H_α emission. Generally the orientation of the plane of polarization coincides with the flare to disk center direction, except for sections where the H_α line has the characteristic form observed in moustaches. We believe that the linear polarization observed in the 16 May 1991 flare was caused by bombardment of the chromosphere by beams of accelerated particles, protons in the main part of the flare and electrons at locations where the H_α line has the characteristic moustache structure.

Key words: Polarization – Flare

1. Introduction

Observation of the linear polarization of spectral lines in solar flares provides unique information on the modes of energy transfer from the corona to the chromosphere during solar flares, and significant properties of the energy transfer process can be derived from the measurement of the line emission polarization vector (Chambe and Hénoux, 1979; Hénoux and Semel, 1981; Hénoux *et al.*, 1983a; Hénoux and Chambe, 1990a; Smith *et al.*, 1990; Aboudarham *et al.*, 1992).

The first observations of impact linear polarization integrated over the profile of the 1437Å S I line in solar flares were made by Hénoux *et al.* (1983b) using the UVSP SMM spectropolarimeter. Observations of linear polarization in the H_α line in flares made with a narrow-band filter were published by Chambe and Hénoux (1979), Hénoux *et al.* (1990b) and Hénoux (1991), and observations of the linear polarization across the H_α line profile were published by Babin and Koval (1983, 1985a,b,c) and Kazantsev *et al.* (1991, 1993). The observed linear polarization cannot be caused by Zeeman or Stark effects. According to Hénoux and Semel (1981), the combination of these two effects, integrated over the H_α line profile, does not exceed 0.5%. Moreover, the wavelength dependence across the H_α line profile of the linear polarization must show the characteristic signatures associated with the Zeeman π and σ components; the observed H_α line polarization profiles do not show such signatures.

Solar Physics **164**: 361–372, 1996.
© 1996 *Kluwer Academic Publishers.*

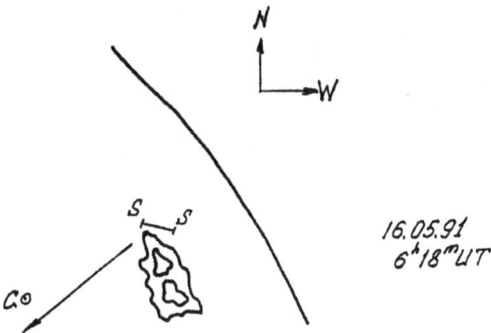

Fig. 1.   A sketch of the active region No.6619. *SS* is the spectrograph slit position. The arrow indicates the direction towards the center of the solar disk.

Linear polarization is found in approximately 30 % of the observed H_α flare spectra. Normally the degree of polarization reaches 3-5 % , but in some cases it exceeds 10 % . In many cases the highest polarization signal is not observed in the brightest regions of the flare. The direction of the plane of polarization in flares basically coincides with the flare to disk center direction. As shown in Hénoux (1991), the analysis of other independent observations leads to the conclusion that the observed polarization is caused by the bombardment of the chromosphere either by proton beams (with a minimum energy for an individual proton at the coronal acceleration site of 200 keV) or by neutral beams.

Despite significant progress in the observation of impact linear polarization during the last decade, only a few observations have been made. In the present paper we present new observations of linear polarization in the H_α line in a flare observed on 16 May 1991 with the Baikal solar vacuum telescope. The spectrograms have been processed with a CCD in order to find the orientation and the spatial dependence of the polarization vector.

2.   Observational data

2.1.   Description of the observations

Our observations of the 16 May 1991 solar flare in the active region No. 6619 SGD (32N56W) were made with the Baikal solar vacuum telescope of ISTP (Skomorovsky and Firstova, 1995), using a $\lambda/2$-plate and a Wollaston prism inserted behind the spectrograph slit. The flare occured north-west of a large sunspot and was observed from 06:18 UT to 06:20 UT during a break in the clouds. Unfavorable weather conditions precluded the determination of the flare duration and importance. Figure 1 shows a sketch of active region No.

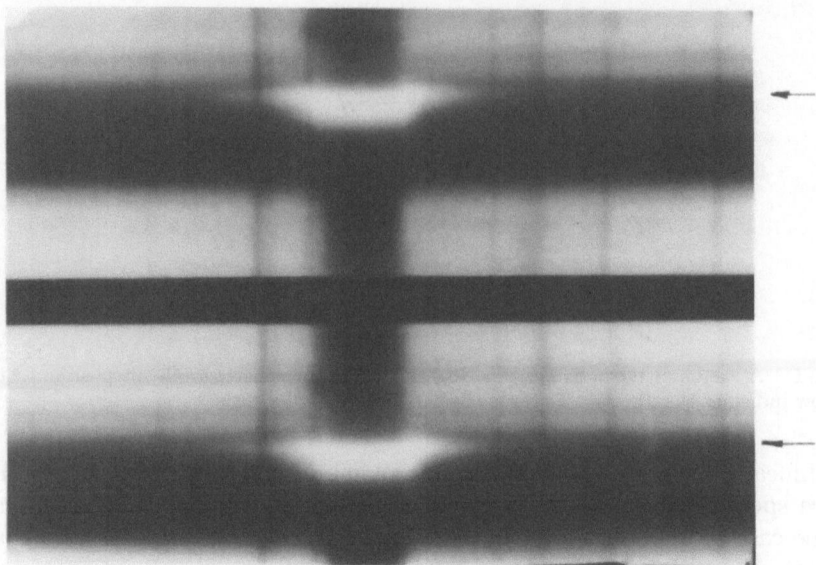

Fig. 2. A flare spectrogram in two orthogonal directions of polarization.

6619 in white light together with the spectrograph slit position at the time of recording and the flare to disk center direction.

The spectra were exposed for different positions of the $\lambda/2$-plate with one of the main axis of the plate making an angle with the direction of the spectrograph slit of $0°$, $22.5°$ and $45°$. These positions correspond respectively to the measurement of the Stokes parameters Q, U and $-Q$. This cycle was subsequently repeated. A total of 12 spectrograms were obtained, each of which consisting of two orthogonally polarized spectral bands. Figure 2 shows a photograph of one of the spectrograms. The height of an individual spectral band is 45 arcsec. Unfortunately the first frame was defocused, and the last six were overexposed. For that reason, only five frames, from the second to the sixth, were fully analysed.

2.2. Technique of analysing spectrograms with a CCD array

The photographic spectrograms were analysed with a CCD system installed at the telescope in 1993. For this purpose, a computer program was developed for constructing the characteristic curve, determining the dispersion, converting photographic densities to intensities, determining the H_α line center, calculating line profiles in units of the continuous spectrum, and for computing the polarization as a function of wavelength.

Selected spectrograms and frames exposed with a step wedge were digitized with a constant gain factor and with the same signal build-up time.
The characteristic curve was constructed using these frames. Subsequently
the dispersion averaged over several sections and common to all spectrogram
frames was determined using reference lines.

It is crucial that the extracted H_α line profiles from the two orthogonal polarization states correspond to the same area on the solar disk. The
sections in each polarization band serving as references were first adjusted
visually, and the remaining sections were then fixed relatively to the reference sections. In order to adjust the reference sections, the whole frame
including the two orthogonal polarization states was displayed on the screen,
and the reference sections were selected at locations were the H_α line has
the same typical form for both directions of polarization. In Figure 2 the
positions of the reference sections in the two spectral bands are indicated
by two arrows. The positions of the other sections were determined by moving in the direction of the spectrograph slit by an equal number of pixels
from the location of the reference sections in the two bands. Thus the po-
sitional accuracy of the sections for the two orthogonal polarization states
depends only on the precision with which the reference sections have been
adjusted.

The beginning of each wavelength scan was selected in order to be as far
as possible from the H_α center, and the value of the continuum was assigned
to the first five points. Subsequently densities were converted to intensities
that were normalized to the value of the continuum. The procedure was then
repeated for every pair of associated sections for each given frame. As a result
we obtained for each section for a set of wavelength positions relative to the
line center, $\pm\Delta\lambda$, the H_α line intensities normalized to the continuum $I_1 =
J_1/J_{cont}$ and $I_2 = J_2/J_{cont}$ in the first and second orthogonally polarized
spectral bands as well as the values of Q/I or U/I given by

$$Q/I \text{ or } U/I = \epsilon(I_1 - I_2)/(I_1 + I_2), \tag{1}$$

where $\epsilon = \pm 1$ depending on the orientation of the $\lambda/2$-plate.

2.3. OBSERVATIONAL ACCURACY

Accurate polarization observations require knowledge of the instrumental
polarization. In addition to the first $\lambda/2$ plate installed in front of the spectrograph slit, a second $\lambda/2$ plate was placed behind the Wollaston prism, so
that the electric vector of the ordinary and extraordinary rays would be at
$45°$ with respect to the grating rulings. In the absence of the first retarder,
the second $\lambda/2$-plate was adjusted by visually balancing the intensities of
the quiet sun in the two spectral bands. If instrumental polarization were
present it would have been impossible to balance the intensities. Furthermore, to improve the accuracy, all measurements of line intensities were

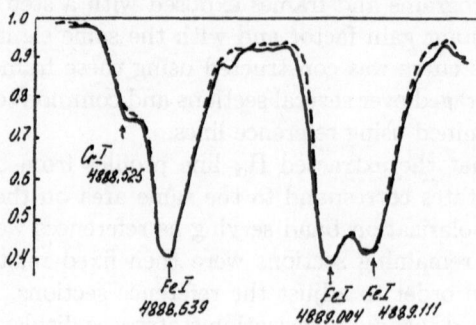

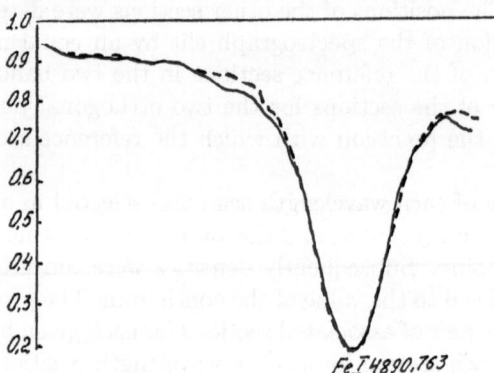

Fig. 3.   Fe I and Cr I line profiles in a quiet region for two orthogonal polarization states.

usually normalized to the continuous spectrum. Figure 3 illustrates profiles of a few unpolarized Fe I and Cr I lines in a quiet region in the two spectral bands. We observe these lines under the same conditions as the H_α spectra, but the solar image was defocused and there should be no intrinsic polarization. A comparison of these lines in the two orthogonal polarization states provides evidence for the absence of instrumental polarization. It is worth noticing that the intensity of the continuous spectrum in the two spectral bands also remained unchanged when inserting the first $\lambda/2$-plate.

The differences that are seen in the line profiles for the two orthogonal polarization states in Figure 3 can be attributed to normal uncertainties of the photographic method. A considerable uncertainty is also present because of the difficulty of adjusting the reference sections. Once they are well adjusted the error in the line intensity between different frames varies from 1% to 8%.

3. Stokes parameters across the H_α line profiles

As mentioned above, five spectrograms taken in an interval of two minutes were processed. The recorded Stokes parameters in these spectrograms were as follows: I (U), II ($-Q$), III (Q), IV (U) and V ($-Q$). In each spectrogram, eight sections across the spectrograph slit were selected. The value $0''$ was assigned to the reference section, and the relative positions of the remaining sections were expressed in arcsec. Figure 4 shows H_α profiles for the two orthogonal polarization states for the section $-3.75''$ in each of the five spectrograms. The solid line shows the H_α profile for the first polarization state, the dashed line corresponds to the second.

In the case of simultaneous exposure of the five spectrograms, the relative intensity variation in the two spectral bands of Figure 4.III (Q) should be opposite to that of Figure 4.II and 4.V ($-Q$), since the direction of polarization has changed by $90°$. This is also what we find, although image motion and intrinsic time variations could have affected the results, since the exposures were not simultaneous. Figure 4 shows the predominance of linear polarization along the slit, rather than along the dispersion direction.

The profiles in Figure 4 were obtained by one recording. Since all sections were recorded twice or three times, the H_α line intensity profiles for each section were averaged and smoothed with the formula

$$I_{k,j} = (I_{k,j} + \frac{2}{3}(I_{k,j-1} + I_{k,j+1}) + \frac{1}{3}(I_{k,j-2} + I_{k,j+2}))/3 , \qquad (2)$$

where $k = 1,2$ is the index of the polarization band, and j is the index of the wavelength coordinate. Figure 5 presents, for spectrogram III, averaged values of I_1 and I_2 in all sections across the spectrograph slit, as well as of the parameter Q/I for these sections. The horizontal bar corresponds to the zero level of the Q/I parameter.

The most systematic behavior of the Stokes parameters is observed at line center. Therefore the data were analyzed in two wavelength ranges around the line center: -1 Å$< \Delta\lambda <+1$Å and -0.5Å$< \Delta\lambda <+0.5$Å. Figure 6a shows the behavior of the mean intensity of the H_α line in the interval -1Å$< \Delta\lambda <+1$Å for the five exposures. The positions of the flare portions along the spectrograph slit are given in arcsec. Figure 6b gives the quantities Q/I, U/I and $-Q/I$ in the same interval; the errors are marked by vertical bars. Crosses show mean values of the Stokes parameters in the wavelength range -0.5Å$< \Delta\lambda < +0.5$Å. To compare time-adjacent values of the Q parameter, the plot in Fig. 6bIII includes the values from Fig. 6bII with reversed sign.

By comparing the Stokes parameters for different spectrograms, the following features emerge. Spectrograms I and IV (Stokes U) agree reasonably well in a relatively quiet region (sections $0'' - 5''$) but are in rather poor

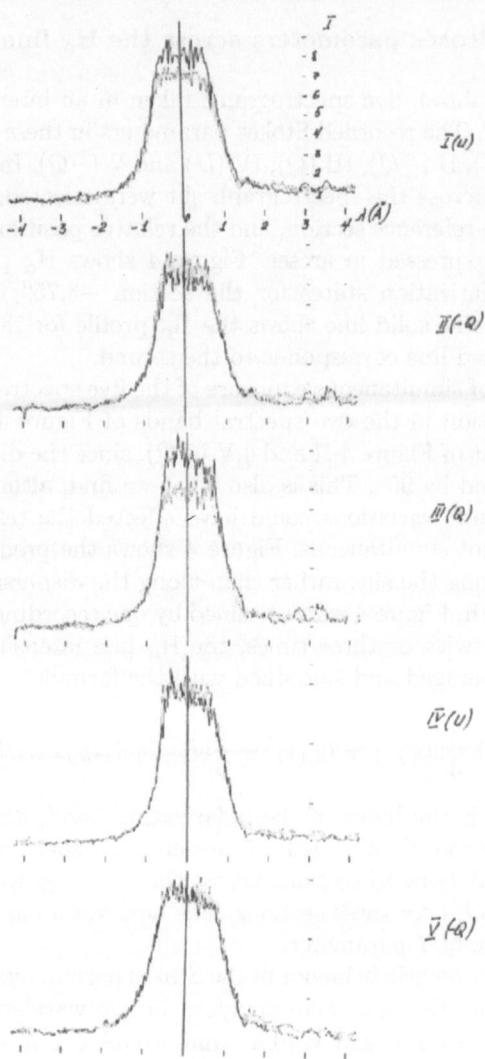

Fig. 4. H_α profiles in the two orthogonal polarization states for the section $-3.75''$ for different orientations of the entrance $\lambda/2$-plate. The full line and dotted line profiles correspond respectively to the directions of polarization numbered 1 and 2 in Eq. (1). Using this equation with $\epsilon = 1$, they provide the Stokes parameters U, Q or –Q.

agreement in the emission part of the flare. As far as Stokes Q is concerned, the situation is the opposite: rather good agreement between $-8''$ and $0''$, and a very strong spread of the values in the relatively quiet region ($0''$ – $5''$).

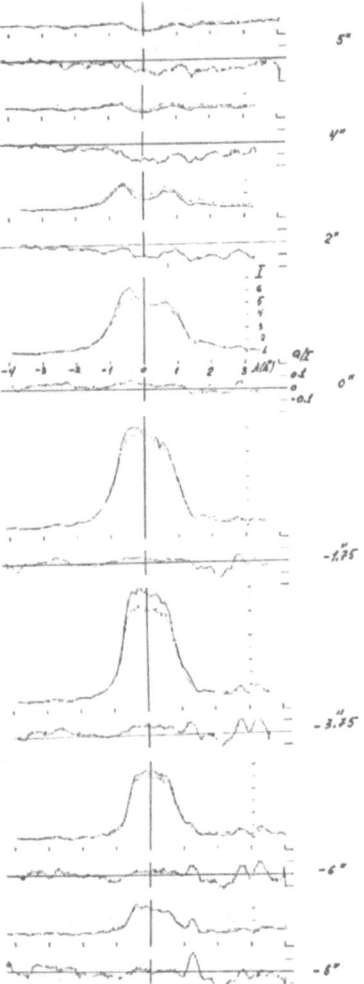

Fig. 5. H_α intensity and Stokes Q profiles for spectrogram III in all sections at various positions along the spectrograph slit.

4. Orientation and amplitude of the polarization vector

To derive the final results, spectrograms IV and V were dismissed, since they were of lower quality. Spectrograms I–III were used to determine the mean intensity of the flare emission along the spectrograph slit. The degree

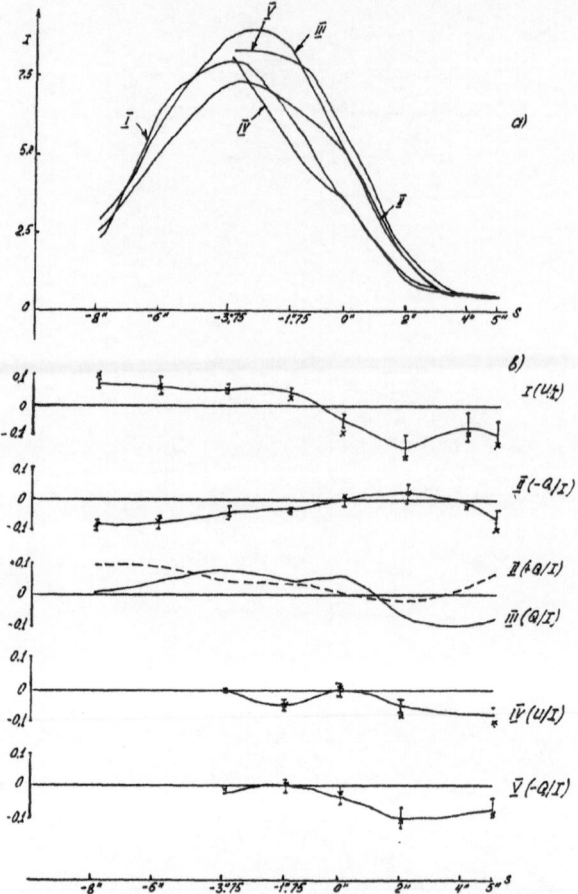

Fig. 6. Variation along the spectrograph slit of the mean H_α intensity (a) and of the mean values of Q/I, $-Q/I$ and U/I (b). Averaging has been made over the interval $-1\text{Å} < \Delta\lambda < +1\text{Å}$. Crosses show mean values averaged over the interval $-0.5\text{Å} < \Delta\lambda < +0.5\text{Å}$.

of polarization P and the azimuth χ of the polarization plane relatively to the direction of the spectrograph entrance slit are given by

$$P = \sqrt{\frac{Q^2}{I^2} + \frac{U^2}{I^2}} \quad \text{and} \quad \cot 2\chi = Q/U. \tag{3}$$

We used spectrograms II, III (providing Stokes Q) and I (Stokes U) to calculate P and χ.

Figures 7a,b,c show a three-dimensional image of I, P and χ versus the wavelength and section number. In spite of multiple averaging and smoothing, the degree of polarization and, to a much larger extent, the polarization

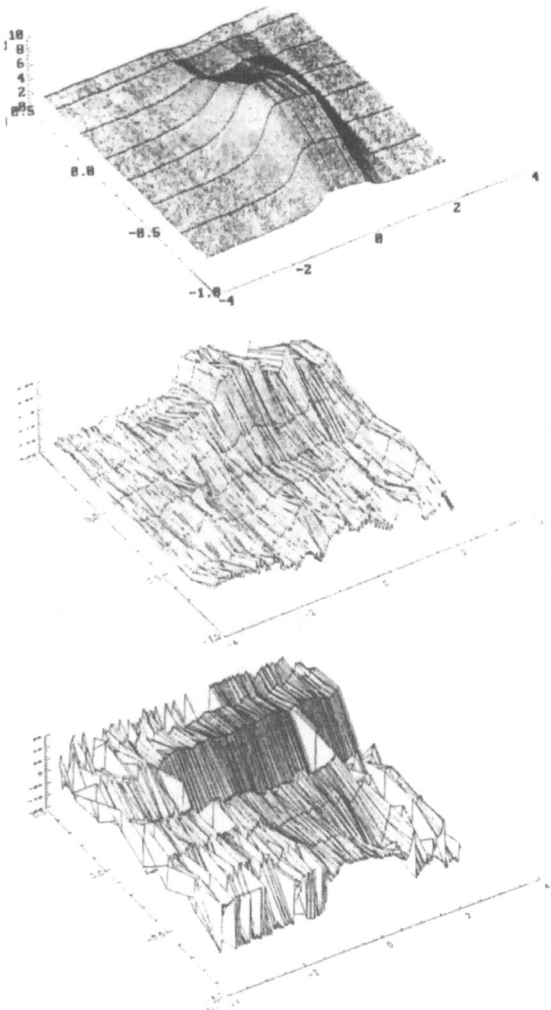

Fig. 7. Three-dimensional representation of I (a), P (b) and χ (c) versus the wavelength and flare position along the spectrograph slit (section number).

azimuth, show a significant spread along the dispersion direction, especially in the line wings.

Figure 8 gives P and χ for the central portion of the H_α line; the photo-metric section with the intensity at H_α line center along the spectrograph slit is also included. The length of each line segment is proportional to P, and its direction is given by the angle χ. The polarization pattern at the

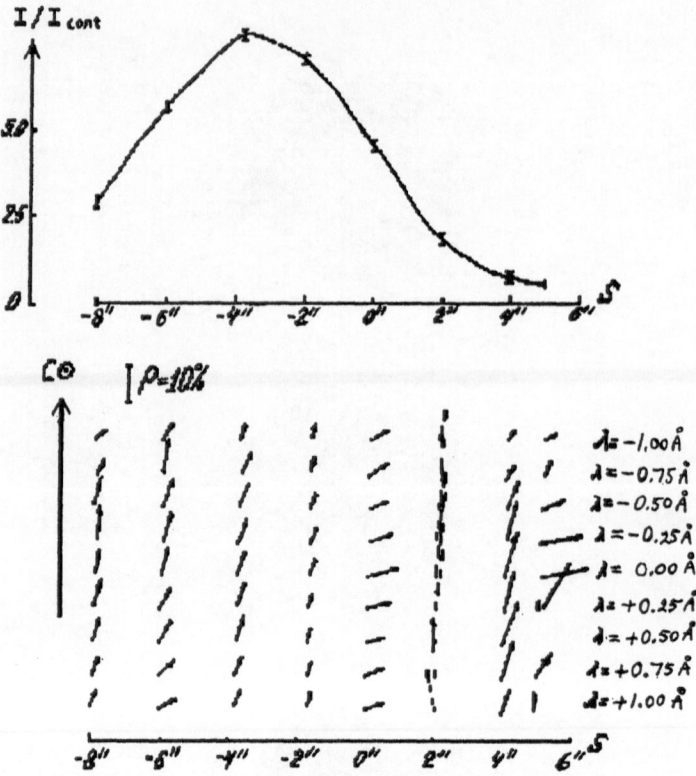

Fig. 8. Linear polarization vector in different sections and in different portions of the central part of the H$_\alpha$ line. The upper curve shows the intensity variation at the H$_\alpha$ line center. The arrow is directed towards the center of the solar disk.

H$_\alpha$ line center is well defined. In sections with high emission (from $-8''$ to $-1.75''$) the polarization vector is generally close to the direction towards the center of the solar disk, and the degree of polarization drops from 8% in section $-8''$ to 4% in section $-1.75''$. In section $0''$, where the H$_\alpha$ emission profile resembles moustaches, the degree of polarization increases again, and the polarization vector is oriented almost perpendicular to the direction toward the center of the solar disk. Moreover, in section $2''$ where the shape of the H$_\alpha$ profile also resembles moustaches, and in sections $4''$ and $5''$, which correspond to an almost unperturbed photosphere, the value of the degree of polarization increases substantially and reaches 20%. The direction of the polarization vector in sections $4''$ and $5''$ is close to the flare to disk center direction. The r.m.s. error of P and χ are 0.007 and $4°$, respectively, for sections within $-6''$ and $+2''$. In section $8''$ and especially in sections $4''$ and $5''$, this error is much larger.

5. Conclusions

The degree of polarization observed in the 16 May 1991 flare is on average 7%, and it reaches 20% in regions with weaker emission and in adjacent areas of the weakly perturbed chromosphere. The plane of polarization is parallel to the direction of the centre of the solar disk. The observed linear polarization vector can be explained (cf. Hénoux, 1991) by bombardment of the chromospheric hydrogen atoms by proton or neutral particle beams.

At flare locations where the shape of the H_α line profile reminds of moustaches (sections $0''$ and $2''$), the direction of the plane of polarization is mainly tangential (perpendicular to the flare to disk center direction). The H_α profiles in these sections agree with calculated H_α profiles excited by accelerated electrons bombarding the chromosphere (Fang et al., 1993), while the H_α line in other parts of the flare corresponds more to profiles, calculated for excitation by proton beams (Hénoux et al., 1993).

Thus we believe that the observed linear polarization in the flare was caused by bombardment of the chromosphere by accelerated particles: protons in the main part of the flare, and electrons in those flare regions, where the H_α line has the shape of a moustache profile.

References

Aboudarham, J., Berrington, K., Callaway, J., Feautrier, N., Hénoux, J.C., Peach, G., and Saraph, H.-E.: 1992, *Astron. Astrophys.* **262**, 302-307.

Babin, A.N., Koval, A.N.: 1983, *Izvest. KrAO* **66**, 89-102.

Babin, A.N., Koval, A.N.: 1985a, *Izvest. KrAO* **72**, 142-153.

Babin, A.N., Koval, A.N.: 1985b, *Izvest. KrAO* **79**, 3-8.

Babin, A.N., Koval, A.N.: 1985c, *Solar Phys.* **98**, 159-161.

Chambe, G., Hénoux, J.C.: 1979, *Astron. Astrophys.* **80**, 123-129.

Fang, C., Hénoux, J.C., Gan, W.Q.: 1993, *Astron. Astrophys.* **274**, 917-922.

Hénoux, J.C., Semel, M.: 1981, *God Solnechnogo Maksimuma*, **1**, 207-210.

Hénoux, J.C., Chambe, G., Heristchi, D., Semel, M., Woodgate, B., Shine, R., Beckers, J.: 1983a, *Solar Phys.* **86**, 115-122.

Hénoux, J.C., Chambe, G., Semel, M., Sahal, S., Woodgate, B., Shine, R., Beckers, J.: 1983b, *Astrophys. J.* **265**, 1066-1075.

Hénoux, J.C., Chambe, G.: 1990a, *J. Quant. Spectrosc. Radiat. Transfer* **44**, 193-201.

Hénoux, J.C., Chambe, G., Smith, D.F., Tamres, D., Feautrier, N., Rovora, M., Sahal-Brechot, S.: 1990b, *Astrophys. J. Suppl. Ser.* **73**, 303-311.

Hénoux, J.C.: 1991, in L.J. November (ed.), *Solar Polarimetry*, NSO/SP Summer Workshop Series No. 11, 285-295.

Hénoux, J.C., Fang, C., Gan, W.Q.: 1993, *Astron. Astrophys.* **274**, 923-930.

Kazantsev, S.A., Firstova, N.M., Gubin, A.V., Lankevich, N.A.: 1991, *Optika i Spektroskopiya* **70**, 990-995.

Kazantsev, S.A., Petrashen, A.G., Firstova, N.M., Hénoux, J.C.: 1993, *Optika i Spektroskopiya* **75**, 644-657.

Skomorovsky, V.I., Firstova, N.M.: 1995, *Solar Phys.* , in press.

Smith, D.F., Chambe, G., Hénoux, J.C., Tamres, D.: 1990, *Astrophys. J.* **358**, 674-679.

FOURIER PARAMETERS AND MOMENTS OF POLARIZATION PROFILES OF MAGNETICALLY ACTIVE LINES. FOURIER VECTOR MAGNETOGRAPH

B. IOSHPA and V. OBRIDKO

IZMIRAN, Troitsk, Moscow Region, Russia

and

I. KOZHEVATOV

NIRFI, Nizhny Novgorod, Russia

Abstract. A new method is proposed to determine all components of the solar magnetic fields using the cumulants of the profile of a magnetic sensitive line. The method is based on polarization measurements in a number of points of the line profile and subsequent calculation of the amplitudes and phases of its two first Fourier–harmonics.

Key words: Solar magnetic field – Fourier–magnetograph

1. Introduction

The methods employed to obtain vector magnetic fields on the Sun are usually based on measurements of the Stokes parameters in magnetoactive lines followed by interpretation with adequate radiative transfer modeling. As a matter of fact, one measures the polarized intensities either in selected regions of the profile (using Babcock–type magnetograph), or all over the profile (using Stokesmeter). The measurements can be carried out simultaneously or by scanning along the profile.

The magnetic field parameters, calculated from the measured polarization, are very sensitive to variations of the line profile due to the physical conditions in the line forming layer (such as temperature, pressure, and inhomogeneities), and strongly depend on the adopted model atmosphere.

The difficulties involved in the methods which employ a single wavelength can be avoided by using integral parameters of the profile (displacement of the gravity center, width, asymmetry etc.). Semel (1970) was one of the first to use relative displacements of the gravity centers of the σ–components of magnetoactive lines as a measure of the longitudinal magnetic field. The present work is, in some sense, an extension of this idea to the vector magnetic field.

Solar Physics **164**: 373–380, 1996.

© 1996 *Kluwer Academic Publishers.*

2. Cumulants and their relations to magnetic fields

As integral parameters of the line we shall take the cumulants determined from the following relations:

$$æ_n = j^{-n} \left[\frac{d^n ln \tilde{I}(w)}{dw^n} \right]_{w=0}, \tag{1}$$

where $\tilde{I}(w) = \int I(\lambda)e^{jw\lambda}d\lambda$ is the Fourier transform of the line depression profile $I(\lambda)$, and $j = \sqrt{-1}$. There exist relationships between the cumulants and the central and starting moments of I (Malakhov, 1978):

$$
\begin{aligned}
æ_1 &= \alpha_1 & &\text{- center of gravity of the line profile} \\
æ_2 &= \mu_2 = \alpha_2 - \alpha_1^2 & &\text{- square of the width of the line profile} \\
æ_3 &= \mu_3 = \alpha_3 - 3\alpha_1\alpha_2 + 2\alpha_1^3 & &\text{- asymmetry of the line profile}
\end{aligned}
$$

where

$$\mu_n = \frac{\int I(\lambda) \, (\lambda - \alpha_1)^n \, d\lambda}{\int I(\lambda)d(\lambda)},$$

are the central moments,

$$\alpha_n = \frac{\int I(\lambda) \, \lambda^n \, d\lambda}{\int I(\lambda)d(\lambda)},$$

are the starting moments, and $I(\lambda)$ is the line depression profile as a function of the wavelength measured from the nominal value. Though the first cumulants coincide with the central moments, we shall rather use the system of cumulants owing to their additivity in the convolution procedure (Malakhov, 1978).

As shown below, the knowledge of the two first cumulants of different states of polarization is sufficient to calculate all components of the vector magnetic field. Let us discuss one of the possible procedures. Let solar emission in a magnetoactive line with Stokes parameters I_0, V_0, Q_0 and U_0 be analyzed by a polarization analyzer consisting of 2 elements: a controlled birefringent plate and a linear polarizer. The operation of the analyzer will be described in the reference frame in which the axes coincide with those of the polarization ellipse, consequently $U_0 = 0$ [*]. The angle between the transverse magnetic field component and the plate axis is denoted by β, and the angle between the beam and the magnetic field vector by γ.

---

[*] In general, the polarization ellipse does not maintain a single orientation along the line profile and, consequently, such a single reference system cannot be found. However, some solutions of the radiative transfer equation do allow such a definition, in particular, those used at the end of this section.

The polarization analyzer has six states:

1) G $(\frac{\lambda}{4}, \beta)$; P $(\beta + 45°)$ – delay between the ordinary and extraordinary beams produced by the plate is $\frac{\lambda}{4}$, angle between polarizer axis and the fast axis of the plate is 45°. The arguments of G and P describe state of the retarder and the polarizer, respectively.

2) G $(-\frac{\lambda}{4}, \beta)$; P $(\beta + 45°)$ – delay is $-\frac{\lambda}{4}$, angles are the same.

3) G $(0, \beta)$; P $(\beta + 45°)$ – delay is zero, angles are the same.

4) G $(0, \beta)$; P (β) – delay is zero, polarizer axis turned by $-45°$.

5) G $(0, \beta)$; P $(\beta - 45°)$ – delay is zero, polarizer axis turned by $-90°$.

6) G $(0, \beta)$; P $(\beta - 90°)$ – delay is zero, polarizer axis turned by $-135°$.

It is easy to show that radiation intensities at the analyzer output for each of its 6 states can be described as follows:

$$I^{(1)} = 0.5 I_o - 0.5 V_o$$

$$I^{(2)} = 0.5 I_o + 0.5 V_o$$

$$I^{(3)} = 0.5 I_o - 0.5 \sin(2\beta) Q_o$$

$$I^{(4)} = 0.5 I_o + 0.5 \cos(2\beta) Q_o \tag{2}$$

$$I^{(5)} = 0.5 I_o + 0.5 \sin(2\beta) Q_o$$

$$I^{(6)} = 0.5 I_o - 0.5 \cos(2\beta) Q_o.$$

Now, let us calculate the relationships between the original Stokes parameters and the cumulants for six states of the analyzer. For the sake of simplicity, assume that I_0 and Q_0 are strictly even functions of λ and V_0 is strictly odd function of λ. By substituting $I^{(1)} - I^{(6)}$ in the expressions for moments we obtain:

$$\mathfrak{x}_1^{(1)} = -\mathfrak{x}_1^{(2)} = -A(H, \gamma) \cdot \cos \gamma$$

$$\mathfrak{x}_1^{(3)} = \mathfrak{x}_1^{(4)} = \mathfrak{x}_1^{(5)} = \mathfrak{x}_1^{(6)} = 0, \tag{3a}$$

$$æ_2^{(1)} = æ_2^{(2)} = B(H,\gamma) - A^2(H,\gamma) \cdot \cos^2\gamma$$

$$æ_2^{(3)} = \frac{B - C\sin 2\beta}{1 - D\sin 2\beta}$$

$$æ_2^{(4)} = \frac{B + C\cos 2\beta}{1 + D\cos 2\beta} \tag{3b}$$

$$æ_2^{(5)} = \frac{B + C\sin 2\beta}{1 + D\sin 2\beta}$$

$$æ_2^{(6)} = \frac{B - C\cos 2\beta}{1 - D\cos 2\beta},$$

where

$$A(H,\gamma) = \frac{\int V_0\lambda d\lambda}{\int I_0 d\lambda} \cdot \frac{1}{\cos\gamma}, \tag{4}$$

$$B(H,\gamma) = \frac{\int I_0\lambda^2 d\lambda}{\int I_0 d\lambda}, \tag{5}$$

$$C(H,\gamma) = \frac{\int Q_0\lambda^2 d\lambda}{\int I_0 d\lambda}, \tag{6}$$

$$D(H,\gamma) = \frac{\int Q_0 d\lambda}{\int I_0 d\lambda}, \tag{7}$$

and H is the magnetic field strength. However, the signals under consideration depend on properties of the atmosphere (pressure, temperature, velocity and magnetic fields, and the variation of these parameters along the line-of-sight). In order to simplify the computation, we adopt the same approximation as Unno (1956). Under this approximation, A, B, C, and D, depend on constant values of H (the field strength), γ (the inclination angle), η_0 (the line-to-continuum opacity ratio), g (the Lande factor), $\Delta\lambda_D$ (the Doppler halfwidth of the line) and the two coefficients which describe the source function. Figure 1 shows calculations based on this approach. The parameters have been chosen to correspond to those of the photospheric Fe I 5250 Å line. The ratio η_0 of opacities at the line center to that in the continuum was taken equal to 10. The field strength is given in units of the Doppler halfwidth ($H = 1$ corresponds approximately to 1000 G). A is expressed in Doppler halfwidths, B and C in square of the Doppler halfwidths.

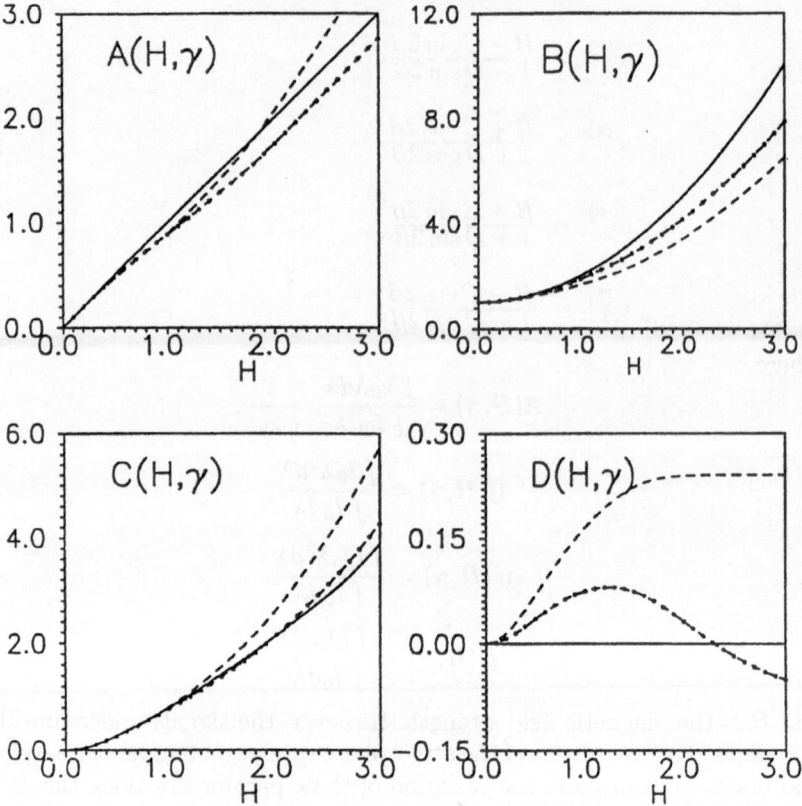

Fig. 1. Dependence of A, B, C, and D on the field strength H. The simulations refer to the line FeI 5250 Å observed in the photosphere, so that $H = 1$ corresponds approximately to 1000 G. The solid line is used for $\gamma = 0$, the dashed line for $\gamma = \pi/2$, and the dash–dotted line for $\gamma = \pi/4$.

3. Calculating cumulants by Fourier–analysis.
Fourier vector magnetograph

Below we describe the method for calculating the cumulants by Fourier–analysis of the line profile. For this purpose, we use expressions connecting the amplitudes and phases of the Fourier harmonics of the line profile with its cumulants (Didkovsky, Kozhevatov and Stepanyan, 1986):

$$\Psi(\Delta\lambda) = æ_1\frac{1}{\Delta\lambda} - æ_3(\frac{1}{\Delta\lambda})^3\frac{1}{3!} + ...,$$

$$ln\frac{\tilde{A}(\Delta\lambda)}{S} = -æ_2(\frac{1}{\Delta\lambda})^2\frac{1}{2!} + æ_4(\frac{1}{\Delta\lambda})^4\frac{1}{4!} + ..., \qquad (8)$$

$$S = \int I(\lambda)d\lambda,$$

where $\Delta\lambda = 2\pi/\omega$ is the period of the Fourier harmonic, and $\tilde{A}$ and Ψ are its amplitude and phase, respectively.

The two first cumulants are enough to determine the magnetic field under our simplifying assumptions (see Section 2). However, the determination of the first three cumulants might allow a more elaborate modeling (for instance, taking into account the asymmetry of the line profiles). Because of this future possibility, and since the determination of two or three harmonics requires a similar computational effort, here we describe how to compute $\neq_1$, $\neq_2$ and $\neq_3$. To determine the three first cumulants, we need to know the amplitudes and phases of the two first harmonics of the Fourier–transform of the line profile (the domain of definition of the line profile is considered as the period of the first harmonic). As seen from Equation (8), the phases of the long–period harmonics (those with small $1/\Delta\lambda$) are proportional to the shift of the center of gravity ($æ_1$). In order to determine $æ_1$ and $æ_3$, we have to know the phases of two harmonics (with small $1/\Delta\lambda_1$ and with somewhat larger $1/\Delta\lambda_2$). A similar procedure is applied to determine the even cumulants from the amplitudes. The period of the first harmonic is large enough, so that the first term dominates the series for the phases. The term $\Delta\lambda^{-3}$ and higher order terms can be neglected, and consequently, $\neq_1$ is determined from the phase of the first harmonic. The second harmonic has a smaller period; in this case the series for the phases has two major terms while the series for the amplitudes has just one term. Since the first cumulant and S are known, the expressions for the phase and amplitude of the second harmonic are used to derive the third and the second cumulants.

According to the Kotelnikov – Shannon criterion (see Bell, 1972), the characteristics of the Fourier harmonics can be determined by making records at discrete points spaced by no more than 1/3 of the harmonic period. Therefore, the intensities measured at six points along the profile are enough to determine the amplitude and phase of the two first harmonics. In this case one condition should be observed: before discretization, higher–frequency harmonics must be eliminated from the line profile to avoid aliasing (see, e.g., Bracewell, 1978). The smoothing can be done by selecting the appropriate resolution of the instrument.

The procedure of determining magnetic fields by the proposed Fourier–method comprises the following stages:

1. Measure intensities at six discrete equidistant points along the line profile at six positions of the polarization analyzer.

2. Calculate the phases and amplitudes of the two first harmonics of the line profile.

3. Use these to find two or three first cumulants of the line profile.

4. Determine the magnetic field parameters from Equations (3)-(7). These expressions yield at least 6 independent equations for 3 unknown values, H, γ, and β. Consequently, one expects no unique solution and, therefore, some kind of least squares best fit has to be sought. This procedure has not been developed yet.

Note that parameters of the Fourier–harmonics can be obtained by a different method than proposed in items 1 and 2 above, for example, directly by using Michelson interferometers (Bell, 1972; Didkovsky, Kozhevatov and Stepanyan, 1986).

4. Conclusions

We believe that the proposed method for determining magnetic field parameters from the cumulants of the line profile offers a number of advantages, compared with the traditional Babcock method, based on the comparison of polarized light intensities at two symmetric points in the line profile.

1. The relation between the cumulants and the magnetic field value does not show saturation or back pass at large field intensities.

2. The magnetic field is determined from integral properties of the line profile, so that the result will not be affected strongly, by the asymmetries of the profile.

3. The use of integral properties will considerably reduce the effect of spatial irregularities (see Rees and Semel, 1979).

4. When determining the magnetic field with a Babcock–type filter magnetograph, the presence of non–compensated line-of-sight velocity may result in significant errors. Our method, in which the effects of the magnetic field and the line–of–sight velocity are separated, does not have this disadvantage, and to a certain extent, it resembles the photographic method. (For detailed description of the photographic method see, for example, Bray and Loughead, 1964)

5. Our method requires measurements at a limited number of points along the line profile.

Acknowledgements

We thank Jorge Sánchez Almeida for valuable comments on the manuscript.

References

Bell, J.: 1972, *Introductory Fourier Transform Spectroscopy*, Academic Press, New York

Bracewell, R.N.: 1978, *The Fourier Transform and its Application*, McGraw Hill, New York

Bray, R and Loughead, R.: 1964, *Sunspots*, Chapman and Hall Ltd., London

Didkovsky, L.V., Kozhevatov, I.E., and Stepanyan, N.N.: 1986, *Izvestia Krimskoy Astrophysicheskoy Observatorii (SU)* **74**, 142

Malakhov, A.N.: 1978, *Cumulant analysis of Random non-Gaussian Processes and their Transformation*, Sov. Radio, Moscow

Rees, D.E., and Semel, M.D.: 1979, *Astron. Astrophys.* **74**, 5

Semel, M.D.: 1970, *Astron. Astrophys.* **5**, 330

Unno, W.: 1956, *Publ. Astron. Soc. Japan* **8**, 108

ASPECTS OF THE ZERO LEVEL PROBLEM OF SOLAR MAGNETOGRAPHS

M.L. DEMIDOV

The Institute of Solar-Terrestrial Physics at the Siberian Division of the Russian Academy of Sciences (ISTP SD RAS) 664033, Russia, Irkutsk, P.O.Box 4026

Abstract. The zero level problem of solar magnetographs is particularly important for observations of large-scale magnetic fields on the Sun. Experiments conducted at the STOP telescope of the Sayan observatory show that, in addition to adjustment errors of the polarization analyzer and the spectrograph focusing, spurious signals of the magnetograph are caused by polarization effects in optical components preceding the polarization analyzer and aberration errors of the spectrograph.

Key words: Solar Magnetic Fields – Solar Magnetographs

1. Introduction

Measurements of large-scale magnetic fields (LSMF) on the Sun have signal levels of several hundred μT for background magnetic fields (BMF) and only several tens of μT for the solar mean magnetic field (SMMF). This corresponds to a degree of circular polarization of $10^{-3} - 10^{-4}$. Diverse instrumental problems manifest themselves as a non-zero magnetograph signal in non-magnetic lines. This raises the issue of the overall reliability of magnetographic measurements and forms the basis for the so-called zero-level problem.

There are currently two reliable methods used to monitor the zero level position: (a) with a non-magnetic line and (b) with a half wave plate. The former method is being used at the Crimean (Severny, 1969) and Stanford (Scherrer et al., 1977; Hoeksema, 1984) observatories, while the latter is being used at the STOP (Solar Telescope for Operative Predictions) telescope of the Sayan Solar Observatory (SSO, see Grigoryev and Demidov, 1987).

STOP is an 18 cm horizontal doublet refractor designed for high precision magnetic field measurements with low spatial resolution (from some dozen arc seconds to observations of the Sun as a star). It has a 30 cm coelostat, a Littrow spectrograph, a Babcock-type longitudinal magnetograph with DKDP crystals running at 500 Hz and a photomultiplier. The zero level is controlled with a $\lambda/2$ plate that is periodically inserted in front of the coelostat. Such a plate reverses the circular polarization. Hence if S_1 and S_2 are measurements with and without the $\lambda/2$ plate, the value of the zero level shift ΔS_0 is

$$\Delta S_0 = (S_1 + S_2)/2, \tag{1}$$

Solar Physics **164**: 381–388, 1996.

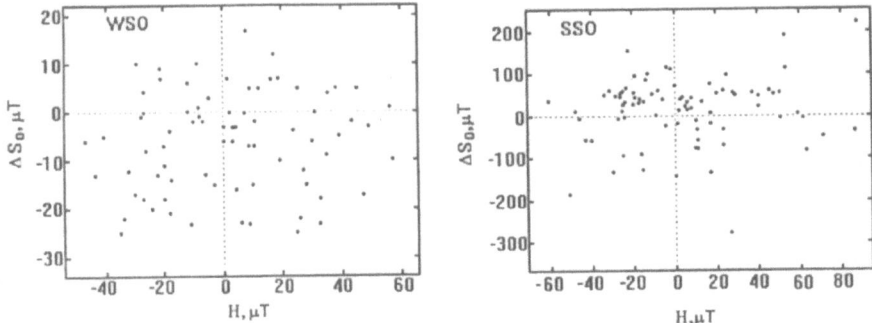

Fig. 1. Left panel: Comparison between the values of the zero level ΔS_0, detected in the Fe I λ512.37 nm line, and the magnetograph signal H in SMMF observations at WSO in December 1993. Right panel: Comparison between the values of the zero level ΔS_0, detected using a $\lambda/2$ plate, and the magnetograph signal H in SMMF observations at the Sayan Observatory in 1993.

which can then be subtracted from the measurements.

The zero level problem was investigated in detail by Duvall (1977) for the WSO (Wilcox Solar Observatory) magnetograph. By measuring in the non-magnetic Fe I line λ512.37 nm and performing analytical calculations, he found that ΔS_0 may be approximated by the formula

$$\Delta S_0 = K \cdot U \cdot F \cdot Q, \tag{2}$$

where K is a constant, U is the Stokes parameter describing the linear polarization with respect to the axes of the electrooptical analyzer (EOA), F is the displacement of the photometer exit slits from the exact focus, and Q is the angle of inclination of the EOA to the light beam incident on it. With perfect focusing ($F=0$), a zero level displacement must, according to Eq. (2), be entirely absent. In practice, however, Duvall was unable to fully eliminate the zero level problem. The left panel of Figure 1 shows the values of the zero level displacements versus the magnetograph signal for December 1993. Each point corresponds to a separate measurement, and usually there are a few such measurements in a day. $|\Delta S_0|$ is typically less than 25 μT. Data for other time intervals behave in a similar way.

The zero level displacement in SMMF measurements at STOP is given by the right panel of Figure 1. It exceeds considerably those observed at Stanford but are comparable to or even smaller than those at Crimea (Kotov, private communication). Thus, observations of LSMF at different observatories demonstrate that, despite careful adjustment of the spectrographs and polarization analyzers, it is still impossible to fully eliminate the zero level problems. Obviously, there are additional mechanisms that have a substantial influence on the generation of spurious magnetograph signals.

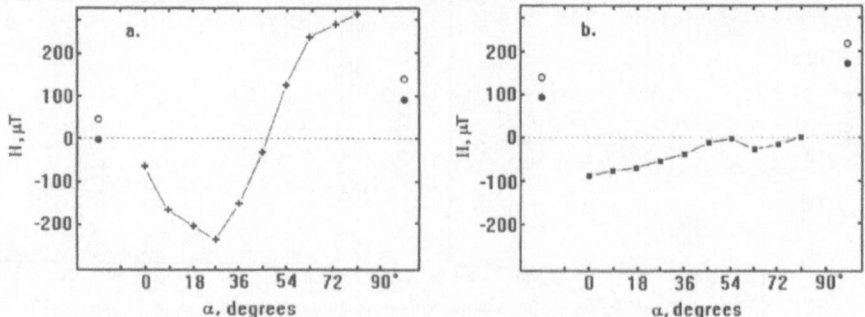

Fig. 2. Variation of the STOP magnetograph signal when rotating the $\lambda/2$ plate in front of the objective lens (a) and behind it (b). Full circles at the beginning and end of each figure show measurements of the SMMF when the $\lambda/2$ plate was installed in front of the coelostat, while open circles show the situation without any plate.

2. Role of the Objective and the Spectrograph Aberrations for the Zero Level Problem

To study the influence of the orientation of the linear polarization on the zero level position, a half wave plate rather than a polarizer was used at STOP because the light from the coelostat mirrors is partially linearly polarized. Inserting such a plate into the light beam changes the sign of the circular polarization, while the rotation of the plate by an angle α causes the azimuth of the linear polarization to rotate by 2α. Relevant measurements with the $\lambda/2$ plate were made both in the magnetic line Fe I $\lambda525.02$ nm and in the non-magnetic line Fe I $\lambda512.37$ nm. The plate was placed at different locations in the optical system: ahead of the coelostat, between the coelostat and the objective lens, and between the objective lens and the polarization analyzer. The rotation of the plate ahead of the coelostat does not affect the zero level position. If, however, the plate is placed behind the coelostat, it leads to an abrupt change in the zero level position, even to the extent of changing its sign. The character of the variation of ΔS_0 caused by rotating the plate depends greatly on whether it is ahead of or behind the objective lens. Figure 2 shows results at STOP in the Fe I $\lambda525.02$ nm line with the wave plate rotated by $18°$ from some arbitrary position in front of and behind the objective lens. If the plate is installed behind the objective lens, a large change of the signal occurs, but rotation of the plate does not lead to variations as it does in front of the objective lens.

This may be explained by the anisotropy of the birefringence of the lens. Different parts of lens may be considered as different retardation plates. When operating the electro-optical modulator this leads (in combination with instrumental polarization of the light incident on the coelostat mirrors)

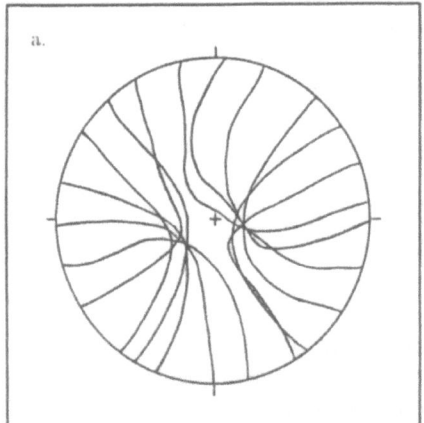 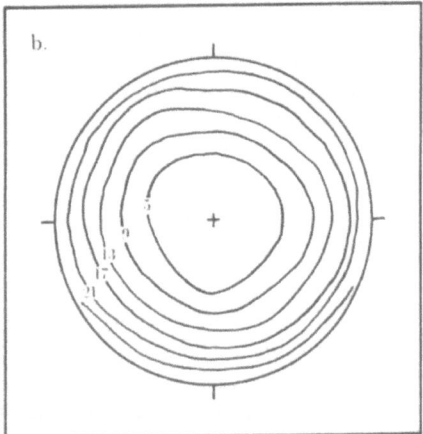

Fig. 3. Polarization properties of the STOP double objective lens: (a) Isocline distribution
map obtained by rotating the polarizers by 15°, when the objective lens is between them.
(b) Birefringence distribution map from measurements with a circular polariscope (in units
of degrees, at a wavelength of λ632.8 nm).

to the complex and variable distribution of brightness across the aperture.
In other words, the system 'objective + coelostat' acts like a kind of polarization analyser that is used for observations of velocities on the Sun by
differential methods. If the plate were placed behind the objective it should
only lead to a change of the sign of the signal, but rotation of the plate
should not influence its value, in accordance with our experimental data.

Measurements show that the STOP objective has birefringence that is
relatively small in the central zone but up to 25° at the edges. Figure 3
shows isoclines of the lens between crossed polarizers and the birefringence.
A significant vertical gradient of the birefringence is present. In combination with the linear instrumental polarization this leads to a considerable
asymmetry of the brightness distribution over the aperture after the light
has passed through the polarization analyzer. For example, with 10 % linear
polarization the brightness values at the center of the objective lens and at
the edges (with a birefringence of 20°) differ by more than 3 % .

As applied to the zero level problem discussed here, it is the deviations
from axial symmetry of the brightness distribution that have the largest
effect. If the brightness across a diameter at opposite edges of the objective
lens varies only by 0.5 % , and if this diameter is perpendicular to the solar
rotation axis, then in two modulation cycles the effective wavelength of an

average spectral line profile at the opposite edges will differ by 1.7×10^{-5} nm, which corresponds to ≈ 10 m/s or a magnetograph signal of ≈ 200 μT. Because of large gradients of the line-of-sight velocities across the field of view, caused by solar rotation and the limb red shift, large zero level displacements in the SMMF measurement may now be due exclusively to the birefringence asymmetry of the objective lens.

The zero level problem also depends on the modulator performance characteristics and the optical properties of the spectrograph. The zero level position is most sensitive to the focusing of the spectrograph. In the presence of spectrograph defocusing the point image turns into a spot of finite size. If the spot includes a brightness gradient in the dispersion direction, the modulator produces a change of the spot's center of gravity, which is perceived as a spurious magnetograph signal (Duvall, 1977).

If the radius of the spot is r and if there is a transverse brightness gradient ∇, the relative variations of the center of gravity with a change of sign of the gradient will be $\Delta x = 2 \cdot \nabla \cdot r/3$. For instance, when $r = 0.04$ mm (which for the STOP spectrograph corresponds to defocusing by only 2 mm), and with a 1 % brightness gradient, the variation of the center of gravity will be 26×10^{-5} mm, which for 0.04 nm/mm dispersion corresponds to 10.4×10^{-6} nm. For the Fe I λ525.02 nm line such a wavelength difference corresponds to a line-of-sight velocity of 6 m/s or to a magnetograph signal of 135 μT.

Blurring of the point image is besides defocusing also caused by many other factors: diffraction, aberration, etc. For example, the size of the Airy diffraction disk with the parameters of the STOP spectrograph objective lens is 0.034 mm (for λ525.0 nm), which corresponds to 1.3×10^{-3} nm. In the case with no diffraction such a broading of the point image corresponds to defocusing by about 1 mm. Since actual errors of the spectrograph optics generally lead to even greater blurring, it is impossible to form a sharp point image. For instance, plane-parallel plates in the spectrograph produce spherical aberration. Therefore the spectrograph should only contain the diffraction grating and the Littrow lens. This is the design of the Stanford magnetograph (unlike the Crimean magnetograph and STOP). It therefore seems likely that also for this reason much smaller values of the zero level displacements occur there.

The energy distributions of a point image for λ525.02 nm and λ512.37 nm differ significantly from each other (see Figure 4, which was calculated under the assumption of geometrical optics). The size of the scattering circle for λ512.37 nm exceeds that for λ525.02 nm. Observations in the magnetic line and zero level monitoring with the non-magnetic line are performed under differing conditions. Furthermore different spectral lines that vary differently across the solar disk differ markedly (Balthasar, 1988).

We have made SMMF and BMF observations at STOP with zero level monitoring using two methods: (a) With the non-magnetic line Fe I λ512.37

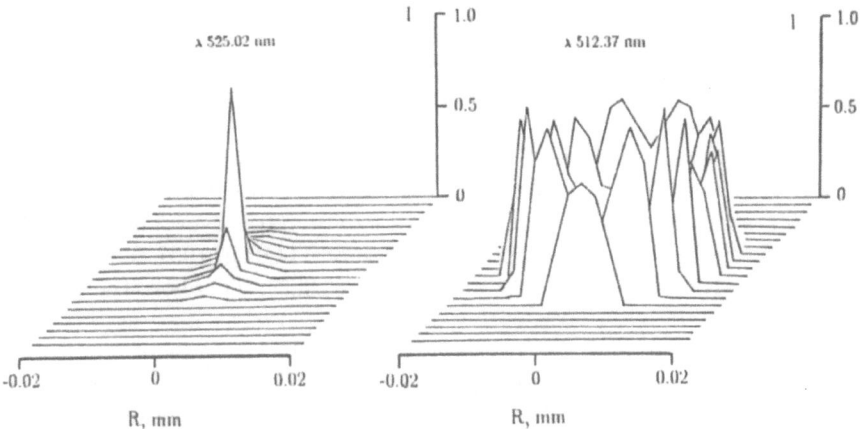

Fig. 4. Results of calculations of the energy distribution across the point image of the STOP autocollimation spectrograph for λ525.02 nm and λ512.27 nm.

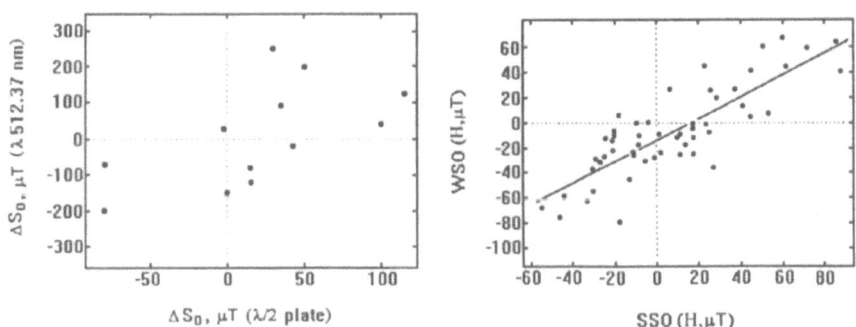

Fig. 5. Left panel: Comparison of zero levels measured with the non-magnetic Fe I λ512.37 nm line and with a $\lambda/2$ plate. The measurements were made in the SMMF observing mode at STOP. Right panel: Comparison of SMMF measurements in 1993 from WSO and SSO (see text).

nm, and (b) with a half wave plate. The values of the zero levels as determined by different methods can differ considerably (by 100 μT or more, see
left panel of Figure 5). Data obtained by monitoring the zero level with the
half wave plate are more reliable.

The right panel of Figure 5 compares SMMF observations from the Sayan
and Stanford observatories for 53 days in 1993. For three of the days when
measurements were available from SSO but not from WSO, values for the
magnetograph signals were obtained by interpolating the observations be-

tween neighboring days. The correlation coefficient is 0.83, and the linear regression equation has the form

$$H_{\mathrm{wso}} = -13.9(\pm 2.77) + 0.87(\pm 0.08) \cdot H_{\mathrm{sso}}. \tag{3}$$

The agreement between Sayan and Stanford SMMF measurements is quite satisfactory and much better than reported in some earlier publications for analysis of both previous Sayan data (Grigoryev and Demidov, 1987) (the situation changed due to redesign of STOP in 1991–92) and data from other observatories (Kotov and Severny, 1983). Therefore the method of monitoring the zero level with a half wave plate, as used at STOP, is sufficiently reliable, and the difference between the methods of monitoring the zero level does not lead to large systematic differences. Nevertheless, it should be recognized that this difference, along with the variation of the polarization properties of objective lenses and instrumental weighting functions in general, is a source of significant discrepancies occurring on individual days.

3. Discussion and Conclusions

In addition to the problems considered by Duvall (1977), other factors are important for the zero level problem: (1) Inhomogeneities of the birefringence of the objective lens or entrance window (due to imperfections or gravitational, thermal and other factors) coupled with instrumental polarization from the coelostat, (2) aberrations in the spectrograph. In observations of the SMMF (or regions of the solar surface with significant line-of-sight velocity gradients along the field of view), factor (1) remains important even under ideal operating conditions.

Based on the preceding considerations the Duvall formula for the displacement of the magnetograph zero level may be modified and generalized:

$$\Delta S_0 \sim K_1 \cdot \Delta A \cdot Q \cdot U + H' + R, \tag{4}$$

where K_1 is a constant, ΔA is the size (in the dispersion direction) of the point image in the plane of the spectrograph exit slits, blurred by defocusing of the spectrograph and by optical aberrations in the spectrograph. Q and U have the same meaning as in Eq. (2). H' is the value of the zero level shift caused by effects of the entrance window. R stands for residual shifts, whose origin is still unclear, and which can be related to operational peculiarities of the polarization modulator.

To minimize the zero level problem in magnetographic observations, it is necessary to eliminate the birefringence and the inhomogeneities of the polarization properties for the optical elements that precede the polarization analyzer. It is more appropriate to use mirror rather than lens objectives. Another recommendation is that it is necessary to achieve as high accuracy as possible for the spectrograph adjustment and to reduce the aberrations. In particular one should use inside spectrographs as few optical components as possible, because even thin plane-parallel plates lead to appreciable distortions in converging light beams.

In practice it is impossible to eliminate all the factors responsible for the zero level shift. Therefore a monitoring of the zero level position is inevitable. Preliminary analysis shows that our method of monitoring the zero level using a half wave plate may be regarded as more advantageous from a conceptual point of view.

Acknowledgements

I am grateful to Drs. V.M. Grigoryev and V.A. Kotov (Crimean Astrophysical Observatory) for helpful discussions on some questions raised in the present paper, and to Dr. P. Scherrer (WSO, Stanford University), who kindly provided me with the Stanford observations and let me use them for the present study. I also thank the referee for a careful reading of the manuscript and for useful comments.

The research described here was made possible in part by Grant NN 3000 from the International Science Foundation and Grant NN 3300 from the International Science Foundation and the Russian Government, as well as by Grant B-002-004 of the ESO C&EE Programme.

References

Balthasar, H.: 1988, *Astron.Astrophys. Suppl.Ser.* **72**, 473
Duvall, T.L.: 1977, 'A Study of Large-Scale Solar Magnetic and Velocity Fields', Ph.D. Dissertation, Stanford University, SUIPR Report No. **724**
Grigoryev, V.M., Demidov, M.L.: 1987, *Solar Phys.* **114**, 147
Hoeksema, J.T.: 1984, 'Structure and Evolution of the Large-Scale Solar and Heliospheric Magnetic Fields', Ph.D. Dissertation, Stanford University, CSSA-ASTRO-84-07
Kotov, V.A., Severny, A.B.: 1983, in 'Mean Magnetic Field of the Sun as a Star. Catalogue 1968-1976', Material from World Data Centre B, Moscow.
Scherrer, P.H., Wilcox, J.M., Svalgaard, L., Duvall, T.L., Dittmer, P.H., Gustafson, E.K.: 1977, *Solar Phys.* **54**, 353
Severny, A.: 1969, *Nature* **224**, 53

SINGULAR POINTS OF THE POLARIZATION TENSOR

M.M. MOLODENSKY and L.I. STARKOVA

IZMIRAN, 142092, Troitsk, Moscow Region, Russia

Abstract. The problem to compute the magnetic field above the chromosphere using data of the vector $\tau = \mathbf{B}_t/B_t$ that gives the projected field direction can be solved with different approximations. The field of direction vectors τ is, however, not the only field accessible to observations. The Stokes parameters, which are components of the radiation tensor, can be measured at each point of the image plane. The directions of the eigenvectors of the radiation tensor define two mutually orthogonal systems of integral curves in the image plane. These families of curves have singular points, which are generally of different type than those of the vector field. When the morphology of $H\alpha$ chromospheric fibrils are used to infer the topology of the magnetic field, a similar problem is met, suggesting that singular points should also be present there.

Key words: Polarization – Fibrils – Chromosphere – Crab Nebula – Solar Eclipse – Singular Points

1. Introduction

To start an analysis of the magnetic field in the solar chromosphere based on data giving the orientation of fibrils, the types of singular points and their arrangement should first be determined (Kulikova *et al.*, 1989, 1990). In the 2–D vector field $\mathbf{B}_t$ (component of the magnetic field tangential to the solar surface) the entire list of singular points is known. They are the node, the focus, the saddle and the center. Each type of singular point of the magnetic field can be related to a well–established chromospheric structure (see Molodensky and Starkova, 1990). The vector nature of the field $\mathbf{B}_t$ and the possibility to relate the field of τ–directions to the orientation of fibrils play an important role in such an analysis. However, the τ–direction is not the only special direction in the image plane. The radiation at each point of the area is, generally speaking, polarized and characterized by the Stokes parameters forming the radiation tensor:

$$\hat{J} = \frac{1}{2}\begin{pmatrix} I+Q & U+iV \\ U-iV & I-Q \end{pmatrix} . \tag{1}$$

The orientation and degree of polarization can be defined in terms of the eigenvectors of tensor (1). The eigenvectors define two special directions in the image plane that are perpendicular to each other.

As was shown by Kulikova *et al.* (1989) for the case of the vector field, it is also possible to find singular points of the field (1) for polarization maps

Solar Physics **164**: 389–396, 1996.

measured, for example, for the Crab Nebula, and to perform a qualitative analysis of this field. This is the aim of the present paper.

2. Singular points of the polarization tensor

The eigenvectors of the polarization tensor are defined by the equation

$$\hat{J}\mathbf{l} = \lambda \mathbf{l} \,. \tag{2}$$

The solution of Equation (2) is

$$\lambda_{1,2} = I \pm (Q^2 + U^2 + V^2)^{1/2} \,, \tag{3}$$

and

$$\begin{aligned} \mathbf{l}_1 &= \{U + iV, \quad Q - (Q^2 + U^2 + V^2)^{1/2}\} \,, \\ \mathbf{l}_2 &= \{U + iV, \quad Q + (Q^2 + U^2 + V^2)^{1/2}\} \,. \end{aligned} \tag{4}$$

Thus

$$(\mathbf{l}_1 \, \mathbf{l}_2) = 2V(iU - V) \,, \tag{5}$$

and for linear polarization we have two special, mutually orthogonal directions $\mathbf{l}_1$ and $\mathbf{l}_2$ at each point of the image plane. If $\hat{J}$ depends smoothly on the coordinates in the image plane, the two families of integral lines $dx/l_{1x} = dy/l_{1y}$ and $dx/l_{2x} = dy/l_{2y}$ form the "grid" of an orthogonal, curvilinear coordinate system. At some points, generally speaking, the polarization becomes zero, and Eqs. (3) and (4) give $\lambda_1 - \lambda_2$, $\mathbf{l}_1 = \mathbf{l}_2$. According to Eq. (4) $\mathbf{l}_1 = \mathbf{l}_2 = 0$ in this case. Thus the directions of the integral curves are lost at some points. By analogy with the singular points of the vector field, these points can be called the singular points of the radiation tensor. These points, however, are different from those of the vector field.

The indices of the singular points of the vector field are always integer (Petrovsky, 1970). This feature is connected with the circumstance that after tracing around a closed path surrounding a singularity, we come back to the starting point, where the direction of the vector coincides with its initial direction. Thus the full angle of rotation is $2\pi n$, where n is an integer.

In the case of linear polarization only the plane of the electric vector can be specified, and the directions $\mathbf{l}_1$ and $-\mathbf{l}_1$ cannot be distinguished. Therefore the singular points have several half-integer indices. These singular points are shown in Fig. 1: half-node (a), half-saddle (b) and half-center (c). The same types of singular points are known to be typical for liquid crystals with long molecules.

It is obvious that if the map of directions of $\mathbf{l}_1$ (ignoring the sign) represents a half-node, the field of $\mathbf{l}_2$ corresponds to a half-center. In the case of

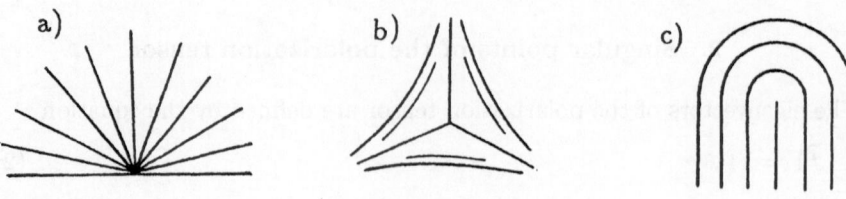

Fig. 1.   Three types of singular points: half-node (a), half-saddle (b) and half-center (c) .

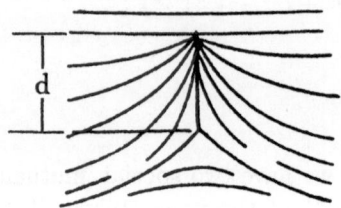

Fig. 2.   Pair of singular points with indices +1/2 and −1/2 .

a half-saddle, however, both l_1 and l_2 represent half-saddles, one of which is rotated by π with respect to the other. It is easy to verify that points of (b) type have index −1/2 and those of the (a) and (c) types correspond to index +1/2
(see Fig. 1).

As in the case of the node-saddle configuration of the magnetic field (see Molodensky and Syrovatsky, 1979) the structure, which cannot be resolved due to the limited spatial resolution d of the telescope, consists of a pair of singular points with indices +1/2 and −1/2.

This is the simplest configuration which turns into an apparently uniform field when the distance between two singular points becomes smaller than d. Figure 2 shows such an elementary configuration.

3. Application

The existence of degenerate singular points having indices 2 (dipole), 3, etc. is possible for a vector field. Correspondingly, the existence of points having

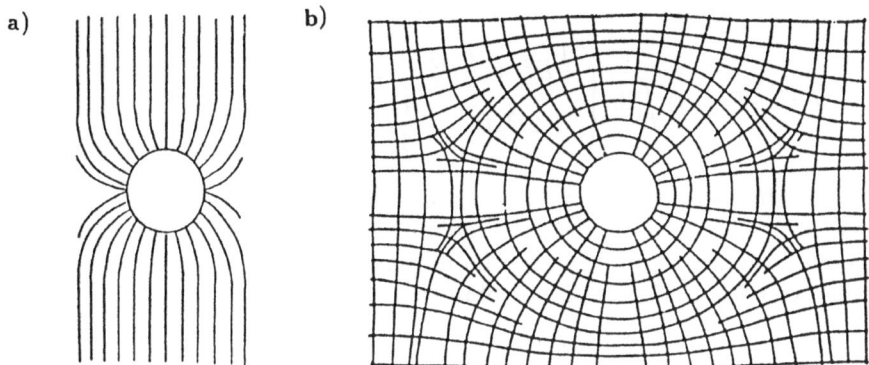

Fig. 3.   a. Polarization map obtained during a solar eclipse. b. Possible structure of the integral lines of polarization.

integer indices is possible also for the field of a radiation tensor, but this is a degenerate case.

Let us consider a polarization map obtained at a solar eclipse as an example. The configuration of the integral lines of polarization was found by Kulidganishvilly (1985). However, points of half-saddle type are not shown in this picture (see Fig. 3a). A possible structure of these lines that turns into a uniform field of polarization is shown in Fig. 3b. A degenerate singular point of center-type is created by the polarized emission of the K corona, and the polarization of the sky is due to the "aurora ring", i.e. to the Rayleigh scattering of surrounding light by the earth's atmosphere. The most favourable conditions to obtain a full picture of both the K+F corona polarization and the polarization of the eclipse sky seem to occur when the sun is at a zenith distance of about $40° - 50°$ (the emission of the ring at larger zenith distances is strongly non-uniform, and at smaller distances the scattered light polarization tends to zero).

In the vicinity of the singular points in Fig. 3 the polarization is close to zero, so direct measurements of the angles determining the direction of the polarization plane become difficult. Knowledge of the types of singularities allows the reconstruction of the corresponding configuration in a self-consistent way. An experimental determination of the positions of the singular points is of great interest in this context.

As shown by Molodensky (1969) the Stokes parameters are related to the Maxwell tensor in the emission area if the radiation is due to the magnetic field. The synchrotron radiation is of greatest interest, because numerous polarization measurements are described in the literature for this case. In the case of synchrotron radiation the equation with a spectral index $\alpha = 1$ takes the form

$$\hat{J} = \frac{1}{2}F(\nu, \gamma) \int dz \begin{pmatrix} \frac{1}{3}B_x^2 + \frac{7}{3}B_y^2 & 2B_xB_y \\ 2B_yB_x & \frac{1}{3}B_y^2 + \frac{7}{3}B_x^2 \end{pmatrix} ,$$ (6)

where the integration is performed along the line of sight z. x and y lie in the image plane, F is a function of the frequency ν and of the spectral index γ. Replacing the integration in Eq. (6) with a multiplication by some characteristic thickness Δ, one obtains:

$$\hat{J} = c(\nu, \gamma) \begin{pmatrix} C_1B^2 + B_y^2 - B_x^2 & 2B_xB_y \\ 2B_xB_y & C_1B^2 + B_x^2 - B_y^2 \end{pmatrix} ,$$ (7)

that is

$$I \sim C_1B^2 , \qquad Q \sim B_y^2 - B_x^2 , \qquad U \sim 2B_xB_y .$$

Thus, the degree of polarization turns out to be zero if

$$4B_x^2B_y^2 + (B_x^2 - B_y^2)^2 = (B_x^2 + B_y^2)^2 = 0 ,$$ (8)

which is possible only if $B_x = B_y = 0$. Hence the singular points of $\hat{J}$ and $\mathbf{B}_t$ coincide in this case. Equations (4) and (7) yield

$$\begin{aligned} l_1 &= \{1, \frac{2B_y^2}{2B_xB_y}\} \sim \{B_x, B_y\} , \\ l_2 &\sim \{-B_x, B_y\} , \end{aligned}$$ (9)

(here we have used the circumstance that the eigenvectors of $\hat{J}$ are defined up to an arbitrary multiplication factor). Two important conclusions follow from Eq. (9):

1. As long as the singular points of $\hat{J}$ and $\mathbf{B}_t$ coincide, and the singular points of $\mathbf{B}_t$ have integer indices, the corresponding indices of the $\hat{J}$–field are also integers. In other words, in the case of the synchrotron radiation the singular points of $\hat{J}$ are degenerate. The latter conclusion is certainly true for a model where Eq. (7) is valid.

2. Taking into account the technique developed for the inverse problem (Kulikova et al., 1989, 1990), we may assume that

$$\tau = l_1/l_1 ,$$ (10)

which turns us back to the problem of the directions of the tangential component of the magnetic field, for which a potential approximation can be used,

Fig. 4.   Polarization map of the Crab Nebula.

$$(\mathrm{curl}B)_z = 0 \ , \tag{11}$$

and the corresponding integrating constant for τ can be derived.

However, from the polarization map we cannot distinguish l_1 from l_2, and condition (11) for l_2 becomes (see Eq. (9)):

$$\mathrm{div}_2\mathbf{B} = 0 \ , \tag{12}$$

which is unacceptable, because the requirement $\partial B_z/\partial z = 0$ corresponds to a rather peculiar geometry of the magnetic field.

Let us turn to the polarization map of the Crab Nebula (Woltjer, 1957) and to the polarization map of the L810 Nebula (Scarrot *et al.*, 1991). From these data the field of directional vectors τ is reconstructed in Fig. 4. One can here see singular points of half-saddle and half-center types. One more half-node is probably present in this map, which is marked with number 1. The polarization map obtained from photometric measurements by Woltjer (1957) differs to some extent from the map of Fig. 4. The main differences are that two half-nodes can be confidently identified in the Woltjer picture, and that the photoelectric data have been obtained with lower resolution, so that the half-centre – half-saddle pairs are not resolved.

It is of special importance to note that no points having integer indices are visible in Fig. 4. This seems to show that the model of an optically thin layer is invalid for the Crab Nebula. At the same time observations are known for which points of this type are present. For example, a polarization map of the L810 Nebula demonstrates a center-type structure at point 2 (see Fig. 5). Also it is worth noting the presence of two points of half-saddle type

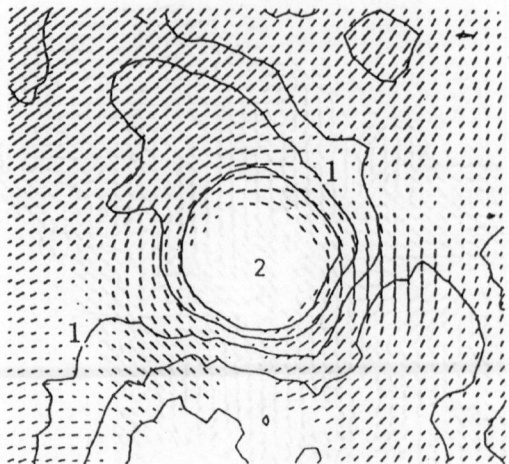

Fig. 5.   Polarization map of the L810 Nebula.

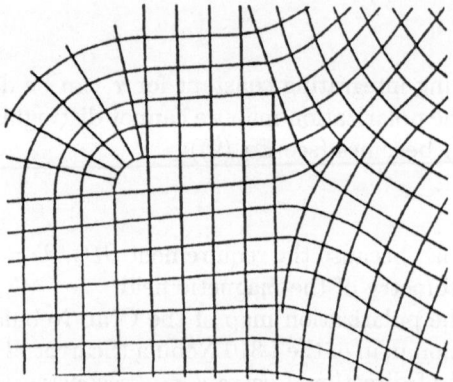

Fig. 6.   Example of a map containing 1/4–saddle and 1/4–center points.

in this figure. The general configuration of the integral lines qualitatively corresponds to the polarization map of the solar eclipse.

4. Singular points with indices +1/4 and −1/4

For a radiation tensor we have actually two families of mutually orthogonal integral curves. Accordingly, in the calculation of the index (the rotation) of the field by tracing around a closed path surrounding some area, we must monitor the rotation of a rectangular cross characterizing the directions l_1 and l_2. Having returned to the initial position, generally speaking, we get the rotation as a multiple of $\pi/2$. This means that in this case indices of singular points may be multiples of 1/4. Figure 6 demonstrates the simplest map containing a 1/4-saddle and a 1/4-center (the latter does not differ from a 1/4-node).

If the distance between them is, as in the previous case, smaller than the angular resolution, the apparent configuration does not differ from a uniform rectangular grid. One can easily see that, similar to the case when the "splitting" of a saddle makes two half-saddles (Fig. 3b), a half-saddle can be substituted by a pair of singular points having indices 1/4. In this case the singular points with half-integer indices represent a degenerate case.

Conclusion

The polarization map of the Crab Nebula contains well identified singular points. The indices of these points have half-integer values. We conclude that the radiation is not connected with a *regular* vector field, including the magnetic vector field.

In the polarization map of the corona two singular points of half-saddle type are found when the zenith distance of the sun during the eclipse is not too large. There is also a possibility of finding new singular points of the polarization tensor with indices $\pm 1/4$.

References

Kulidganishvilly, V.I., Nikolsky, G.M., Stepanov, A.I.: 1985, *Bull. ABAO* **60**, 109
Kulikova, G.N., Molodensky, M.M., Filippov, B.P.: 1989, *Astron. Zhurn. (Russia)* **66**, 271
Kulikova, G.N., Molodensky, M.M., Filippov, B.P.: 1990, *Astron. Zhurn. (Russia)* **67**, 65
Molodensky, M.M.: 1969, *Astron. Zhurn. (Russia)* **46**, 797
Molodensky, M.M., Syrovatsky, S.I.: 1979, *Astron. Zhurn. (Russia)* **54**, 1293
Molodensky, M.M., Starkova, L.I.: 1990, *Astron. Zhurn. (Russia)* **67**, 1309
Petrovsky, I.G.: 1970, *Lectures on the Theory of Ordinary Differential Equations*, M. Nauka
Scarrot, S.M., Rolph, C.D., Tadhunter, C.N.: 1991, *Mon. Not. R. Astr. Soc.* **248**, 27
Woltjer, L.: 1957, *Dissertation*, The Netherlands

DETERMINATION OF POLARIZATION OF VACUUM-ULTRAVIOLET RADIATION BY FLUORESCENCE AND PROBE-BEAM TECHNIQUES

V.YU. BACKMAN, S.V. BOBASHEV and O.S. VASYUTINSKII

Ioffe Physico-Technical Institute, Russian Academy of Sciences, SU-194021, St.-Petersburg, Russia

Abstract. Polarization analysis of vacuum ultraviolet (VUV) radiation, especially the analysis of elliptical polarization is of considerable scientific interest in many areas of physics including solar physics. Polarization measurements in the VUV region are much less developed than those in visible and infrared spectral regions. It is due to lack of efficient phase shifting elements and small reflection coefficients of polarizers for VUV radiation. We propose two techniques, to determine the arbitrary polarization of VUV radiation with photon energies in the range 10 – 100 eV, which seem to be more efficient and precise than the known techniques, especially to study the polarization of atomic or molecular spectral lines.

Key words: Polarization – Atomic fluorescence – Laser radiation

1. Introduction

Determination of polarization of VUV radiation in the wavelength range λ < 105 nm by ordinary techniques is a difficult problem (see Elleume, 1989; Bahrdt, 1992). One of the known experimental methods for polarization measurements in the region of 10 eV to 30 eV photon energies is based on analysis of the intensity of radiation reflected from a surface for which optical constants are accurately determined. Sufficient progress has been recently achieved in this field (see Glushkin, 1992; Koide, 1992), but the problem is far from being solved. Our proposal for experimental investigation of polarization parameters of VUV radiation, is based on the suggestion reported by us elsewhere (see Bobashev and Vasyutinskii, 1992; Backman *et al.*, 1994), that to our knowledge, has not yet been explored experimentally. We present two techniques to determine the arbitrary type polarization of radiation, in the wavelength range 10 – 100 nm. These techniques are more efficient and precise than the known techniques, especially to study the polarization of atomic or molecular spectral lines. The essence of the techniques is, to produce by the initial VUV radiation, nonequilibrium population of excited atomic or molecular magnetic sublevels by orientation and alignment (see Blum, 1981), to measure these orientation and alignment and then, to calculate polarization of the initial VUV radiation using well known expressions. The measurements can be done in the infrared, visible, or near UV spectral

Solar Physics **164**: 397–401, 1996.
© 1996 *Kluwer Academic Publishers.*

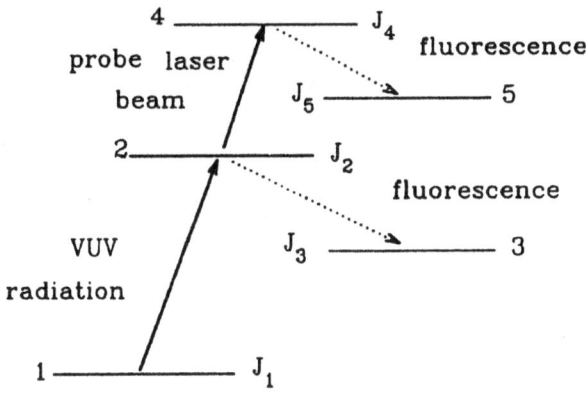

Fig. 1.   Energy level diagram showing the transitions discussed in the present paper.

regions, detecting either fluorescence polarization, or dichroic absorption of a probe laser light from the excited atomic or molecular states. The proposed techniques are specially applicable for determination of polarization of VUV radiation emitted by excited atomic (for example H or He) gases, and therefore can be successfully used in astrophysics. We also discuss the use of atomic Ne, Ar, and molecular CO as the targets that can widen the spectral range of polarization measurements.

2. Fluorescence Technique

If atomic gas is illuminated by VUV radiation, the radiation whose frequency is in resonance with one of the atomic transitions (say, the transition $1 \rightarrow 2$ in Figure 1), will be absorbed and can excite atomic fluorescence in the transition $2 \rightarrow 3$ in UV, visible, or IR spectral range, which are convenient for polarization measurements. The degree of fluorescence polarization is determined uniquely by the degree of VUV light polarization and by the angular momenta of the states involved in the transition (see Mitchell and Zemansky, 1934). Therefore analysis of the Stokes parameters for the fluorescence light allows to recover those of the original VUV radiation by well-known formulae (see Bobashev and Vasyutinskii, 1985; Mitchell and Zemansky, 1934).

3. Results and Discussion

A schematic diagram of the experimental setup, to determine the degree of circular polarization of the initial VUV radiation, is shown in Figure 2. The

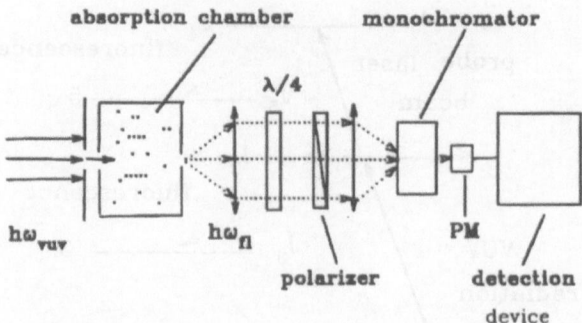

Fig. 2. Experimental setup to determine circular polarization of the incident VUV radiation.

VUV radiation that is going to be investigated passes through monochromatization elements, and then interacts with atomic target in the absorption chamber. To avoid a systematic depolarization of the excited atoms due to collisions, a typical gas density of $n < 10^{14}$ cm^{-3} should be used. Absorption chamber should be screened from an external magnetic field, to prevent the fluorescence depolarization due to Hanle effect. The fluorescent light should be analyzed by usual means, as shown in Figure 2 using quarter-wave phase plate, linear polarizer, monochromator and photodetector. For example, to determine the polarization of VUV radiation at 53.7 nm the following photoreaction can be used:

$$\text{He } (1^1S_0) + h\nu_{vuv} \rightarrow \text{He } (3^1P_1) \rightarrow \text{He } (2^1S_0) + h\nu_{fl} . \tag{1}$$

Here the polarization of fluorescence at 501.57 nm should be detected, and normalized Stokes parameters (see Shurcliff, 1962 for details) of the fluorescence radiation are equal to those of the VUV radiation (see Bobashev and Vasyutinskii, 1992). The accuracy of our method is close to 3% if the VUV photon flux is not less than $N=10^7$ phot/sec eV.

Using other He resonance transitions, the polarization measurements can be carried out also at 58.43 nm, 52.22 nm, 51.21 nm, etc. Ne and Ar atomic gases and ions such as $^4\text{He}^+$, $^{14}\text{N}^{+3}$, and $^{14}\text{N}^{+4}$ can also be used, which make it possible to cover the spectral region 10 – 90 nm.

4. Probe-Beam Technique

The essence of this technique is in the recording of dichroism of atomic or molecular gas previously irradiated by VUV radiation under study, followed by a calculation of initial VUV radiation Stokes parameters. The energy level diagram that elucidates the photoinduced process is shown in Figure 1 (dash

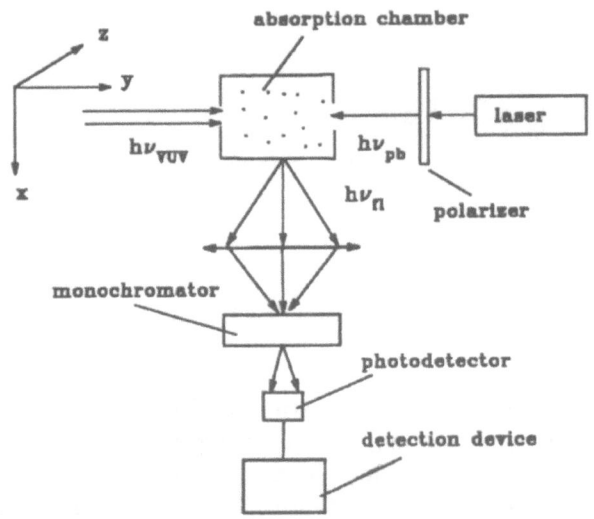

Fig. 3.   Experimental setup for measuring dichroic absorption of the probe-beam.

lines). It is assumed that, at first the atoms are excited by VUV radiation from the ground state 1 with angular momentum J_1 to the state 2 with angular momentum J_2, and then the excited atomic or molecular vapour is probed by polarized probe laser radiation that is in resonance with the $2 \rightarrow 4$ transition. Circular and linear dichroic absorption of this probe beam is measured by LIF technique (see Demtroder, 1982) detecting the intensity of fluorescence from level 4 to level 5. The required experimental setup is shown in Figure 3. To determine the polarization of VUV radiation at 58.43 nm one can use the following reaction:

$$\text{He } (1^1S_0) + h\nu_{vuv} \rightarrow \text{He } (2^1P_1) \, , \tag{2}$$

$$\text{He } (2^1P_1) + h\nu_{pb} \rightarrow \text{He } (4^1S_0) \rightarrow \text{He } (3^1P_1) + h\nu_{fl} \, . \tag{3}$$

Here the wavelength of the probe and fluorescence radiation are $\lambda_{pb} = 504.8$ nm and $\lambda_{fl} = 2113.20$ nm respectively. In order to determine the Stokes parameters of initial VUV radiation, four measurements should be done. In the first one, probe radiation linearly polarized in the vertical direction with Stokes parameters $P_1 = 1$, $P_2 = P_3 = 0$ should be used. In the second and third measurements also, the probe radiation polarization should be linear but rotated with respect to the **Y** - axis (see Figure 3), by angles $\pi/2$ ($P_1 = -1$, $P_2 = P_3 = 0$) and $\pi/4$ ($P_2 = 1$, $P_1 = P_3 = 0$) respectively. In the fourth measurement, the probe beam should be circularly polarized with the Stokes parameters $P_3 = 1$, $P_1 = P_2 = 0$. The expressions for Stokes parameters of the initial VUV radiation can be presented as follows (see Backman *et al.*, 1994):

$$I_1 = \frac{11}{3} \frac{S_1 - S_2}{S_1 + S_2} , \tag{4}$$

$$I_2 = \frac{11}{3} \frac{S_1 + S_2 - 2S_3}{S_1 + S_2} , \tag{5}$$

$$I_3 = \frac{11}{5} \frac{S_1 + S_2 - 2S_4}{S_1 + S_2} , \tag{6}$$

where four quantities S_1, S_2, S_3, and S_4 are the results of the four measurements. To determine the polarization of VUV radiation in other wavelength regions, some other lines of He, Ar, Ne, or ionic resonance lines can be used.

This probe-beam method has distinct advantages over the fluorescence detection. First, it has a very high analysing power, which is much more than that of the other technique. Second, this method can transfer the excitation and polarization information to a more convenient wavelength for efficient detection. Third, if the probing is pulsed with a short pulse laser, the effects of collisions, depolarization of excited state 2 due to cascade population, and hyperfine depolarization, on the measured polarization are greatly suppressed.

Acknowledgements

The work was carried out with financial support from the Russian Foundation for Fundamental Research, Project No. 93-02-14269, and Grant No. R 65300 from International Science Foundation.

References

Backman, V.Yu., Bobashev, S.V., and Vasyutinskii, O.S.: 1994, *Sov. Tech.Phys.Lett.* **20**, 772
Bahrdt, J., et.al.: 1992, *Rev.Sci.Instrum.* **63**, 339
Blum, K.: 1981, *Density Matrix Theory and Application*, Plenum Press, New York
Bobashev, S.V., and Vasyutinskii, O.S.: 1992, *Rev.Sci.Instr.* **63**, 1509
Bobashev, S.V., and Vasyutinskii, O.S.: 1985, *Sov. Tech. Phys. Lett.* **11**, 599
Demtroder, W.: 1982, *Laser Spectroscopy*, Berlin
Elleume, P.: 1989, *Rev.Sci.Instrum.* **60**, 1830
Glushkin, E.S.: 1992, *Rev.Sci.Instrum.* **63**, 1523
Koide, T., et al.: 1992, *Rev.Sci.Instrum.* **63**, 1458
Mitchell, A.C.G., and Zemansky, M.W.: 1934, *Resonance Radiation and excited Atoms*, Cambridge University Press, Cambridge
Shurcliff, W.A.: 1962, *Polarized Light: Production and Use*, Cambridge University Press, Cambridge

BALLOON-BORNE POLARIMETRY

The Flare Genesis Experiment

D.M. RUST, G. MURPHY and K. STROHBEHN

The Johns Hopkins University, Applied Physics Laboratory, Laurel, MD 20723, USA

and

C.U. KELLER

National Solar Observatory, P.O.Box 26732, Tucson, AZ 85726-6732, USA

Abstract. For about two weeks in 1995, the balloon-borne Flare Genesis Experiment will continuously observe the Sun well above the turbulent, image-blurring layers of the Earth's atmosphere. The polarization-free 80 cm telescope will supply images to a liquid-crystal based vector magnetograph, which will measure magnetic features at a resolution of 0.2 arcsec. An electrically tunable lithium-niobate Fabry-Perot provides a spectral resolution of about 0.015 nm. In a follow-up series of Antarctic balloon flights, the Flare Genesis Experiment (FGE) will provide unprecedented details about sunspots, flares, magnetic elements, filaments, and the quiet solar atmosphere.

Key words: Stratospheric balloon – Antarctica – Vector polarimetry – Flares

1. Introduction

To better understand solar activity, the Applied Physics Laboratory (APL) of the Johns Hopkins University has launched an effort to obtain spatially highly resolved pictures of the solar atmosphere over a substantial period of time. These images will reveal magnetic fields and motions in the solar photosphere and chromosphere with an advanced 80 cm telescope that circles the South Pole. The project makes use of modern technologies, including devices that had to be designed for this particular program and some that were developed during the Cold War and are no longer needed by the military. In fact, the end of the Cold War is presenting solar physicists with opportunities to use high-technology equipment that would otherwise be beyond the means of solar research programs. It is reminiscent of the fruitful period following World War II, when, for example, scientists inherited the V-2 rockets.

2. Ballooning in Antarctica

The Flare Genesis Experiment (FGE) seems to be a retreat to old technology, because a balloon instead of a rocket will carry it above the atmosphere.

Solar Physics **164**: 403–415, 1996.
© 1996 *Kluwer Academic Publishers.*

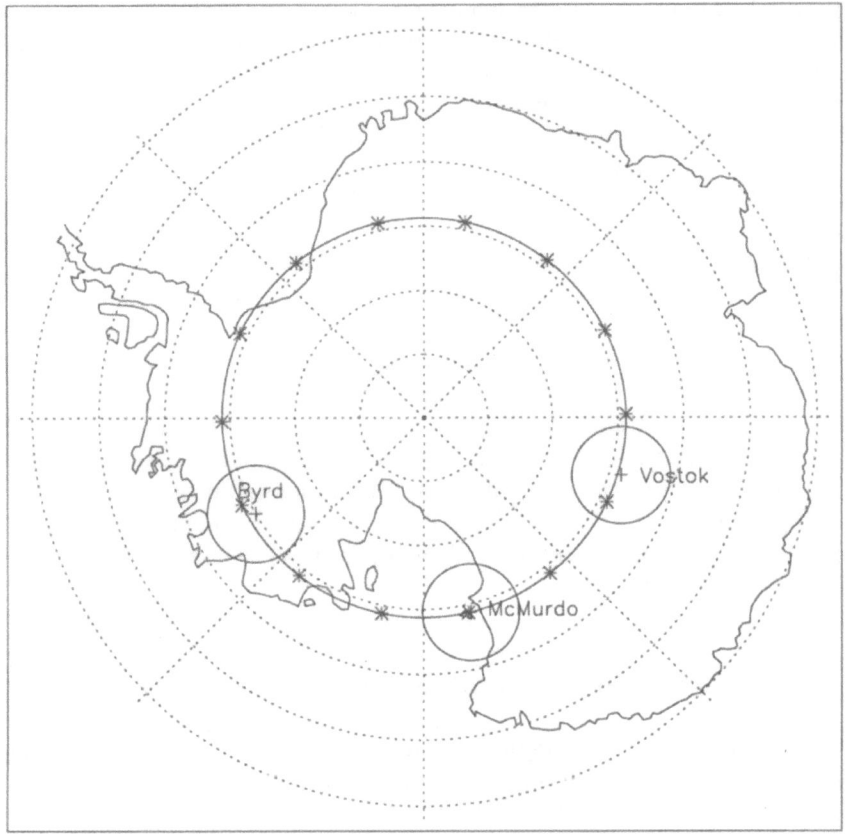

Fig. 1. The projected flight path of the Flare Genesis Experiment in Antarctica, starting from McMurdo base, passing over Russia's Vostok station and Byrd station, over the Ross Ice Shelf and finally back to near McMurdo. Actual flight paths have deviated by hundreds of miles from the ideal 78-degree South Latitude circle shown here. The small circles show the maximum range of radio contact with the experiment.

But space-based experiments are still about a factor of 100 more expensive as compared with comparable balloon-based experiments. Furthermore, flight opportunities to space are rare. Ballooning in Antarctica is new and offers special advantages over satellites in near-earth orbits, because the sun can be studied without interruption during the Austral summer for periods of usually about 14 days, and 28-day flights are possible.

Long-duration ballooning in Antarctica began in 1988 when Carl A. Rester of the University of Florida flew a gamma-ray telescope that had been grounded by the suspension of Space Shuttle flights. Shortly after Supernova 1987A appeared, Rester realized that the gamma rays from the element-building processes in it might be detectable with his telescope. To detect the gamma rays, Rester had to get above most of the atmosphere and stay there for a week or more. Since the supernova was in the Southern Hemisphere, a stratospheric balloon flight in Antarctica would provide the perfect viewing platform. Vortical winds would carry it on a slow circum-navigation of the South pole (see Figure 1). But no one had ever done it before. Worse, the kind of helium-filled balloon he needed had not been a very reliable vehicle in the 1980's.

Rester, with support from the Air Force Geophysics Laboratory and many other agencies, decided anyway to try to get a balloon up in the 1987-88 Austral Summer. The National Science Foundation (NSF) joined the consortium of sponsoring agencies. The NSF's Division of Polar Programs supports U.S. research in Antarctica from its base at McMurdo Sound. Rester's mission was a phenomenal success (Hughes, 1988). Rester not only detected the gamma rays and verified the theory of element production in supernovae, he and his consortium also opened Antarctica to long-duration scientific ballooning. Since then, balloons have flown successfully many times, thanks to NASA's National Scientific Ballooning Facility and the development of highly reliable balloon film.

3. Scientific Objectives of the Flare Genesis Experiment

The principal aim of the FGE is to understand how the fibrous magnetic fields at the solar surface emerge, coalesce, unravel, and erupt in solar flares by measuring the vector magnetic field in flaring regions over long periods and at the highest possible spatial resolution. It is important because magnetic fields are the root cause of flares and other solar activity phenomena. All solar activity arises from the action of magnetic fields, but much better observations of the magnetic fields are needed. The potential of the SVMG and other ground-based instruments for probing fundamental flare physics is limited by atmospheric blurring. Sharper images, produced with the large FGE telescope flown above most of the atmosphere, will provide much better sensitivity.

Although the FGE's primary scientific objective is flare research, successful balloon flights should contribute broadly to a better understanding of many other features on the sun even when solar activity is low. The FGE vector magnetograph can measure the radial and vertical distribution of magnetic fields in a sunspot and the relationship between the magnetic field and fine-scale features such as the dark fibrils of the penumbra. The FGE can answer the questions of whether subsurface flows or visible surface flows control the migration of magnetic features and how large-scale flows are related to magnetic energy build-up. Furthermore FGE is well suited to study oscillations at the smallest scales.

There will be several X-ray telescopes in space during the balloon flights. By comparing the FGE images with pictures of the sun in X-rays, it should be possible to determine precisely how the X-ray features relate to the underlying vector magnetic field structures. Joint observations of fields and flows with the FGE and of the X-ray emission with a satellite-borne telescope could show how energy concentrates in the solar corona. They could settle the important and long-standing question about the origin of coronal heating.

4. Ground-Based Prototype

The plan to fly a balloon-borne telescope had its origins in 1986, when APL received a University Research Initiative grant from the Air Force Office of Scientific Research to start a center at APL for solar research. APL designed and built a solar vector magnetograph (SVMG) that is now operated at the National Solar Observatory at Sacramento Peak, New Mexico (Rust et al., 1988). It is the prototype of the FGE vector magnetograph.

At Sacramento Peak, the scientific objective is to infer enough information from vector magnetograms at about 2 arcsec resolution to give useful physical insight into the structure and evolution of the magnetic fields in sunspot regions. Magnetic flux emergence and disappearance and regions with twisted magnetic fields need to be better understood if substantial progress in understanding solar flares should be made. A central problem is how to identify the build-up of magnetic energy for flares in an active sunspot region.

We have found that imaging through a narrow-band filter (Rust et al., 1986) and analyzing the results with the weak-field approximation (Jefferies and Mickey, 1991) can give useful temporal and spatial data on magnetic fields (Cauzzi et al., 1993), but the limited spatial resolution imposed by atmospheric blurring means that interpretation of the measurements can be ambiguous.

5. Instrument Description

5.1. DESIGN STUDY

Drawing on the experience with the magnetograph at Sacramento Peak, APL initiated a design study of a balloon-borne solar vector magnetograph. The study showed that a balloon-borne instrument is feasible, but presents significant challenges. Principal difficulties, such as thermal control and precise pointing, can be overcome by modern technology. Because the experience of other ballooning groups, particularly those at the Harvard Smithsonian Center for Astrophysics (Nystrom et al., 1992) and the Air Force Phillips Laboratory, is available, the FGE will acquire many images with unprecedented resolution and achieve a sufficient polarization sensitivity.

The design study focused on the effects of image motion, gravity, and temperature on a balloon-borne telescope to determine whether the FGE will maintain the desired resolution of about 0.2 arcsec. The principal conclusions of the study were that (1) with an Image Motion Compensator (IMC), sufficient image stability can be maintained at the FGE focal plane to allow magnetic mapping at 0.2 arcsec resolution, despite the motion of the gondola and the harsh environment; (2) a complete FGE, including on-board data recorders, can be built within the customary power and weight limits for balloon flights; (3) the needed optical components either exist or are commercial items that can be obtained in less than six months; (4) most of the needed mechanisms are similar to those used in the SVMG at Sacramento Peak, and their incorporation into the balloon-borne instrument should pose no serious problem if they are operated in a temperature-controlled, pressurized chamber; (5) the FGE can achieve the required polarization sensitivity (Rust et al., 1990).

5.2. OVERVIEW

The gondola consists of an aluminum framework, 1.5 m wide by 1.8 m deep by 4 m in height. On the top of the structure, there is an angular-momentum dump unit, which consists of a fly wheel that is either connected to the gondola or to the balloon. The gondola is rotated (moved in azimuth) by spinning up or down the flywheel while its motor is connected to the gondola. To prevent the flywheel from having to spin too fast, the angular momentum can be dumped into the balloon by loosely coupling the flywheel to the balloon cables. Special care has been taken to minimize friction in the bearings. The telescope is moved in elevation by a motor whose axis goes through the center of gravity of the telescope. Figure 2 shows a drawing of the modified telescope in the gondola.

The telescope is mounted in a cylindrical frame that is surrounded by thermal blankets. In front of the telescope, a diaphragm shields the telescope structure including the mount of the secondary mirror from direct sunlight. Thermal shields on the inside of the telescope prevent direct sunlight hitting the secondary mount in the case that the telescope is not tracking accurately. The inside of the telescope tube is protected by a reflective layer.

Attached to the rear of the telescope is a temperature-controlled, pressurized vessel that contains the post-focus optics. Therefore, standard, laboratory components can be used. This approach avoids the cost of qualifying parts for the harsh environment at the flight altitude. Batteries, computers, tape drives etc. are located in separate pressurized vessels below the azimuth drive unit. Solar cells are attached to both sides of the main gondola frame and provide electricity at 28 V for the whole payload. Batteries provide power for up to 12 hours during the launch, and if tracking of the sun should be lost.

The Target Selection Telescope (TST) is mounted on the side of the main telescope. The beam from this 10 cm refractor is also sent into the pressure vessel on the back of the main telescope. The TST is used to derive guiding signals for the azimuth and elevation drives and provides a full-disk image through the Fabry-Perot filter. Either of the two beams, from the TST or the main telescope, can be sent through the etalon filter to the CCD camera. The full-disk Hα images are used to autonomously select regions of interest.

The polarization modulator housing is located between the main telescope and the pressure vessel containing the post-focus optics. It contains a collimating lens, an entrance filter that blocks most of the light to prevent damage to two liquid crystal variable retarders and a linear polarizer. These are oriented at 0 and 45 degrees with respect to the analyzing polarizer. The modulators are in a temperature-stabilized oven that is wrapped in thermal blankets. The polarization analyzer is directly behind the prime focus where the optics are still cylindrically symmetric. This minimizes instrumental polarization. The collimating lens forms an image of the pupil on the IMC. This array of three mirrors is the only optical element after the telescope prime focus that is not protected in a heated and/or pressurized compartment. The collimated and stabilized beam is then sent into the optical pressure vessel.

Table I outlines the basic properties and capabilities of the FGE.

5.3. NEW TECHNOLOGIES

The Strategic Defense Initiative Organization transferred the former Starlab telescope to the project, a space-qualified telescope that was intended to be operated on the Space Shuttle. The primary mirror is 80 cm in diameter. The Cassegrain telescope has been modified to meet the requirements of solar observing.

The Starlab telescope body is made of graphite-epoxy fiber and the mirror is made of ultra-low-expansion glass. The dimensions of both materials are stable over wide temperature ranges. Therefore, it is expected to maintain diffraction-limited performance despite the potential differences between the temperature at which the telescope will operate and laboratory temperature. If the thermal design is correct, the operating temperature at 38 km in the Antarctic Stratosphere will be 20°C.

TABLE I

FLARE GENESIS EXPERIMENT CAPABILITIES

Spatial resolution	limited by diffraction to 0.2" (145 km at solar disk center)
Spectral resolution	0.015 nm passband tunable over spectral line profiles with repeatability to 1×10^{-4} nm
Wavelength range	610 - 660 nm
Field of view	100" x 150"
Detector	Kodak Megaplus 1.6 10 bit CCD camera, 1024 x 1534 pixels
Data products	time series of vector magnetograms at various wavelengths, vector velocity and intensity in the photosphere and chromosphere
Data storage capacity	100 Gbytes in two 10-Exabyte-cassette loaders
Telemetry downlink	1 Mbit/s image data, 1 Kbit/s for commands and status checks
Spectral lines	1) 630.25 nm, g = 2.5 Fe I
	2) 656.28 nm, g = 1.045 H I (Hα)
	3) other lines selectable between 610 nm and 660 nm

Figure 3 is a schematic drawing of the modified telescope optics and the optical instruments behind the focal plane. The key high-technology elements are the light-weight honeycomb, ultra-low-expansion primary mirror, the single-crystal silicon secondary mirror, the graphic-epoxy mirror cell and tube, a tunable lithium-niobate Fabry-Perot filter (Rust, 1994), an image motion compensator, and the on-board FORTH-language Reduced Instruction Set Computer (FRISC, Hayes et al., 1987).

5.3.1. *Ultra-Low-Expansion Materials*

The lightweight mirror and graphite-epoxy structure provide high thermal stability to maintain the optical figure of 1/25 of a wave over a wide temperature range. Another important advantage of the lightweight structure is ease of recovery. Although the one-million cubic meter helium balloons used for long-duration flights can lift about 2500 kg to the required altitude of 38 km, there are still important advantages in keeping the payload light and compact. Two men can easily recover the lightweight telescope if the gondola lands in the Antarctic highlands where only small airplanes can set down. The mirror weighs 50 kg – about one-third the weight of a solid mirror. Both the primary and the secondary mirrors are coated with silver and protective layers so that they reflect about 97 % of the incident solar energy throughout most of the spectrum. About 500 W strike the primary, secondary, and heat-dump mirrors. The heat absorbed by the primary is radiated out of the back. The heat absorbed at the secondary is removed with heat pipes. The heat-dump mirror reflects 90 percent of the sunlight back out of the telescope and is cooled by conduction to the telescope framework. A single crystal of silicon is used for the secondary mirror material because silicon

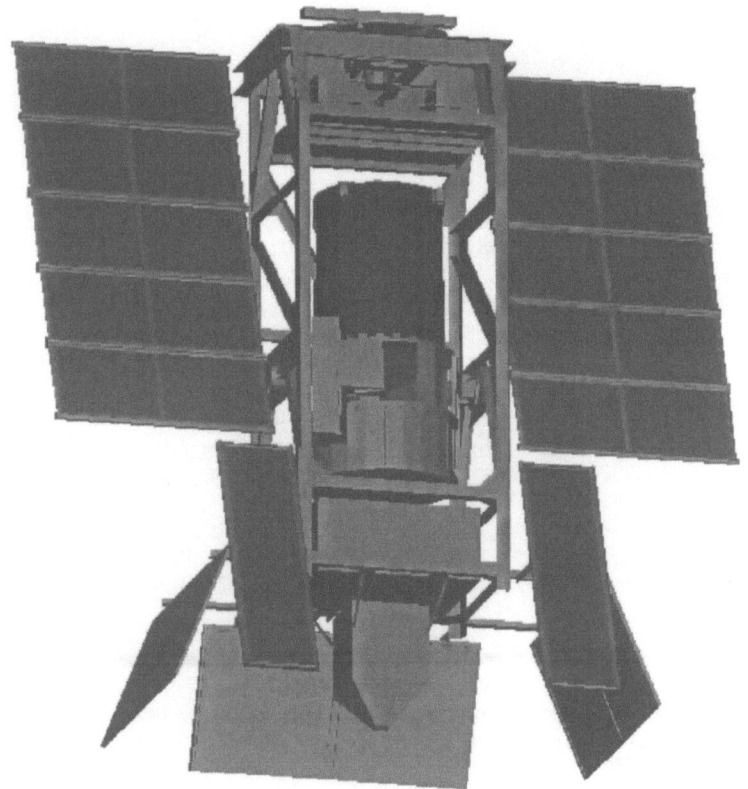

Fig. 2.   A computer-aided design drawing of the Flare Genesis Experiment. The height of the gondola is 4 m; the main telescope has a 80 cm aperture and a 3 arcmin field of view; the target selector telescope has a 10 cm aperture and a one degree field of view.

takes a very smooth polish and because it has a high ratio of conductance to thermal expansion. This facilitates removal of the absorbed solar energy with minimum mirror distortion.

5.3.2. *Fabry-Perot Etalon*
The etalon consists of a 75-mm diameter wafer of lithium niobate polished to a flatness of about 1/100 of a wave. Indium-tin-oxide coatings on the crystal faces make them electrically conductive. An electric field can then alter the index of refraction of the lithium niobate which, in turn, alters the optical path in the Fabry-Perot cavity. A consequence is that the wavelength of light passed by the filter is proportional to voltage. The filter has a passband of about 150 mÅ, and it can be tuned to any wavelength between 610 nm and 660 nm. Unwanted orders of the Fabry-Perot are suppressed by narrow-

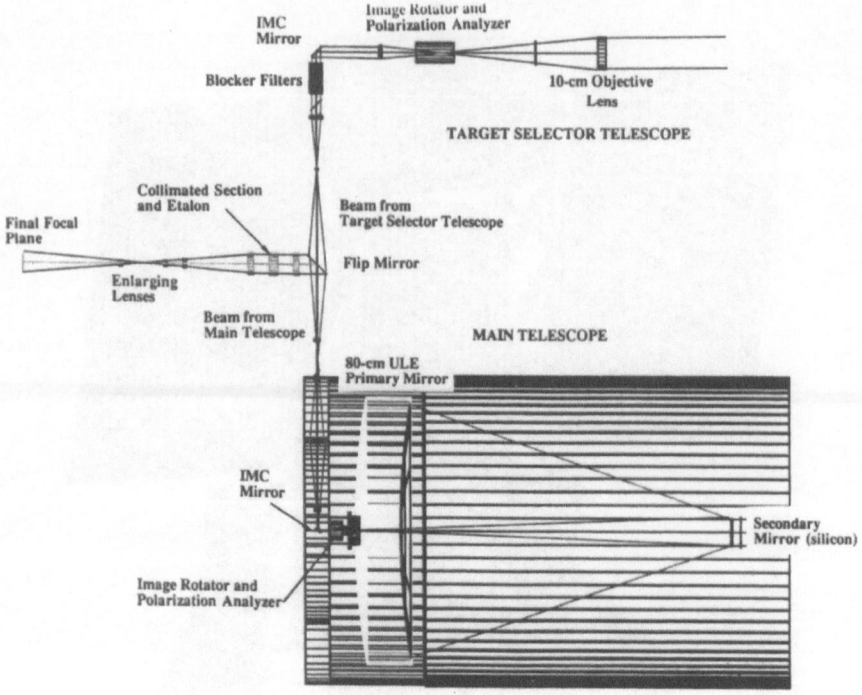

Fig. 3. Optical schematic of the Flare Genesis Experiment (some fold mirrors are not shown). The image rotators are not used on the first science flight in December 1995.

band interference filters with a passband of slightly more than 0.1 nm. The etalon as well as the filter wheel with the interference filters are mounted in temperature stabilized ovens.

5.3.3. *Image Motion Compensator and Silicon Retina*

The IMC senses translational motion of the images and reduces or removes it with an agile tip-tilt mirror. For the FGE, the closed-loop bandwidth of the IMC must be > 50 Hz to provide adequate disturbance rejection. We have chosen an analog servo-loop approach based on a novel image position sensor. The analog approach has significant advantages over digital systems in terms of weight, power, simplicity, reliability, and performance.

Many image motion sensors require a bright or dark feature before they lock up to track and remove motion at the focal plane. The main solar telescope does not provide an image of the whole solar disk to track on the limb, nor can we expect to find conveniently placed sunspots to track. An analog servo approach is still feasible because of a series of novel image position sensors. These sensors combine analog computations with signals from a photo-detector array to continuously compute image position. All image sensing and computations are carried out in a single microchip which produces a video image as well (Strohbehn et al., 1993). The video output has proven to be very helpful during system debugging and integration. This correlation-based position sensor tracks subtle, low-contrast features such as solar granulation by using neuromimetic analog current-mode computations. The detector includes a silicon retina for image processing.

The tip-tilt mirror developed for the FGE uses magnetostrictive actuators. This tip-tilt mirror has a wide bandwidth (first bending mode at 200 Hz) and achieves a throw of ±1200 arcsec in the main telescope. Tests on the ground show that the actual jitter of the telescope should not exceed ±10 arcsec, but the tip-tilt mirror operates at a small image of the entrance pupil, and angular deviations are magnified by about 25 times. Another advantage of the relatively new technology of magnetostrictive actuators is that they can be driven with only ±15 volts, so they will be completely immune to arcing, which might be a problem with high-voltage piezoelectrical actuators.

6. Flight Plan

The first flight of the FGE tested the telescope pointing scheme. This test flight took place in New Mexico on January 23, 1994 and lasted eleven hours. Only the Target Selector Telescope was aboard. A wooden facsimile represented the main telescope, which did not have to face the rigors of flight until the first scientific mission. This test flight showed that the gondola, pressurized vessels, optical positioning mechanisms, light modulator, target selector telescope, motors, on-board and ground-based computers, telemetry links, and power distribution system all functioned within specifications.

In December 1995 the entire experiment will fly in Antarctica. Except for the first 20 hours, it will be out of radio contact and it will have to react automatically to solar activity according to the predefined science priorities. The images are stored on-board on Exabyte tapes. Computer software will direct the FGE to examine the state of the sun periodically and prioritize the science objectives. For example, if a large sunspot is at the center of the solar disk, then a code to map the motions and magnetic fields in the penumbra might be chosen automatically.

Table II outlines the autonomous observing program. Many details, such as initial pointing and check-out of the telescope have been omitted from the table, but virtually all operations must be foreseen. Resources in Antarctica are so limited, that it might not be possible to have any contact with the experiment except to get minimum health and safety data. But the au-

TABLE II

AUTONOMOUS OBSERVING PLAN

1. select area for study on basis of full-disk images from Target Selector Telescope
2. select timing sequence on the basis of science program, e.g., flares, magnetic field evolution, oscillations, granule proper motion, etc.
3. select wavelength: adjust blocker filter and voltage to the Fabry-Perot
4. determine optimum exposure time
5. move main telescope to desired azimuth and elevation
6. activate Image Motion Compensator and check contrast (focus) of IMC images
7. select voltages for the two liquid-crystal devices in the polarization analyzer on the basis of science, temperature, and apparent angle of the solar axis
8. command dark-current and gain exposures by CCD camera
9. start exposure sequence
10. dump data to Exabyte tapes
11. if science program allows, interrupt exposure sequence once per hour to verify pointing and record full-disk images from the Target Selector Telescope
12. end sequence, obtain new darks and flats
13. verify focus and instrument health
14. review science priorities in light of solar activity, completed observations, and expected time and tape remaining. Go to step 1

tonomous observing program is able to select sunspots, flares, and other features from typical solar images (Figure 4).

After one or possibly two circumnavigations of the South Pole, the FGE will drop to the ice under a parachute. A landing point will be chosen that is accessible to helicopters or LC-130 Hercules aircraft, which are capable of bringing the entire payload back to McMurdo in one piece. If the experiment lands at a high elevation, a Twin Otter airplane equipped with skis will land near it and only the most important things can be retrieved. The data tapes have first priority, the primary mirror the second.

7. Conclusions

The FGE currently is undergoing final adjustments and calibrations at the Air Force Phillips Laboratory facilities at Kirtland AFB in Albuquerque, New Mexico. After the Mission Readiness Review in Palestine, Texas, in mid August, it will be shipped to McMurdo Sound, Antarctica. A crew of up to nine persons will check out the instrument starting in mid-November and will launch the FGE in mid-December. There will be a high-speed digital link during the first 20 hours in the flight that will provide real-time digital

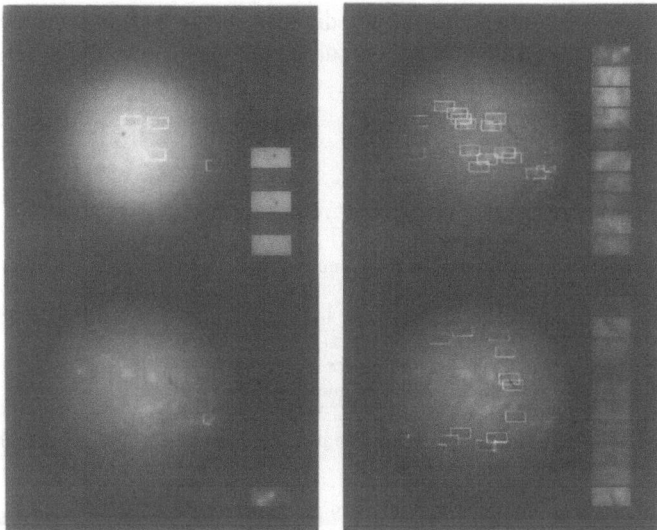

Fig. 4. (upper left) An image of the sun showing the photosphere, where sunspots are the most prominent features. The autonomous observing program has identified five sunspot regions and listed them in order of decreasing sunspot area. (upper right) This is an $H\alpha$ image. Most of the bright features are plages - regions of enhanced magnetic fields, but the brightest feature is a flare. (lower left) The autonomous observing program has removed the natural darkening of the solar disk near the edges and has discovered the flare. (lower right) The program has identified solar filaments. At a site that will flare soon, a filament there will darken and twist about, so the program will select the darkest filament and point the main telescope at it and collect data until a flare occurs.

images at the ground station. It is likely that a LC-130 Hercules aircraft will be equipped with computers and antennas to fly under the balloon and down-load data and check-out the payload every few days. In conclusion, the flight in Antarctica has an excellent chance to achieve major breakthroughs in solar research. Future flights will collect more data during the rise of the next solar cycle.

Acknowledgements

The Flare Genesis Project is supported by NSF, AFOSR, NASA, the Air Force Phillips Laboratory, and APL internal funds. We are grateful to Steve

Keil, John Hayes, Harry Eaton, Russ Cain, Jack Colson, Jeffrey Blanchette, and Ashok Kumar for help with the experiment. Phil Wiborg created the software for the autonomous observing program and the transmitter. Jeff West, John Ground, and Lt. Col. David Williamson of the Air Force Phillips Lab provided valuable advice and assistance with the telescope and gondola. The National Solar Observatory is a division of the National Optical Astronomy Observatories, which are operated by the Association of Universities for Research in Astronomy, Inc. (AURA) under cooperative agreement with the National Science Foundation.

References

Cauzzi, G., Smaldone, L. A., Balasubramaniam, K. S., Keil, S.L.: 1993, *Solar Phys.* **146**, 207

Hayes, J. R., Fraeman, M. E., Williams, R. L., Zaremba, T.: 1987, in Proc. 2nd Intl. Conf. Architech. Support Prog. Lang. & Oper. Syst., IEEE, p. 42

Hughes, D.: 1988, *Aviation Week and Space Technology* March 7, 44

Jefferies, J. T., Mickey, D. L.: 1991, *Astrophys. J.* **372**, 694

Nystrom, G, Cheimets, P., Couvault, C., Grindlay, J., Coyle, L., Licata, F., Kuosmanen, V., 1992, in Symp. Thirty Years of Scientific Ballooning in India, Hyderabad

Rust, D.M.: 1994, *Opt. Eng.* **33**, 3342

Rust, D. M., Burton, C. H., Leistner, A. J.: 1986, *SPIE* **627**, 39

Rust, D. M., O'Byrne, J. W., Harris, T.: 1988, *Johns Hopkins APL Tech. Dig.* **11**, 77

Rust, D. M., Murphy, G. A., Strohbehn, K., Hochheimer, B, Henshaw, R., Hayes, J.R., Lohr, D. A., Harris, T. J.: 1990, *Design Study of a Balloon-borne Solar Vector Magnetograph*, APL Space Physics Tech. Rept. (unpub.)

Strohbehn, K, Rust, D. M., Andreou, A. G., Jenkins, R. E.: 1993, in R.R. Radick (ed.), *Real-Time and Post-Facto Solar Image Correction*, Proc. of the 13th Sacramento Peak Summer Workshop, National Solar Observatory, Sunspot, NM, p. 32

ZEEMAN-DOPPLER IMAGING OF SOLAR-TYPE STARS: MULTI LINE TECHNIQUE

M. SEMEL and J. LI *

Observatoire de Paris, URA 326 (CNRS), F 92195, Meudon, Cedex, France

Abstract. In this work, a multi-line spectropolarimetric detection using an Echelle spectrograph is described. The polarization of Zeeman effect is detected by the use of more than 200 lines observed in the solar type star, HR1099. Using the statistics analysis in a sample of 200 lines, we found on the average a polarization signal of about 3×10^{-4}.

Key words: Solar type stars – Magnetic fields – Spectropolarimetry

1. Introduction

The early observations of Zeeman-Doppler Imaging (ZDI), were made with a single order spectrograph using three lines. Detection was achieved for four stars, (Donati et al., 1992). Zeeman polarization in the spectra of solar-type stars was found in general to be very weak (less than a percent). As suggested in paper I (Semel, 1989), the use of many lines may increase the signal to noise ratio. Now, the technique of adding the signals from different lines is not necessarily trivial and may require a careful radiative transfer calculation. Moreover most of the spectral lines are blended due to Doppler broadening related to the rapid rotation of the star. This makes spectral analysis even more difficult. It is also important to avoid spurious signals due to crosstalk from the intensity I to the Stokes V parameter and from the Doppler to the Zeeman effects. In this paper, an attempt to apply a very simple procedure is described. Most of the technical details of ZDI are given in paper III (Semel et al.1993). Here, we list only the important modifications.

2. Observations

The first attempt in this direction was performed during 19-21 of August 1991 at the Anglo-Australian Telescope (AAT), and by using the cross dispersion spectrograph UCLES (Walker and F. Diego, 1985). Using the 79 grooves/mm grating and Thomson CCD detector with 1024×1024 pixels of 19μ size, we could observe 14 orders of $\simeq 40$ to 55 Å– length. This allows us to measure add and later add together few hundreds spectral lines. The

* Present address 602 Kukuiula Loop, Honolulu, HI 96825, U.S.A.

Solar Physics **164**: 417–428, 1996.
© 1996 *Kluwer Academic Publishers.*

polarimeter and the attending optics described in paper III were designed to observe a single order at a time and therefore not achromatic. It was therefore necessary to introduce a few modifications in the instrumentation.

They were : 1) replacement of the aberration free beam splitter by one with a smaller separation ($\simeq 300\mu$) ; 2) Use of double optic fibers of 150μ diameter ; 3) t modification of the double image slicer to form 3 slices for each beam and to fit the f ratio required for the spectrograph.

For illustration, a sample of the polarimetric spectrum of the Moon obtained with UCLES, is shown in Figure 1. Only six orders out of the 14 are shown. Note that each order contains two spectra, right and left, corresponding to the two states of polarisation. Each of them is composed of three bands in the direction of the dispersion, (vertical in the figure). They are due to the image slicer at the entrance of the spectrograph. In short, the device transforms the output circular image of each optical fiber to a long and narrow image. This helps to increase spectral resolution and still allow all the light to enter the spectrograph. The usual way of increasing spectral resolution by closing the slit of the spectrograph would have cost us a very significant loss of light. In the present observation each fiber was "sliced" to three parts.

The observing procedure was similar to the one explained in paper III: successive exposures with alternation of the state of polarisation in each beam. However, this time we proceeded by a cycle of four exposures instead of two, in order to increase symmetry and reduce time dependant effects.

3.   Data reduction

3.1. THE MULTILINE TECHNIQUE.

In this method one has to add the polarisation signals originating from different spectral lines as function of the coordinate (common to all lines) $x = (\lambda - \lambda_i)/\lambda_i$ where λ_i is the central wavelength of the line i and λ is the wavelength in the line profile where the polarisation is measured. (for more details, see paper I)

3.1.1.   *Wavelength resampling.*

The beginning and the end of each order are removed, and the start and end wavelengths are determined in the following way : Denote by λ_{sk} and

Fig. 1. A portion of a spectropolarimetric image of the solar spectrum, as recorded by observing the Moon, From left to right the six orders : 38,39..., and 43 (where one may easily recognize the Fe I line 5250.02Å). Each order appears twice corresponding to left and right circularly polarised spectra. The three vertical bands in each of them correspond to the three slices created by the image slicer. See explanation in the text.

λ_{sm} the start wavelengths of the orders k and m, respectivly : they were chosen to satisfy the condition :

$$k \times \lambda_{sk} = m \times \lambda_{sm}. \tag{1}$$

This holds for all the orders and beams. Similarly λ_{ek} and λ_{em} the end wavelengths for the orders k and m satisfy :

$$k \times \lambda_{ek} = m \times \lambda_{em}. \tag{2}$$

For each beam in each order k the spectrum extracted in the range $\lambda_{sk} < \lambda < \lambda_{ek}$ is rebinned to 1024 new pixels. This is important as will be explained in the next section.

3.1.2. Calculation of the circular polarisation

There are two beams per order and they allow us to calculate two spectra S_l and S_r observed simultaneously at each exposure. These two spectra correspond to two orthogonal states of polarisation. As described in paper III the procedure of observation requires that a $\lambda/4$ plate takes $\pm 45°$ orientations relative to the axes of the beamsplitter. Thus, in a cycle of four exposures, the $\lambda/4$ plate is orientated successively to the states $+ - -+$.

For the *ideal case*, we anticipate that :

$$S_{l1} = I + V, \; S_{l2} = I - V, \; S_{l3} = I - V, \; S_{l4} = I + V,$$

$$S_{r1} = I - V, \; S_{r2} = I + V, \; S_{r3} = I + V, \; S_{r4} = I - V,$$

where S_{lj} and S_{rj} are the normalised spectra for the exposure j and for the beams l and r respectively.

For the beams l and r we calculate :

$$P_l = \frac{S_{l1} + S_{l4} - S_{l2} - S_{l3}}{S_{l1} + S_{l4} + S_{l2} + S_{l3}}, \tag{3}$$

and

$$P_r = \frac{S_{r1} + S_{r4} - S_{r2} - S_{r3}}{S_{r1} + S_{r4} + S_{r2} + S_{r3}}, \tag{4}$$

For the *ideal case* one may substitute $I \pm V$ for S_{lj} and S_{rj} in Eqs. (3-4) to obtain $P_l = V/I$ and $P_r = -V/I$.

Finally, we calculate the polarisation P from :

$$P = \frac{1}{2}(P_l - P_r), \tag{5}$$

and, for the *ideal case*, $P = V/I$.

This procedure solves the problem of pixels sensitivity and the difficulties of producing a satisfactory flat field calibration.

We wish to stress again that we intend to measure a polarisation that is weak relative to a lot of effects in the first order. Take for instance the transmission of the optics (the optical fibers etc.), it may vary by 20%. However, we assume reasonably that it is achromatic within the range of each of the spectral orders. Thus, we apply a "practical" normalisation, i.e. divide the signal of each pixel by the sum of those of all of the pixels in the same order. This removes transmission effects, and by the way, also the crosstalk from I to V. Next, the sensitivity of each pixel can not be determined to much better than 1%. Therefore, we first compare the successive signals of the same pixel while the state of polarisation changed in time. It is a kind of modulation, where we anticipate that the pixel's sensitivity was constant during the observation cycle - about one hour.

The spurious signals due to time variations are considerably reduced, as well, by this procedure. First - the modulation cycle, $+--+$, removes the effect of linear component of the time variations, second - we assume that the time variations are the same for the two beams and are removed through Eq. (5).

Some more detailed discussion on the removal of spurious signals can be found in (Semel,1994), where it is proposed to proceed by combining double exposures and double beams, i.e., four spectra. Here, we used four exposures, i.e., eight spectra and proceeded by subtractions, Eq. (3-5) instead of divisions, as in Eq. (8) in (Semel,1994). Formally, the two kinds of algebra, subtraction or division, give practically the same result for weak polarisation signals. However, the fragility of each equation is not necessarily the same.

3.1.3. *Addition of spectral lines*

The polarisation of the line i, $P_i(x)$ as a function of the variable $x = (\lambda - \lambda_i)/\lambda_i$ is extracted as follows : the wavelength of the line λ_i allows to determine the order and the corresponding pixel. Let y be the running pixel number in the order and y_i the pixel number corresponding to the wavelength λ_i. We now extract a reduced spectral domain of 120 pixels around y_i. The variable x is obviously :

$$x = y - y_i \qquad\qquad\qquad (6)$$

The polarisation $P_i(x)$ of the line i as a function of the variable x (common to all spectral lines) is now determined, and we can proceed to the addition of the polarisation of all the lines as described in paper I :

$$\overline{P}(x) = \frac{1}{N} \sum_{i=1}^{N} P_i(x), \tag{7}$$

where $\overline{P}(x)$ is a straight forward average of the observed N lines.
Eventually one may give a statistical weight to the signal $P_i(x)$, the contribution of the line i , its the Lande factor g_i or its solar equivalent width w_i or their product $g_i w_i$, to obtain the weighted polarisations P_g, P_w or P_{gw}, respectively :

$$\begin{aligned}
P_g(x) &= \sum g_i P_i(x)/\sum g_i \, , \\
P_w(x) &= \sum w_i P_i(x)/\sum w_i \, , \\
P_{gw}(x) &= \sum g_i w_i P_i(x)/\sum g_i w_i \, .
\end{aligned} \tag{8}$$

The same procedure may be applied to the normalised line intensity profile, $I_i(x)$. For instance by using the statistical weight $g_i w_i$ for the line i one obtains,

$$I_{gw}(x) = \sum g_i w_i I_i(x)/\sum g_i w_i. \tag{9}$$

3.1.4. *Spectral lines selection*

How to select the stellar spectral lines ? The answer to this question may be not simple, a complete spectral analysis of a rapidly rotating double star system is certainly too heavy. The compromise done here was to list all the spectral lines, well seen in the solar spectrum as observed through the moon light, and with the same setup. Their spectral identifications is easy. The assumption is then made that the same lines should be dominant also in the spectrum of the solar type star. This approached proved to be efficient as reported in the following. The solar spectral lines were selected by the use of the table Moore,1966).

4. Results

4.0.5. *The selected objects*

To apply and test the present method, we selected a cycle of four exposures of the moon, expected to exhibit no circular polarisation.The sun as a star has certainly very weak magnetic field. Magnetic detection was expected on the RSCVn star HR1099. (Donati, 1992).

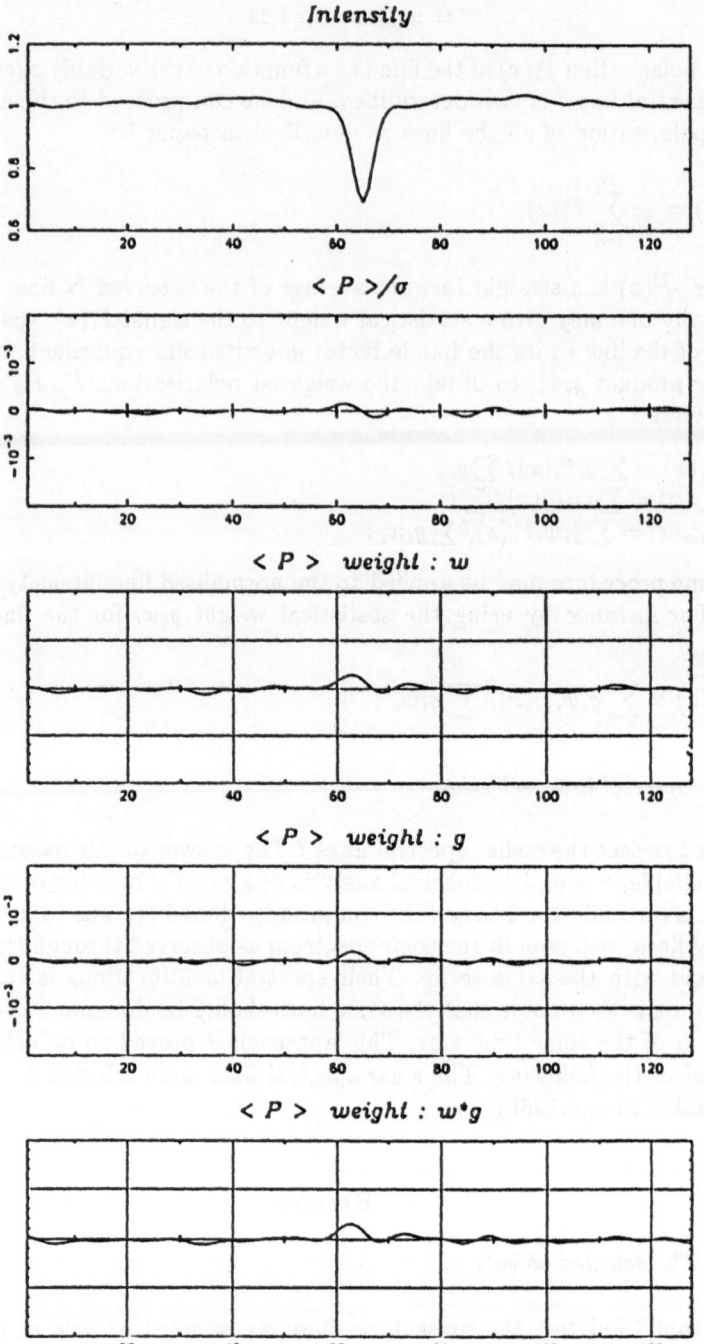

Fig. 2. The lunar observation - the results of adding 202 spectral lines. From top to bottom are shown: Intensity, $\overline{P}$, P_w, P_g and P_{gw}, respectivly.

4.0.6. *Spectral resolution*

One pixel in the Thomson CCD corresponds to 2.45 km/sec. The actual spectral resolution in this first experiment was certainly not less than 5 km/sec. The new pixel created corresponds to 2.17 km/sec. All the computations, like line additions, were made with the new pixels. Thus during the data reduction there was no deterioration of spectral resolution. However, a Fourier smoothing was applied to reduce noise in the presentation of the calculated polarisation. Thus, the spectral resolution in the figures corresponds to $4 \times 2.17 = 8.7$ km/sec, or, in other words equal to $\simeq 35000$.

4.0.7. *Residual polarisation in the solar spectrum*

To test for spurious signals, crosstalks from I to V, from Doppler to Zeeman effects and any other possible spectroscopic effects, we observed the moon. The spectrum of the sun as a star, as reflected from the moon should show very little circular polarisation, if at all, in the Fraunhofer lines. Thus, the circular polarisation found in the lunar observation was very weak, about 2×10^{-4} and served as indication that spurious signals were almost absent. (see figure 2).

Moreover the addition of all the line profiles resulted in a well determined and sharp "Doppler" line profile with very little wiggly continuum. Note that the "Doppler" line is measured in km/sec and has not in wavelength units. Its position and shape may disclose the Doppler shift of the observed object as well as its Doppler broadening. This in itself was a fair argument in favour of this particular approach of line addition.

4.0.8. *Circular polarisation in the binary star HR1099*

The addition of the intesity profiles of all the selected lines (202), results in two "Doppler" lines representing the two stars (see Figure 3 - top). The relative positions of these lines is related to their relative motion.
The stellar system is assumed to rotate nearly as a rigid body. The $v \sin i$ for each star is therefore proportional to its radius, and in the Doppler diagram it explains the width of the respective line :
1) the primary star - a very wide one, and
2) the secondary star whose radius is about 4 times smaller - a narrow line.
The circular polarisation appears mainly in the blue side of the profile of the primary star and has only a slight trace in the secondary star - a possible detection for a marginally weak field.

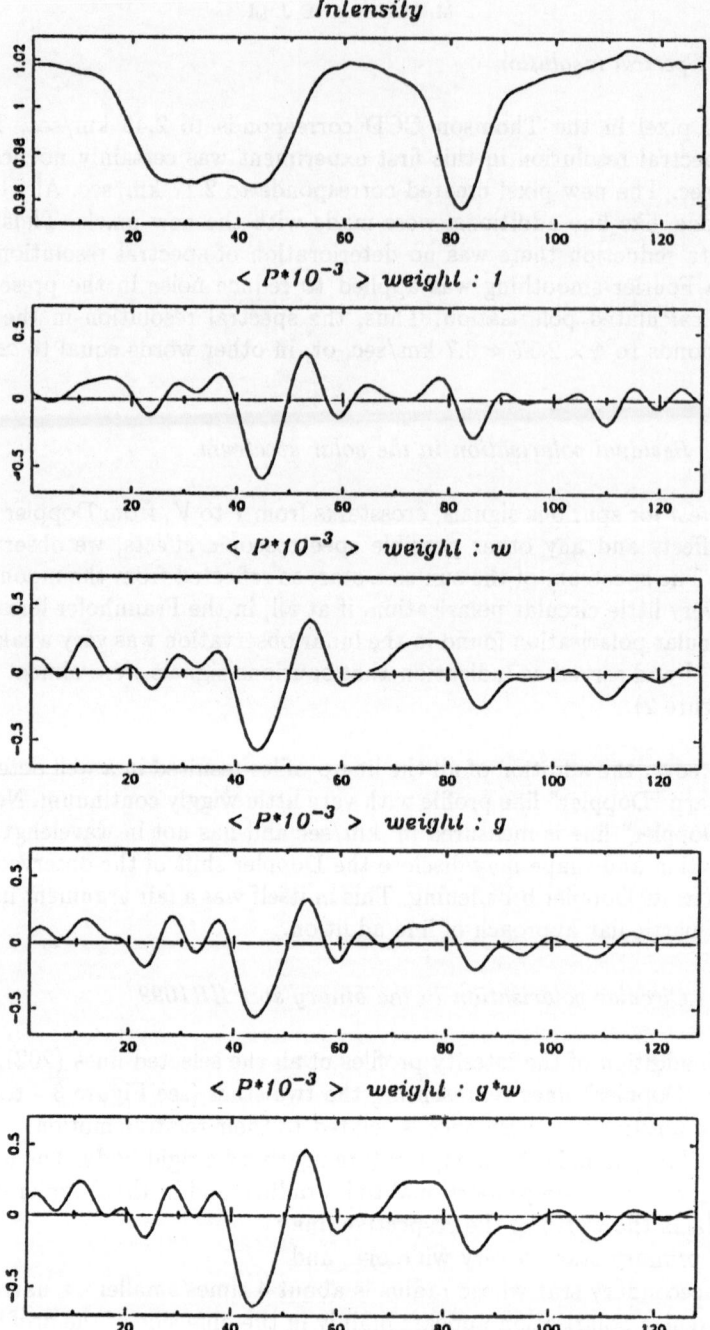

Fig. 3.   HR1099 a binary star - the results of adding 202 spectral lines. From top to bottom are shown: Intensity, $\overline{P}$, P_w, P_g and P_{gw}, respectivly.

We may reasonably assume that the circular polarisation in a spectral line is fairly proportional to the g factor of the line and also to the line strength. We may proceed by a linear regression :

$$P_i(x) = a_W(x) + b_W(x) \times W_i, \tag{10}$$

where W_i is the weight for the line i, for instance g_i, w_i and $g_i w_i$ already discussed before. By least square fit the coefficients $a_W(x)$ and $b_W(x)$ were determined and compared to the $\sigma_W(x)$ obtained as well for of the cooefficients through the minimization of the sum of the squares. In Figure 4, we show the result for HR1099 with the weight $g_i w_i$.

5. Summary

5.1. THE MULTI LINE TECHNIQUE

It seems that the first attempt of the method is promising. A simple technique of adding spectral lines proves to work and improves the signal to noise ratio. The noise, after integrating over all the lines, is at the level of 10^{-4}. A small residual polarisation, of the order 2×10^{-4} due to spurious signals and crosstalk is still seen. Improvements in the instrumentation as well as in the data reduction are anticipated.

5.2. CIRCULAR POLARISATION IN HR1099

An apparent signal of circular polarisation can be detected in the blue side of the primary star at the level of 3×10^{-4} (see Figures 3. and 4.).The correlation of the circular polarisation with the Landé factor discloses the Zeeman origin. The circular polarisation is weaker than the average of the earlier observations. This star is known for its variability and in this period, August 1991, the magnetism of HR1099, as can be seen by ZDI, was fairly weak. This is also in agreement with the early magnetic observation of this solar type star, where in some periods, the Zeeman signal could not be detected.

Acknowledgements

We are grateful to the director - Russel Cannon, Peter Gilingham and Jason Spyromilio who made this experiment possible. We thank Keith Shortridge for helping us to use Figaro. We also thank the editors and the referee for their careful reading of the paper and for their useful comments.

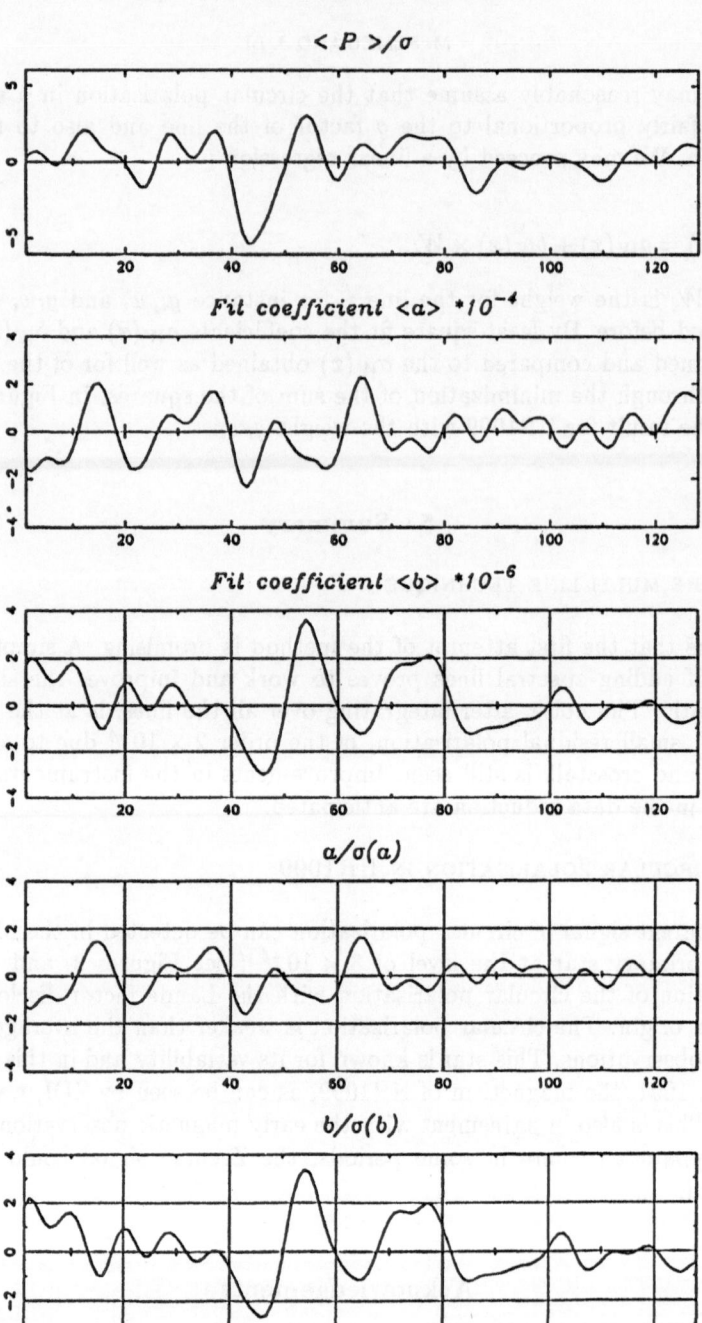

Fig. 4.   HR1009 - Linear regression of the circular polarisation as function of the weight $w_i g_i$. From top to bottom are shown: P_{gw} divided by σ, the coefficients a_{gw} and b_{gw}, and finally the same coefficients divided by their respective error bars.

References

Semel M.: 1989, *Astron. Astrophys.* **225**, 456 - Paper I

Donati J. -F. , Semel M., and Rees D.E. : 1992, *Astron. Astrophys.* **265**, 669 - paper III

Semel M., Donati J.-F. and Rees D.E. : 1993, *Astron. Astrophys.* **278**, 231

Moore C.E.: 1966, *The Solar Spectrum 2935 Å to 8770 Å*, NBS Monograph 61

Walker D.D. and Diego F.: 1985, *Mon. Not. R. astr. Soc.* **217**, 355

Semel M.: 1994, in R.J. Rutten and C.J.Schrijver, (eds.), *Solar Surface Magnetism*, Kluwer Academic Publishers, p. 509.

AUTHOR INDEX

LIST OF PARTICIPANTS

AUSTRALIA
D.E. Rees   CSIRO Division of Radiophysics, Epping

CZECH REPUBLIC
P. Heinzel   Astronomical Institute, Ondřejov

ESTONIA
T.F.Viik   Inst. of Astrophys. and Phys. of Atmosphere, Tartu

FRANCE
V. Bommier	Observatoire de Paris, Meudon
C. Briand	Institut d'Astrophysique Spatiale, Orsay
M. Faurobert-Scholl	Observatoire de la Côte d'Azur, Nice
H. Frisch	Observatoire de la Côte d'Azur, Nice
J.-C. Henoux	Observatoire de Paris, Meudon
M. Semel	Observatoire de Paris, Meudon

GERMANY
H. Domke	Astrophysikalisches Institut Potsdam
D. Engelhardt	Dr. Remeis Sternwarte, Bamberg
F. Kneer	Universitäts-Sternwarte Göttingen
K. Muglach	Kiepenheuer-Institut für Sonnenphysik, Freiburg
J. Staude	Astrophysikalisches Institut Potsdam
O. Steiner	Kiepenheuer-Institut für Sonnenphysik, Freiburg

INDIA
K.N. Nagendra   Indian Institute of Astrophysics, Bangalore

ITALY
E. Landi Degl'Innocenti	University of Florence
M. Landolfi	Osservatorio Astrofisico di Arcetri

SPAIN
J.H.M.J. Bruls	Instituto de Astrofisica de Canarias, Tenerife
J.C. del Toro Iniesta	Instituto de Astrofisica de Canarias, Tenerife
J. Sánchez Almeida	Instituto de Astrofisica de Canarias, Tenerife
J. Trujillo Bueno	Instituto de Astrofisica de Canarias, Tenerife

RUSSIA
A.V. Boulatov	Institute of Solar-Terrestrial Physics, Irkutsk
M.A. Demidov	Institute of Solar-Terrestrial Physics, Irkutsk
N.M. Firstova	Institute of Solar-Terrestrial Physics, Irkutsk
G.B. Gelfreikh	Pulkovo Observatory, St. Petersburg
Yu.N. Gnedin	Pulkovo Observatory, St. Petersburg
S.A. Grib	Pulkovo Observatory, St. Petersburg
B.A. Ioshpa	IZMIRAN, Moscow
V.V. Ivanov	St. Petersburg University
V.N. Karpinsky	Pulkovo Observatory, St. Petersburg
A.M. Kasaurov	St. Petersburg University
E.V. Kononovich	Sternberg Institute, Moscow
V.M. Loskutov	St. Petersburg University
V.I. Makarov	Pulkovo Observatory, St. Petersburg
D.I. Nagirner	St. Petersburg University
Yu.A. Nagovitsyn	Pulkovo Observatory, St. Petersburg
Yu.I. Poytanen	St. Petersburg University
L.I. Starkova	IZMIRAN, Moscow
A.V. Stepanov	Pulkovo Observatory, St. Petersburg
G.F. Vyalshin	Pulkovo Observatory, St. Petersburg

SWITZERLAND
P. Bernasconi	Institute of Astronomy, ETH Zurich
I. Rüedi	Institute of Astronomy, ETH Zurich
S.K. Solanki	Institute of Astronomy, ETH Zurich
J.O. Stenflo	Institute of Astronomy, ETH Zurich

UKRAINE
D.N. Rachkovsky	Crimean Astrophysical Observatory

USA
C.U. Keller	National Solar Observatory, Tucson
A. Skumanich	High Altitude Observatory, Boulder